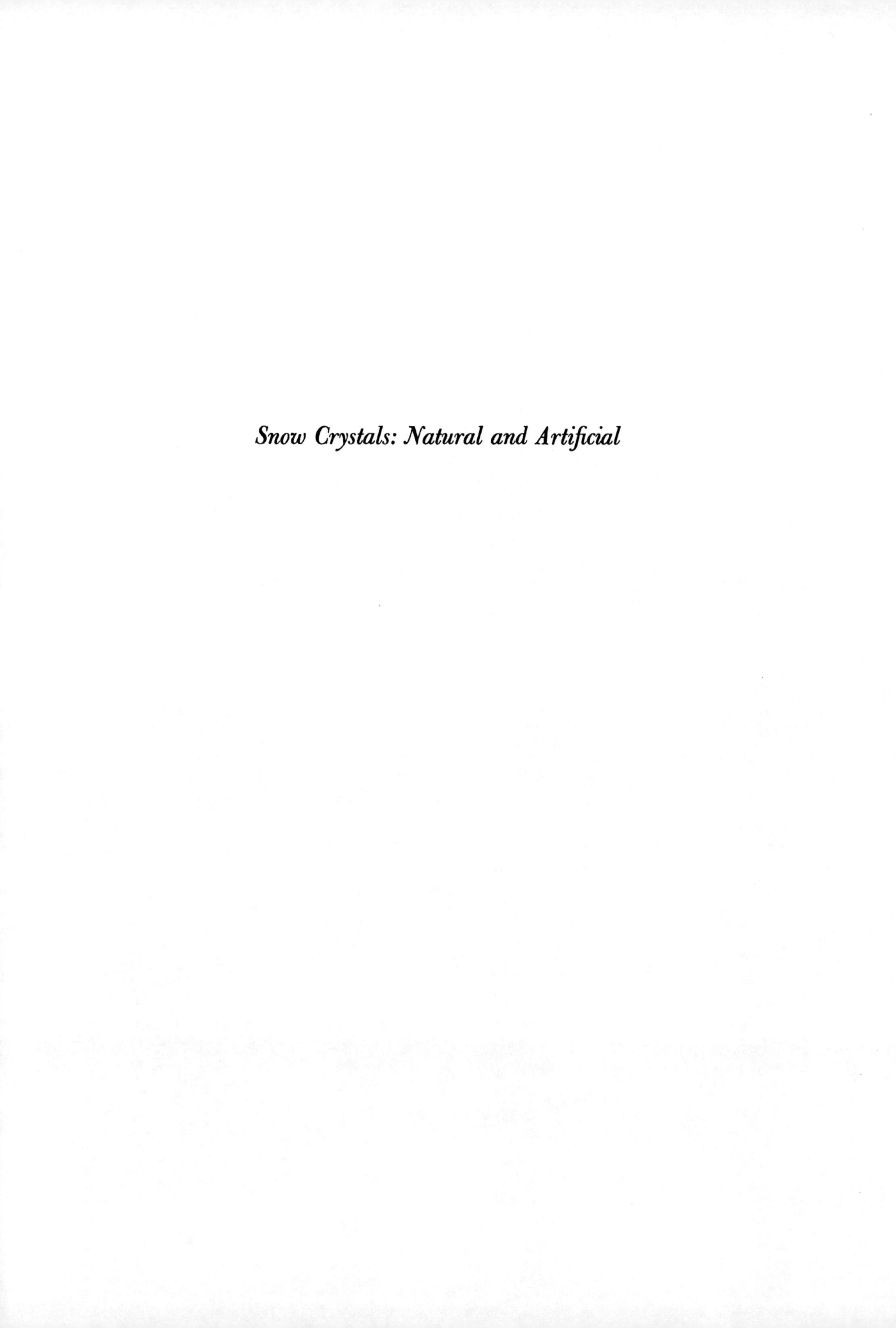

Snow Crystals: Natural and Artificial

Snow Crystals

NATURAL AND ARTIFICIAL

Ukichiro Nakaya

HARVARD UNIVERSITY PRESS · CAMBRIDGE · 1954

DISTRIBUTED IN GREAT BRITAIN BY
GEOFFREY CUMBERLEGE · OXFORD UNIVERSITY PRESS · LONDON

LIBRARY OF CONGRESS CATALOG CARD NUMBER 52–5038

PRINTED IN THE UNITED STATES OF AMERICA

Preface

When the Faculty of Science was established at Hokkaido University, Sapporo, Japan, in April 1930, I was offered and accepted a position in the Department of Physics and immediately took up residence in Sapporo. My experiences during two snowy winters in Sapporo led to the idea of conducting physical investigations on snow, and I began the study of snow crystals toward the end of 1932 as my first project in the field of snow investigation.

I spent the first winter in taking photomicrographs of snow crystals at our University laboratory in Sapporo. During the next few winters I changed my place of study to the *Hakugin-sō* (Silvery Villa), a dormitory of forest guards of the Hokkaido Prefectural Office. The villa was located at a height of 1030 m above sea level on a slope of Mount Tokachi, near the center of the main island of Hokkaido. During three winters there, I endeavored to gather as many photomicrographs of various kinds of crystals as possible. In the observations carried out during the first two winters, we found numerous varieties of snow crystals – not only every type thus far reported, but also some crystals hitherto quite unknown in the literature. We were therefore able to make a general classification of snow crystals which was extremely broad in its coverage.

In parallel with these observations at Mount Tokachi, the work of taking photomicrographs of crystals at Sapporo was also continued. Later we spent two winters at a mountain hut located on a slope of Mount Asari near Sapporo. Photographs of crystals taken at this spot added considerably to our collection of crystal pictures. I was ill at the time, and Professors Seiji Kaya and Jumpei Harada, two of my colleagues, took my place in directing the observation squad sent to the hut. I express my gratitude here to these professors.

In addition to the work on a general classification scheme, we commenced (in 1934) the study of the physical nature of each type of crystal. This undertaking, chiefly conducted at Mount Tokachi, included measurements of the mass and velocity of fall of individual crystals, for the purpose of finding the relation between these quantities and the size and form of crystals. The electrical nature of each snow particle was also studied. Lastly, the frequency of occurrence of each crystal form was studied, the observations for this purpose being carried out for a whole winter. These observations indicate that plane crystals of hexagonal symmetry, which have been believed to be the representative form of snow crystals, constitute only a small part of natural snow crystals and that in actuality irregular types of crystals are more abundant.

These studies constituted the first part of our investigations on snow. The results obtained were comprehensive, but we could not as yet begin to explain the simplest and most primary problem, why snow crystals show such a complicated variation in form and structure. With regard to this problem, many studies had been made in Europe and America on the relation between crystal forms and meteorological conditions. We also worked on this subject in Hokkaido. However, inasmuch as the snow crystals are formed high up in the atmosphere and meteorological conditions usually recorded are the results of observations on the surface of the earth, we could not expect any simple relation between them. As we had anticipated, the actual observations yielded little or no pertinent information.

It was felt that a solution of this problem should be obtainable by producing these crystals artificially in the laboratory. Thus, by studying the relation between the crystal form and the conditions under which it is formed, the basic question could be answered unambiguously. However, we could not foresee the difficulties that lay before us in solving this problem, since no work on the artificial production of snow crystals had been reported at that time. So a beginning was made with the artificial production of frost crystals. The form and structure of frost crystals may be regarded as very similar to those of a single branch of a snow crystal, some forms of frost crystals corresponding to those of snow. Hence, the conditions of snow formation may be inferred, to some extent, on the basis of the knowledge acquired in producing various types of frost crystals.

Even the early experiments on the artificial production of frost proceeded smoothly and yielded useful data, but it was natural that the crystals of frost which developed on the cooled surfaces of solid bodies were in various ways different from snow crystals formed afloat in the air.

At about this time, construction of the Low Temperature Laboratory was begun at Hokkaido University; the new laboratory was completed in February 1936. The experiments on the artificial production of frost were moved into the cold chamber of the new laboratory, in which the temperature was usually kept at −35°C. In this chamber we found that we could produce frost crystals without using a cooling surface. Soon thereafter, owing to the improvement in equipment, we succeeded in producing ice crystals afloat in the air – in other words, snow crystals – by making an ice crystal develop on a point of thin rabbit hair.

Since then, our studies on the artificial production of snow crystals have been continued, and it is felt that the basic problem is nearing complete solution. Today we can produce every type of natural snow crystal in the laboratory, and the conditions of their formation have been determined in almost all cases.

September 1946 U. Nakaya

Four years have already elapsed since I finished the last page of the manuscript of the Japanese edition of this book. Meanwhile, the printing work was making slow but steady progress under the adverse conditions of the War, when unfortunately the printing office was bombed and the whole copper type and original type of the text were burnt. At that time it was impossible to reset the type and I gave up hope of publication.

In the early spring of 1949, I was greatly encouraged by Dr. H. C. Kelly of the Economic and Scientific Section, GHQ, SCAP, to again prepare the photographs for publication. The authorities at Hokkaido University also rendered assistance by making available a special research fund. Nothing surpasses the author's happiness as he looks forward to the birth of this book from a copy of the proof which survived the war. Its publication has been made possible by the kindness and sympathy of various persons, especially by the courtesy of Dr. Charles F. Brooks, Director of Blue Hill Meteorological Observatory, Harvard University, and of Dr. B. C. Dees of the Economic and Scientific Section, GHQ, SCAP, who gave much advice on the preparation of this English edition. The author is also much indebted to Mr. Harold M. Lane, who took the trouble to read through the manuscript. The most sincere gratitude is expressed to the American Academy of Arts and Sciences, which made the grant of $1500 from The Permanent Science Fund necessary to supplement the funds that the Harvard University Press was in a position to allocate to the manufacture of this book, and to Dr. Brooks, by whose kind assistance this grant was obtained. Although the progress of research during the past few years might make advisable additions or deletions in some places, the author decided to present the book in the form of 1946, being afraid of losing the opportunity again. Nevertheless, the last chapter summarizes the recent research in cloud physics, including work done in our laboratory.

February 1953 U. Nakaya

Acknowledgments

This work was accomplished by the coöperation of staff personnel of my division in the Department of Physics, Hokkaido University, as well as of undergraduate students who did their undergraduate research theses under my charge.

The author takes this opportunity to express his thanks to Assistant Professors Toichi Terada and Masando Hanajima, to Messrs. Tamakichi Takano and Ikuzo Kagami, who worked with him for many years under the terrible chill, and to Messrs. Tsuneo Iijima, Katsuji Hashikura, Isonosuke Satō, Yataro Sekido, Motoichi Tada, Yasuaki Toda, Shuzo Maruyama, Yasushi Sumi, Shun Inada, Yukio Miyazaki, Hiroshi Endō, and Chōji Magono, who undertook parts of the work as their graduation experiments.

The success in the most difficult experiments in the latter half of the study on artificial snow is especially due to the efforts of Assistant Professor Hanajima, who continued the experiments for many years in the low-temperature room. The author herewith presents his thanks and expresses his admiration for the persistent exertion of this splendid colleague.

This series of researches was aided by the Education Ministry subsidy for scientific research and by grants from the Japan Society for Promotion of Scientific Research and from the Hattori Hōkōkai. In addition, the progress of our study owes much to the benevolence of Mr. Kōichi Hanajima. Herewith I express my hearty thanks to these contributors.

Contents

Part II: Artificial Snow

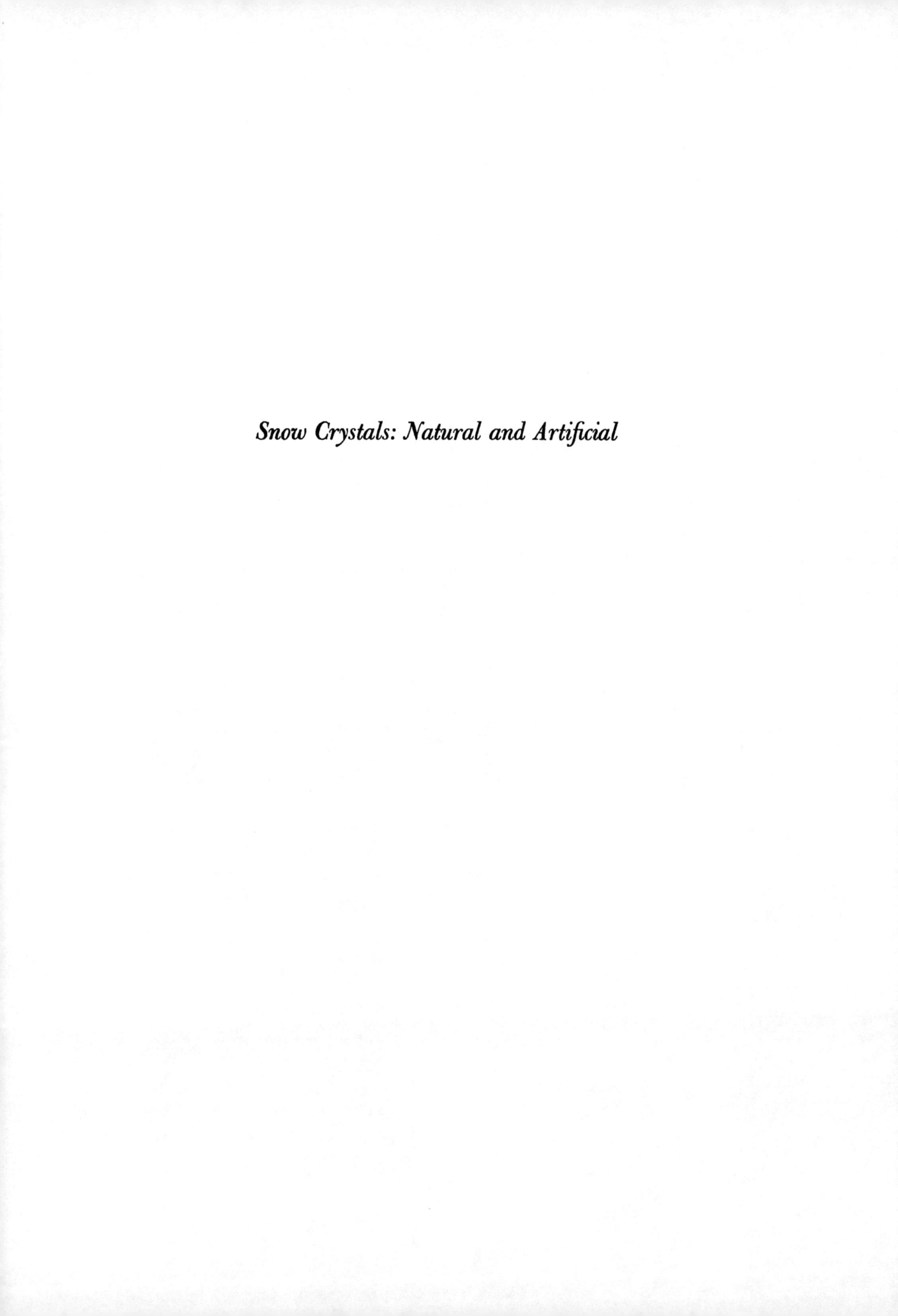

Snow Crystals: Natural and Artificial

Introduction

Snow consists of ice crystals, made of water vapor condensed on some sort of nucleus by sublimation at low temperature. Usually, the snow crystal has the shape of a hexagonal plane or column of delicate structure, and has almost innumerable variations in form.

The fact that snow consists of ice crystals of such complicated forms was first recognized with the naked eye; in Europe the fact was first reported sometime about the middle of the thirteenth century. The history of the study of snow crystals in Europe is detailed by Hellmann.[1]

According to Hellmann, the first sketch of snow crystals observed with the naked eye was made by Olaus Magnus, the Archbishop of Uppsala, about 1550. The sketch by Magnus does not indicate that snow crystals have hexagonal symmetry. It is said that it was the famous Kepler who first pointed out that snow crystals belong to the group of hexagonal crystals. Later (in 1635), Descartes made observations on snow crystals and published his sketches in Amsterdam.[2] Figure 1 reproduces some of his sketches. These are thought to be the first scientific records of snow crystals. We find in these sketches a drawing of a crystal composed of a hexagonal column with plane crystals developed on both ends. It is surprising that a sketch of this rather rare type of crystal should be contained in this first scientific report in the history of snow-crystal investigation. There are a few later studies made by various persons, most of whom observed snow crystals with the unaided eye; the use even of simple magnifying glasses was quite rare.

When the microscope was invented in the latter half of the seventeenth century, the observation of nature with this new instrument made rapid progress. Robert Hooke, who is famous as a discoverer of plant cells, published in *Micrographia* in 1665 various sketches of natural things observed with his microscope. This work contains a number of sketches of snow and frost crystals. He observed that in hexagonal snow crystals all twigs extending from a main branch are parallel to the neighboring main branch, thus correcting the inaccuracy or error in the observations made prior to his researches.

Hooke was followed by Friedrich Martens, a German traveler, who sailed from Spitzbergen to Greenland, and published an account of his experiences during his trip. In his report, he wrote about the snow crystals he observed in the arctic regions. In particular, we find a discussion on the relation between the crystal

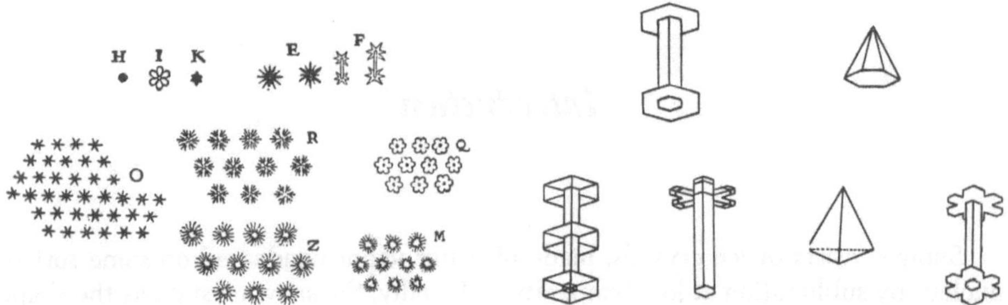

FIG. 1. Snow crystals by Descartes.

FIG. 2. Snow crystals by Scoresby.

forms and weather conditions – the first discussion of this subject in the literature.

In 1681, Donat Rossetti, a mathematician and priest at Revorno, Italy, collected the sketches of 60 snow crystals, and made the first attempt to classify them.

These observers opened the way for the study of snow crystals, but, with the exception of Descartes, they confined their observations to hexagonal plane crystals. Besides the well-known plane types, snow crystals of hexagonal column and pyramid forms are quite common. The first detailed description of such forms was made by William Scoresby – a whale fisher of England – in 1820. In his book describing the history and status of whale fishing in the arctic [3] he described the snow crystals he had observed, and clarified the structure of new types which had been overlooked, including various types of columns and more complicated ones formed through combinations of columnar and plane crystals.

Shortly after Scoresby's work was published, there appeared (1832) in Japan the *Sekka Zusetsu* (Illustration of snow blossoms) by Toshitsura Ōinokami Doi, a feudal lord of Koga, Shimousa. He used a "Dutch glass" in his observations of snow crystals and 86 sketches thus obtained were published in the book. The book also contained 12 sketches by Martinet, which were reproduced from *Kakuchi Mondō* (Catechism on Nature). According to Yajima [4] this *Kakuchi Mondō* is a translation of *Katechismus der Natur* by J. F. Martinet, which contains, besides the sketches of snow crystals reproduced in *Sekka Zusetsu*, a number of sketches of frost crystals. Although some of the sketches of *Sekka Zusetsu* apparently show inaccuracies in observation, most of them are supposed to be the results of quite faithful observation.

Toshitsura Doi published the *Sekka Zusetsu, continued* in 1839; this work contains 97 sketches of similar snow crystals. The sketches in these two books are superior to most of those published at the time. For instance, the sketches by Doi are very nearly equal in quality to the illustrations of snow crystals by the English meteorologist James Glaisher, published in 1855 (in other words 23 years later

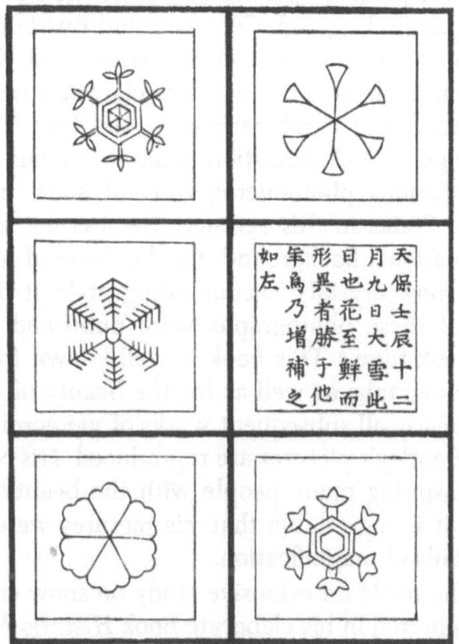

Fig. 3. Snow crystals from *Sekka Zusetsu* (Illustration of snow blossoms).

than the *Sekka Zusetsu*), which are considered to be the most accurate observations published before the development of photomicrography. There are reasons for this sudden appearance of such an achievement at a corner of Musashino in the Tokugawa Era. The chief retainer of Doi, Senseki Takami, was a scholar of Dutch science studying under Nobutō Kawaguchi, one of the pioneer scientists of Japan. Takami materially aided Doi's studies, and the two men worked on their observations on snow crystals for many years, from about 1812 to 1832. In 1802, ten years before these observers began their work, Hoshū Katsuragawa wrote about the use of the microscope by order of the Bakufu (Tokugawa Shōgunate Government). It seems probable that Doi and his group used a compound microscope in their observation. In fact, there was a scholar, Ranzan Ono (1729–1810),[5] who used a compound microscope in observing snow crystals prior to Doi. Although he left no publication of his snow sketches and there is no indication of when he made his study, he is considered to be the first to make observations of snow crystals in Japan. With the *Sekka Zusetsu, continued* the study or observation of snow crystals in Japan was completely terminated until well into the twentieth century. Even the modern Western sciences imported after the Meiji Restoration left no trace of any study of snow crystals in Japanese scientific literature until recently.

In Europe, with the development of photomicrography, research on snow crystals made remarkable progress, particularly since the end of the nineteenth century. Many meteorologists in England, Germany, and Russia took photomicrographs of snow crystals. Among them, Hellmann of Berlin [6] and Nordenskiöld of Stockholm [7] are well known. Both of these workers classified snow crystals into three kinds – planar, columnar, and combinations of the two. The basic idea of this system remains in the general classification of snow crystals to the present day.

A series of very famous photomicrographs of snow crystals was made by Bentley of the United States in this century. He was not a professional scientist with an orthodox educational background, but he devoted his whole life to taking photomicrographs of snow crystals, accumulating at least 6000 before his death. Approximately 3000 of these photographs were published in 1931 (the year of Bentley's death) in book form.[8] This book is well known for its unprecedentedly large collection of photographs as well as for the beauty of those pictures. It may safely be said that in nearly all subsequent works of meteorology in the world dealing with this subject, Bentley's pictures are reproduced. His collection made a great contribution toward inspiring many people with the beauty of snow crystals, but as scientific material, it is regrettable that his pictures were retouched and were published without details of magnification.

Another scholar who made an extensive study on snow crystals is Dobrowolski. His observations are reported in his elaborate book *Historja Naturalna Lodu* (Warsaw, 1922), but this book seems to be not well known because of the language in which it was printed.

During the period from the time when Bentley and Dobrowolski did their work to the time when the author and his colleagues started their study in 1932 in Hokkaido, there was no conspicuous report in this field except the work of Stüve,[9] who tried to find some relation between crystal types and the conditions of the upper atmosphere.

The author and his colleagues started their study of snow toward the end of 1932, with extensive observations on natural snow. The observations were conducted at Sapporo and Mount Tokachi (which is located near the center of Hokkaido Island). From the results of these observations, we made a general classification of snow crystals, and then carried out some investigations on the mass, speed of fall, electrical nature, frequency of occurrence, etc., of each type of crystal.

Along with the study on natural snow, we had worked in the laboratory of our University in Sapporo to find a method of producing snow crystals artificially. The first success in the artificial production of snow crystals was achieved in 1936. Subsequent experiments were chiefly concentrated on the production of artificial snow crystals, and by 1944 we were able to produce every type of snow crystal artificially, and were therefore able to determine clearly the conditions of their formation. This book is a summary of the work of the author and his colleagues between 1932 and 1944.

PART I

Natural Snow

CHAPTER 1

Observation of Snow Crystals

1. Method of taking photomicrographs

No difficulty is encountered in taking photomicrographs of snow crystals when the temperature is below −5°C. Ordinary methods may be employed with a duly cooled microscope placed in a shady place. When the temperature is above −5°C, it is necessary to have cooled the glass plate before placing snow crystals on it. When the temperature is above 0°C, needless to say, the microscope must be well cooled. In such cases, however, a part of the crystal is often melted before the observation is begun and it is therefore impossible to take good photographs.

Photographs are usually taken by transmitted light. As the source of illumination, the light of the sky was used in the daytime, and at night an electric lamp was used. When the temperature is below −10°C, no arrangement for absorbing heat rays from the lamp is necessary; thus the lamp light may be used without any adaptation. Photographing by transmitted light is advantageous in getting a clear picture of the boundary and internal structure of the crystal, or, in other words, the crystal pattern. However, ordinary transmitted light leaves something to be desired in that it does not show clearly the topography of the surface.

In order to get a white image of the crystal on a black background, reflected light must be used. Figure 4 shows two crystal microphotographs taken by reflected light with the Ultropak microscope. This method increases the beauty of the crystals, outlining the white image clearly against the black background, but the delicate structure inside the crystals is not revealed.

Bentley took his photographs using transmitted light, and he attempted to show white images against a black background. For this purpose, he cut away the background part of the film on duplicate negatives along the rim of the crystal image and printed them. This process was employed in all of the 3000 photographs contained in his book, four examples of which are reprinted in Fig. 5. Some people

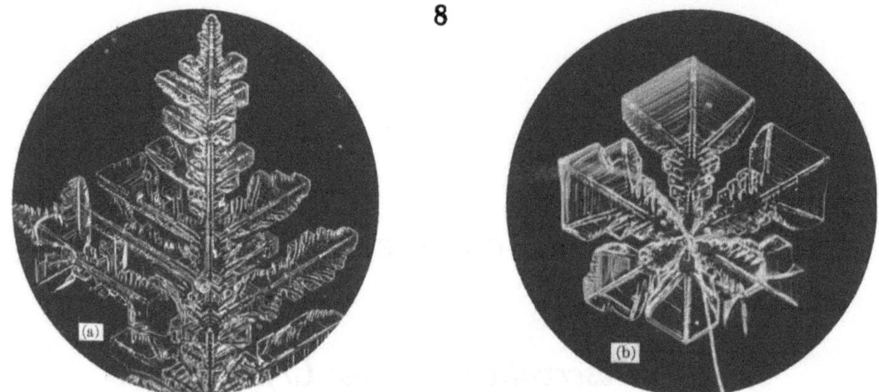

FIG. 4. Photomicrographs made by reflected light; (*a*) × 32; (*b*) × 44.

FIG. 5. Snow crystals by Bentley. [From W. A. Bentley and W. J. Humphreys, *Snow crystals* (McGraw-Hill Book Co., Inc., New York and London, 1931), pp. 78, 192.

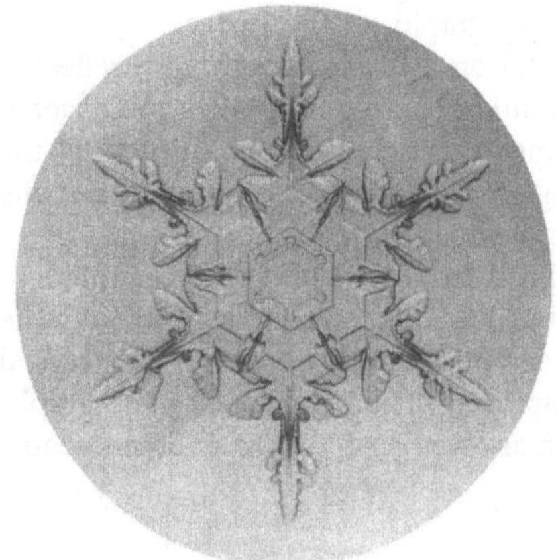

FIG. 6 [1]. Photomicrograph by oblique illumination (× 28).*

FIG. 7 [2]. Photomicrograph by oblique illumination (× 35).

* The photomicrographs here reproduced are numbered serially. The number is given in brackets in the legends to figures in the text. Numbers that do not appear there will be found in the plates.

say that this method of reproduction has an "effect like a diamond placed on velvet." Others, like Hellmann, accuse this operation of depriving the photograph of its value as scientific material. In examining the detailed structure of crystals of such complicated forms as are shown in the lower part of Fig. 5, photographs reproduced by such a process cannot be used to advantage. We therefore gave up the attempt to show white images on black backgrounds, and chiefly used transmitted light.

The best method of revealing the slight ruggedness of the crystal surface is to use a little oblique illumination. Prior to our work, Sigson had employed this method in taking his photographs,[10] and Okada also used the same method.[11] However, if the light is slanted too much in order to get a clearer picture of the surface ruggedness, a part of the visual field becomes excessively dark. The adjustment of the angle of light has to be done by hand, and the best light angle is the one that slightly darkens one side of the visual field. Two examples of photographs taken by this method are reproduced in Figs. 6 and 7. As is illustrated in these photographs, it is possible by this means to get clear pictures of the minute ruggedness of the surface as well as the boundary shape and distinct images of the internal structure of crystals. We therefore employed this method almost exclusively in our series of investigations.

2. Observations at Sapporo and Mount Tokachi

The mean temperature at Sapporo during the winter is −3.2°C for December, −6.3°C for January, −5.4°C for February, and −1.7°C for March. Throughout these four months snow falls very frequently and the whole district is covered continually by snow drifts. It is therefore possible to take photographs of crystals almost every day by exercising care. Observations at Mount Tokachi were conducted at a cottage called Hakugin-sō (Silvery Villa), which was located on the slope half way up Mount Tokachi, the altitude of the cottage being 1030 m above sea level. The temperature at this spot in the beginning (and end) of winter usually ranged between a maximum of −3°C and a minimum of approximately −10°C. In midwinter the maximum was approximately −10°C, the minimum approximately −20°C. We thus found that the temperature there was very suitable for the study of snow crystals. We could manipulate snow crystals as if they had been delicate glassware. In addition, the spot where the cottage is situated was protected from strong winds by the surrounding ridges, and beautiful crystals were observed in large quantity, thus greatly facilitating our collecting of photographs.

The microscope used was one made by Leitz. The photographic apparatus was of a very common type, consisting of a bellows and a quarter-plate frame attached to it. The plates used were ordinary orthochromatic plates, which were found to be suitable for getting the minute surface ruggedness. Process plates, which had been often used by many workers in this line, were not adopted, because they were found to be too contrasty.

In the case of plane crystals all scattered light should be eliminated. Therefore, a hollow cylinder coated black on the inner surface was placed on the microscope slide over the crystal and the object glass. With crystals having a solid structure, however, we could take better photographs by adding a small amount of reflected light to the transmitted light. When a large crystal of solid structure was chosen, we could obtain a good photograph with considerable depth of focus by using the microscope objective having an iris (such as the Miral objective of Leitz).

In the observations at Mount Tokachi, where the temperature was usually below −10°C, we could investigate the crystals by manipulating them in various ways; for example, it was found possible to cut a crystal in a desired manner and to take a photomicrograph of a side view by erecting the crystal on a slide of the microscope. In this case there was no danger of the crystal's melting, but we had to pay attention to the loss of fine pattern and the decrease in size due to sublimation, which occurs even when the temperature is below −10°C. Figure 8 shows the sublimation of a natural dendritic crystal. The original fine structure of the crystal is shown in (*a*); this crystal was placed on a glass plate at about 7 o'clock in the evening, when the temperature was approximately −15°C. In (*b*) it can be seen that the crystal has lost most of its fine structure within the twigs; this picture was taken 10 min after the crystal was exposed to the atmosphere. After an additional 20 min, the crystal had become considerably smaller, as shown in (*c*); (*d*) shows the state of sublimation after an additional hour. In these cases there was little or no wind. The process of sublimation is greatly accelerated by wind, and increases still faster as the temperature rises toward 0°C. As photographs are usually taken at temperatures between −5°C and −10°C, quick operation is necessary for obtaining good pictures. It is a well-known fact that when a crystal has a relatively high value of equilibrium vapor pressure, molecules evaporate from the part with larger curvature (i.e., projecting parts) and condense onto indented parts, owing to the surface energy. Ice has the relatively high vapor pressure of 4.58 mm-of-mercury at 0°C; it is 3.02 mm at −5°C, 1.96 mm at −10°C and 0.78 mm even at −20°C. The rapid sublimation of ice is thus accounted for. However, it seems that this point is often overlooked. For example, some of the photographs of Bentley (especially a number of those of the dendritic crystals), show crystals that have already undergone sublimation to a considerable degree. In his case sublimation must have taken place in the atmosphere near the earth's surface, before the crystals were brought under observation.

Snow crystals are thus so apt to sublime that the particles in snow kept on the ground as snow cover do not long retain their original crystal form. The structure

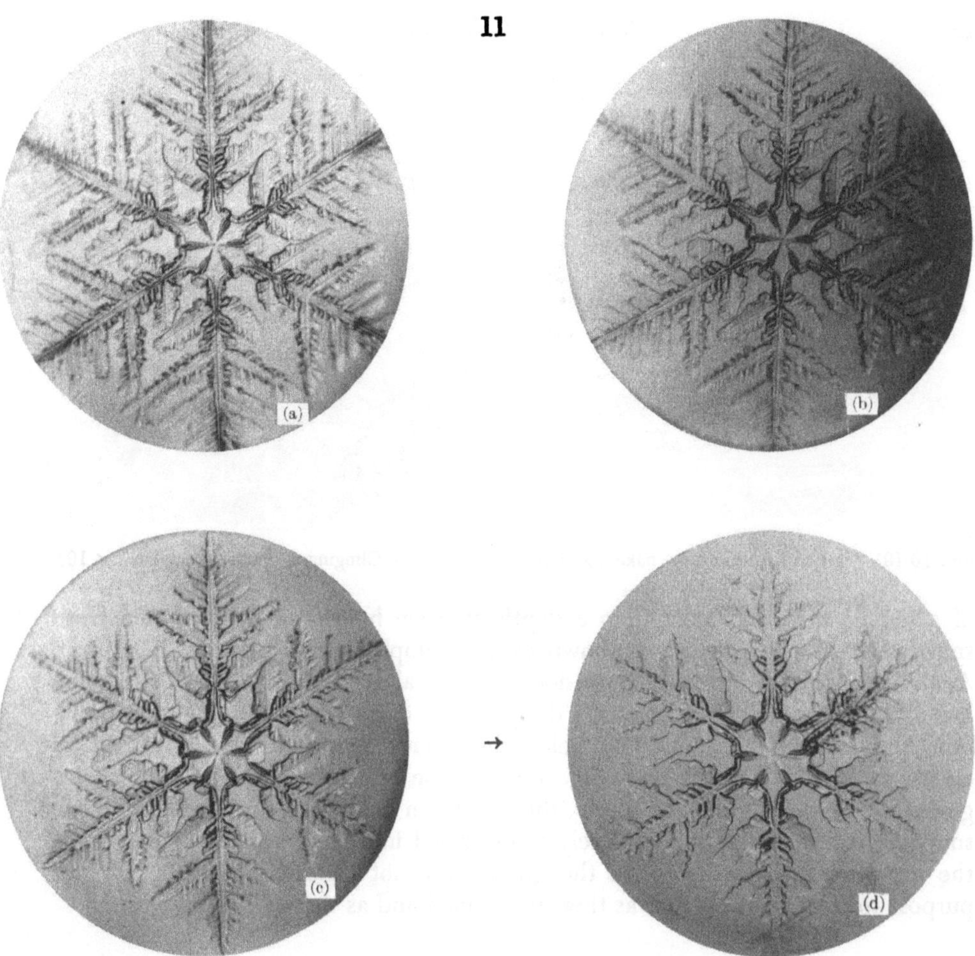

FIG. 8 [3–6]. Sublimation of a snow crystal (all × 16).

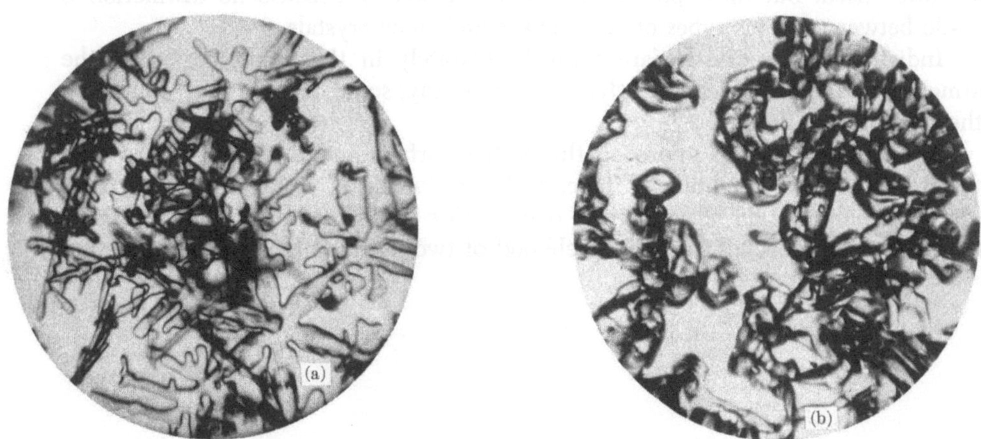

FIG. 9 [7, 8]. Particles of ground snow.

FIG. 10 [9]. Part of a cotton-snow flake (× 12).

FIG. 11 [10]. Clinging of dendritic crystals (× 10).

of a sample of snow, taken from a depth of 5 cm below the surface of a fresh snowfall at Mount Tokachi, is shown as an example of this point in Fig. 9(*a*). As the temperature in this locality does not rise above −5°C in this season, the loss of the fine structure and delicate design in the crystal is clearly due to sublimation. Furthermore, the snow 30 cm below the surface retains no trace of its original crystal form, as is evident from Fig. 9(*b*). These crystals eventually become ice granules after a long time. Similarly, the snow 1 m below the surface of this same snowfall had been almost completely transformed into ice granules. In this book, the problems of snow cover on the ground are not considered, since it is our purpose to treat snow crystals as they are formed and as they fall to the ground.

3. General view of snow crystals

To distinguish the kinds of snow, the words "cotton snow" or "powder snow" are often used. But these phrases are very ambiguous, because no distinction is made between various types of snow flakes and snow crystals.

Individual snow crystals are formed separately in the upper region of the atmosphere and fall independently. Needless to say, some of them may collide on their way to the ground.

As the snow crystals approach the earth's surface and the temperature rises toward 0°C, the temperature of the crystals also increases. As long as ice particles are very cold, they do not adhere to one another when in contact; but when the temperature approaches 0°C, the likelihood of two particles being joined together when they collide increases markedly. Thus, as they collide the crystals gradually form a cluster. We shall call this cluster of crystals a "snow flake." What is commonly called "cotton snow" is made up of snow flakes consisting of many crystals; one "flake" may consist of ten or thirty or perhaps even thousands of crystals. Figure 10 shows a part of a cotton-snow flake.

When the temperature is very low all the way down to the earth's surface, crystals generally do not make snow flakes but fall separately. However, even at the low temperature of $-10°C$ or $-15°C$, crystals of complicated dendritic form sometimes cling to each other by the points of their twigs and make flakes. An example of this is shown in Fig. 11. Snow flakes of this type are extremely light, so that ground snow consisting of such flakes sometimes is found to have a specific gravity of 0.01 or even less, when measured shortly after snowfall. In many cases, individual crystals are mixed with these flakes. At Mount Tokachi we often observed this extremely light snowfall when the temperature was below $-10°C$ and there was little or no wind.

In Hokkaido, needle crystals are often observed at the beginning and end of winter. One example of these needle crystals is shown in Fig. 12. This kind of crystal, when received upon a sleeve of a black overcoat, appears to consist of numerous thin silk threads cut into 1- or 2-mm lengths and scattered.

Needle crystals are said in Europe and America to be a kind of snow very rarely observable, but they are not so rare in Hokkaido. Every year we could observe them several times in one winter.

Figure 13 is an example of columnar crystals and bullet types. When these crystals fall, they appear to the naked eye like very fine scattered powder. These crystals are formed when the crystal was developed along the principal axis of the hexagonal system of crystallization. They are not so rare, and are observed at Mount Tokachi more frequently than needle crystals.

The representative snow crystal has hitherto been thought to be the dendritic crystal developed in one plane – hexagonal plane crystal. However, dendritic crystals do not always develop in one plane but sometimes have a solid structure. Figure 14 shows an example of dendritic crystals with radiating extensions from a center, taking a form like a chestnut burr. We rarely see photomicrographs of this kind of crystal in the literature from Europe and America, but it falls so frequently in Hokkaido that most of the dense snow consists of this type together with a similar type of spatial assemblage of dendritic branches with a stellar base crystal, which will be described in a later section.

If we continue careful observation of actual snowfall, we notice that irregular combinations of various types of crystals are observed more frequently than the abovementioned rather regular crystals; an example is shown in Fig. 15. More unevenly shaped ones, to be called irregular, also fall very often.

Figure 16 is a photomicrograph of the so-called "flour snow." During the coldest

FIG. 12 [11]. Needle crystals (× 18).

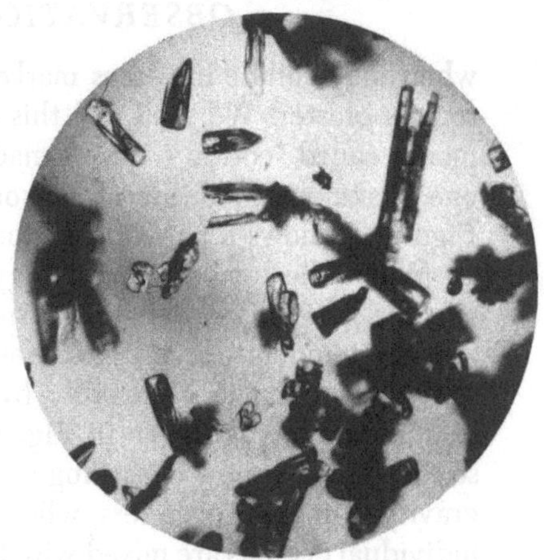

FIG. 13 [12]. Columns and bullets (× 20.5).

FIG. 14 [13]. Burrlike crystals (× 12).

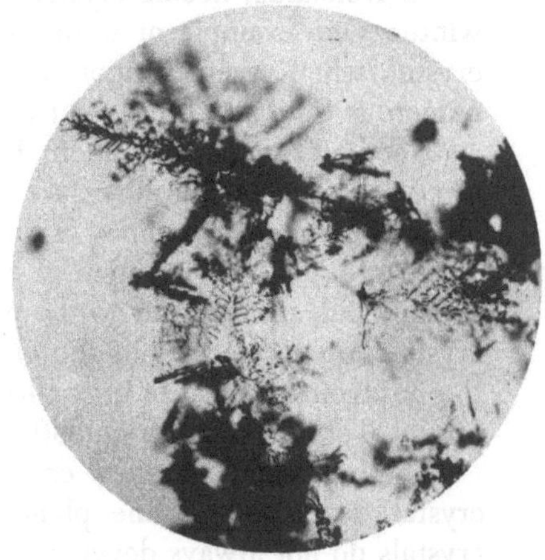

FIG. 15 [14]. Irregular crystals (× 10.5).

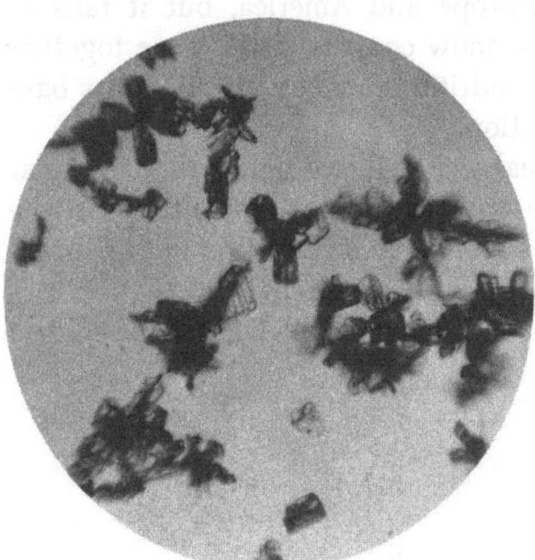

FIG. 16 [15]. "Flour snow" (× 19).

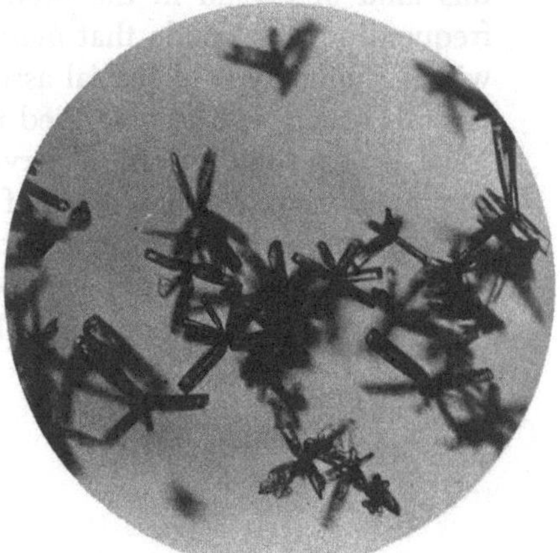

FIG. 17 [16]. Combination of bullets (× 17.5).

part of winter we often notice upon the summits of high mountains white but lusterless snow that looks like chalk powder or flour. The name comes from its appearance, but the structure is, as shown in the photograph, an assemblage of small columns with small plates attached to them. These crystals are formed when the temperature is low and the supply of water vapor is scanty.

A group of columns with pyramids, bullet type or rock-crystal type, is shown in Fig. 17. This kind of crystal belongs to a very rare type in the existing foreign literature, but we could observe it quite frequently at Mount Tokachi.

It is evident that there are many types of snow crystals besides the hexagonal plane crystal, which has hitherto been regarded as the representative one. And these crystals not merely satisfy our curiosity but in actuality fall very frequently.

Detailed descriptions of each of these types will be given in the following sections.

4. Classification of hexagonal plane crystals

The hexagonal crystal is the one that develops within the base plane of the hexagonal system of crystallization, showing six-pointed symmetry. From the scientific point of view, it would better be called a "regular plane crystal with hexagonal symmetry," but for convenience' sake, we call it a "hexagonal plane crystal." All crystals belonging to this type have hitherto been considered by many workers in this line as having six-pointed symmetry, but, as is mentioned later, irregular development is found no less frequently in this plane type. We call these irregular types "malformed" and exclude them from the hexagonal plane crystals.

The hexagonal plane crystal that has hitherto been most familiar to us is the one that has branches of dendritic extension. Models of these types with other plane crystals are shown in Fig. 18. The simplest form of the dendritic type is the crystal with six straight branches extended from the center, called the stellar type [Fig. 18(*e*)]. This type, when further developed, sends out twigs from the six main branches; it is then called the regular dendritic, or simply the dendritic [Fig. 18(*f*)]. The characteristics of the dendritic crystal are best seen in the one that has well-developed twigs arranged parallel with each other; we give this type the special name of fernlike crystal [Fig. 18(*g*)].

It has been usually believed in the science of crystallography that the dendritic development occurs when the degree of supersaturation in the atmosphere is high and that the plate form is obtained when it is low. But we came to notice, as a result of the study of artificial snow described in Part II, that temperature plays the more important role in the dendritic formation. When the degree of supersaturation of water vapor becomes low and the temperature goes beyond the limit of dendritic development, the plate crystal develops [Fig. 18(*a*)]. The sector-type

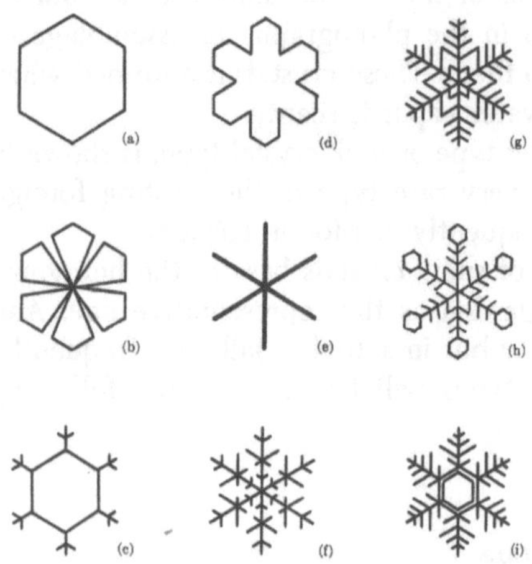

FIG. 18. Classification of regular plane crystals: (*a*) simple plate; (*b*) branches in sector form; (*c*) plate with twigs; (*d*) broad branches; (*e*) simple stellar form; (*f*) dendritic form; (*g*) fernlike form; (*h*) dendritic form with plates; (*i*) plate with dendritic extensions.

FIG. 19 [17]. Fernlike crystal (× 16.5).

crystal [Fig. 18(*b*)] is the one very close to the plate type. We can think of the broad-branched dendritic type [Fig. 18(*d*)] as an intermediate type between the plate and the dendritic.

Next, there is a combination of the plate or sector type and dendritic branches. It sometimes happens that the plate or sector crystal is formed in the upper region of the atmosphere, and as it nears the ground it passes through an atmospheric layer that is suitable for the dendritic development. Depending on the thickness of this atmospheric layer and the weather conditions, the crystal may become a plate with simple extensions at the corners [Fig. 18(*c*)], or a large dendritic type with a plate at the center [Fig. 18(*i*)].

There is a converse type to this plate with dendritic branches, namely, the dendritic crystal with plates at the point of its six branches. We call this the dendritic with plates [Fig. 18(*h*)]. In our observations in Hokkaido, this type was rather rare.

The afore-described crystals are classified into nine types as shown in Fig. 18. However, hexagonal plane crystals have almost uncountable variations so that they can not be definitively classified into these nine types. As a matter of fact there are many successive intermediate forms, but for the convenience of description, such intermediate forms are included in this book under some one of these nine types.

5. Forms of hexagonal crystals

(i) *Fernlike hexagonal crystal.* The fernlike hexagonal crystal is the largest of the snow crystals. The diameter of a large one is 8 to 9 mm, and the form and structure of such large crystals can easily be studied by the naked eye. Crystals of this type with complete development which could serve as models were often observed at Mount Tokachi. One of the best examples of this type that we found is shown in Fig. 19.

The crystal of Fig. 19 is an especially beautiful one. Crystals of this type are found to be always very thin in our climate. Viewed from the side, they have the appearance shown in Fig. 20, which is a side view of a crystal similar to that in Fig. 19. The central part is somewhat thick and the crystal becomes thinner toward the points. To take a side-view photograph of this type, the crystal must be erected vertically on its side upon the slide. We found the following method convenient for this purpose: Wet an end of a matchstick with saliva and make a small stain of saliva upon the glass by tapping it with the stick. The saliva remains liquid for a short time in a supercooled condition. Break another matchstick and hang a crystal from the ragged end of the broken match. Carefully bring the crystal near the saliva stain. As soon as the point of a crystal branch touches the supercooled saliva, it suddenly freezes and the crystal is fixed vertically on the glass.

In measuring the average thickness of crystals of this type, we first measured the weight of a crystal by the method detailed in Sec. 34, Chapter 3, and then measured its area upon an enlarged photograph of the plane figure with a planimeter. From these data we could calculate the mean thickness, and we found that the average value of the mean thickness of the fernlike type and regular dendritic type is approximately 0.01 mm. This figure applies to the crystals with a complete dendritic development as shown in this book. Hexagonal plane crystals which are observed in Manchuria are reported to be thicker. In view of the fact that Humphreys [12] meant by the term "plane crystals" those whose thickness did not exceed one tenth of the diameter, thicker crystals seem to be often observed in the United States.

The sharp outline and distinct internal structures of a crystal like this show that it has been continuously developed under the conditions most suitable for dendritic development until just before the observation. Usually the growth of the crystal is temporarily stopped at a place near to the ground either by an increase of temperature or the lack of sufficient water vapor. In such cases, evaporation due to sublimation occurs and the fine structure of the crystal branches tends to disappear. Figure 21 shows an example of such a crystal. As a principle, crystals a part of which has been evaporated by sublimation are excluded from the photographs of this book.

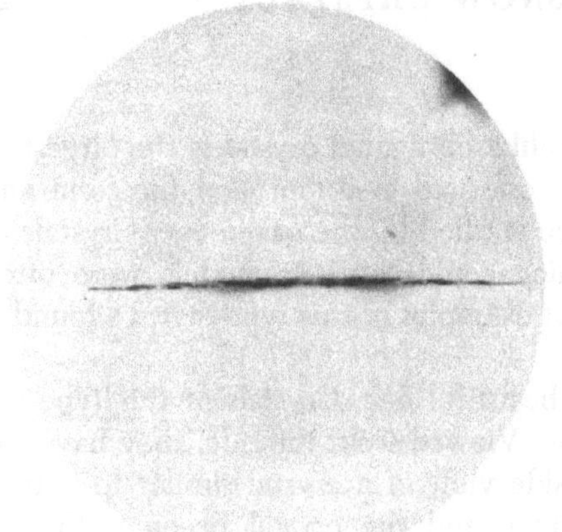

Fig. 20 [18]. Side view of a fernlike crystal (× 17).

Fig. 21 [32]. Slightly sublimed crystal.

Fig. 22 [23]. Crystal showing partial asymmetry (× 21).

Fig. 23 [61]. Ordinary dendritic crystal (× 19).

Fig. 24 [70]. Ordinary dendritic crystal (× 27).

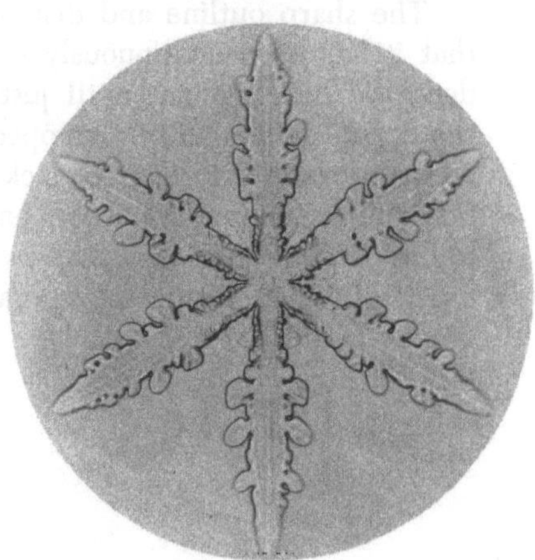

Fig. 25 [76]. Ordinary dendritic crystal, partially sublimed.

Even the fernlike hexagonal crystals of a quite regular shape do not show complete symmetry when observed minutely. The crystallization system can decide only the angles of branches but not the external form. Therefore some of the branches naturally have different forms. For example, the crystal of Fig. 22, representing a hexagonal symmetry as a whole, shows a marked deviation from perfect symmetry. Crystals No. 21, 22, 24, 25, 27, and 33, in Plates 2, 3, 4, and 7, show other examples. However, crystals that can be considered to have hexagonal form as a whole are included in this section and those that have remarkably irregular forms are included in the malformed type mentioned later. A sharp distinction between the two is impossible. Other crystals belonging to this group are reproduced in Plates 7–12.

(ii) *Ordinary dendritic crystals.* Two examples of ordinary dendritic crystals are shown in Figs. 23 and 24. Figure 23 is a representative example of this type, and Fig. 24 is a somewhat simpler form. Crystals with fewer twigs are also included in this group, as shown in Fig. 25. The form of this type is closely connected with the stellar type with a small number of twigs, and the distinction of these two is made for the convenience of description. There is another group of crystals, the form of which cannot be clearly distinguished from the simpler form of fernlike type. Many intermediate types between this and the fernlike one are observed. One example of a slightly irregular shape is shown in Fig. 26. This crystal is to be regarded as an intermediate type between the present one and the broad-branched type to be described later. Figure 27 is closer to the sector type with dendritic extensions. Thus there are various intermediate forms between the various types. They are shown in Plates 13–24.

(iii) *Dendritic crystals with plates.* This type is a dendritic or stellar crystal with plates attached to the ends of branches. Primarily it means such a crystal as is shown in Fig. 28. Crystals of this type are rarely observed in Hokkaido. Similarly shaped crystals are the dendritic or stellar ones with the tips of branches slightly broadened. A good example is shown in Fig. 29. This type is also included in this group. Other similar types are those with extremely short stellar parts, as shown in Nos. 129 and 130, Plate 27. Other examples of these crystals are collected in Nos. 115–132, Plates 25–27.

(iv) *Stellar crystals.* A model example of stellar crystals is shown in Fig. 30. Such a complete stellar crystal is very rarely observed in Hokkaido. Most examples of this type observable in our climate have short twigs from stellar branches, and are to be considered as the intermediate type between this and the dendritic one. Stellar crystals with small plates at the center, like Fig. 31, are often seen. Other crystals of this type are collected in Nos. 135–152, Plates 28–30.

(v) *Crystals with broad branches.* Dendritic crystals with broadened branches are named the broad-branched type. Two representative crystals of this kind are shown in Figs. 32 and 33. Besides these types, there is an intermediate type be-

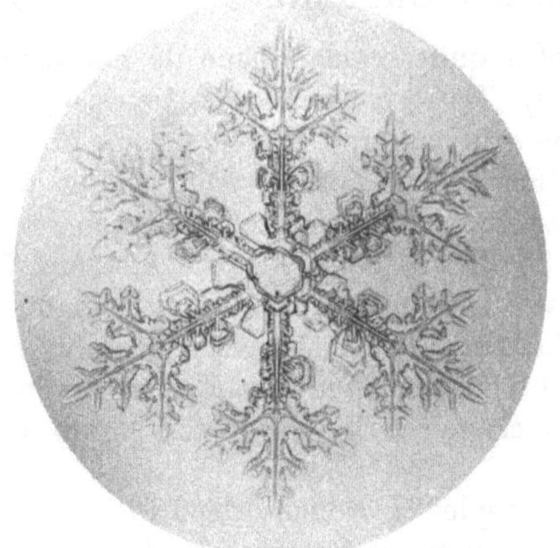

Fig. 26 [73]. Intermediate state of dendritic and broad branches (× 20.5).

Fig. 27 [74]. Sector type with dendritic extensions (× 21).

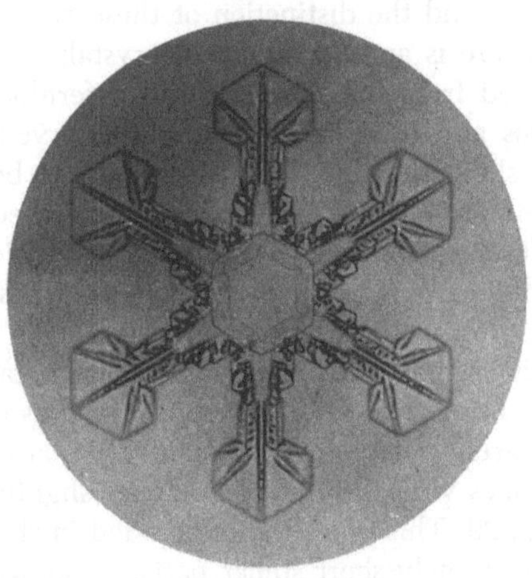

Fig. 28 [125]. Stellar with plates at the ends (× 25).

Fig. 29 [113]. Stellar with branching at the ends (× 20).

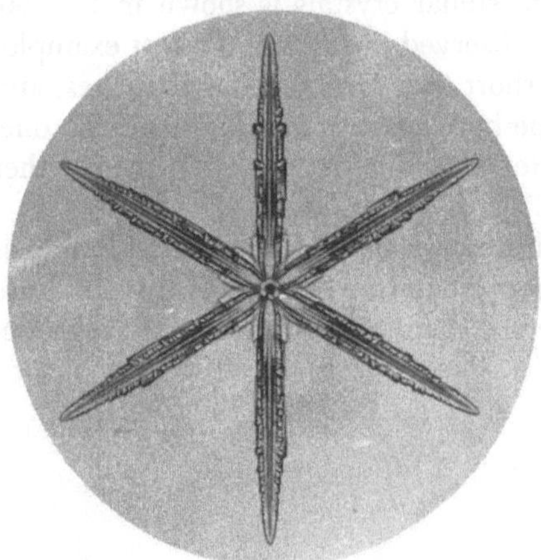

Fig. 30 [133]. Stellar crystal (× 33.5).

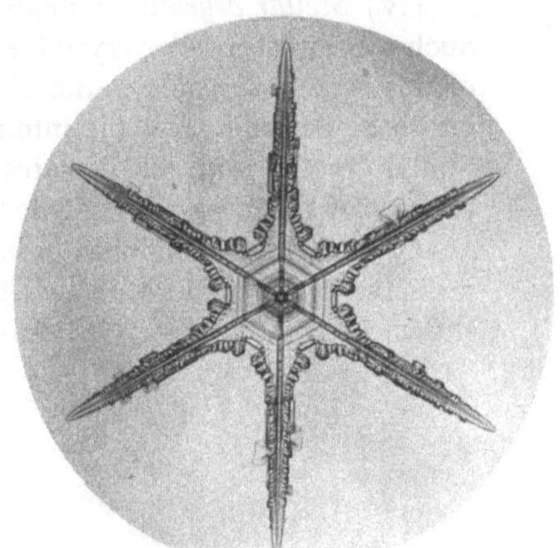

Fig. 31 [134]. Stellar crystal with a plate at the center (× 27.5).

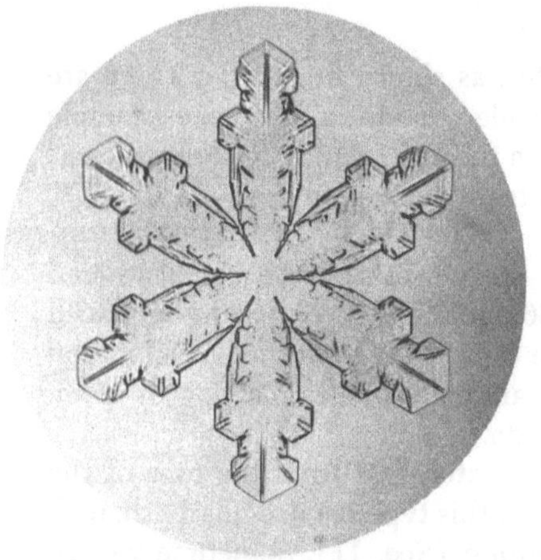

FIG. 32 [153]. Broad branches (× 26).

FIG. 33 [156]. Broad branches (× 39).

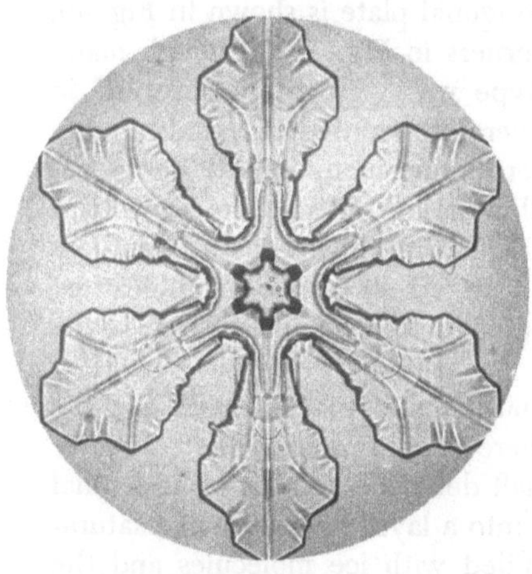

FIG. 34 [162]. Intermediate type of broad branch and plate (× 48.5).

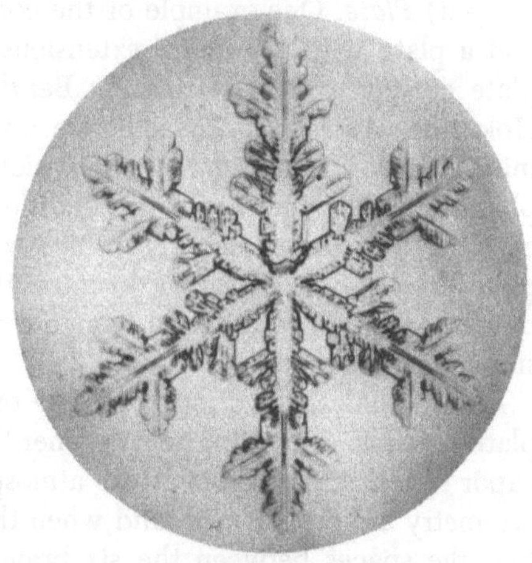

FIG. 35 [163]. Intermediate type of dendritic and broad branch (× 25).

FIG. 36 [178]. Sector form (× 37).

FIG. 37 [179]. Sector form (× 37).

tween the above-mentioned dendritic with plates, as shown in Fig. 34. There are other crystals that are closer in form to the regular dendritic ones, one example being reproduced in Fig. 35. These are included in this type simply for convenience' sake. The other examples belonging to this category are seen in Nos. 159–176, Plates 31–33.

(vi) *Sector-form crystals.* The sector-form crystal consists of six sectors gathered to a center. Two examples are shown in Figs. 36 and 37. This type is to be placed between the broad-branched type and the plate type, both from the consideration of the condition of its formation and from the standpoint of classification by form. Not many of this type were observed in Hokkaido.

Some sector crystals have dendritic or stellar extensions from the tops of the six sectors, as shown in Fig. 38. Often the sectors of this type are irregularly shaped. These crystals are included in the malformed group (Sec. 11). Sometime we see sectors developing at the corners of a hexagonal plate. One good example is shown in Fig. 39.

(vii) *Plate.* One example of the complete hexagonal plate is shown in Fig. 40, and a plate with a trace of extensions at the corners in Fig. 41. Although many plate crystals were reported by Bentley, this type was very rarely observed in Hokkaido. At Mount Tokachi, we observed several times many crystals in the initial stage of formation, among which there were found many minute hexagonal plates. But the meteorologic condition that allows plate crystals to grow to a considerable size is very rarely met with in Japan. The dimensions of Bentley's plate crystals are unknown, because they lack a statement of the magnification, but they seem to be different from our minute plate which is the initial stage of the snow crystal proper.

It is usual that various patterns of hexagonal symmetry are seen inside the plate crystals. According to Wegener,[13] when there is an ample amount of water vapor in the upper layers of the atmosphere, small dendritic crystals of hexagonal symmetry are first formed, and when they come into a layer with less supersaturation the spaces between the six branches are filled with ice molecules and the crystals become plates. When the plates come down to the next layer with much supersaturation, branches develop from the plate corners. This plate with dendritic branches becomes a larger plate again in the next drier layer. This theory gives a pretty clear explanation of the cause of the development of various symmetrical patterns inside the plate crystals. So Wegener's theory has been admitted by many people in this line. But we came to know, as a result of our experiments on artificial snow described in Part II, that the spaces between dendritic branches are rarely filled up by ice molecules when the degree of supersaturation becomes less. Plate crystals are not frequently observed. Examples are shown in Nos. 191 to 196, Plate 35.

(viii) *Plate with dendritic extensions.* Plates with dendritic extensions at the

FIG. 38 [184]. Sector form with dendritic extensions (× 36).

FIG. 39 [182]. Plate with sector extensions.

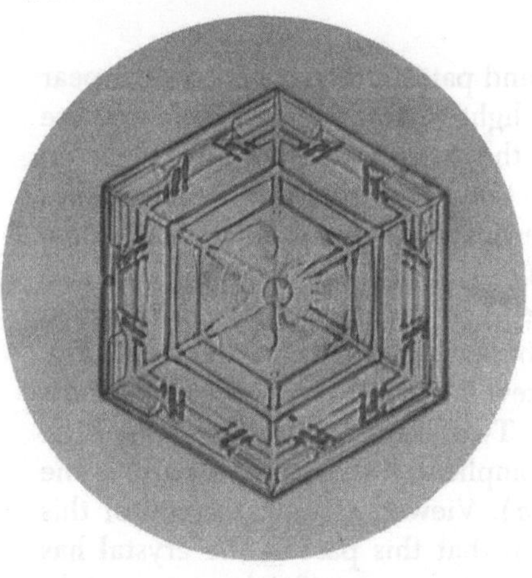

FIG. 40 [189]. Hexagonal plate (× 38.5)

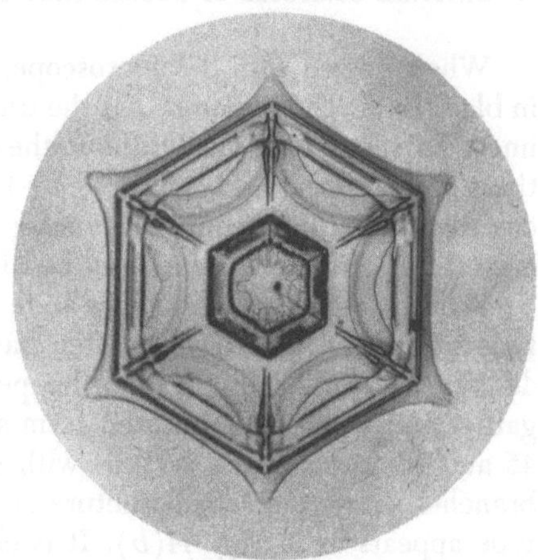

FIG. 41 [190]. Hexagonal plate with a trace of extensions (× 45).

FIG. 42 [197]. Plate with fernlike extensions (× 24).

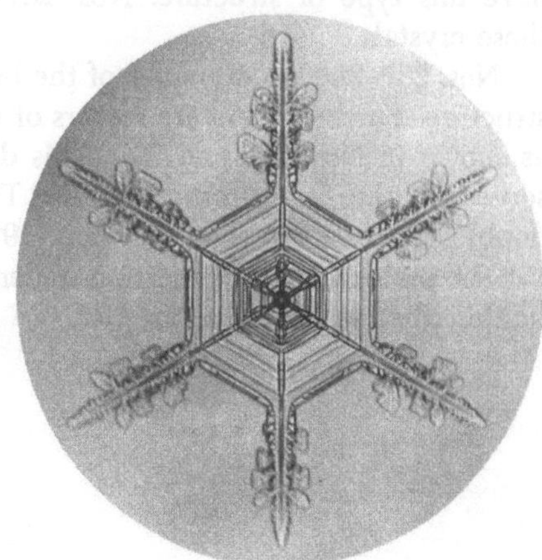

FIG. 43 [203]. Plate with simple dendritic extensions (× 20).

corners are already well known because of their beauty in form and structure. Two representative examples of this type are shown in Figs. 42 and 43. There are varied complexities in the dendritic extensions; in an extreme case the extensions may be the straight branches as shown in No. 281, Plate 54. Another example is the one with six small plates attached to the central plate as shown in No. 211, Plate 42, and No. 267, Plate 52. Most other forms take intermediate shapes between these two, with considerable variation. In any case, when the upper layer of atmosphere is the one in which plates are formed and there are various different meteorological conditions nearer to the earth's surface, this type of crystal is developed. Examples of this type are collected in Nos. 198–284, Plates 36–54.

There are many crystals of this type in Bentley's photographs, and Shedd [14] based his theory of the relation between the growth of the crystal and meteorological conditions upon this material, but it has no experimental basis.

6. Crystals composed of double sheets

When viewed with the microscope, the rim and pattern of snow crystals appear in black lines. The reason is that the transmitted light is totally reflected along those lines. To speak specifically about the pattern, the narrow ditches and ridges on the surface appear in black lines by total reflection. The structure of the ditches or canals along the dendritic branches was investigated in detail on artificial snow crystals. The result will be described in Part II.

When the dendritic branches were broadened and became the broad-branched type mentioned before, they often have the characteristic pattern shown in Fig. 44(*a*). In natural snow, it often happens that several of the same sort of patterns gather and yield a complicated form structure. Two examples are shown in Figs. 45 and 46. Even in the crystals with such a complicated structure, a part of the branches shows the same structure as Fig. 44(*a*). Viewed aslant, a crystal of this type appears as in Fig. 44(*b*). It is clearly seen that this part of the crystal has two sheets. We first noticed this double-sheet structure on artificial snow crystals, but our later observation revealed that there are many natural snow crystals that have this type of structure. Nos. 287–304, Plates 55–57, are other examples of these crystals.

Nos. 287–294 are examples of the broad-branched type which have this double structure. Further, there are sectors of the same structure developing from a plate, as shown in Nos. 296 and 297. This double structure was often very clearly observed in the sector-form crystals. The examples of sector-form crystals with double structure are shown in Nos. 299–304, Plate 57.

The meteorological condition under which this double structure is formed will be described in detail in Part II.

7. Hexagonal crystals with a small plate at the center

In many cases, the central part of the dendritic hexagonal plane crystals, when closely observed, shows the existence of a small hexagonal plate. Examples are shown in Nos. 309–314, Plates 58 and 59. Small plates are also seen in the crystals mentioned heretofore, oftener than expected. Such crystals comprise a considerable percentage of all the hexagonal plane crystals. Examples in which the existence of a plate is clearly recognized are Nos. 1, 2, 17, 32, 43, 44, 46, 47, 48, 49, 51, 56, 69, 71, 77, 82, 84, 87, 88, 92, 97, 98, 101, 103, 109, 115, 117, 125, 126, 127, 136, 137, 138, 148, 155, 156, 158, 166, 167, 169, 171, 172.

In our study of artificial snow, observation of the initial stage of crystal formation revealed the fact that first a small skeleton crystal in the form of a short column is formed and then one of the bases alone is often developed into a hexagonal plane crystal proper. In such a case, the growth of the other base is stopped and it remains as a small hexagonal plate. A side view of such a crystal has the form **ⵊ**; viewed from above, the small hexagonal plate appears at the center of a dendritic plane crystal. The small plate does not lie in the same plane as the dendritic crystal, but a little above it.

Later we investigated on natural snow whether this central plate exists in the same plane as the dendritic extensions, by altering the focus of the microscope. With most such crystals the result was just as we had expected. We could also take two photographs – plan view and side view – of one crystal of this type, as shown in Figs. 47 and 48. Another example will be seen in Nos. 305, 306, Plate 58. The two small plates at the center of the crystal of Fig. 47 correspond to the plates above and below the hexagonal plane crystal proper, as seen in the side view of Fig. 48.

In some examples one base of the skeleton, without stopping at the stage of minute plate, grows to a small hexagonal dendritic extension, as shown in Nos. 26, 28, 39, 58, 59, 72, 74, 76, 79, 114, 142. Such crystals are naturally to be expected, according to our knowledge on the course of formation of a snow crystal. Detailed descriptions of the development of a crystal from the initial stage of the skeleton type will be given in the later article dealing with the artificial snow.

8. Combinations of dendritic crystals

Even when the temperature is as low as $-10°C$, the dendritic crystals often fall in the form of several crystals combined. One example is given in Fig. 49. In many such cases, crystals are combined at a considerable height and then fall through the same path, thus resulting in the same form of development. Figure

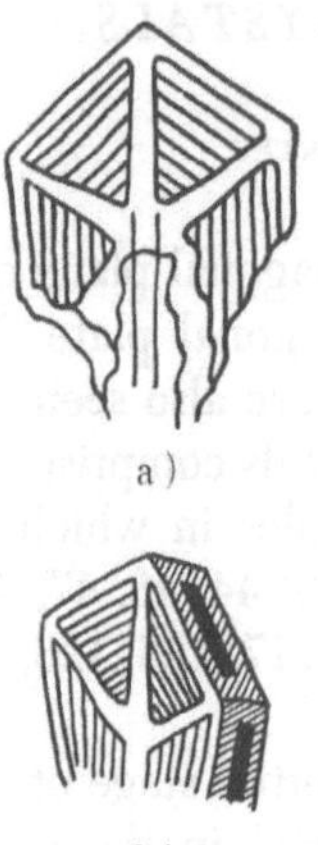

Fig. 44. Double-sheet structure

Fig. 45 [285]. Dendritic crystal showing double-sheet structure (× 22).

Fig. 46 [286]. Dendritic crystal showing double-sheet structure (× 25).

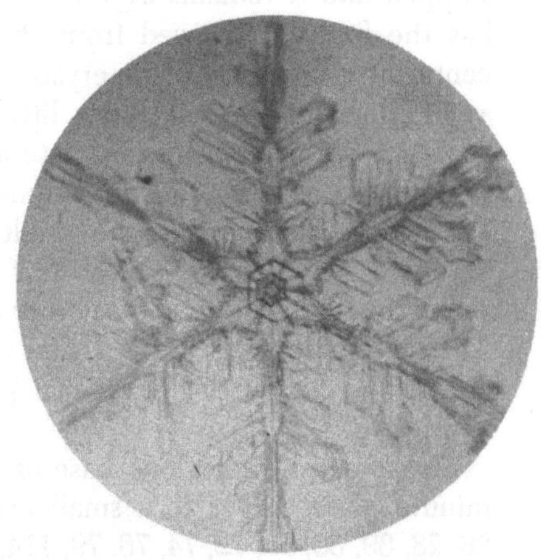

Fig. 47 [307]. Plan view of a dendritic crystal (× 19).

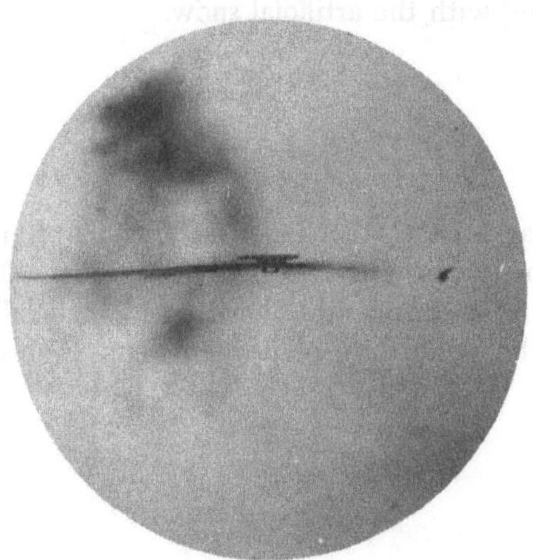

Fig. 48 [308]. Side view of the crystal of Fig. 47 (× 19).

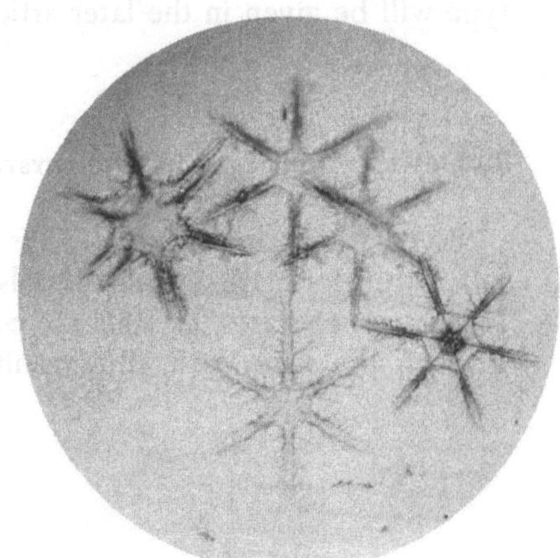

Fig. 49 [324]. Combination of dendritic crystals (× 10.5).

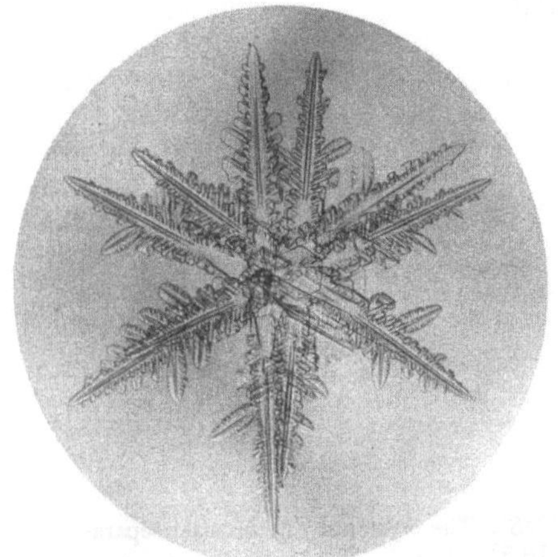

FIG. 50 [332]. Combination of two dendritic crystals (× 21).

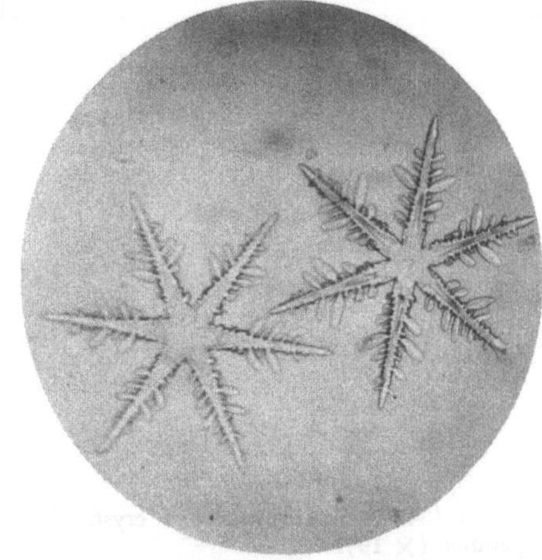

FIG. 51 [323]. Combination similar to Fig. 50, separated into components (× 18).

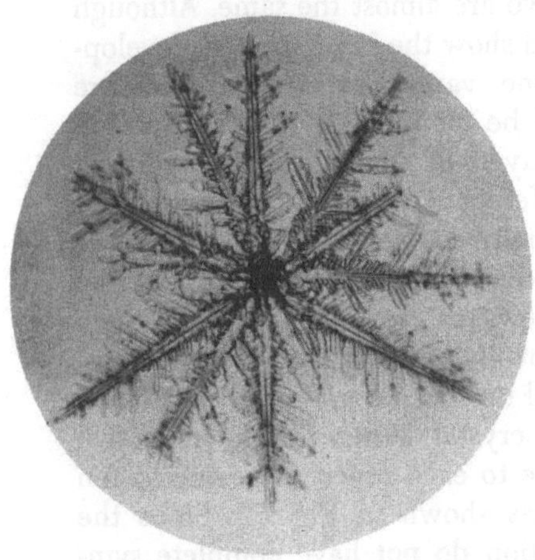

FIG. 52 [330]. Irregular twelve-branched crystal (× 18).

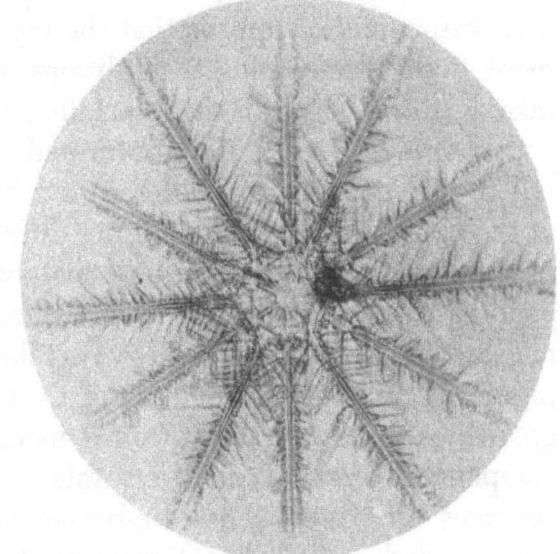

FIG. 53 [335]. Twelve-branched crystal, fernlike branches (× 16).

FIG. 54 [338]. Twelve-branched crystal, broad branches (× 28.5).

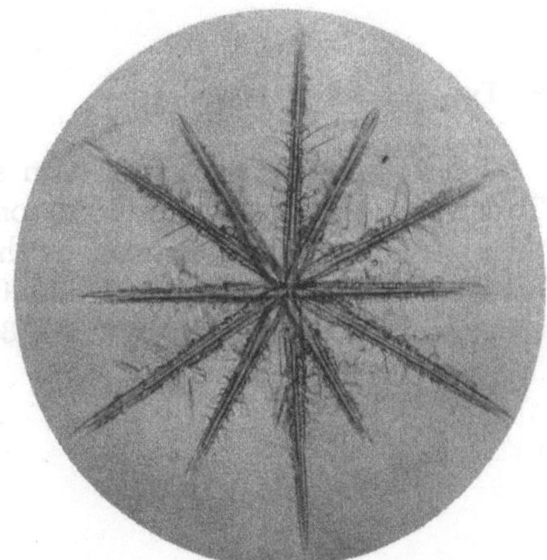

FIG. 55 [344]. Twelve-branched crystal, long and short branches (× 20.5).

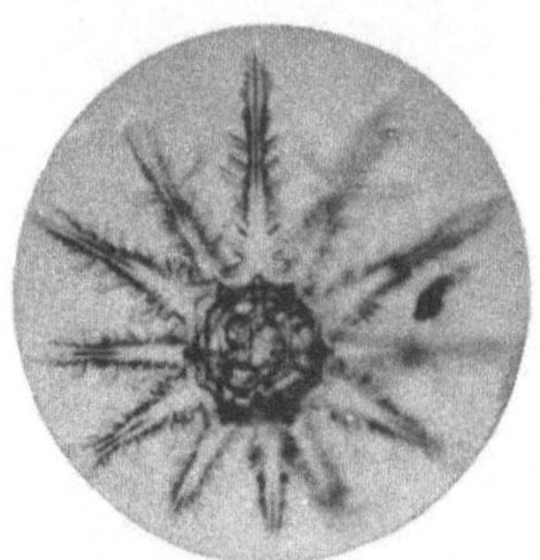

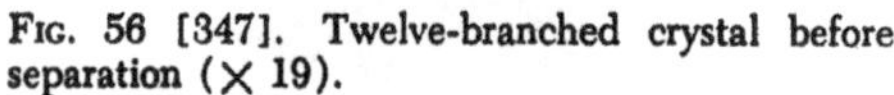

FIG. 56 [347]. Twelve-branched crystal before separation (× 19).

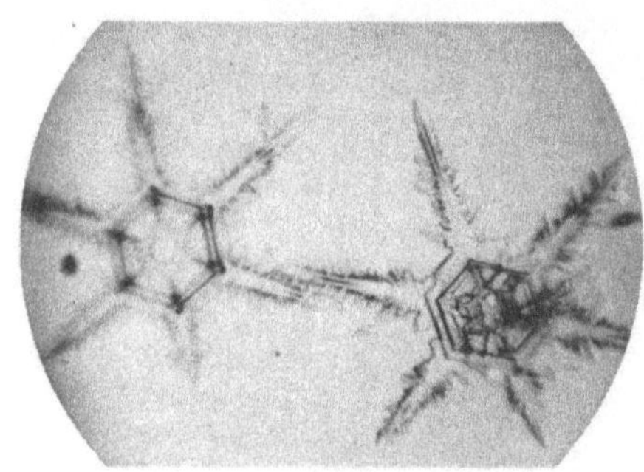

FIG. 57 [348]. The crystal of Fig. 56 after separation (× 18.5).

50 is an example of two crystals combined. A similar one separated into its components is shown in Fig. 51. The shapes of the two are almost the same. Although it is reasonable to suppose that the crystals should show the same form of development if the meteorologic conditions are the same, various types of crystals are mixed in one snowfall at a certain time, as will be mentioned later. The reason is that different types are first formed in various layers of the atmosphere and then fall at different speeds. Therefore, a mixture of different types of crystals is observed at the earth's surface. The minute differentiation in one type of crystals observable in one snowfall at a certain time may be accounted for by the fact that the external conditions are quite different, in a microscopic view, for each of the crystals. Such microscopic fluctuations are always taking place, owing to disturbance in the atmospheric layers. Only in the case of these combined crystals do we see the identical growth of form and structure for more than one crystal. When the nuclei of two components of this type of crystals are very close to each other, the combination becomes an irregular twelve-branched crystal, as shown in Fig. 52. Since the twelve-sided crystals mentioned in the next section do not have complete symmetry, the distinction between the irregularly shaped twelve-sided ones and the crystals of this type is not clear.

9. TWELVE-SIDED CRYSTALS

Twelve-sided crystals have been so long known that the *Sekka Zusetsu* by Toshitsura Doi published in 1832 contains seven sketches of this type. Most of them have dendritic extensions, as shown in Fig. 53, but rarely we find broad-branched ones, as shown in Fig. 54. Other examples of twelve-sided crystals are collected in Nos. 331, 334, 336, 337, 339–343, 345, 346, 352, Plates 63–66.

TABLE 1. Angles between branches of twelve-branched crystals (see Fig. 59).

No.	I	II	III	IV	V	VI
335 *	27.0°	31.5°	28.0°	34.0°	27.0°	32.0°
336 *	27.0	32.0	28.0	32.5	28.5	32.0
337 *	31.0	31.0	29.0	30.5	30.0	28.5
338 †	26.0	34.5	26.0	34.5	25.0	34.5
339 *	28.5	30.5	30.0	30.5	30.0	30.5
340 †	21.0	39.0	20.5	40.0	20.5	40.0
341 †	25.5	34.0	27.0	35.0	26.0	33.5
342 ‡	16.0	44.5	15.5	46.0	15.5	43.0
343 †	29.0	31.5	28.5	31.0	29.5	31.0
344 †	29.0	31.0	30.0	32.0	27.0	31.0
345 ‡	34.0	27.5	34.0	25.5	31.5	29.5
346 †	24.0	34.5	25.5	36.0	25.5	34.5
347 †	31.0	31.0	28.0	30.5	29.5	30.0
352 ‡ left	15.5	45.0	14.5	45.5	14.5	45.5
352 ‡ right	29.0	32.5	31.0	29.5	31.0	28.5

In many of the twelve-sided crystals, six alternate branches are a little longer than the remaining six between them. The difference in length is hardly conspicuous in general, as is seen in Figs. 53 and 54, but rarely the difference is remarkable, as shown in Fig. 55. As all of the twelve-sided crystals can be separated into two crystals, we can see that every other branch belongs to a different crystal. In other words, two hexagonal plane crystals, one attached upon the other, make the twelve-sided crystal.

According to Shedd,[15] Bentley already knew that the twelve-sided crystal can be separated into two hexagonal crystals. We also tried the separation on many crystals, one example of which is given in Figs. 56 and 57. Figure 56 is the crystal before separation. Figure 57 shows the two crystals obtained by separating the crystal of Fig. 56. No. 349, Plate 66, is another twelve-sided crystal after separation.

A photograph of one of the crystals obtained by separating a twelve-sided crystal is shown in Fig. 58. No. 351, Plate 66 is a side view of the similar crystal. Both of them have a small hexagonal plate at the center. Shedd considers that the twelve-sided crystal is two hexagonal crystals connected by a short column, but it seems to us that most of them consist of two hexagonal crystals overlapped, each of which had developed from a base of a skeleton column in the initial stage of crystal formation.

The problem is why the branches of one crystal are so situated as to divide evenly the angle formed by two branches of the other crystal. Some have attributed it to the electrical charge of the two crystals, but the theory has no reliable basis. The actually measured angles between the branches are not always constant. The angles between branches were measured for Nos. 335–352 and the results are shown in Table 1.

Those marked by an asterisk (*) in the table have twelve branches radiating

TABLE 2. Angles between branches of components of twelve-branched crystals (see Fig. 59).

No.	θ_1	θ_2	θ_3	θ_1'	θ_2'	θ_3'	Distance OO' between centers (mm)
338	60.0°	61.0°	59.0°	60.0°	59.5°	60.5°	0.14
340	59.5	60.5	60.0	59.0	60.0	61.0	.12
341	59.0	62.0	59.0	61.0	60.0	59.0	.10
343	60.0	60.0	60.0	59.5	60.5	60.0	.07
344	58.5	61.5	60.0	60.0	60.5	59.5	.12
346	59.0	61.0	60.0	60.5	61.0	58.5	.09
347	59.0	62.0	59.0	60.0	60.0	60.0	.08

from one point at the center. A dagger (†) shows the crystals whose two centers are not exactly coincident, as shown in Fig. 59. Measurement of the angles between the branches of each of the two component crystals in the latter case, namely, $\theta_1, \theta_2, \ldots, \theta_3'$ of Fig. 59, reveals that each angle is approximately 60°, as shown in Table 2. This fact shows that two crystals are attached one upon the other with their centers a little apart. The distance between the centers, OO', is about 0.1 mm, as seen in the last column of Table 2.

The crystals marked with a double dagger (‡) in Table 1 are those one or both of whose component crystals have not a complete hexagonal development with six branches radiating from one point. It being impossible to find two centers in this case, these crystals will be excluded in our following analysis. In the case of those marked † the angles between branches do not change even if one crystal is displaced without rotation and placed upon the other. Therefore, if we take the mean value of the angles I, III, V, and of angles II, IV, VI, respectively, we can find out whether they are the regular twelve-sided crystals or not. Deviation of the mean angles from 30° is shown in Table 3.

If deviations up to 1° can be regarded as within the error of observation, five of the eleven examples can be considered as the regular twelve-sided crystals. As for the four examples that show a remarkable deviation, namely, Nos. 338, 340, 341, and 346, attention ought to be called to the fact that the centers of the two component crystals are considerably apart.

From the above results, it was clarified that in the twelve-sided crystals with coincident centers branches of one crystal have the tendency to divide equally the angles between branches of the other crystal, which has direct relation to the crystal structure, and that when the centers are not coincident, deviation occurs in the angles.

Sometimes an eighteen-branched crystal is observed, which is made up by overlapping of three component crystals. An example is shown in Fig. 60. The mode of overlapping of three component crystals may be clearly seen in the side view, Fig. 61.

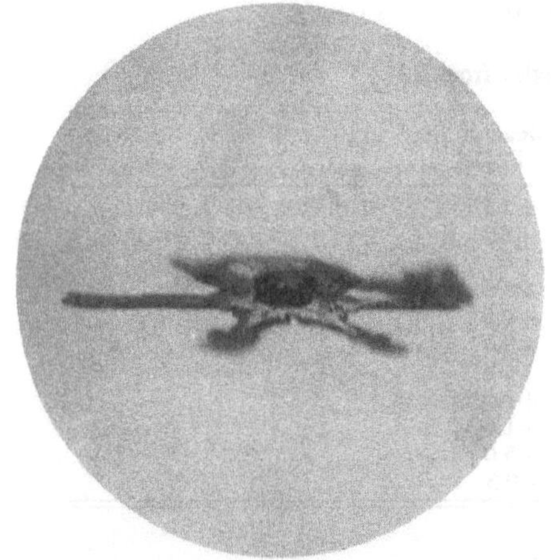

FIG. 58 [350]. Side view of a component of a twelve-branched crystal (× 28).

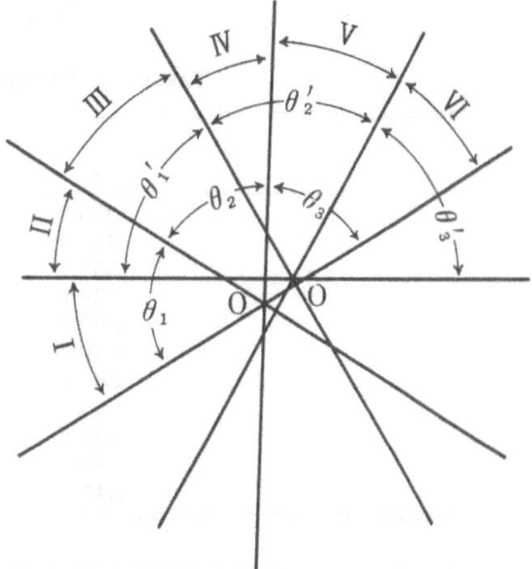

FIG. 59. Angles between branches of twelve-branched crystal.

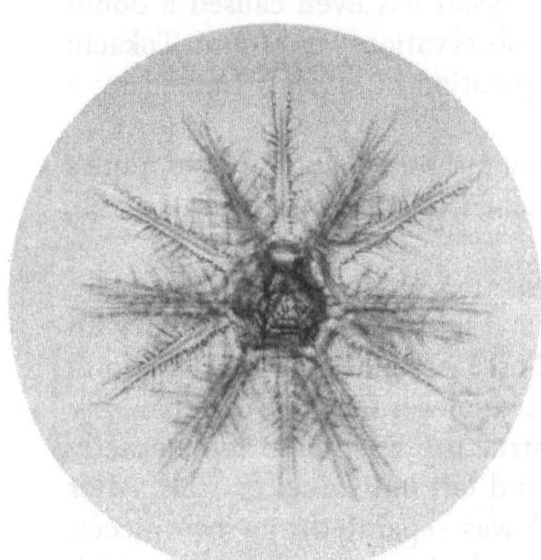

FIG. 60 [353]. Eighteen-branched crystal (× 20).

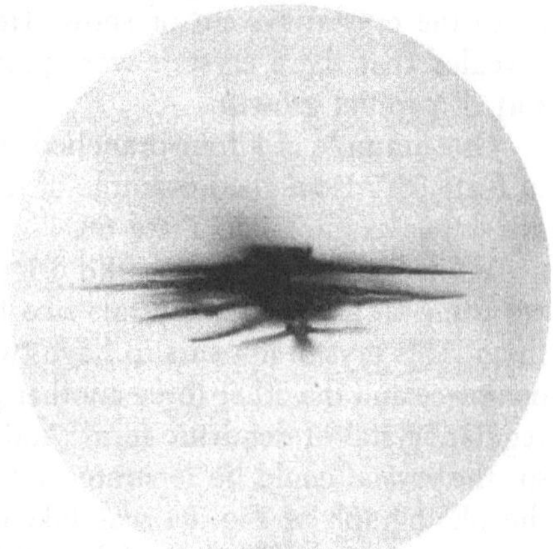

FIG. 61 [354]. Side view of Fig. 60 (× 21).

FIG. 62 [356]. Four-branched crystal (× 18).

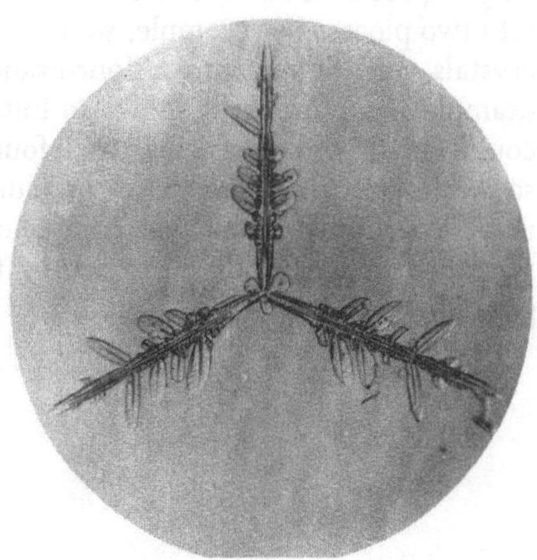

FIG. 63 [357]. Three-branched crystal (× 18).

TABLE 3. Deviation of mean angles from 30°.

No.	Mean of I, III, V	Mean of II, IV, VI
335 *	−2.5°	+2.5°
336 *	−2.2	+2.2
337 *	0	0
338 †	−4.5	+4.5
339 *	−0.5	+0.5
340 †	−9.5	+9.5
341 †	−4.0	+4.0
343 †	−1.0	+1.0
344 †	−1.0	+1.0
346 †	−5.0	+5.0
347 †	−0.5	+0.5

10. THREE- OR FOUR-BRANCHED CRYSTALS AND THE LIKE

It is recorded that three- or four-branched crystals were occasionally observed. In fact, the extraordinary form of four-branched crystal has even caused a doubt as to the crystal system of snow. However, our observations at Mount Tokachi revealed that these crystals were produced by separation of crystals formed by a sort of parallel growth.

One example of a four-branched crystal is shown in Fig. 62, and other examples in Plate 68. Also a good example of the three-branched crystal is shown in Fig. 63, and other examples in Plate 68.

While we were studying the origin of the formation of these sorts of crystals, we came to know that crystals like that shown in Fig. 64 often fell mixed with them. This crystal appears to have two centers. Three branches on the right form one piece and the other three another piece, these two pieces forming an apparently regular crystal of dendritic form. And the actual structure was found to be exactly so: the crystal could be separated into two right and left individual crystals. After the photograph of Fig. 64 was taken, the crystal was separated into two pieces, as shown in Fig. 65. A crystal of this type is already known and a photograph similar to Fig. 64 is reproduced in Dobrowolski's book.[16]

Likewise, we examined various hexagonal crystals which appeared to be ordinary regular ones and found that a considerable number of them could be separated into two pieces. For example, some of them are separated into two three-branched crystals, and others into a four-branched crystal and a two-branched one. An example of the former is shown in Fig. 66 and the latter in Fig. 67. In this way we could obtain three-branched and four-branched crystals in an artificial manner, so we inferred that the three- or four-branched crystal which had long been an unsettled problem was one component of these combined crystals which had separated in a natural course during the fall or at the moment they reached the ground.

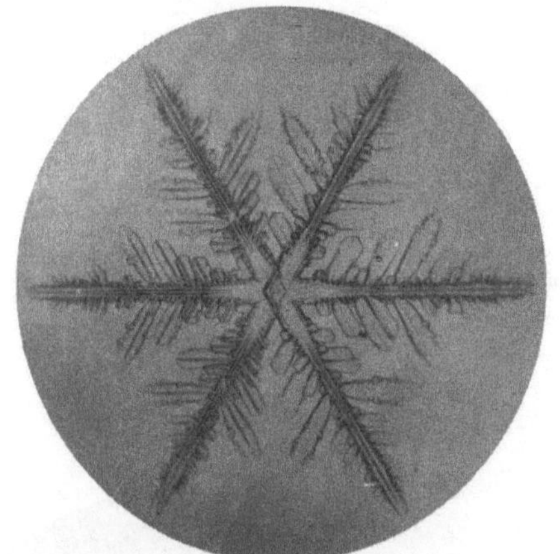

FIG. 64 [359]. Two three-branched crystals, before separation (× 19).

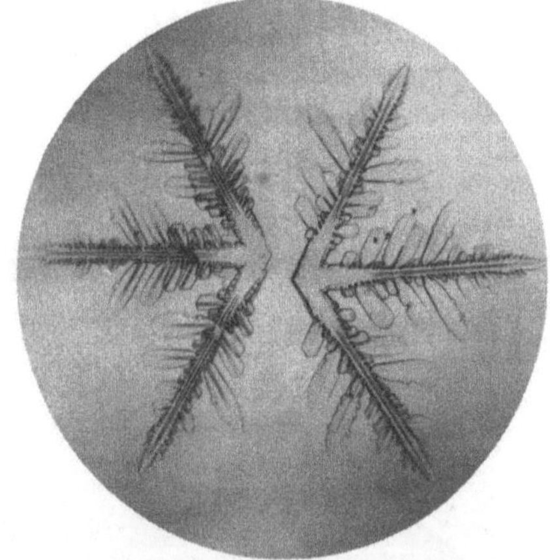

FIG. 65 [360]. The crystals of Fig. 64 after separation (× 19).

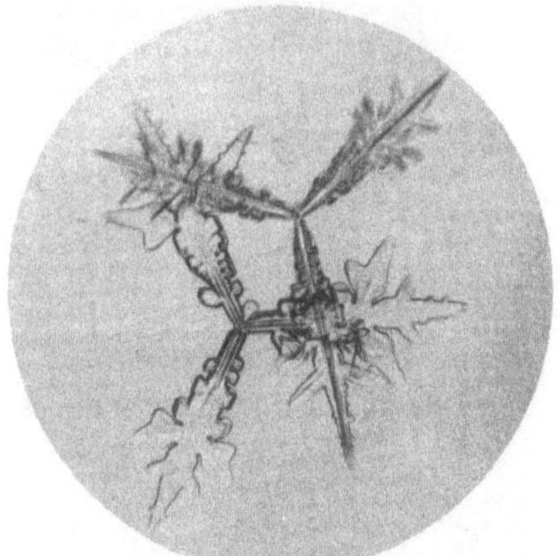

FIG. 66 [361]. Two three-branched crystals (× 16.5).

FIG. 67 [362]. Two-branched and four-branched crystals (× 18.5).

However, the problem still remains of the central structure of these separable crystals. If we examine the photograph closely, we can recognize that both the three-branched crystal of Fig. 63 and the four-branched one of Fig. 67 belong actually to the hexagonal type. The branches that appear to be lacking are so short and thin that they are apt to be overlooked. It can readily be imagined that these thin branches sublime very easily and they often disappear before they are brought to microscopic observation. In case of the genuine four-branched and

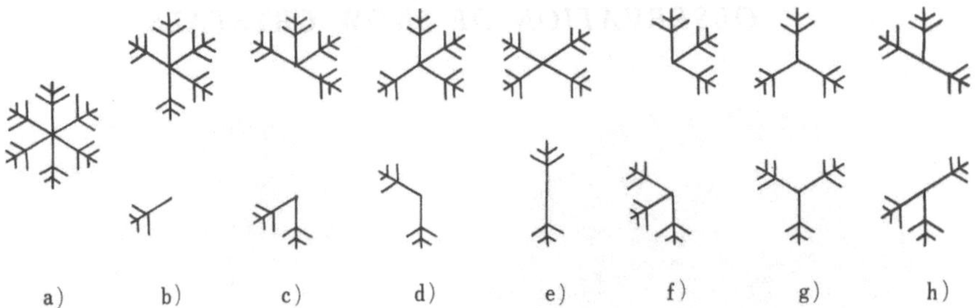

Fig. 68. Modes of combination of two components.

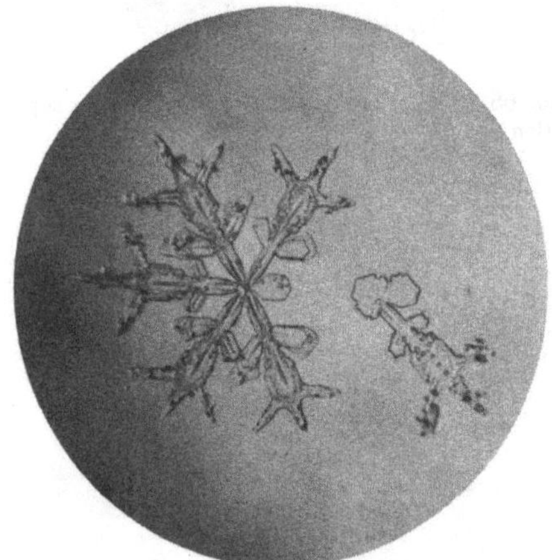

Fig. 69 [363]. Type (*b*) of Fig. 68 (× 49.5).

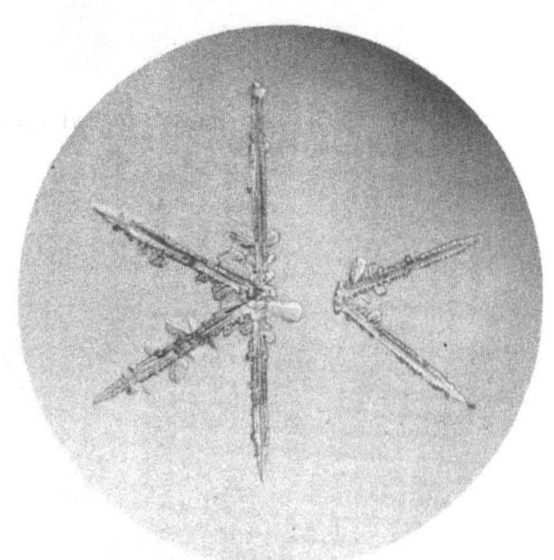

Fig. 70 [364]. Type (*c*) of Fig. 68 (× 19).

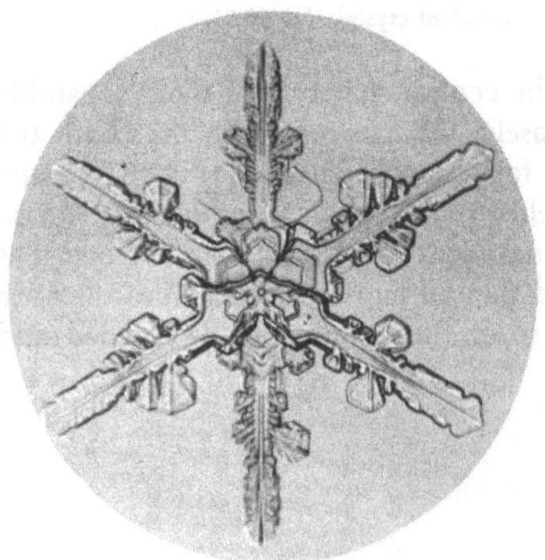

Fig. 71 [425]. Type (*d*) of Fig. 68 (× 36).

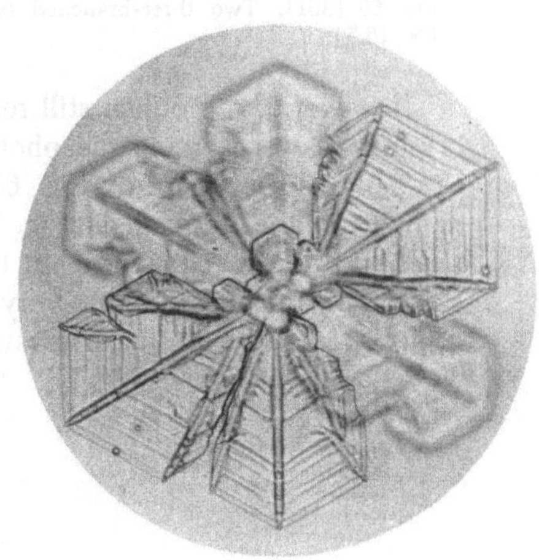

Fig. 72 [519]. Type (*h*) of Fig. 68 (× 42).

three-branched crystals shown in Fig. 62, and No. 358, Plate 67, such minute branches would have already sublimed before observation or would not have been developed at all.

This fact leads us to infer that crystals of this type have two centers, one lying upon the other, and that a few branches accidentally grew out of one center and the remaining branches from the other, both groups jointly forming an apparent hexagonal development.

The origin of two three-branched crystals such as that shown in Fig. 66 will be explained as follows. At the initial stage of the growth of a crystal, three alternate branches belonging to one center and the other three branches of the other center begin by chance to develop respectively just a little earlier than the corresponding ones belonging to the opposite nucleus. Then the supply of water vapor from the surrounding atmosphere to the branches of crystal is chiefly attracted to the elongated branches, so that three branches belonging to one center are fully developed, while the branches of the other center just underneath these developed ones remain in a shriveled state, and vice versa. A twin crystal grown in this manner must be composed of two components which can be easily separated by a slight external force, thus yielding the three-branched crystals in question.

From the standpoint of this two-center theory, seven pairs of component crystals can be expected, if it is assumed that the initial development of some of the branches is determined by chance. The seven pairs are shown schematically in Fig. 68. As a matter of fact, all these component crystals have been actually observed, in considerable numbers. The examples are given as follows: (*a*) An apparently regular hexagonal crystal; (*b*) a separated single branch is sometime observed; Fig. 69 shows this type of crystal after the separation; (*c*) Fig. 70; (*d*) Fig. 71, before the separation; (*e*) Fig. 67; (*f*) Figs. 64 and 65; (*g*) Fig. 66; (*h*) this rather extraordinary form of component crystal is also observed, as shown in Fig. 72, the distance between the two centers being actually observable in this case.

In the next place, if the central part has the double structure, there may be two ways to explain the structure. First, it is possible that crystals may be developed from both of the bases of the primitive column of skeleton structure. The best example is Fig. 73. It will be clearly seen in the photograph that three branches are extended from one base of the minute hexagonal column of skeleton structure and the remaining three, from the other base. Similar examples are seen in Nos. 388, 392, and 397, Plates 71, 72, and 73. In this case the two crystals cannot be in the same plane. But the primitive column in the initial stage of crystal formation is generally so short that the two crystals are photographed in one focal plane under this degree of magnification. Some of these crystals have pretty long primitive columns. In such crystals it is clearly recognized that the two components

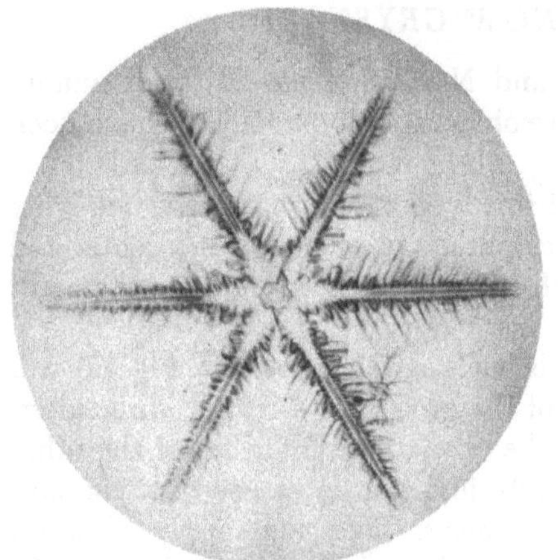

Fig. 73 [379]. Example of parallel growth (× 19).

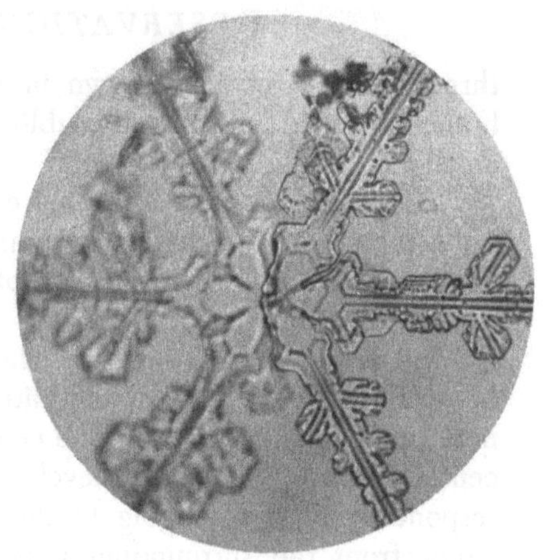

Fig. 74 [521]. Two components in different planes.

Fig. 75 [371]. Crystal with appendant nuclei (× 24).

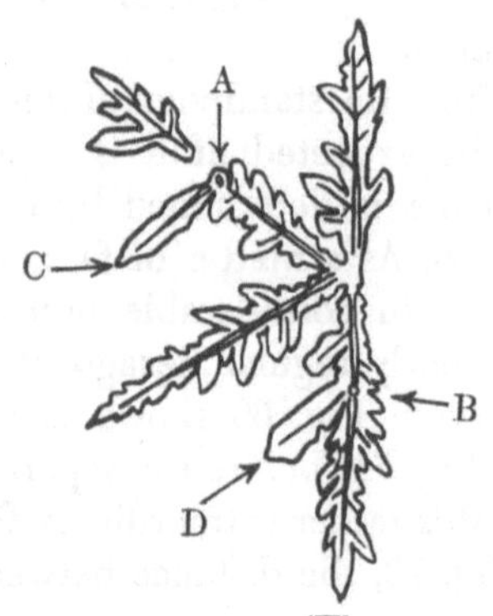

Fig. 76 [372]. Sketch of the crystal of Fig. 75.

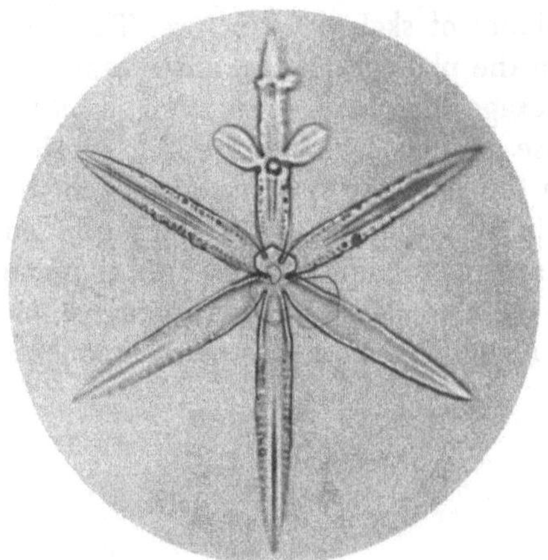

Fig. 77 [373]. Twigs caused by attachment of nucleus (× 49).

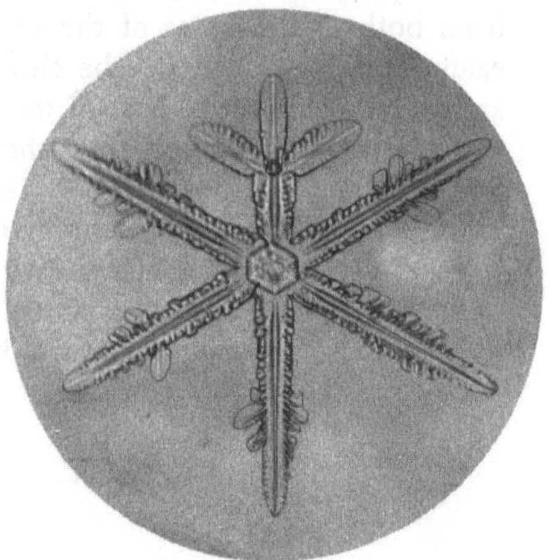

Fig. 78 [374]. Twigs caused by attachment of nucleus (× 45).

do not develop in one plane. One good example of such a crystal is shown in Fig. 74, while Fig. 72 is another example. Although these crystals have rather irregular shape, there is no doubt that the former corresponds to (*f*) of Fig. 68, and the latter, to (*h*). Although it is not so clear on the photograph, Nos. 388, 392, and 404, Plates 71, 72, and 74 also show that the two developments are not in the same plane. Hence, without doubt many of the three- or four-branched type of crystals are the result of parallel growth of the crystal from a hexagonal column of skeleton structure in their initial stage and branches are developed from both its bases.

In the second place, we can imagine a case of two nuclei, one placed upon the other. Although we do not consider the nuclei themselves in this book, they are not always placed at the center, but sometimes are found in the branches. Figure 75 shows a crystal of type (*c*) of Fig. 68 in the separated state. There is an appendant nucleus at the middle of one of its left hand branches, from which a twig is extended, as if it were attached at the point. As is illustrated by the sketch of Fig. 76, a nucleus is attached at the point *A*, so that the branch beyond the point *A* can be separated without being destroyed.

An interesting fact is that the development of the crystal is accelerated at the point where an appendant nucleus is attached. The crystal shown in Fig. 76 has another appendant nucleus at the point *B*. The twigs *C* and *D* extending from *A* and *B* are remarkably larger than the other twigs. Representative examples showing this nature are the crystals shown in Figs. 77 and 78. The influence of nuclei is quite conspicuous in these examples.

Although it is difficult to find the crystals that are formed on two nuclei, one placed upon the other, we tentatively include the crystals of Figs. 62, 63, 64, 67, and others like them, in this type, that is, having no trace of development from a column of skeleton structure. Further investigation along this line is difficult, since there is no definite theory as yet about the nuclei of snow crystals nor is their substance known, although there are various arguments.

A collection of two- or four-branched crystals and the like is added in Plates 69–77, among which Nos. 375–378 and 380–394 belong to type (*f*) of Figure 68; Nos. 395–400 to type (*c*); Nos. 401–407 and 409–411 to type (*e*); Nos. 412–417 to type (*g*); Nos. 418–420 and 422 to type (*b*); and Nos. 424 and 428 to type (*d*).

11. Malformed crystals

As is mentioned above, plane crystals do not always show perfect hexagonal symmetry or regular dendritic development. Although it is now acknowledged that all snow crystals belong to the hexagonal system of crystallization, this means only that the arrangement of atoms has hexagonal symmetry. The form and pattern of the crystal do not necessarily become those of regular hexagonal symmetry. The problem of the form of crystal, that is, the crystal habit, has not been studied much in the past.

Crystals other than the regular hexagonal type are collected and considered together in this section as malformed crystals. Photographs are shown in Figs. 79–102 and Plates 78–97. The malformed crystals have a high frequency of occurrence and are of such variety that a distinct classification is impossible. Some of them, however, have comparatively clear characteristics, on which basis we attempted a rough classification, or rather a sort of grouping. For instance, Fig. 79 shows a slight deviation from the regular crystal form; Fig. 80 is characterized by its remarkable asymmetry; Figs. 92 and 93 are several plane crystals, one overlapping with the other; and Figs. 81 and 82 have nuclei attached during their development. From these characteristics the malformed crystals will be roughly classified as follows.

(i) *Dendritic crystals with branches grown asymmetrically.* To account for the extraordinarily symmetric nature of the six branches of some snow crystals has long been a problem. A typical example is shown with some discussion in Sec. 33. Crystals with asymmetric branches, however, are not less often observed. Considering the frequency of occurrence, these asymmetric crystals may be more common than the symmetric ones. Of course the question of symmetry is a matter of degree. Such crystals as those shown in Plate 82 are at present considered as symmetric in this book, although a close examination of the photographs reveals that the mode of distribution of twigs differs slightly for different branches. The degree of asymmetry is more marked in large dendritic crystals. The criterion of regular crystals and asymmetric ones is rather arbitrary. The crystal of No. 449, Plate 82, is an example which shows a further branching of a twig, the mode of this secondary branching being asymmetric for two sides of one main branch. Most of the complicated crystals of this sort show some deviations from a symmetric form, and a perfectly symmetric crystal is very rarely observed.

Plane crystals fall, as hydrodynamics shows, in a nearly horizontal position, turning round the axis of a vertical line and taking a spiral course. This revolving movement helps the symmetric growth of branches. This symmetry, however, is apt to be broken, owing to the nuclei attaching to some point on the branch.

As described in Sec. 10, the acceleration of development of a crystal by the attachment of nuclei is one of the causes of the asymmetric growth of branches. The crystals shown in Figs. 77 and 78 are of course to be grouped with the asymmetrical crystals. Among the more complicated fernlike crystals we see very often the asymmetric development of branches due to the attached nucleus. One example is shown in Fig. 79.

Another characteristic type of asymmetry is a marked elongation of one branch or a few branches over the other ones. A remarkable photograph was obtained of

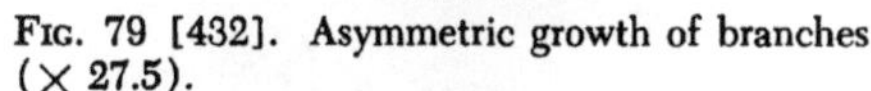

FIG. 79 [432]. Asymmetric growth of branches (× 27.5).

FIG. 80 [435]. Remarkably asymmetric growth of branches (× 16.5).

a crystal which is the best example of this kind, as shown in Fig. 80. In this case there is little doubt that the water vapor was chiefly supplied to the crystal in one direction (downward in the photograph). Similar examples are shown in Nos. 452–469, Plates 83–85.

(ii) *Malformation by attachment of nuclei.* As described in the preceding section, when a nucleus is attached to a point on the branch, the crystal makes a malformed development at that point. There are various kinds of malformation, which are collectively shown in Figs. 81–84 and Plates 86 and 87.

Figure 81 is a remarkable example, in which one sees that another dendritic crystal develops from the attached nucleus. This sort of crystal can be separated at the point of nucleus into two parts, as shown in Fig. 82. Plate 86, No. 472, shows a similar crystal, whose structure is clearly seen in the photograph after separation, No. 473. Plate 86, Nos. 474 and 475, are the photographs of similar crystals after separation.

Figure 83 exhibits the influence of nucleus attachment most clearly. It shows that a nucleus was attached to a corner of the central hexagonal plate when it grew to a certain extent, and that plane development occurred both from the former crystal and from the new nucleus. In Fig. 84, one sees that in an earlier stage, when the main crystal was a small hexagonal plate, another nucleus became attached to it and in the subsequent course the main and the appendant crystals passed through the same stage of development. The crystal of Fig. 85 seems to have formed such an irregular shape, because a similar phenomenon occurred several times.

FIG. 81 [470]. Malformed crystal before separation (× 17.5).

→

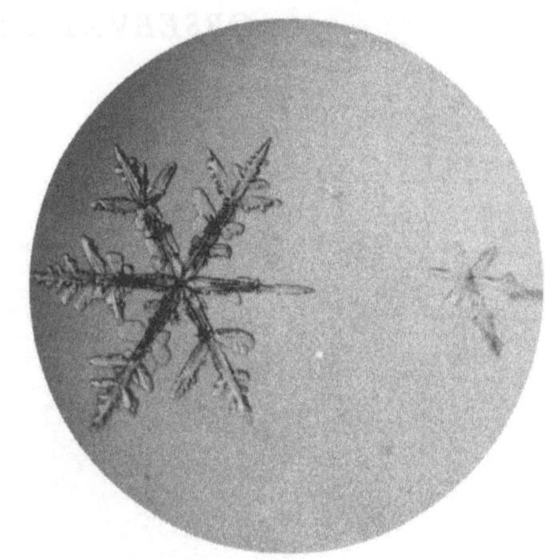

FIG. 82 [471]. The crystal of Fig. 80 after separation (× 14.5).

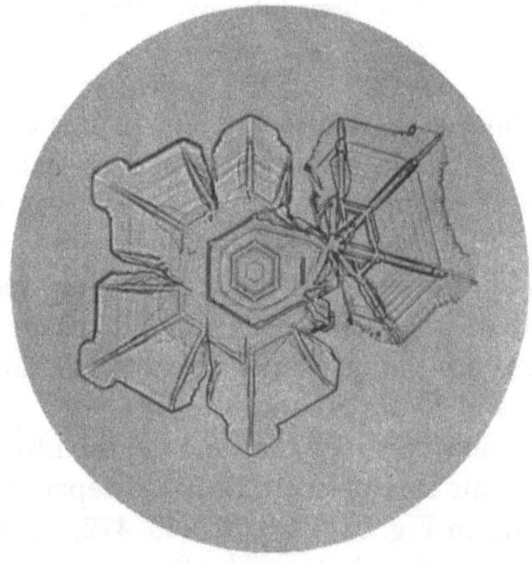

FIG. 83 [480]. Malformation by attachment of a nucleus (× 48).

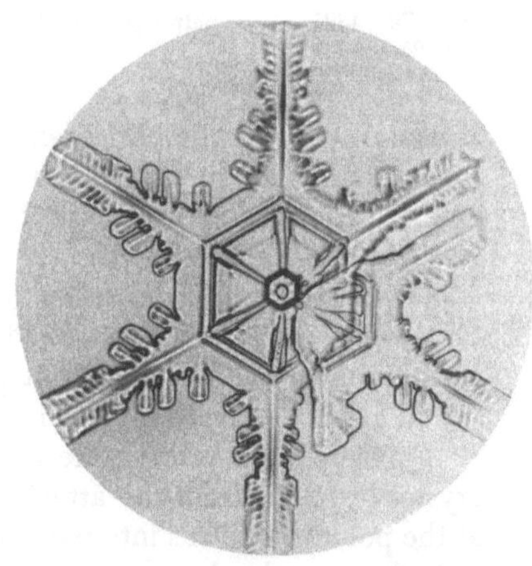

FIG. 84 [421]. Malformation by attachment of a nucleus (× 50.6).

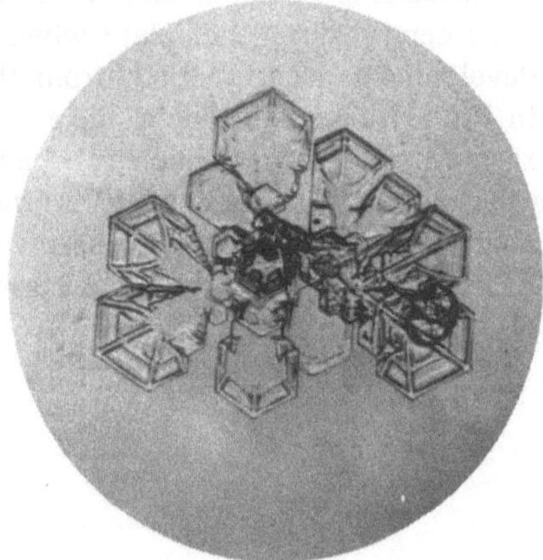

FIG. 85 [481]. Malformation by attachment of several nuclei (× 30).

FIG. 86 [483]. Malformed crystal of four-branched type (× 29).

(iii) *Malformed crystals of the three- or four-branched type.* The prototype of Figs. 86 and 87 and Nos. 482, 484–486, and 489, Plates 87 and 88, seems to be the four-branched type, but their branches are developed asymmetrically. The crystal of Fig. 86 seems to be similar to that of Fig. 88, with its two-branched part detached. The state of separation is very well shown in Fig. 89. Before separation this crystal showed an appearance like that of No. 482, Plate 87. Figure 87 and Nos. 495 and 496, Plate 89, are malformed crystals of the four-branched type developed in a sector-form.

Figure 90 and Nos. 501–503, Plate 90, are malformed crystals of the three-branched type.

(iv) *Crystals having bilateral symmetry.* Besides the malformed crystals of the four-branched type, there is another type which shows bilateral symmetry. Nos. 512–515, Plates 91–92, belong to this category and it is rather surprising that crystals of such a unique form were observed again and again.

(v) *Crystal with components developing in different planes.* The above-mentioned are malformed crystals that are chiefly due, for whatever reason, to the asymmetric development of branches. Besides these, there are some that ought to be treated as malformed crystals because their two component crystals are not in the same plane. Examples of such crystals are shown in Fig. 91, and in Figs. 72 and 74 already cited.

(vi) *Overlapping of several planes.* Sometime crystals are made up of two or more plates, one overlapping the other. Figures 92 and 93 are examples. Similar crystals are reproduced in Nos. 524 and 527, Plate 93. The origin of this structure of overlapping of planes was clarified in our experiments on artificial snow, to be described in Part II.

(vii) *Malformed plates.* Plates are no less frequently malformed than regular dendritic crystals. The most characteristic one has a V-shaped notch on one side. One example is shown in Fig. 94. The crystal assumes various complicated forms as the number of these notches increases. More complicated crystals of this sort are reproduced in Figs. 95 and 96. One can see that the essential feature of malformation is in the V-cut of the plates in these cases. Therefore, if the cause of the formation of this notch is clarified, the cause of such malformation will be understood. In the artificial production of plate crystals, this notch is often formed. The cause is not yet clear, but surely it is connected with the problem of the transition of sectors into a plate. In the example shown in Fig. 97, one part of the crystal is a plate form and the other part is made of sectors.

The V-shaped notch is clearly seen also in the photographs of Plates 94 and 95.

(viii) *Miscellaneous.* Besides what are classified above, there are still more various types, which are collectively shown in Figs. 98–103 and Nos. 546, 548, 550–555, Plates 96 and 97. Among them, Figs. 98 and 99 may be regarded as examples of malformation from one-branched or two-branched crystals. Nos. 551

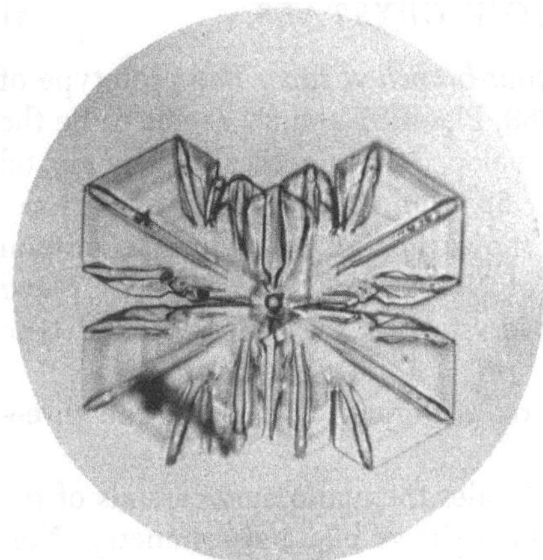

Fig. 87 [494]. Malformed crystal of four-branched type (× 65).

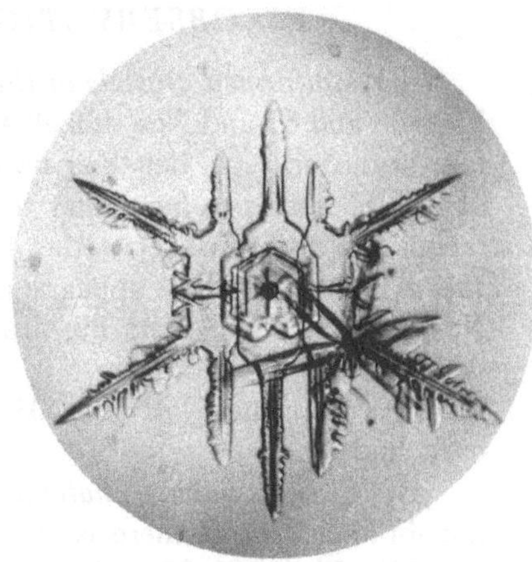

Fig. 88 [488]. Malformed crystal of type of Fig. 68(*e*), before separation (× 26).

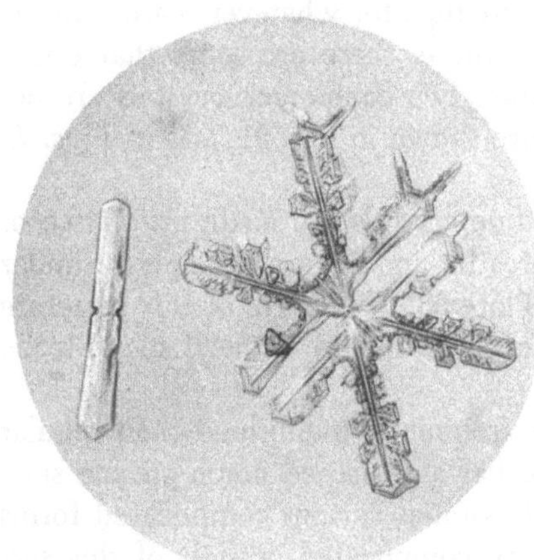

Fig. 89 [487]. A crystal similar to that of Fig. 88, after separation (× 19).

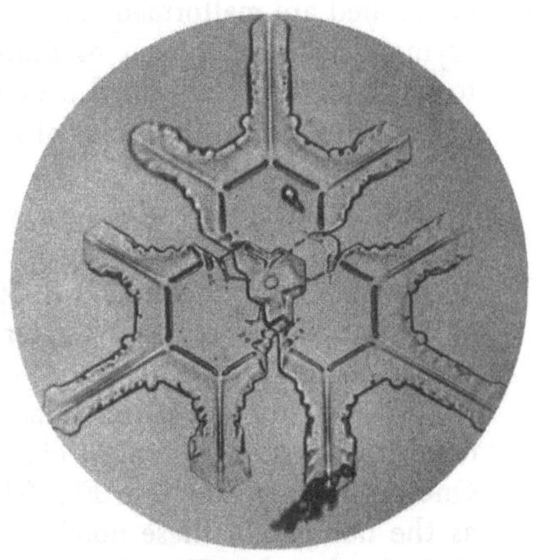

Fig. 90 [500]. Malformed crystal of three-branched type (× 43).

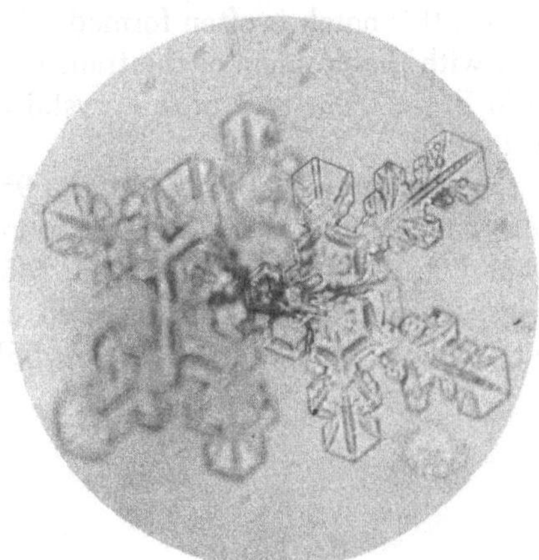

Fig. 91 [523]. Components developing in different planes.

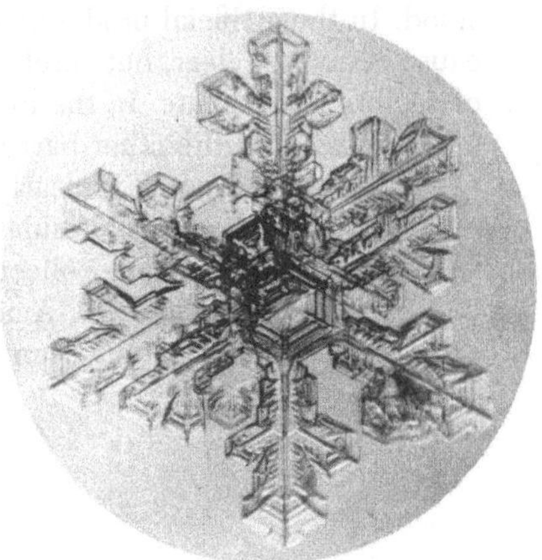

Fig. 92 [436]. Overlapping of several planes (× 21).

Fig. 93 [437]. Overlapping of several planes (× 19.5).

Fig. 94 [531]. Malformed plate (× 28).

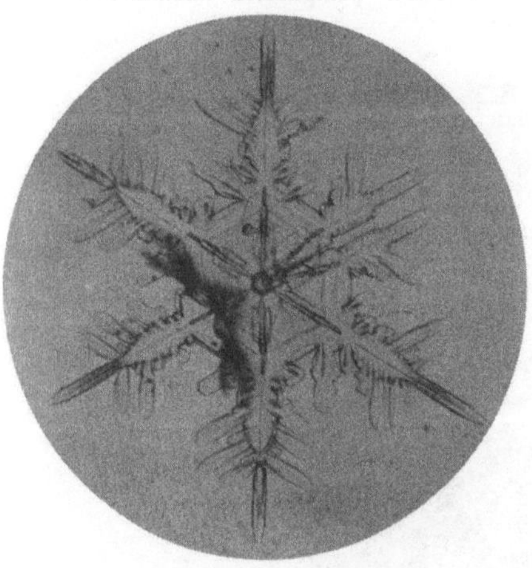

Fig. 95 [534]. Malformed plate with dendritic extensions (× 23).

Fig. 96 [535]. Malformed plate with dendritic extensions (× 24.5).

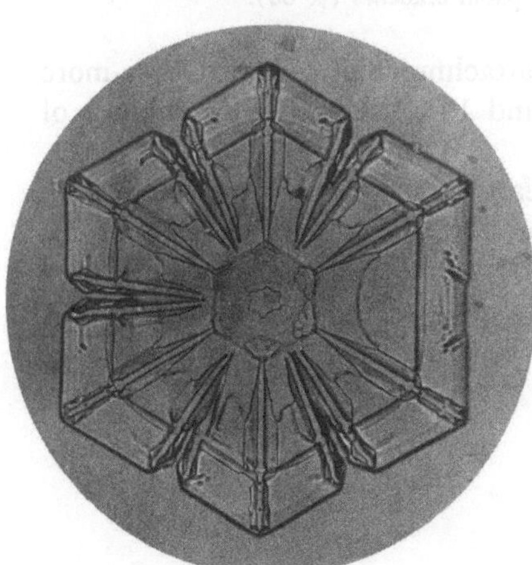

Fig. 97 [542]. Malformed sectors (× 57).

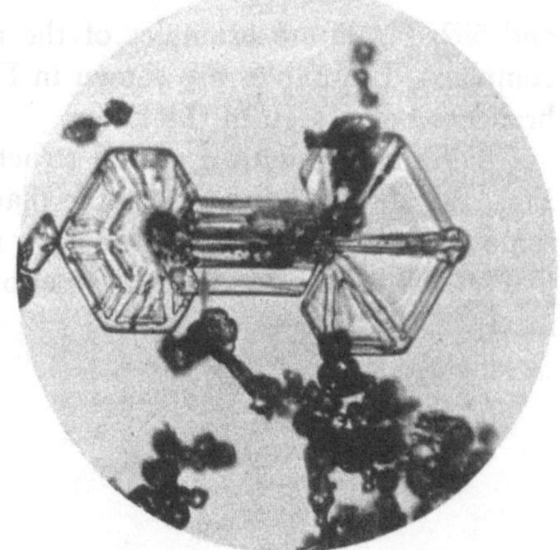

Fig. 98 [547]. Malformed crystal of two-branched type (× 50).

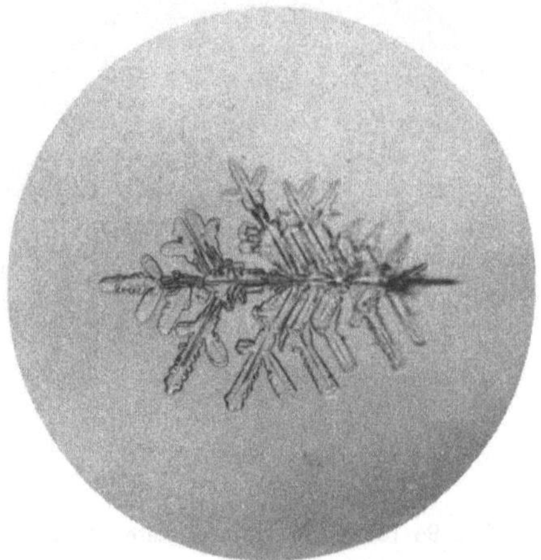

FIG. 99 [549]. Malformed crystal of one-branched type (× 21).

FIG. 100 [525]. Complicated malformation (× 25).

FIG. 101 [556]. Malformed crystal with spatial branches (× 38).

and 552, Pl. 96 are examples of the repeated attachment of nuclei. Other more complicated examples are shown in Figs. 100 and 101, the spatial assemblage of branches being seen in the latter.

Malformation with a spatial structure is clearly seen in the crystal shown in Figs. 102 and 103; Fig. 102 is the plan view and Fig. 103 is the side view of the same crystal. The dark vague image in Fig. 102 is the upward extension shown in Fig. 103. The side view shows also the overlapping of two plane crystals very nicely.

12. Spatial dendritic crystals, with a stellar base

The crystals mentioned so far are the ones developed in one plane, but they often gather spatially and form a solid structure. Photographs of crystals of this kind have hitherto been rarely, if ever, published, not because such crystals do not occur, but perhaps because they do not attract attention. In our observation at Hokkaido, crystals of this type were observed to exist in greater quantity than the plane hexagonal crystals. When such a dense snow falls that one cannot see 20 m ahead, the snow consists of crystals of this type.

Spatial dendritic crystals may roughly be classified into two types. The one has a hexagonal dendritic crystal at the base and many dendritic extensions growing out of various points on each branch. The other has dendritic branches growing out of the center in every direction, giving an appearance like a chestnut burr. We named the former one the spatial assemblage of dendritic branches with a stellar base, or simply the "spatial hexagonal type," and the latter the radiating type or the "burr type." Of the two, this section will be devoted to the spatial hexagonal type.

Examples of spatial hexagonal crystals are shown in Figs. 104–110 and Nos. 561, 567–578, 581–585, 587, 589–594, 596–598, 600, 602–606, Plates 98–103. Let us note Fig. 104, which is a representative example. The focus of this photograph is adjusted on the branches that extend upward, the image of the hexagonal base being vague. Seemingly, a small hexagonal plane crystal is first formed in the upper layer of the atmosphere and while it falls many nuclei are attached to various points on its branches, from which spatial branches are developed. From the result of experiments on artificially grown snow crystals, we learn that the hexagonal plane crystal must be of small hexagonal form from the initial stage of its formation.

Figure 105 shows three photographs of the same crystal: (*a*) is focused on the hexagonal base crystal, (*b*) on the spatial branches, and (*c*) is a side view. The appendant upward branches are very well developed in this crystal. Usually the spatially distributed branches are not so long as in this case. Figure 106 shows a rather common form of this type of crystal, in which (*a*) is a plan view and (*b*) a side view of the same crystal. Nos. 567–572, Plate 98, are similar photographs; the photographs on the left were taken from above, those on the right are the side views of the same crystals. As these examples show, many of the spatial branches grow out of one surface of the hexagonal base. The reason may be that many nuclei are attached to the lower surface of the plane crystal while it falls in a nearly horizontal position, and the supply of water vapor is more abundant on that surface. A similar phenomenon is seen in crystals with cloud particles attached, to be mentioned later.

Nos. 573–578, Plate 99, are photographs of three crystals taken with different

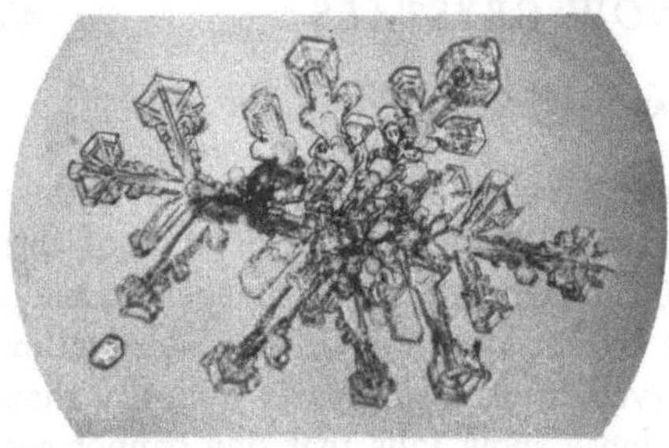

FIG. 102 [558]. Plan view of a crystal with spatial structure (× 19).

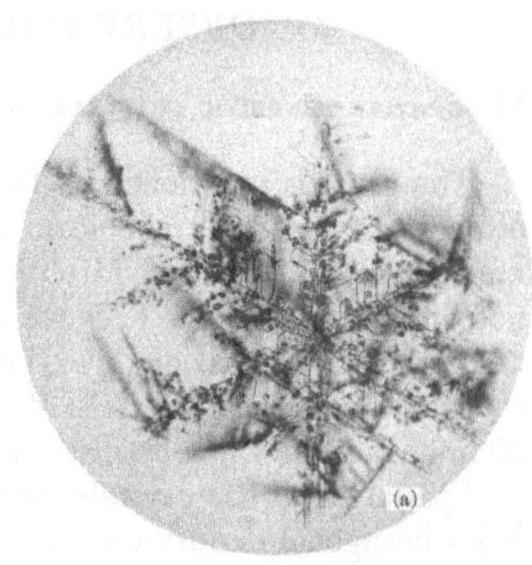

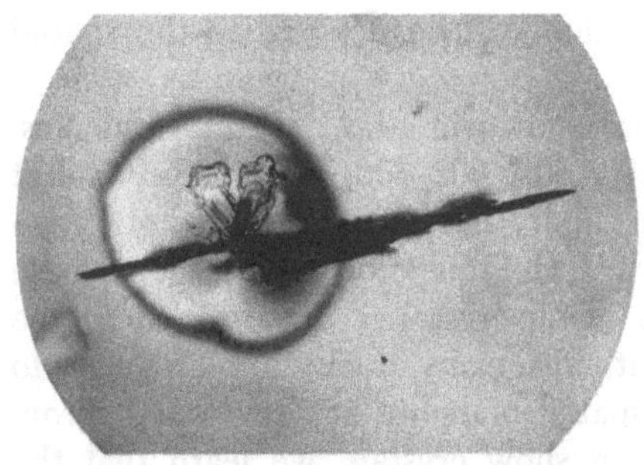

FIG. 103 [559]. Side view of the crystal of Fig. 102 (× 19).

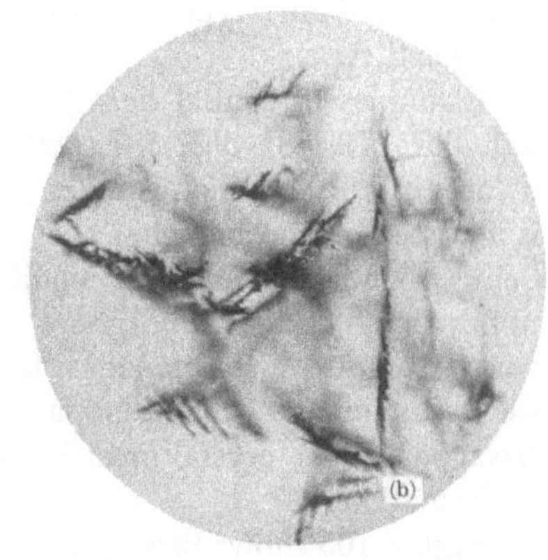

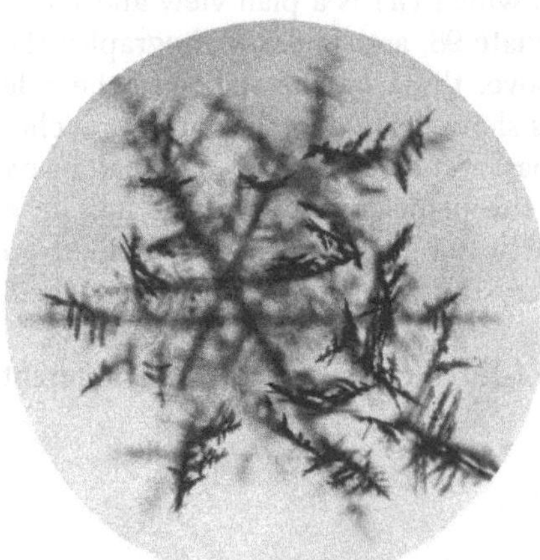

FIG. 104 [560]. Spatial hexagonal crystal (× 18).

FIG. 105. Spatial hexagonal crystal: (*a*) [562] focused on the hexagonal base; (*b*) [563] focused on the spatial branches; (*c*) [564] side view (× 16.5).

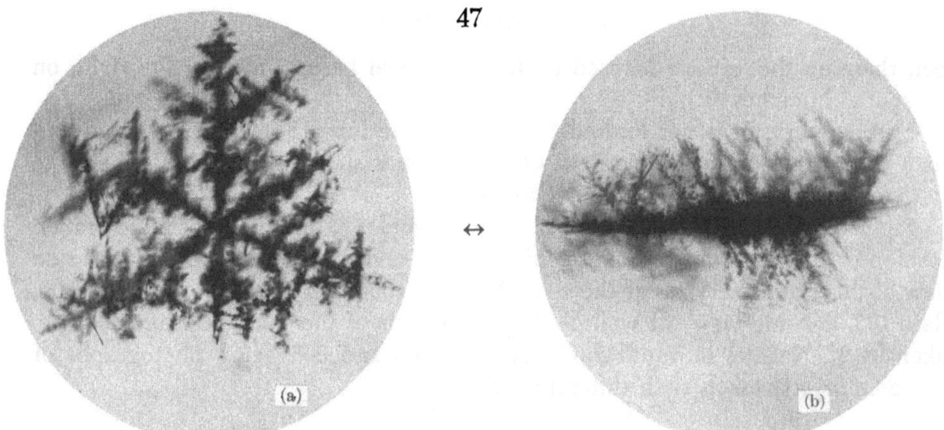

FIG. 106. Spatial hexagonal crystal: (*a*) [565] plan view; (*b*) [566] side view (× 19.5).

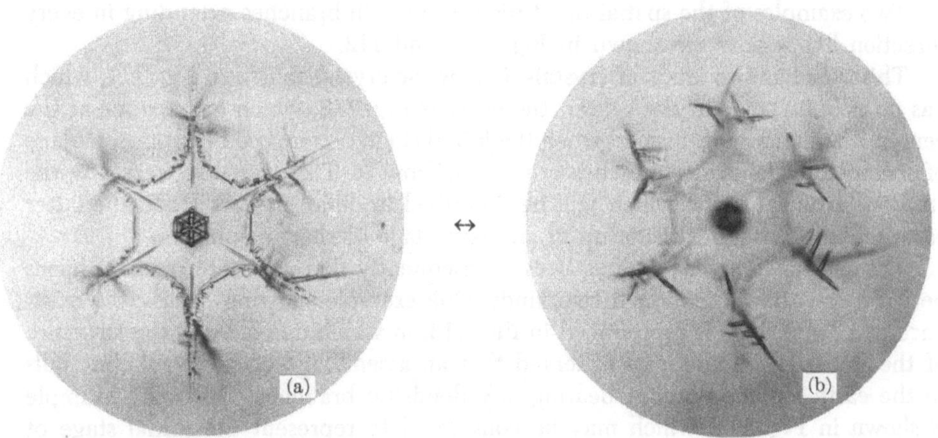

FIG. 107. Plate with spatial branches: (*a*) [579] focused on the base; (*b*) [580] focused on the spatial branches (× 19).

FIG. 108 [595]. Crystal with one spatial branch (× 21.5).

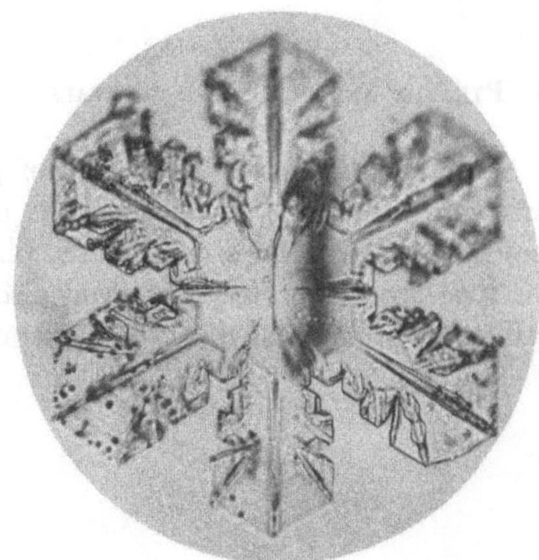

FIG. 109 [599]. Crystal with one spatial branch (× 36).

foci; those on the left are focused on the hexagonal bases, those on the right, on the spatial branches.

The base crystal is not limited to the dendritic form. Sometimes it may be a plate with extensions at the corners. One example is shown in Fig. 107, where the two views were taken with different foci.

Figures 108 and 109 and Plate 102 show examples of the type with one branch developed upward from the center of the base crystal. One example of the side view of this simpler crystal of the spatial type is shown in Fig. 110. Nos. 604–606, Plate 103, are side views of well-developed crystals of the spatial hexagonal type, taken by K. Nakata, in South Sakhalin, Nos. 605 and 606 being photographs of the same crystal taken with different foci.

13. Spatial dendritic crystals, radiating type

Two examples of the spatial dendritic crystal with branches extending in every direction like a burr are shown in Figs. 111 and 112.

This type has two kinds of crystals. One is the crystal shown in Fig. 111, which has no definite stone at the center; the other is Fig. 112, which has a stone at the center. The first kind is formed when the initial stage is composed of an assemblage of sectors and dendritic branches grown out from it. This was discovered by the study of artificial snow, which will be described in detail in Part II. In the latter case, the initial stage is made up of an assemblage of short columns.

The central stone has various sizes. Consequently there are intermediate forms between the above-mentioned two kinds. One example showing clearly the existence of a large stone is reproduced in Fig. 113, in which one can see the structure of the stone. It is sometimes observed that an assemblage of short columns falls to the earth surface without bearing any dendritic branches. One such example is shown in Fig. 114, which may be considered to represent the initial stage of this sort of crystal. In any case, there must be an atmospheric layer of considerable thickness suitable for dendritic development, with slightly different conditions in the higher atmospheric layer for each type.

14. Pyramid and columnar crystals

Scoresby discovered the crystal of pyramid type in the arctic and sketched it. But there are very few photographs of this type. Figures 115 and 116 are side views of the pyramid-type crystals, photographed at Mount Tokachi. Crystals of this type are very rarely observed also in Japan. However, we succeeded in the artificial production of this type rather easily, as will be described later.

Fig. 110 [601]. Side view of a crystal similar to those of Figs. 108 and 109 (× 27).

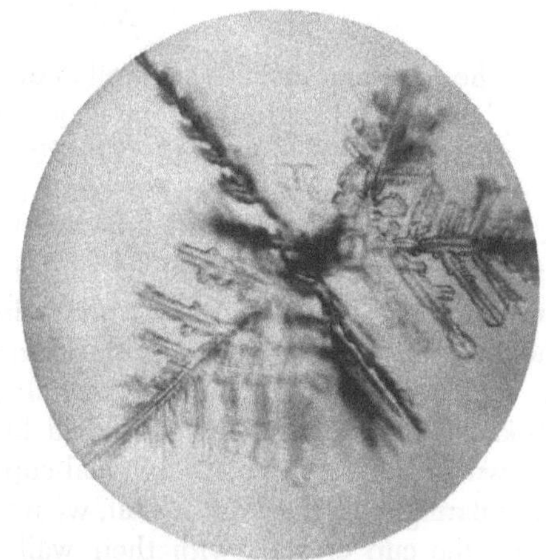

Fig. 111 [613]. Burr type without a stone (× 19.5).

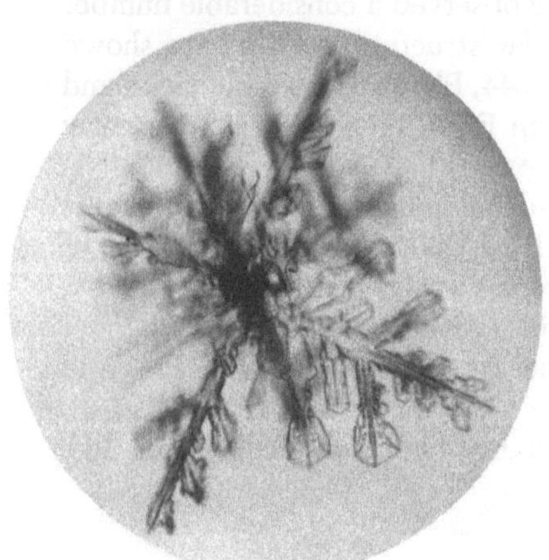

Fig. 112 [614]. Burr type with a small stone (× 20.5).

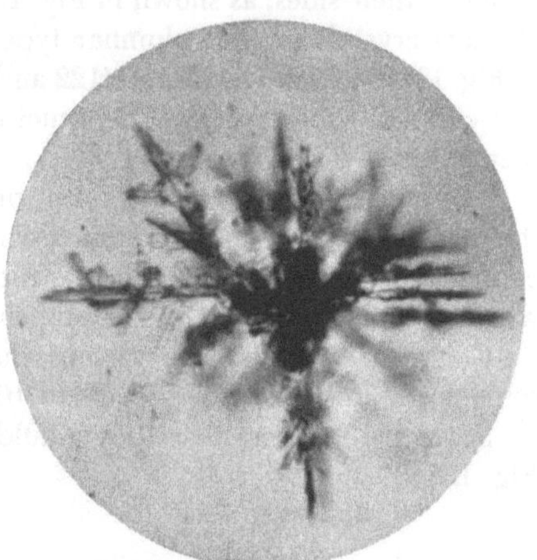

Fig. 113 [615]. Burr type with a large stone (× 27).

Fig. 114 [616]. Structure of a stone (× 66).

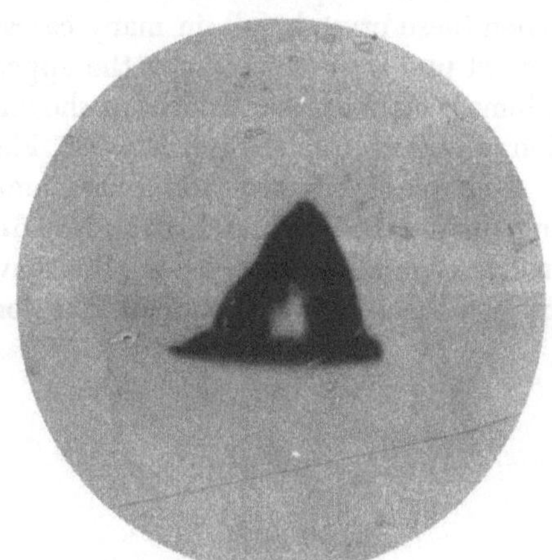

Fig. 115 [631]. Pyramid (× 55).

The columnar crystals are well known and have hitherto been frequently photographed. In Hokkaido this type is observed at least five or six times in one winter. Examples are shown in Figs. 117–120 and 122–124 and in Plates 106–109. Almost without exception, columnar crystals have hollows on both bases like the kick of a wine bottle. In many photographs, these hollows appear cone-shaped, as shown in Figs. 117 and 118. Most of the photographs of columnar crystals that have hitherto been taken by many workers in Europe and America show these conelike hollows. But this is due to the transformation of the original shape by sublimation or melting. We found out, by examining a column with complete shape at Mount Tokachi, that this hollow is shaped like a hexagonal whisky glass, as is clearly shown in Fig. 119. This hexagonal-cup type is often found among frost crystals. Comparing with the frost crystal, we infer that the columnar crystals are developed from the cup crystals with their walls gradually thickened. Figure 120 may be considered as representing the intermediate state to show this transition.

It is frequently observed that many of the cup crystals of frost have rolled parts in one of their sides, as shown in Fig. 121 (*a*). We observed a considerable number of snow crystals of the columnar type with similar structure, such as are shown in Fig. 121 (*b*) and (*c*). Figure 122 and Nos. 639–644, Plate 107, are side views and end views of such crystals. The structure shown in Fig. 121(*b*) and (*c*) is clearly seen in these photographs.

The cup crystal of ice is a result of skeleton development, and we can expect the structure like a minute staircase in the wall of the hollows in the bases of columnar crystals. This skeleton structure is very clearly seen in Fig. 123(*b*), which is the end view of the crystal shown in Fig. 123(*a*).

Columnar crystals have various shapes. Some are slender, like Fig. 124, and some are sturdy like No. 656, Plate 109.

Some are so short that they would more suitably be called thick plates, like Fig. 125.

15. Bullet crystals and their combination

The bullet type means a column with a pyramidal head like a rock crystal. When these crystals fall, in many cases the snow consists of only this kind. They present under the microscope the appearance shown in Fig. 126. Mixed with the columnar crystals, they appear as shown in Fig. 13. Crystals consisting of combination of bullets, as shown in No. 662, Plate 109, have a similar appearance.

Examples of bullets and their combination are reproduced in Figs. 127–132 and in Nos. 664–667, 669, 671, 672, 674–677, 679–682, 684–691, Plates 110–113. In this type also, the crystals often have kicks in their bases; Fig. 127 is a good example. Humphreys [17] explains the formation of haloes of unusual radii on the

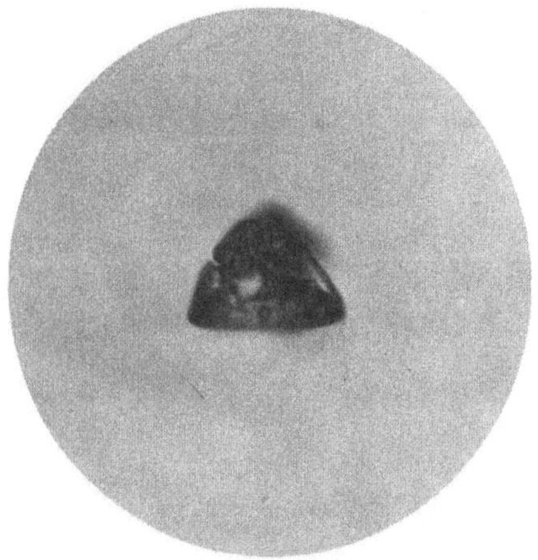

FIG. 116 [632]. Pyramid (× 110).

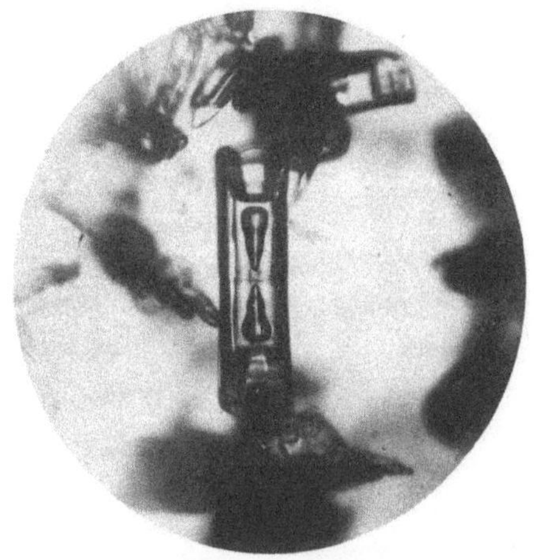

FIG. 117 [659]. Degenerate column (× 48).

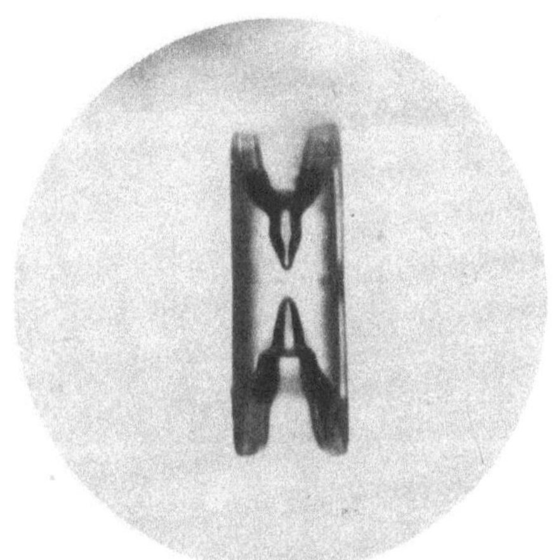

FIG. 118 [660]. Degenerate column (× 95).

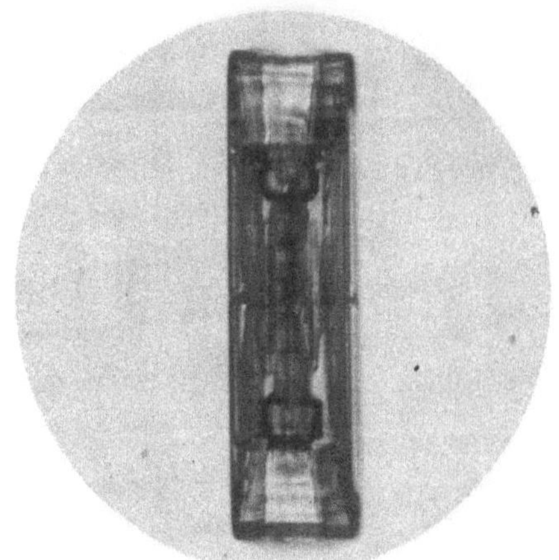

FIG. 119 [633]. Column in complete form.

FIG. 120 [654]. Transition from cup to column (× 58).

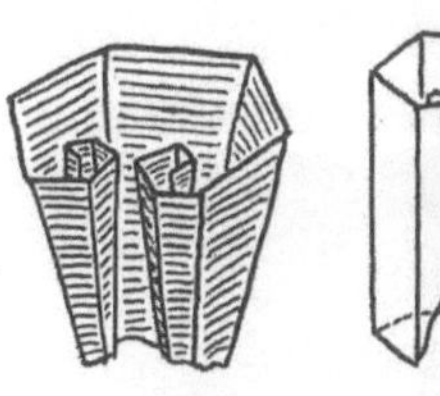

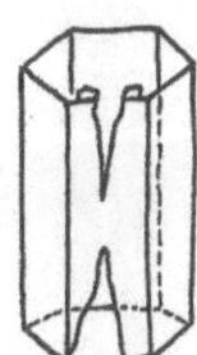

FIG. 121. Sketches of cup and column.

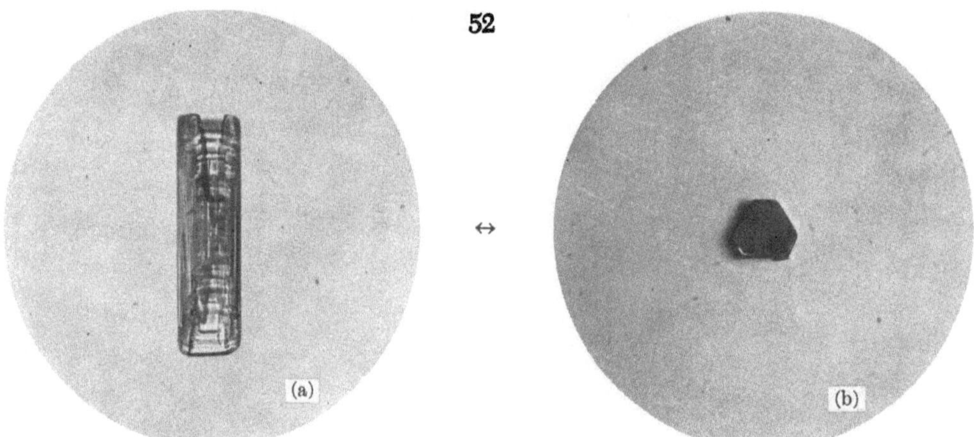

Fig. 122. Column with one side rolled up: (*a*) [637] side view; (*b*) [638] end view (× 35).

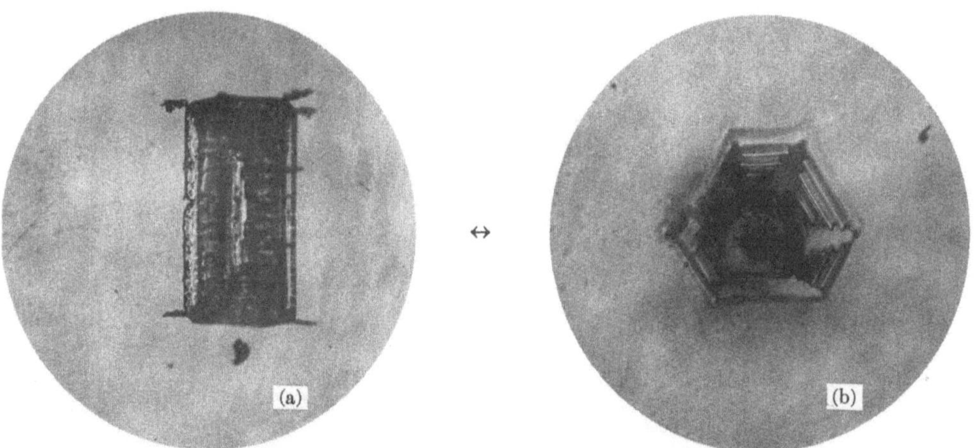

Fig. 123. Column showing skeleton structure: (*a*) [647] side view (× 39); (*b*) [648] end view (× 54).

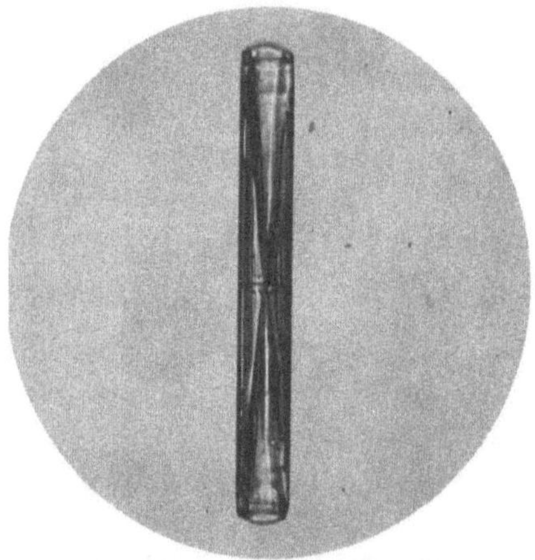

Fig. 124 [649]. Slender column (× 42).

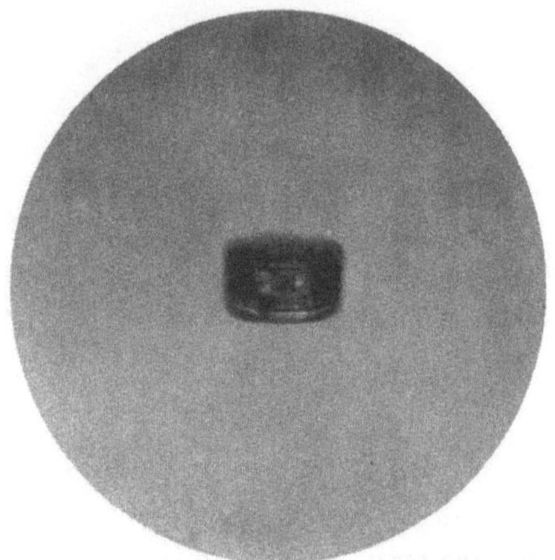

Fig. 125 [658]. Short column (× 54).

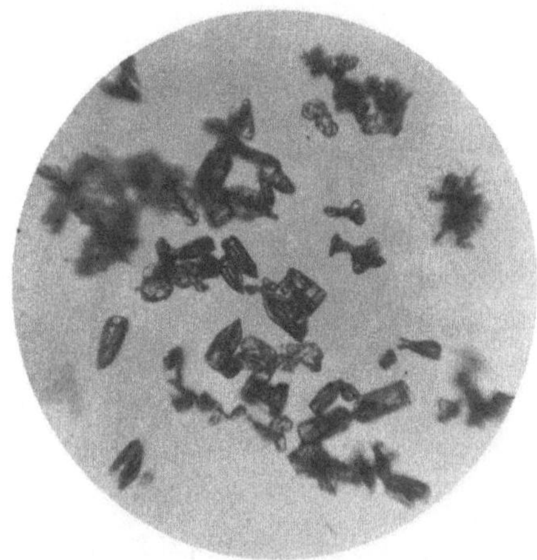

Fig. 126 [661]. Snow composed of bullets (× 16.5).

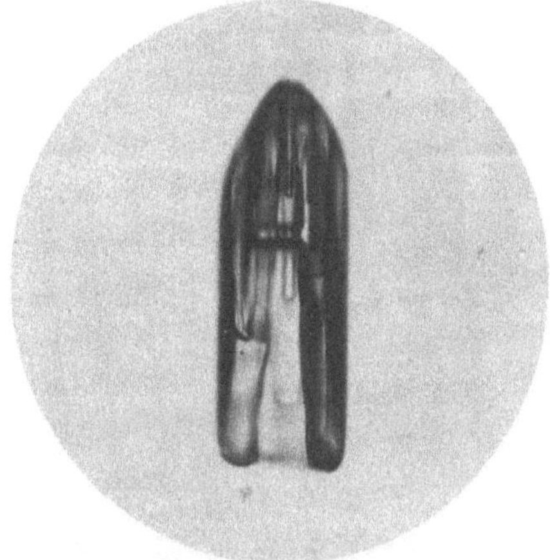

Fig. 127 [663]. Bullet-type crystal (× 86).

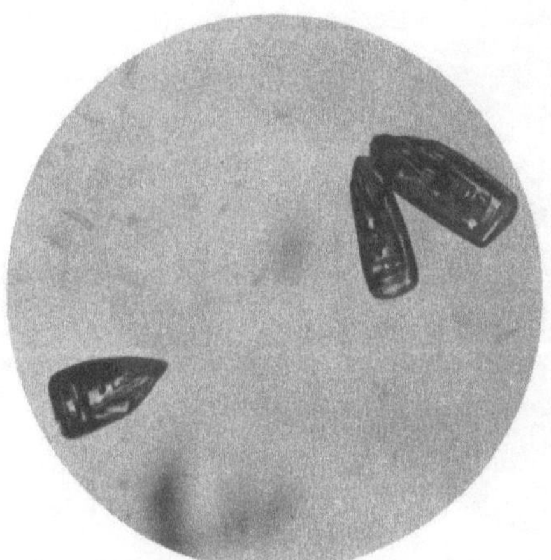

Fig. 128 [668]. Three bullets (× 42).

hypotheses that ice crystals which constitute the upper clouds have forms similar to this bullet type, and the halo is produced by the refraction of solar light by these crystals.

The bullet-type crystals tend to gather at their heads. Examples of this combination are shown in Figs. 129–132. A photomicrograph of a similar crystal composed of many columns is given in Dobrowolski's book.[18] Nos. 686–688, Plate 113, show other types of crystal development added to the combination of bullets.

Stüve [19] says that when a solid dust particle with many corners becomes the

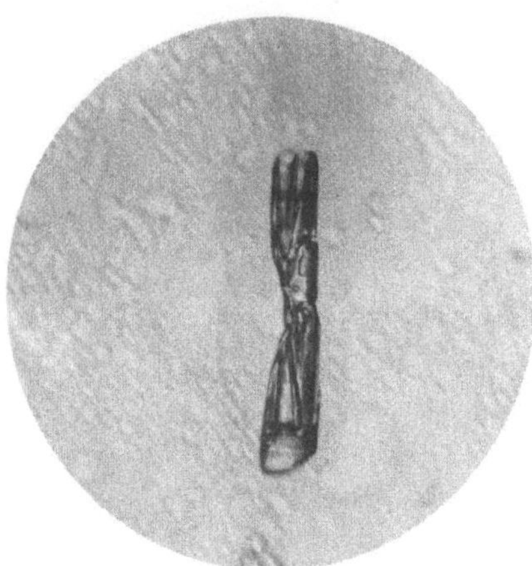

Fig. 129 [670]. Combination of two bullets (× 51).

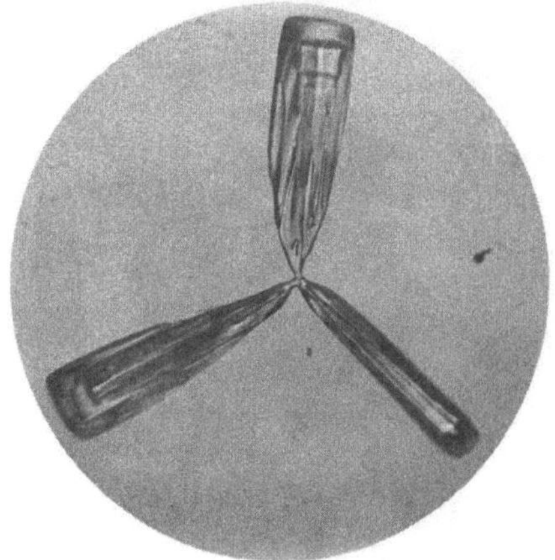

Fig. 130 [673]. Combination of three bullets (× 43).

Fig. 131 [678]. Combination of four bullets (× 72).

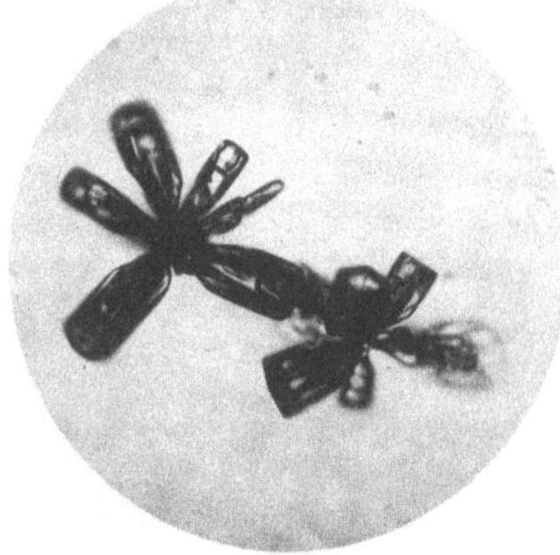

Fig. 132 [683]. Combination of several bullets (× 37).

nucleus and cup crystals extend from each of the corners, this combination of bullets is formed. His theory does not go beyond imagination at present, but it seems worth further examination.

16. Needle crystals

Needle crystals of snow have hitherto been regarded as one of the rarest types. This type seems to be very rare in Europe and America. Wegener [20] says that the opportunity for observation was so rare that he could not decide whether this type is simply a slender columnar crystal or a part of skeleton. However, it is more frequently observed in Hokkaido. In Sapporo we could see this type an average of four or five times during one winter. We have even met an occasion at Mount Tokachi, when a quite dense snow consisted almost entirely of needles and continued to fall for nearly an hour. Therefore, we were enabled to investigate not only the structure of the crystals, but also their mass and speed of fall. Photographs of these crystals are reproduced in Figs. 133–136 and Nos. 694–703, 705, 707–709, 712–715, Plates 114–116.

Needle crystals generally have a structure like a bundle of extremely thin simple needles. Good examples of this structure are shown in Figs. 133 and 134. Very rarely simple needles fall individually. The structure of this simple needle is shown in Fig. 135. This photograph indicates that the simple needle also has holes on both ends extending in the direction of the axis. To this extent, this type resembles the columnar type in structure, but from the point of view of the conditions of formation they are utterly different. So we consider that the needles should be regarded as another type different from slender columnar crystals. This difference in the conditions of formation was clarified by the study of artificial snow.

Often two needles are combined, making a cross. Photographs of this cross type are reproduced in Figs. 133 and 136. The shape of the cross being always the same, the simple assumption that two needles were glued together while they fell cannot be accepted.

Figure 137 and Nos. 712–715, Plate 116, show hitherto unknown crystals. They have vertical needles extending from hexagonal plane bases. Figure 137(*a*) shows the plan view of a hexagonal plane crystal of broad branches with needles extended vertically from the base; Fig. 137(*b*) shows the side view of the same crystal. In No. 707, Plate 115, plan views of two crystals and a side view of one crystal are seen. No. 708, Plate 116, is an example of a *tsuzumi*-type crystal to be described in the next section, with needle extensions from one of its plane crystals. Today since it is possible to produce these crystals artificially in the cold chamber, the mechanism and conditions for the formation of this extraordinary crystal are well known to us.

17. Tsuzumi type

Occasionally plane crystals are developed on both bases of a columnar crystal. Former snow researchers called this crystal a wheel or a tea-table type, but we named this a *tsuzumi* (Japanese tomtom) type because, when the plane crystals on both bases are plates, the crystal assumes a shape exactly like a *tsuzumi*, a photograph of which is shown in Fig. 138. Plate development, however, is rare in

Fig. 133 [706]. Group of needle crystals (× 20).

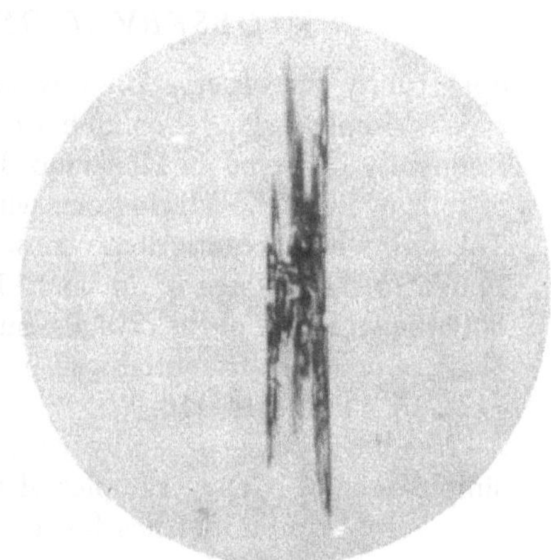

Fig. 134 [692]. Needle crystal (× 34.5).

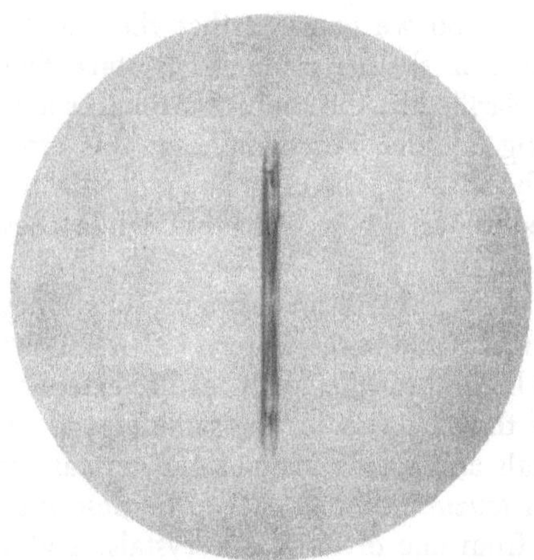

Fig. 135 [693]. **Simple needle** (× 48).

Fig. 136 [704]. Cross combination of needles (× 20.5).

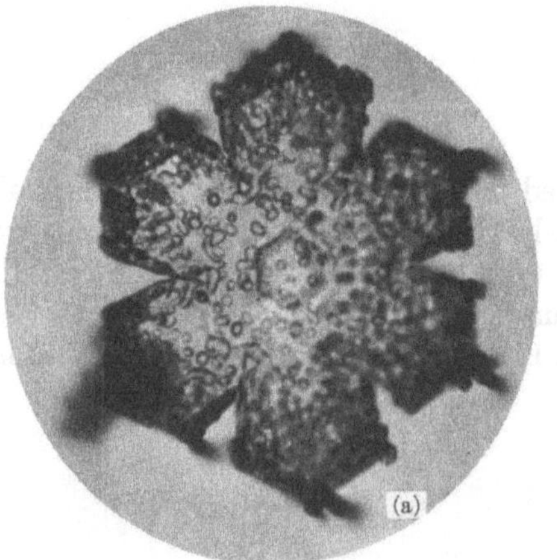

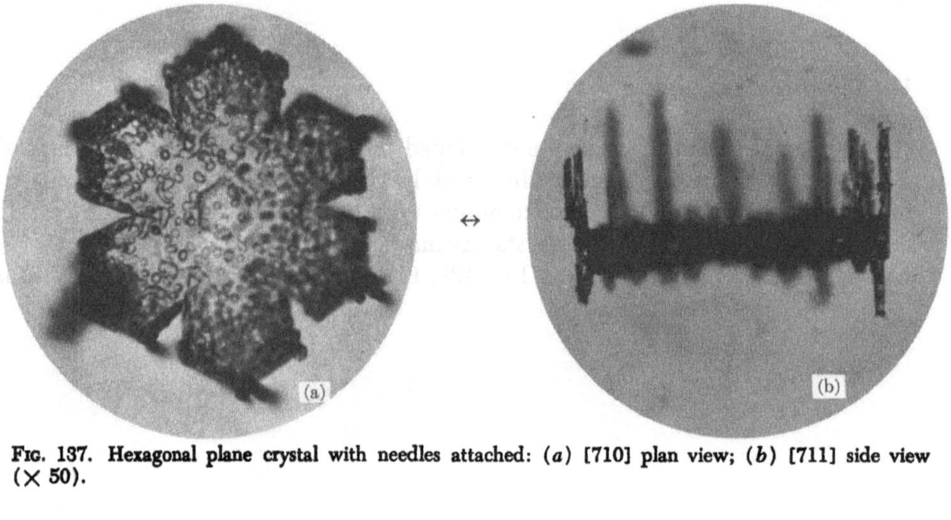

↔

(b)

Fig. 137. Hexagonal plane crystal with needles attached: (*a*) [710] plan view; (*b*) [711] side view (× 50).

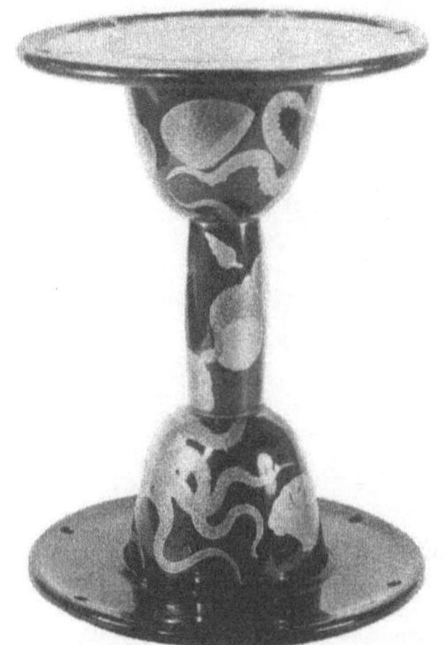

Fig. 138. Japanese tomtom, *tsuzumi*.

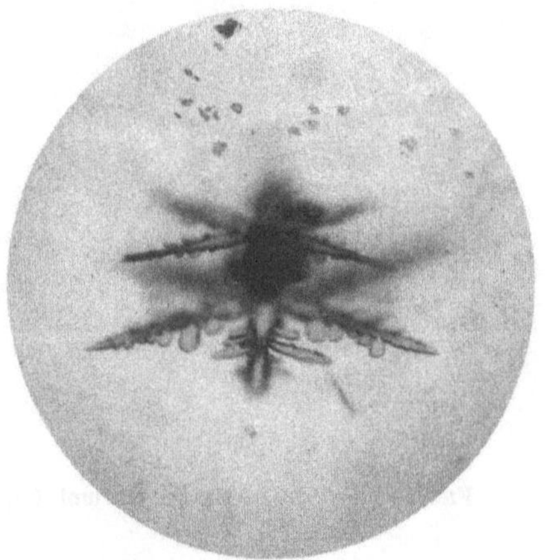

Fig. 139 [718]. Dendritic tsuzumi (× 61).

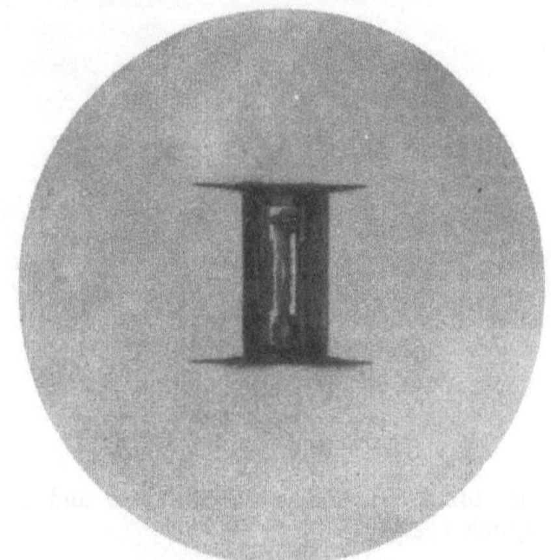

Fig. 140 [725]. Plate tsuzumi (× 50).

our climate and usually this type has dendritic hexagonal development on both bases. When two or more tsuzumis are connected, the crystal becomes a complicated shape with several plane developments on both ends and also between the ends. We called this complicated type a sectioned tsuzumi. In some cases, spatial hexagonal crystals develop on the bases of a column. The crystals of tsuzumi type are collectively shown in Figs. 139–149 and Nos. 719–724, 726–729, 732–738, 740–755, 757, 763, 768–777, 779–781, 783, Plates 117–125.

Figures 139 and 140 are side views of ordinary tsuzumi-type crystals, the end

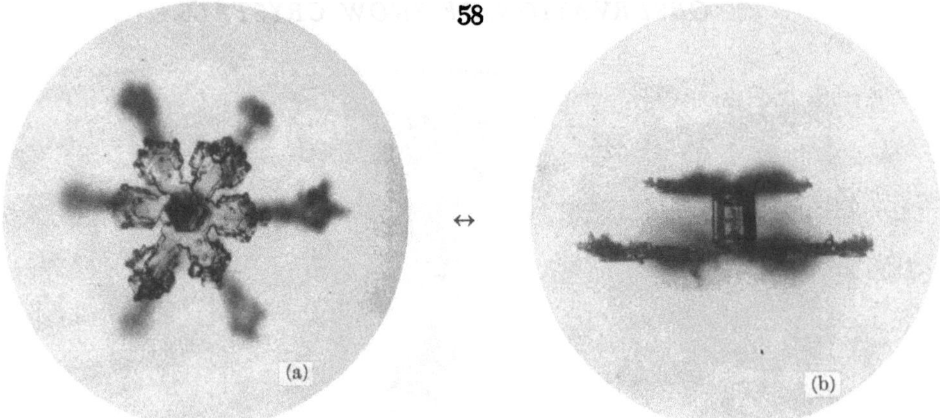

Fig. 141. Tsuzumi with different planes: (*a*) [730] plan view; (*b*) [731] side view (× 29).

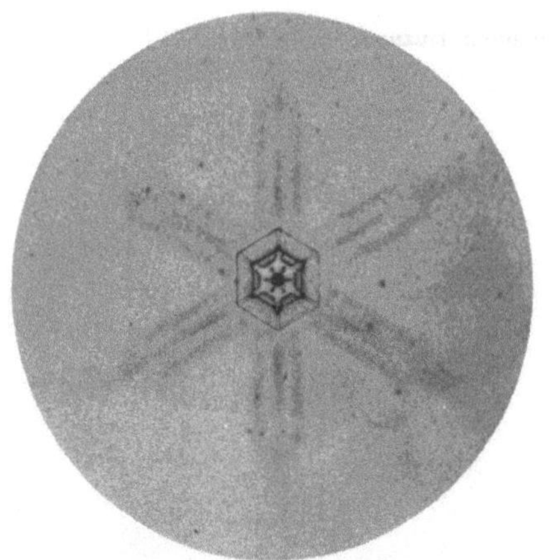

Fig. 142 [739]. Tsuzumi with plate and stellar crystals (× 44).

Fig. 143 [756]. Sectioned tsuzumi (× 32.5).

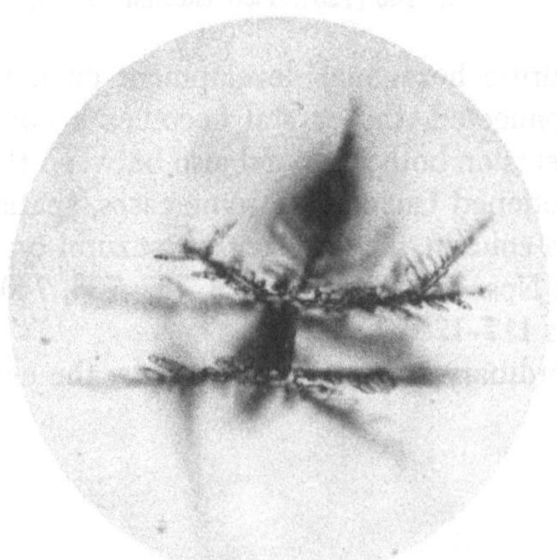

Fig. 144 [778]. Tsuzumi with a spatial dendritic crystal (× 21).

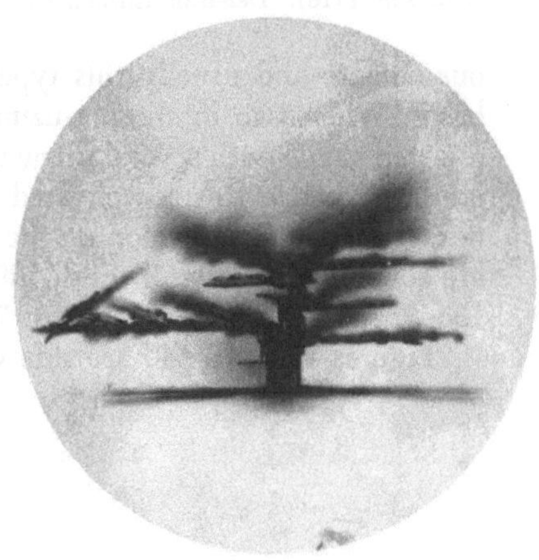

Fig. 145 [716]. Sectioned tsuzumi with a spatial dendritic crystal (× 18.5).

Fig. 146 [782]. Spatial dendritic tsuzumi with cloud particles attached (× 28).

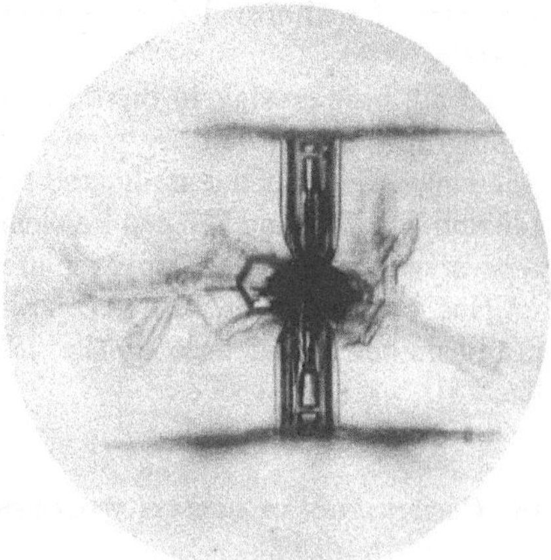

Fig. 147 [717]. Tsuzumi with appendant crystals.

crystals being of dendritic form in the former and of plate form in the latter. The crystal shown in Fig. 141 is an example of a crystal in which upper and lower end plane crystals are of different size, (*a*) being the plan view and (*b*) the side view of the same crystal. In Nos. 732 and 733, Plate 118, the focus is adjusted on the upper and lower hexagonal development of the same crystal. Nos. 734 and 735, Plate 119, are similar examples.

Figure 142 is a comparatively rare example which has a small hexagonal plate on one end and a well-developed stellar crystal on the other end.

Examples of the sectioned-tsuzumi type, namely, crystals composed of many columns with end plates and also one or more intermediate plates, or several tsuzumis connected, are shown in Fig. 143 and Nos. 754–757, 759, Plate 122.

Sometimes the end crystals of tsuzumi or sectioned-tsuzumi crystals develop in spatial dendritic form. One example of a tsuzumi with spatial dendritic crystals is shown in Fig. 144, and a sectioned tsuzumi is shown in Fig. 145. The crystal shown in Fig. 145 is a beautiful example. A crystal of this structure has hitherto been regarded as a rare kind. For instance, Kassner [21] made a sketch of a crystal of this type viewed under a magnifying glass and reported it as a new species of snow crystal. Figure 146 is an example of this type with numerous cloud particles attached.

Figure 147 is an extraordinary crystal of this type with other dendritic branches extending from the middle point of a column. In this case the column is composed of two bullet crystals joined together at the pointed ends, the structure being very clearly observed from the forms of the hollows inside the bullets.

Figures 148 and 149, and Nos. 760–763, 768–775, Plates 123 and 124, are examples of malformed development of ordinary tsuzumi and sectioned-tsuzumi crystals.

Of all these crystals in these pictures, plan view and side view are reproduced. Photographs on the left are plan views and those on the right are corresponding side views. Even such complicated crystal structures as are shown in Figs. 148 and 149 and Nos. 768 and 770 can be clearly analyzed in this way. No. 776 is a tsuzumi-type crystal from whose upper plate short needles extend.

Nos. 780 and 781, Plate 125, are photographs of one crystal, focus being adjusted on the upper base in No. 780 and on the middle part of the column in No. 781.

18. Combination of bullets with dendritic crystals

In some cases, dendritic branches extend from the base of a bullet crystal. Figures 492–495 are good examples of this type, but such examples are rather rare. Many of the crystals of this type are combinations of a group of bullets with dendritic crystals attached to the bases. A beautiful example is represented in Fig. 150. A crystal similar to Fig. 150 but composed of four columns has already been observed and its photograph reproduced in Dobrowolski's book.[22] When many crystals of this type fall, they look like No. 785, Plate 126.

The combination of a bullet and a plate is rarely observed in our climate. Figure 151 is one of these rare examples. In these crystals, when further developed, the plates tend to become dendritic extensions. The series of aforementioned photographs, Figs. 492–495, illustrates very nicely the course of development of plates into dendritic forms.

A combination of three pyramids, each having a dendritic crystal at the base, is represented in Fig. 152. This is one of the rarest cases in our climate.

19. Spatial assemblage of plates and "powder snow"

Although spatial combination of minute plates is sometimes observed in our climate, it rarely develops to a combination of big plates. One example of this type is shown in Fig. 153. This is one of the rarest crystals observable in our country, and seems to be very seldom observed in Europe and America. A similar example, composed of four plates, is shown by Bentley.[23]

Figure 154 shows a crystal similar to Fig. 153, crushed on a glass plate to lay the component plates in one plane for the purpose of photography. This crystal is composed of nine plates.

Spatial combination of minute plates is almost always accompanied by irregular

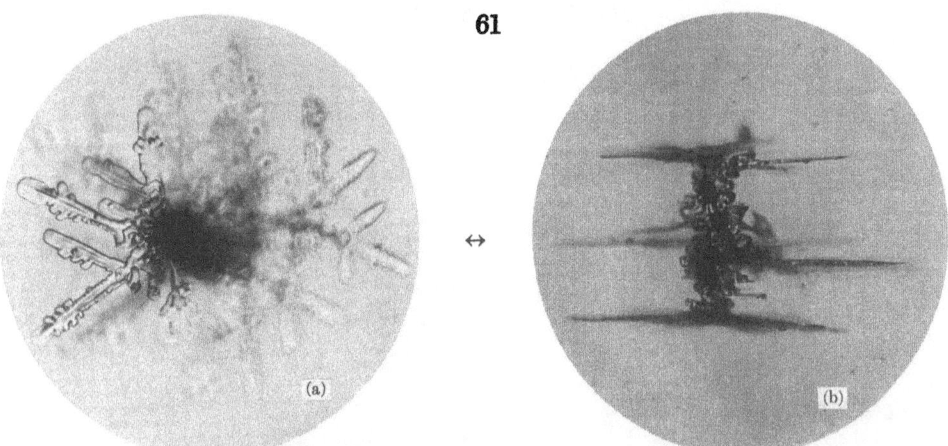

Fig. 148. Malformed sectioned tsuzumi: (*a*) [766] plan view; (*b*) [767] side view (× 32).

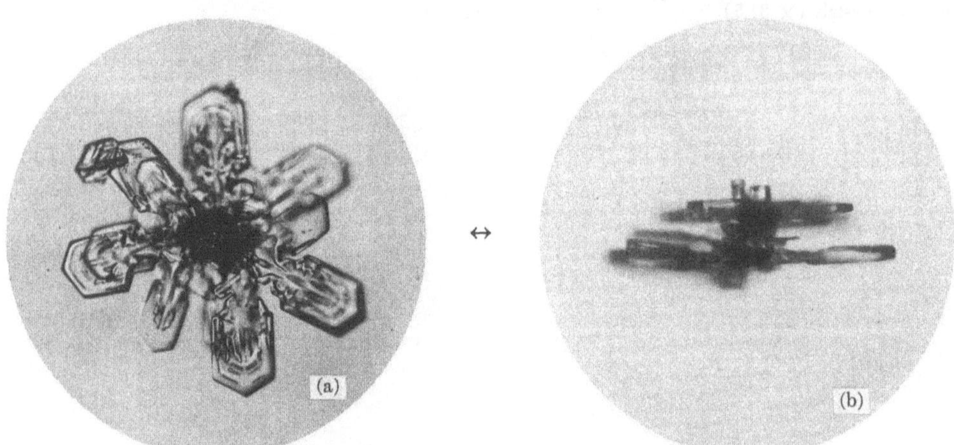

Fig. 149. Malformed tsuzumi: (*a*) [764] plan view; (*b*) [765] side view (× 37).

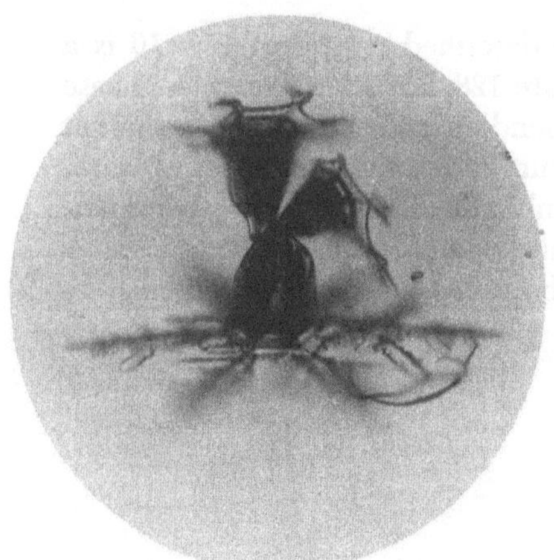

Fig. 150 [784]. Combination of bullets with dendritic crystals (× 61).

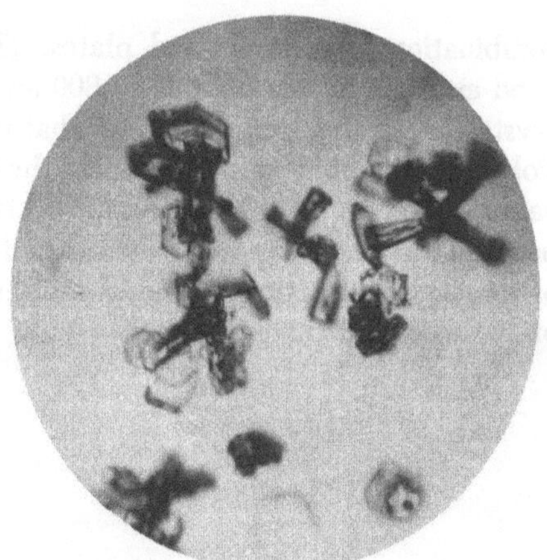

Fig. 151 [793]. Combination of bullets with plates (× 25).

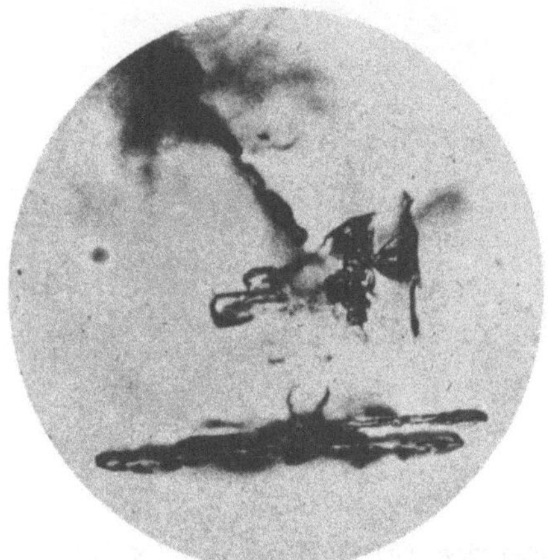

FIG. 152 [787]. Combination of pyramids with dendritic crystals (× 31.5).

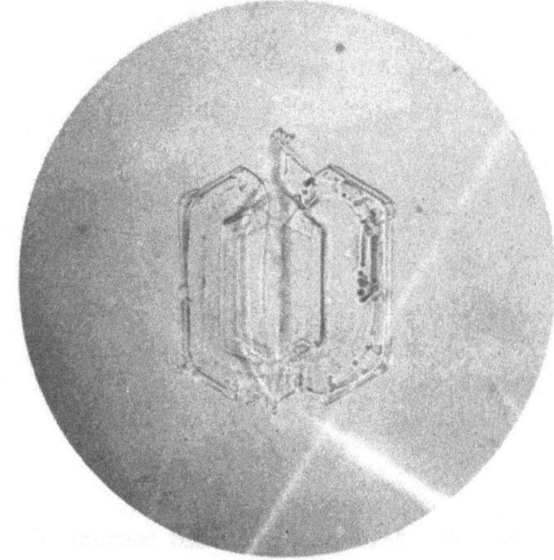

FIG. 153 [798]. Spatial combination of plates (× 35).

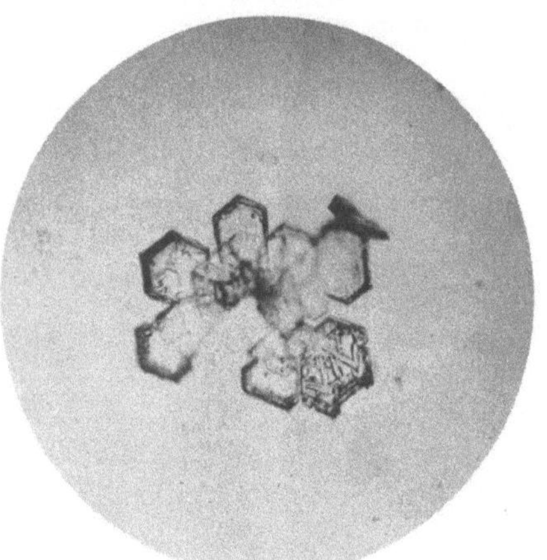

FIG. 154 [799]. Combination of nine plates (× 26).

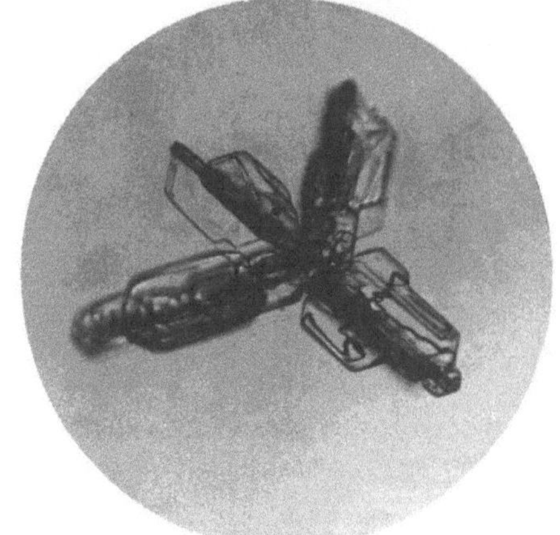

FIG. 155 [804]. Columnar crystal with extended side planes (× 38).

combination of columns and plates. The above-described crystal of Fig. 16 is a good example of this type. Nos. 800 and 801, Plate 128, are other examples. These crystals make the "powder snow" that is often found around the summit of Mount Tokachi. This snow is formed when the temperature is low and the supply of water vapor is scarce. The "powder snow" is commonly called "flour snow" by skiers, because the whole aspect of the heap of snow quite resembles that of flour. In view of the crystal structure, it would more suitably be called an irregular combination of columns and plates.

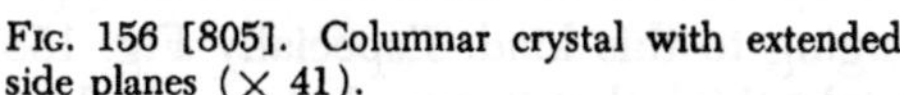

FIG. 156 [805]. Columnar crystal with extended side planes (× 41).

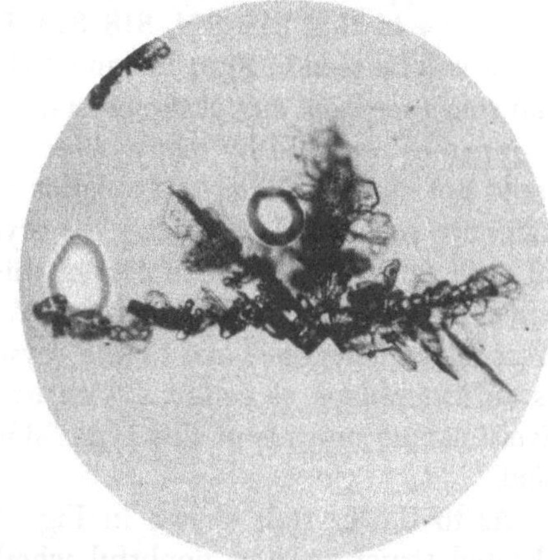

FIG. 157 [815]. Complicated side-plane crystal (× 28).

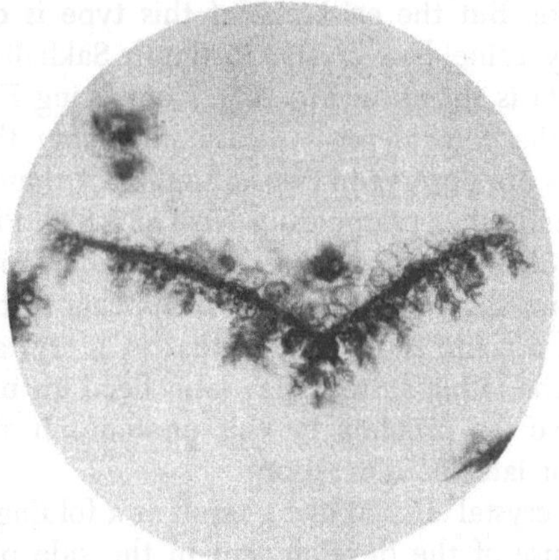

FIG. 158 [816]. Very complicated side-plane crystal (× 6.5).

20. COLUMNAR CRYSTAL WITH EXTENDED SIDE PLANES

Toward the evening of 16 January 1935, we experienced a strange snowfall at Sapporo, consisting solely of crystals of curious type quite unrecorded in the former literature. These crystals were irregular combinations of columns and plates. Photographs taken at that time are reproduced in Figs. 155 and 156 and Nos. 806–811, Plate 129. Since then we have kept a lookout for this type and found crystals of the same type two or three times at Mount Tokachi. Photographs of those crystals are shown in Nos. 812–814, 818–823, Plates 130 and 131.

As will be seen in Figs. 155 and 156, they are quite different in both appearance and structure from any of the crystals that have been hitherto observed by authors in Japan or reported by many observers in various parts of the world. At first we could not classify the crystals of this type, because their structure was unknown. Later we succeeded in making the crystal shown in Fig. 262, in our experiments on artificial frost. The plate part of this crystal appears to be the development of one of the side planes of the column. We infer that crystals classified in this section are substantially the same as the crystal of Fig. 262. We therefore regard them as an assemblage of columns with irregularly developed side planes, and tentatively named crystals of this type "side-plane crystals," to be classified as a new kind.

As to the crystals shown in Figs. 157 and 158, which have extremely complicated structure, it is doubtful whether they should be included in this type, but we tentatively classified them as belonging to this kind. Especially Fig. 158 has such an extraordinary shape that one might think that it was photographed by some incidental mistake. But the existence of this type is confirmed by the discovery of a completely coincident crystal in South Sakhalin by K. Nakata. The crystal found by Nakata is shown in Fig. 159. Comparing Figs. 158 and 159, one will be surprised by the coincidence. Occasionally only the skeletons or bony frames of these crystals are observed. Two examples of these skeletons are shown in Figs. 160 and 161, and other examples in Nos. 827–829, Plate 131.

A crystal of unexpected shape was observed at Mount Tokachi, of which front, back, and side views are shown in Fig. 162. It appears to be a crystal of square form. The existence of a cubic snow crystal, that is, a crystal that belongs not to the hexagonal but to the cubic system, has long been an unsettled problem. Although we cannot solve the problem by this photograph alone, we present this example as a datum for later consideration.

Figure 163 shows a crystal shaped like a scroll or a folding screen. This peculiar form is obtained because of the development in the side planes only. This type being produced artificially in the cold chamber, we shall describe it in detail in Part II.

21. Initial stage of snow crystals

In photograph No. 20, Plate 1, a small hexagonal crystal is seen beside the large fernlike crystal, about one-seventh the size of the latter. When such well-developed large crystals fall, very small crystals of a similar type often mingle with them. This phenomenon was especially well observed at the spot 1030 m

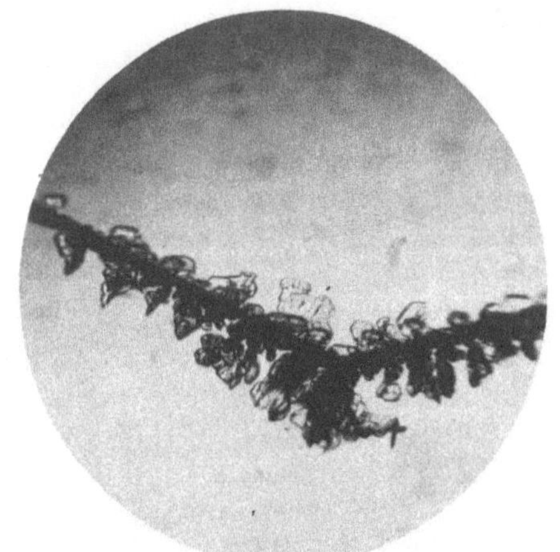

Fig. 159 [817]. Crystal similar to Fig. 158, observed by Nakata in South Sakhalin.

↕

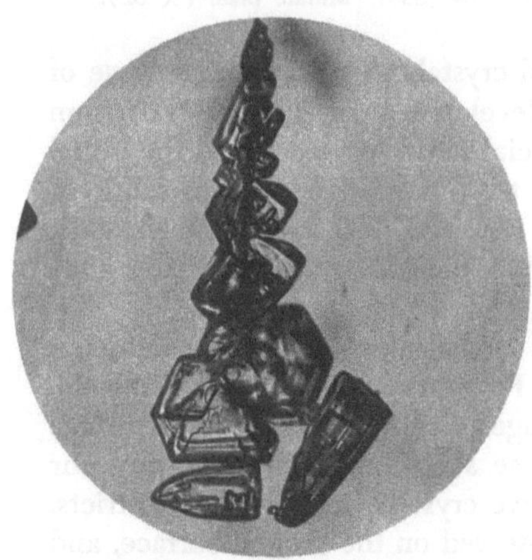

Fig. 160 [825]. Peculiar crystal (× 43).

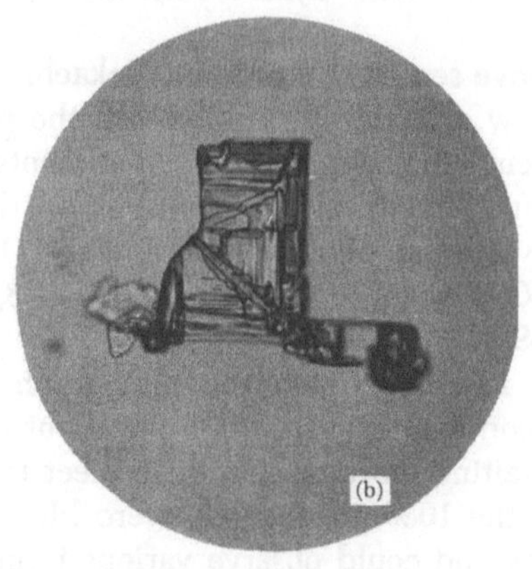

↕

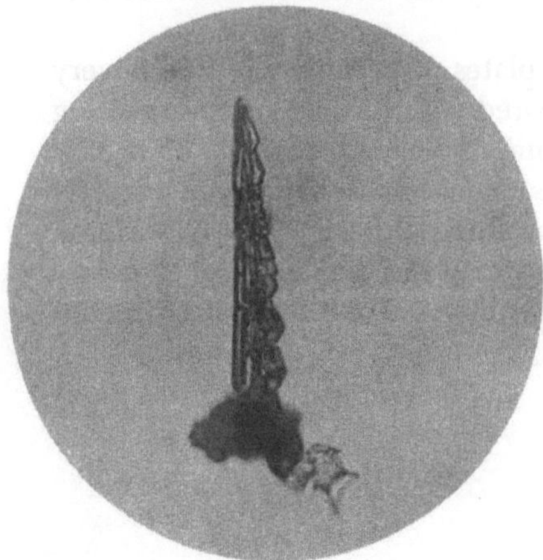

Fig. 161 [826]. Peculiar crystal (× 32.5).

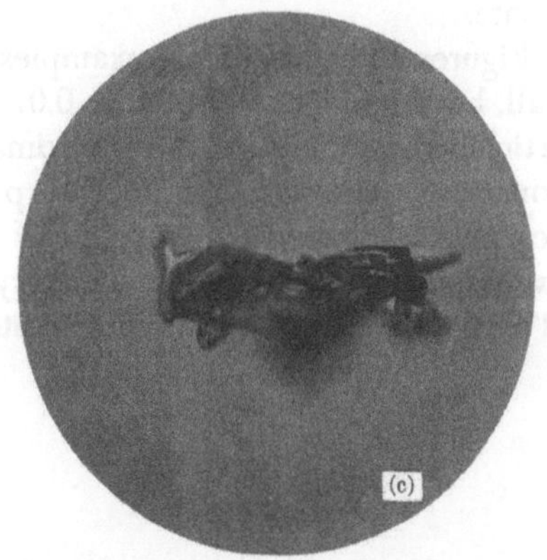

Fig. 162. Snow crystal of apparently square form: (*a*) [830] front view; (*b*) [831] back view; (*c*) [832] side view (× 38.5).

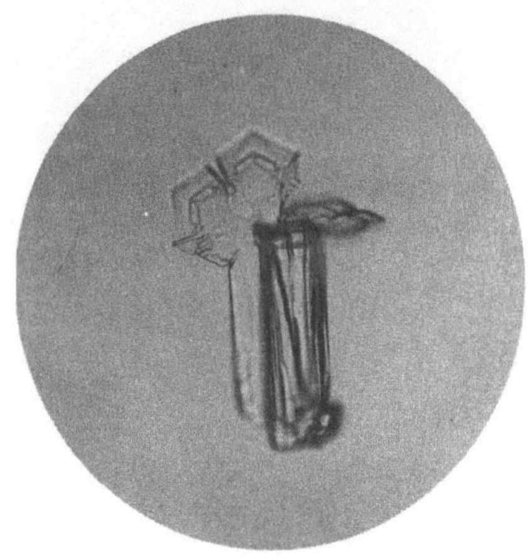
FIG. 163 [835]. Crystal of scroll form (× 41).

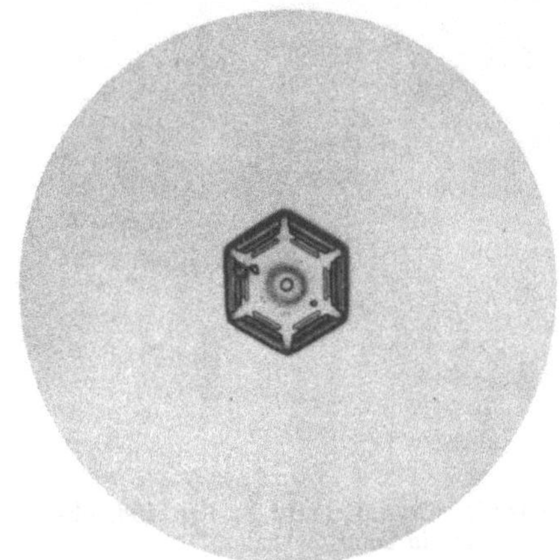
FIG. 164 [838]. Minute plate (× 52).

above sea level on Mount Tokachi. These small crystals show the initial stage of snow crystals. We could learn the process of development of snow crystals from their initial stage, in our experiments with artificial snow. We examined the initial stage of natural snow crystals on the basis of this knowledge. Various kinds of crystals are shown in Figs. 164–171 and Nos. 836, 837, 839, 840, 842, 843–848, 850, 851, 853–856, 858–863, 866–868, 870, 872, 873, 875–879, 882, 883, Plates 134–139.

In order to take the photographs of crystals in the initial stage, we conducted a continuous observation for about a month in February 1940 at Mount Tokachi, awaiting the opportunity to meet the initial stage of snow at the ground surface of the 1030-m spot. We were blessed with three such opportunities during our stay and could observe various forms of primitive crystals. In very cold districts, the so-called ice fog or ice crystals are often observed on the ground surface, and when lighted by sunshine they are seen floating with the naked eye as many minute brilliant particles. Some people call them "diamond dust." The primitive crystals mentioned here are the same as these ice crystals observed in the coldest climate.

Figures 164 and 165 are examples of minute plates. Especially Fig. 165 is very small, having a diameter of only 0.07 mm. Compared with the size of a normal fog particle, whose diameter is approximately 0.03 mm, the crystal of Fig. 165 is only about twice as large. Such minute plates as this one seem to have been scarcely known in the past. In the crystal of Fig. 164 the internal design due to skeleton development is clearly seen. This crystal is also small, the diameter being nearly 0.2 mm. Generally, these minute crystals, not larger than 0.2 or 0.3 mm in diameter,

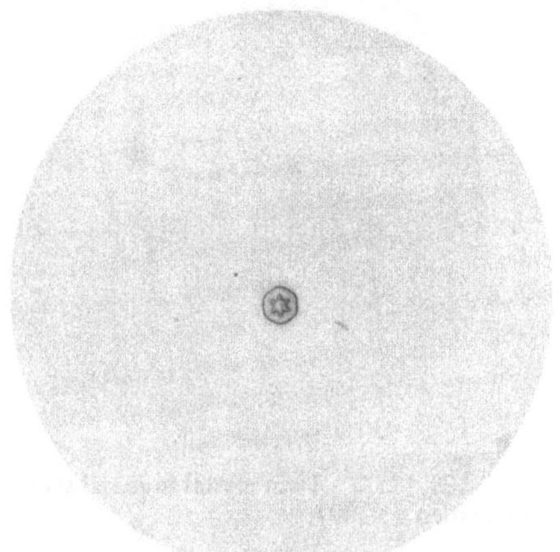

FIG. 165 [841]. Minute plate (× 65).

FIG. 166 [849]. Minute stellar crystal (× 70).

have comparatively simple internal structure. But when they grow to 0.5 mm or more in diameter, various patterns begin to appear, as shown in No. 847, Plate 134, which indicates that they have come through several stages of development. We classified crystals up to 0.5 mm in diameter as crystals in the initial stage.

Some of the primitive crystals of the same size as the minute plates above mentioned have already made hexagonal development. Examples are shown in Fig. 166 and in Nos. 848, 850, 851, 853–856, 858, Plates 134 and 135. The crystal of Fig. 166 is 0.3 mm in diameter and Nos. 850 and 851 are 0.5 mm. A small tsuzumi crystal is also observed in an initial stage. One example is given in Fig. 167. This minute tsuzumi is often observed in the initial stage of artificial snow and will be described in detail in Part II, Sec. 71. An interesting fact is that the aforementioned three- or four-branched type appears even in this stage, as Fig. 168 and Nos. 851, 856, and 858, Plate 135, show. Figure 168 is a photograph of a small hexagonal crystal 0.5 mm in diameter after separation. In the initial stage of artificial snow we could obtain minute plates and hexagonal crystals even smaller than these primitive crystals of natural snow. Their photographs will be introduced later.

Figure 169 shows plan and side views of a thick plate which is developed in the skeleton structure. Nos. 862 and 863, Plate 136, are similar crystals. The pattern appearing in No. 862, which is the unit of patterns often seen in the branches of larger crystals, is due to the skeleton structure. The thickness of this sort of crystal is of the order of magnitude indicated by Fig. 169(*b*).

Minute columns and bullets are frequently observed in the initial stage. Usually they have a strong tendency for the skeleton development, the wall being very thin

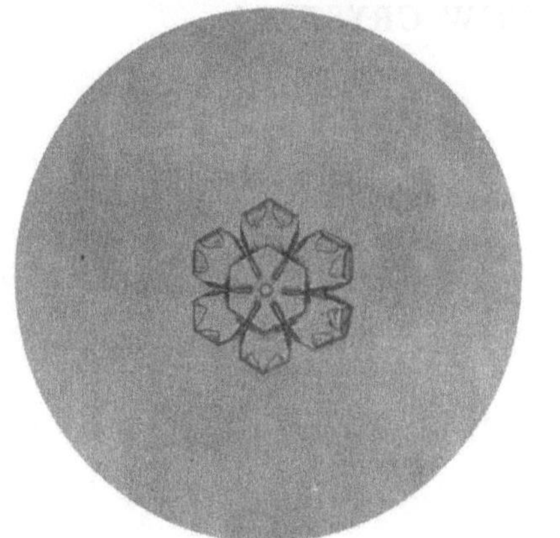

Fig. 167 [852]. Tsuzumi in initial stage (× 34).

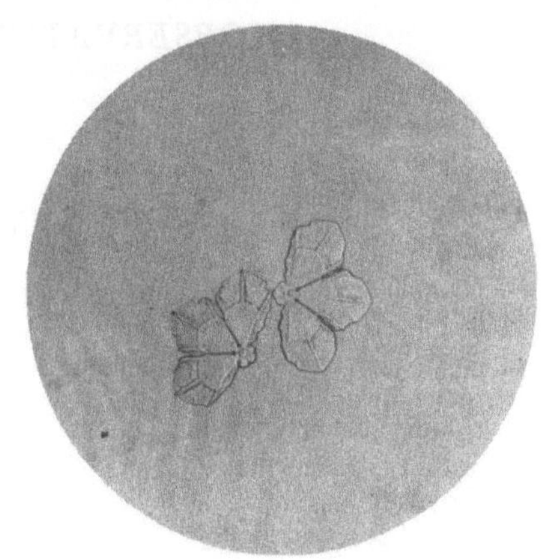

Fig. 168 [857]. Twin crystal in initial stage, after separation (× 39).

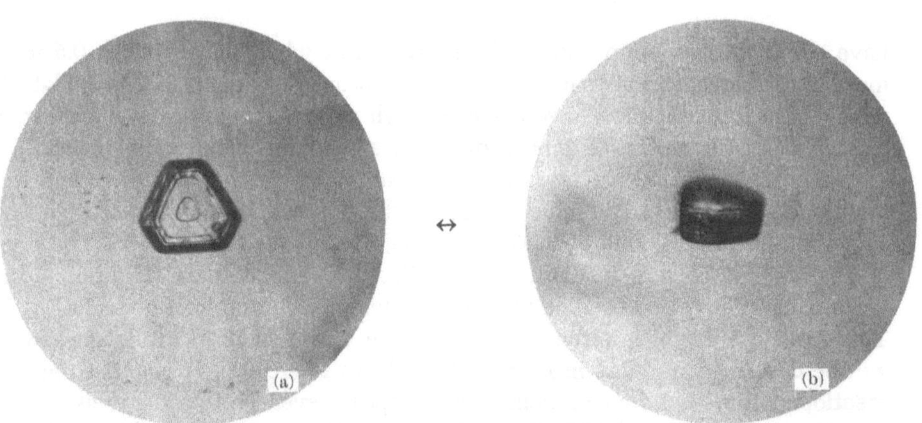

Fig. 169. Minute thick plate: (*a*) [864] plan view; (*b*) [865] side view (× 52).

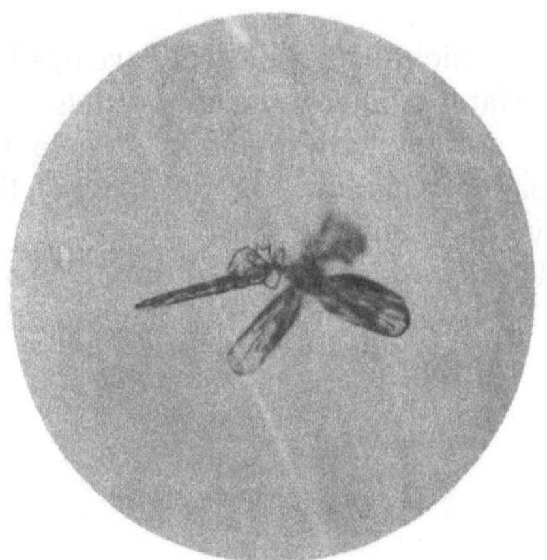

Fig. 170 [871]. Bullet with thin wall (× 34).

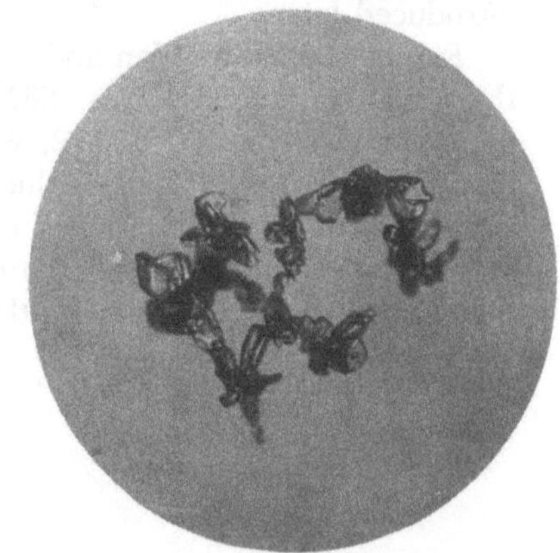

Fig. 171 [874]. Initial stage of irregular form (× 34).

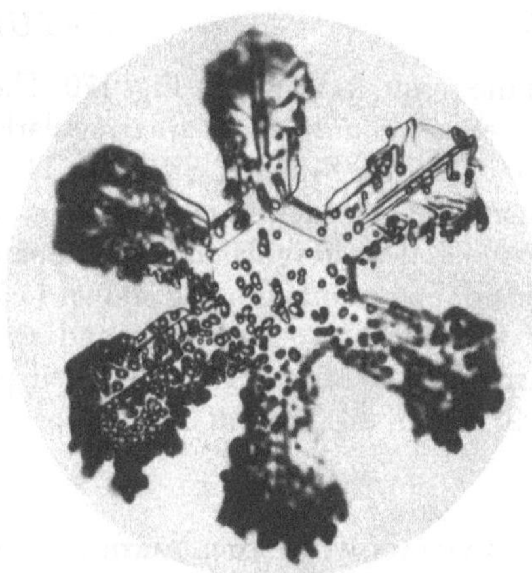

FIG. 172 [898]. Plane crystal with cloud particles (× 21).

FIG. 173 [895]. Plane crystal with cloud particles (× 40).

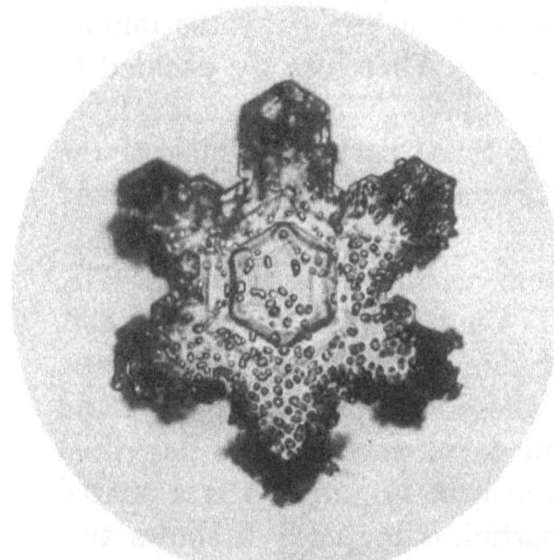

FIG. 174 [903]. Plane crystal with cloud particles (× 19).

FIG. 175 [911]. Spatial dendritic crystal with cloud particles (× 34).

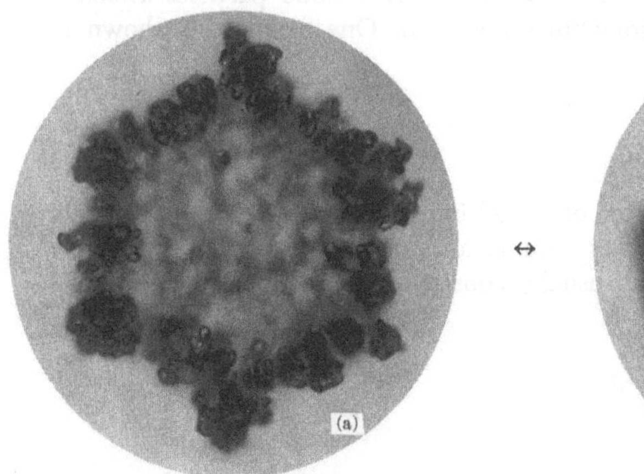

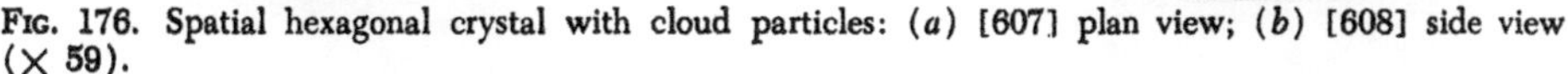

FIG. 176. Spatial hexagonal crystal with cloud particles: (*a*) [607] plan view; (*b*) [608] side view (× 59).

as the result, as shown in Fig. 170. They often form complicated combinations and sometimes minute plates are irregularly attached to the combination. Nos. 866, 868, 870, 872, 873, 875–879, 882, 883, Plates 138 and 139, are examples of such combinations, but their classification is impossible as yet owing to the shortage of material. The irregular crystal, apparently of amorphous form, is sometimes observed, one example being shown in Fig. 171. Mixed with these ice crystals, rare crystals such as cup crystals and scroll-type crystals of natural snow were also discovered at Mount Tokachi. Their photographs are reproduced in Figs. 498 and 499, and an explanation of their formation is given in chapter 7.

22. Crystals with cloud particles attached

Bentley states that he occasionally observed crystals with small water drops attached to them. Our observation in Hokkaido revealed the fact that crystals with small water drops are more frequently observed throughout the winter than ordinary crystals without droplets. As is fully described in Sec. 37, these minute water drops were found to be particles of cloud or fog. Therefore we decided to name these crystals "crystals with cloud particles attached." When there is a layer of cloud or fog consisting of supercooled droplets, those particles attach to the crystals while they fall and are frozen to them. Crystals of this type are formed in this way.

Crystals with cloud particles attached are found, as shown in Figs. 172–178, in almost all kinds of crystal types. Figures 172–174 and Nos. 884–887, 897, 899, 900–902, 904–906, Plates 140–143, are examples of plane crystals such as dendritic or plate crystals with cloud particles attached. Of these examples, in Fig. 174 and in Nos. 887 and 906, needle development or scroll development, to be described later, is seen in the especially dark part upon the rim. Figure 175 is a spatial dendritic crystal of radiating type with cloud particles. Figure 176 shows two views of a spatial hexagonal crystal with cloud particles.

As needles fall in many cases when the temperature is near 0°C, they more often show attached cloud particles. Figure 177 and Nos. 889–892, Plate 139, are examples of needles with cloud particles. Columns with cloud particles attached are comparatively few, but are sometimes observed. One example is shown in Fig. 178.

23. Thick plane crystals

Very thick plane crystals are often observed in our climate. At Mount Tokachi it sometimes occurs that the air is very humid and a dense fog covers the whole district. In such a case the snowfall usually consists of these thick plane crystals

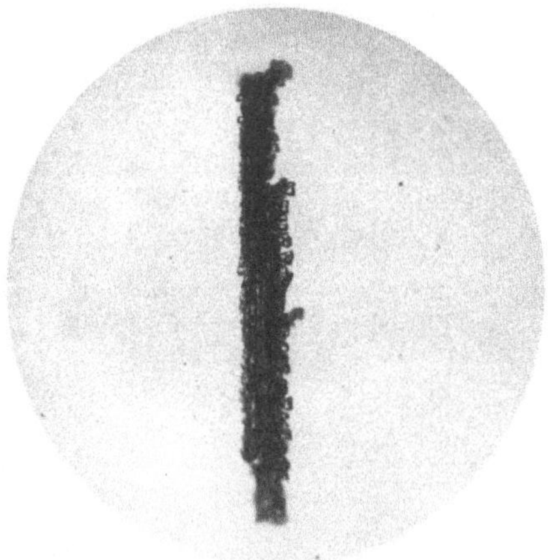

Fig. 177 [888]. Needle with cloud particles (× 19).

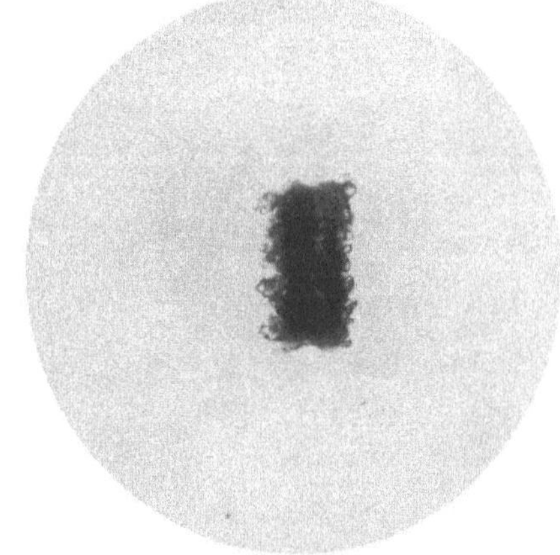

Fig. 178 [893]. Column with cloud particles (× 36).

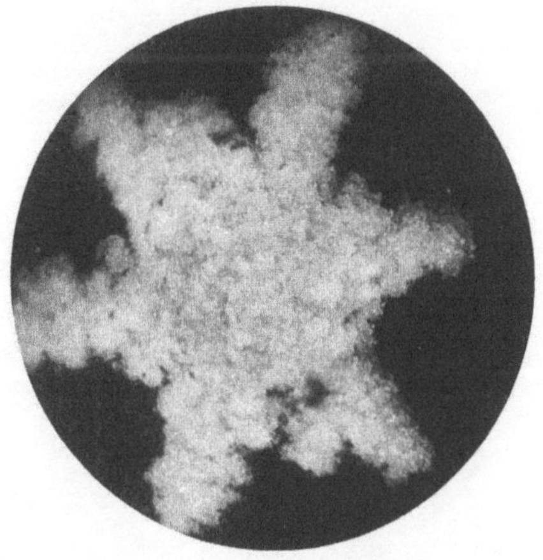

Fig. 179 [912]. Thick plate by reflected light (× 9.3).

Fig. 180 [913]. Section of a thick plate (× 19).

and the graupellike snow to be described in the next section. These thick plane crystals appear under a microscope by reflected light as many white particles attached to one another, as shown in Fig. 179, and by transmitted light the whole crystal appears dark. We succeeded in cutting a crystal of this type into two parts and in taking a photograph of its section under a microscope. It has a structure, as shown in Fig. 180. As this photograph shows nicely, cloud particles are attached on the surface of a plane crystal in many layers and form a thick plane crystal as a whole. We named crystals of this type "thick plane crystals."

Fig. 181 [924]. Spatial hexagonal crystal with cloud particles (× 12).

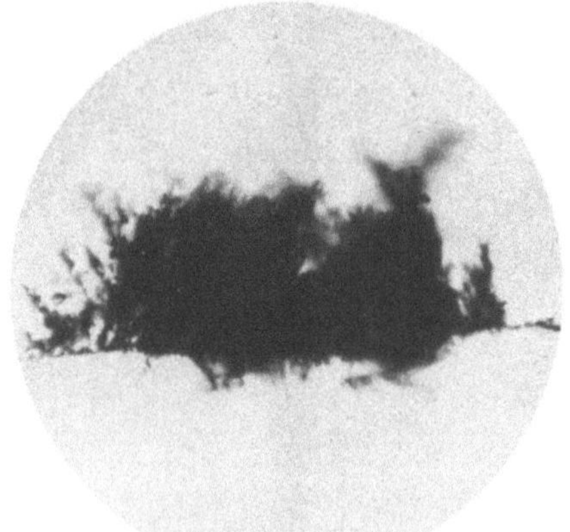

Fig. 182 [925]. Graupellike snow, hexagonal type (× 12).

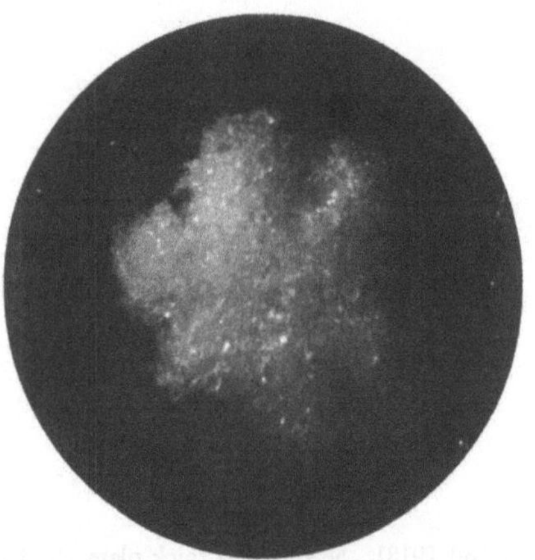

Fig. 183 [926]. Hexagonal graupel (× 15).

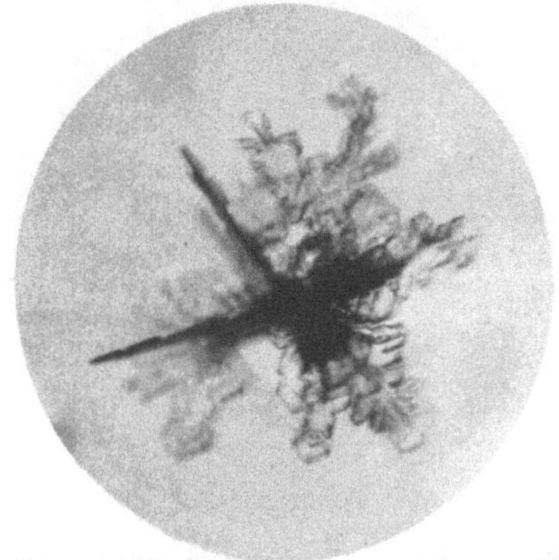

Fig. 184 [927]. Burr-type crystal with cloud particles (× 19).

Thick plane crystals have attached cloud particles chiefly on one surface alone. It seems that while a plane crystal falls, in an almost horizontal position, cloud particles become attached to its lower surface. It is well known that in the case of rime frost, seen on high mountains in winter, fog particles attach themselves chiefly to the weather side of an exposed object. Crystals of this type may be considered as snow crystals with rime frost attached. If the cause of formation is as mentioned above, each type of plane crystal must have the corresponding thick plane one. This is actually so. Photographs of various thick plane crystals are reproduced in Nos. 914–923, Plates 144 and 145. Although very few thick plane crystals are recorded in former literature on the subject, this type is very often observed both at Mount Tokachi and at Sapporo in Hokkaido.

24. Origin of graupel (snow pellets)

In Hokkaido, graupel or pellett-type snow falls very frequently not only at the beginning of winter but also in the middle of winter. Graupel in Hokkaido is rather small in size. Graupel in midwinter is mingled with crystals with cloud particles or thick plane crystals in most cases.

It is now generally accepted that the graupel is formed in a lower atmospheric layer where the supercooled water droplets are abundant, but the mechanism of its formation is not clearly known. The author had many chances at Sapporo and at Mount Tokachi to observe snow crystals that seemed to be in intermediate states between snow and graupel. There are three types of graupel as already noticed by many observers: lump graupel, the conelike one, and another which shows a trace of hexagonal symmetry. The author proposes to explain the lump graupel as a modified form of a spatial dendritic crystal of radiating type and the last-mentioned type as developed from a spatial dendritic one with a stellar base crystal.

In our climate, snow crystals with numerous water droplets attached are very frequently observed. When many water droplets attach themselves to a plane dendritic crystal of snow, it turns into a thick plane crystal, as described in the preceding section. These droplets are likely to attach themselves also to the spatial dendritic crystals and in that case these snow crystals seem to become graupel.

A series of photographs supporting this theory is reproduced in Figs. 181–183. Figure 181 shows a side view of a spatial dendritic crystal with a stellar base, to which many droplets are attached. Figure 182 is a side view of a similar crystal, the only difference of which from the one in Fig. 181 being that the attached cloud particles are so numerous in this case that the whole crystal presents a semisolid appearance. It is proposed to call snow crystals in this stage graupellike snow. When the number of attached cloud particles increases still further, this graupellike snow becomes more compact in structure and turns into the usual graupel. The appearance of the graupel belonging to this series, therefore, shows a trace of hexagonal symmetry. One example of this type of graupel is shown in Fig. 183. Thus the origin of this special form of graupel is explained.

Another series of photographs, showing the course of formation of lump graupel, is given in Figs. 184–186. Figure 184 is a spatial crystal of radiating type, Fig. 185 is a similar crystal with numerous cloud particles attached, and Fig. 186 is a lump graupel which seems to have been developed from this kind of graupellike snow. Thus the origin of a lump graupel is explained.

FIG. 185 [928]. Graupellike snow, lump type (× 20).

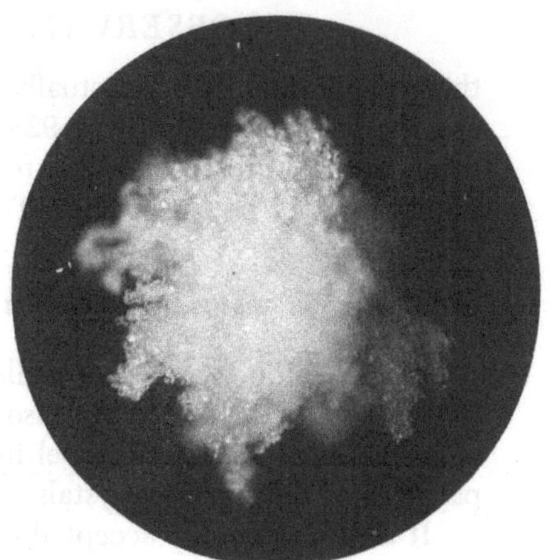

FIG. 186 [929]. Lump graupel.

FIG. 187 [935]. Graupellike snow, burr type (× 17.5).

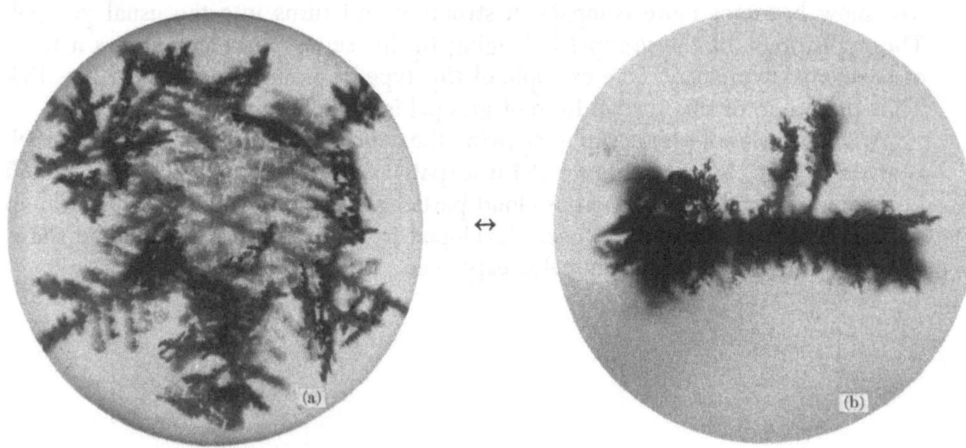

FIG. 188. Graupellike snow, hexagonal type: (*a*) [942] plan view; (*b*) [943] side view (× 12).

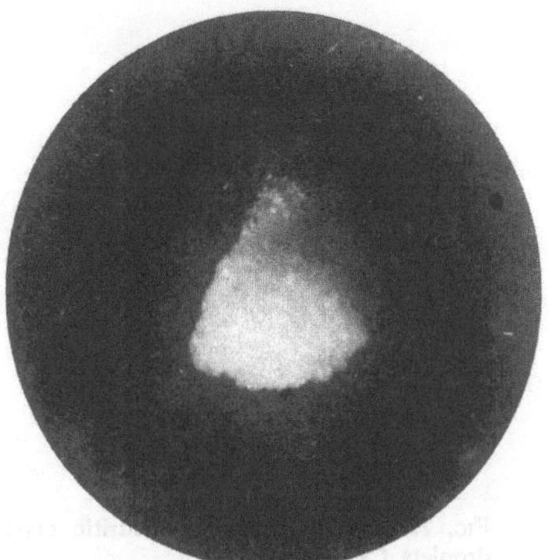

FIG. 189 [948]. Conelike graupel.

FIG. 190 [952]. Lump graupel (× 15.5).

The origin of the graupel of a conelike form remains unclarified. The rotation of the graupel that occurs while it is falling may be a cause, as described in Czermak's paper,[24] but at present the author refrains from touching this question.

Crystals in an intermediate state between snow and graupel were also noted by European observers. Barkow [25] proposed a theory similar to the present author's on the formation of the graupel. Some parts of his paper must be corrected as pointed out by K. Wegener,[26] but the author considers that this theory, which assigns the formation of graupel to the attachment of numerous cloud particles to a snow crystal, is in the main correct. The point of importance is that any sort of snow crystal is observable, the amount of cloud particles attached to which is varied continuously from snow to graupel. There is no reason why a graupel pellet must be considered separately from a snow crystal with many cloud particles attached.

The condition favorable for the production of graupel is the existence near the earth's surface of a thick cloud layer of supercooled droplets, the temperature in which is a few degrees below 0°C, and a snow layer over it. Any upward current will be very favorable for the attachment of an innumerable number of cloud particles to the original crystal.

25. Graupellike snow

As mentioned in the preceding section, graupellike snow has already been recorded in European literature; for instance, in the report of Greim [27] the cause of its formation, structure, etc. are described.

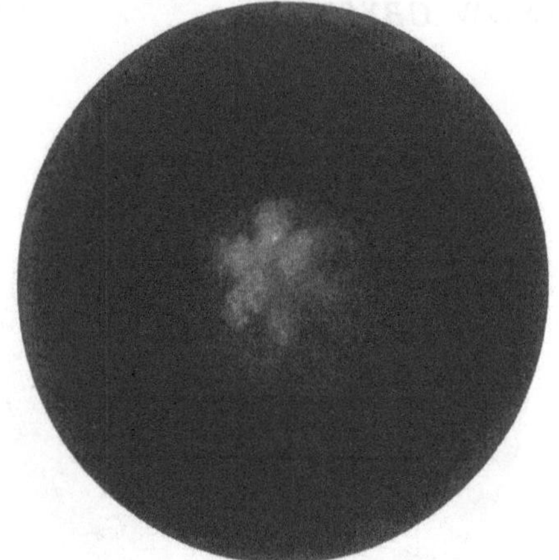

Fig. 191 [956]. Hexagonal graupel.

Fig. 192 [961]. Irregular dendritic crystal with droplets (× 23.5).

Fig. 193 [966]. Irregular dendritic crystal with droplets.

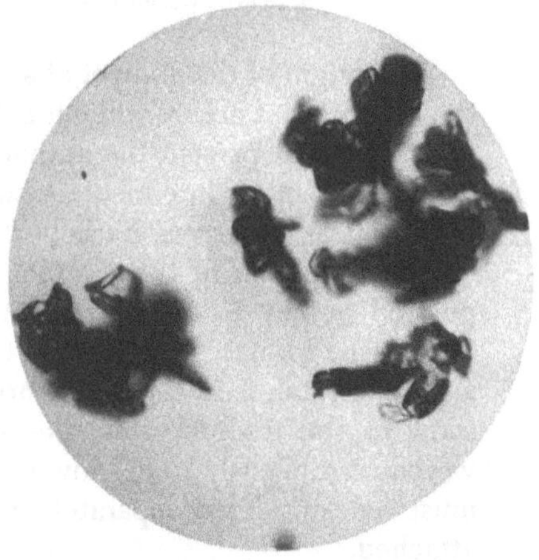

Fig. 194 [968]. Ice particles (× 35).

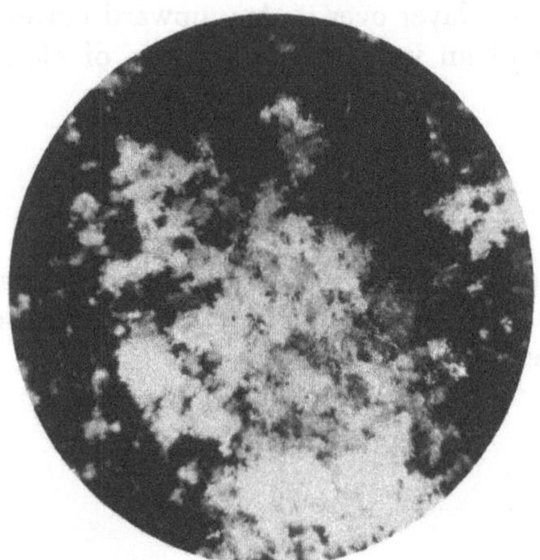

Fig. 195 [969]. Amorphous particles with droplets (× 13).

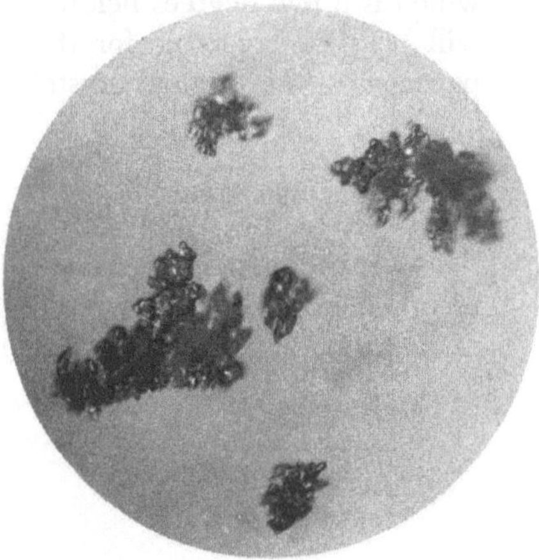

Fig. 196 [976]. Granular snow (× 33).

Figure 187 and Nos. 930–934, 936, 937, Plates 146 and 147, are examples of graupellike snow developed from spatial dendritic crystals of radiating type. Figure 188 and Nos. 944–947, Plates 147 and 148, are those developed from spatial hexagonal ones, plan views and side views being shown of each of three crystals. Of these No. 946 is a rare example of a hexagonal plate with cloud particles attached around its edges. Cloud particles often attach themselves in this way, which is occasionally observed in the minute structure of rime frost deposited on trees in high mountain districts.

26. Graupel (snow pellets)

Of the three kinds of graupel (snow pellets) an example of conelike graupel is shown in Fig. 189. An example of lump graupel is shown in Fig. 190. Small prominences on the rim of this lump graupel, which are the traces of spatial dendritic branches forming the skeleton of this pellet, can be clearly observed in Fig. 190. Graupel pellets of a complete shape as seen in these figures is observed only when the temperature is quite low. When the temperature is close to 0°C, the number of surfacial prominences becomes less as shown in No. 954, Plate 149.

An example of hexagonal graupel is reproduced in Fig. 191. It will be clearly seen in the photograph that the skeleton of a fully developed hexagonal graupel pellet like this example is a spatial hexagonal snow crystal. Other examples of graupel are shown in Nos. 949–951, 953–955, 957–959, Plates 148 and 149.

27. Irregular snow particles

It sometimes happens that many of the snow crystals which actually fall cannot be classified in any of the above-mentioned types, because of their indefinite shape. We named them collectively "irregular snow particles."

This irregular type falls more frequently than one might suppose. If we come to know the details of the characteristics of these snow particles, further classification will become necessary. At present, we tentatively divide them into the irregular assemblage of dendritic branches with numerous cloud particles attached, the assemblage of ice particles, and the one with so many cloud particles that it appears as a lump of frozen droplets.

Two examples of the type of irregular dendritics with cloud particles are shown in Figs. 192 and 193. A representative example of an assemblage of ice particles is shown in Fig. 194. Figures 195 and 196 are good examples of the type that appears like a lump of frozen droplets. Sometimes a trace of dendritic branches can be seen inside the particle, but usually the skeleton is indistinguishable. This irregular snow has hitherto been completely neglected, but in future snow of this kind will have to be classified more scientifically.

CHAPTER 2

General Classification of Snow Crystals and Their Frequency of Occurrence

28. Method of classification

As is detailed in the foregoing chapter, snow crystals observed in Hokkaido take many forms. The hexagonal plane crystal, which has hitherto been commonly regarded as the representative type of snow crystal, was found to comprise only a part of natural snow. In actuality, spatial crystals and irregular types fall in greater quantity, contrary to the general assumption. One of the reasons for the past popularity of the photographs of regular hexagonal crystals may be that those crystals make beautiful photomicrographs. Before the method of photomicrography was fully developed, crystals with spatial structure were often reported in the literature, but they were usually more or less neglected by the workers using photomicrography. In this sense, if we may borrow the words of Wegener, we could say that the development of the photomicrograph obstructed the study of snow crystals. We therefore, in attempting our general classification, have tried to attach always equal importance to every type of crystal that was actually observed in nature.

It is so far accepted, although there is some argument, that the snow crystal belongs to the hexagonal system of crystallization. When it develops chiefly along the principal axis, it becomes a hexagonal column or pyramid, and when the crystal grows in the base plane, it becomes a plane crystal showing a hexagonal symmetry. The persons who noticed this and attempted the first scientific classification are Hellmann [28] and Nordenskiöld.[29] The classifications of these two scientists are the same in principle. They classified the snow crystals into the columns, the plates, and their combinations, making subdivisions of each type. The principle of this classification remains to the present day. The classification of Bentley and

Humphreys [30] is a recent attempt, which has five types. But from the structural viewpoint, their classification has a tendency to confuse some of the types.

29. General classification of snow crystals

We took approximately 2200 photographs during our main observations covering three years and the supplementary observations covering the succeeding two years. In taking these photographs, we tried to include as many kinds as possible, discarding our subjective prejudice in the selection. The result is tabulated in Table 4, as the general classification of snow crystals, based on structure and cause of formation.

Our general classification is more or less similar to that of Hellmann and Nordenskiöld, but the distinguishing feature of our classification lies in the addition of new types and the alteration in some items of classification on the basis of crystalline structure. Although needles are the same as columns in that they are developed along the principal axis, we classified them into a separate type because we discovered in our experiments on artificial snow that the conditions for their formation are quite different from those of columns. We also attached due importance to the hitherto neglected types, such as those of spatial structure, graupel-like snow, irregular crystals, etc., making our classification extensive as a whole. A chart of this general classification is shown in Fig. 197, and photographs illustrating it are presented in Fig. 198. Classification numbers in the following brief descriptions refer to Fig. 197.

I. Needle crystal, *N*. Considering the results of the experiments on the artificial production of needle crystals, this had better be taken as a kind distinct from the elongated column.

1. *Simple needle.* Usually this type of crystal shows a structure like a bundle of several thin pillars growing parallel to each other (*N1b*). The component needle is rarely observed (*N1a*).

2. *Combination of needles.* A combination of two needles giving an *X*-shaped appearance is usually observed (*N2*).

II. Columnar crystal, *C*.

1. *Simple column.* Crystals growing in the direction of the principal axis of crystallization are grouped in this type.

a. Pyramid. This is the least frequently observed (*C1a*).

b. Bullet type. A hexagonal column with one end pyramidal has an appearance like a bullet (*C1b*).

c. Hexagonal column. This is a hexagonal cylinder with both ends plane, and is considered to be a twin crystal composed of two bullet crystals (*C1c*).

2. *Combination of columns.*

TABLE 4. General classification of snow crystals.

I	*N*	Needle crystal	1. Simple needle	*a.* Elementary needle *b.* Bundle of elementary needles
			2. Combination	
II	*C*	Columnar crystal	1. Simple column	*a.* Pyramid *b.* Bullet type *c.* Hexagonal column
			2. Combination	*a.* Combination of bullets *b.* Combination of columns
III	*P*	Plane crystal	1. Regular crystal developed in one plane	*a.* Simple plate *b.* Branches in sector form *c.* Plate with simple extensions *d.* Broad branches *e.* Simple stellar form *f.* Ordinary dendritic form *g.* Fernlike crystal *h.* Stellar crystal with plates at ends *i.* Plate with dendritic extensions
			2. Crystal with irregular number of branches	*a.* Three-branched crystal *b.* Four-branched crystal *c.* Others
			3. Crystal with twelve branches	*a.* Fernlike crystal *b.* Broad branches
			4. Malformed crystal	Many varieties
			5. Spacial assemblage of plane branches	*a.* Spacial hexagonal type *b.* Radiating type
IV	*CP*	Combination of column and plane crystals	1. Column with plane crystals at both ends	*a.* Column with plates *b.* Column with dendritic crystal *c.* Complicated capped column
			2. Bullets with plane crystals	*a.* Bullets with plates *b.* Bullets with dendritic crystals
			3. Irregular assemblage of columns and plates	
V	S	Columnar crystal with extended side planes		
VI	*R*	Rimed crystal (crystal with cloud particles attached)	1. Rimed crystal	
			2. Thick plate	
			3. Graupellike snow	*a.* Hexagonal type *b.* Lump type
			4. Graupel	*a.* Hexagonal graupel *b.* Lump graupel *c.* Conelike graupel
VII	*I*	Irregular snow particle	1. Ice particle	
			2. Rimed particle	
			3. Miscellaneous	

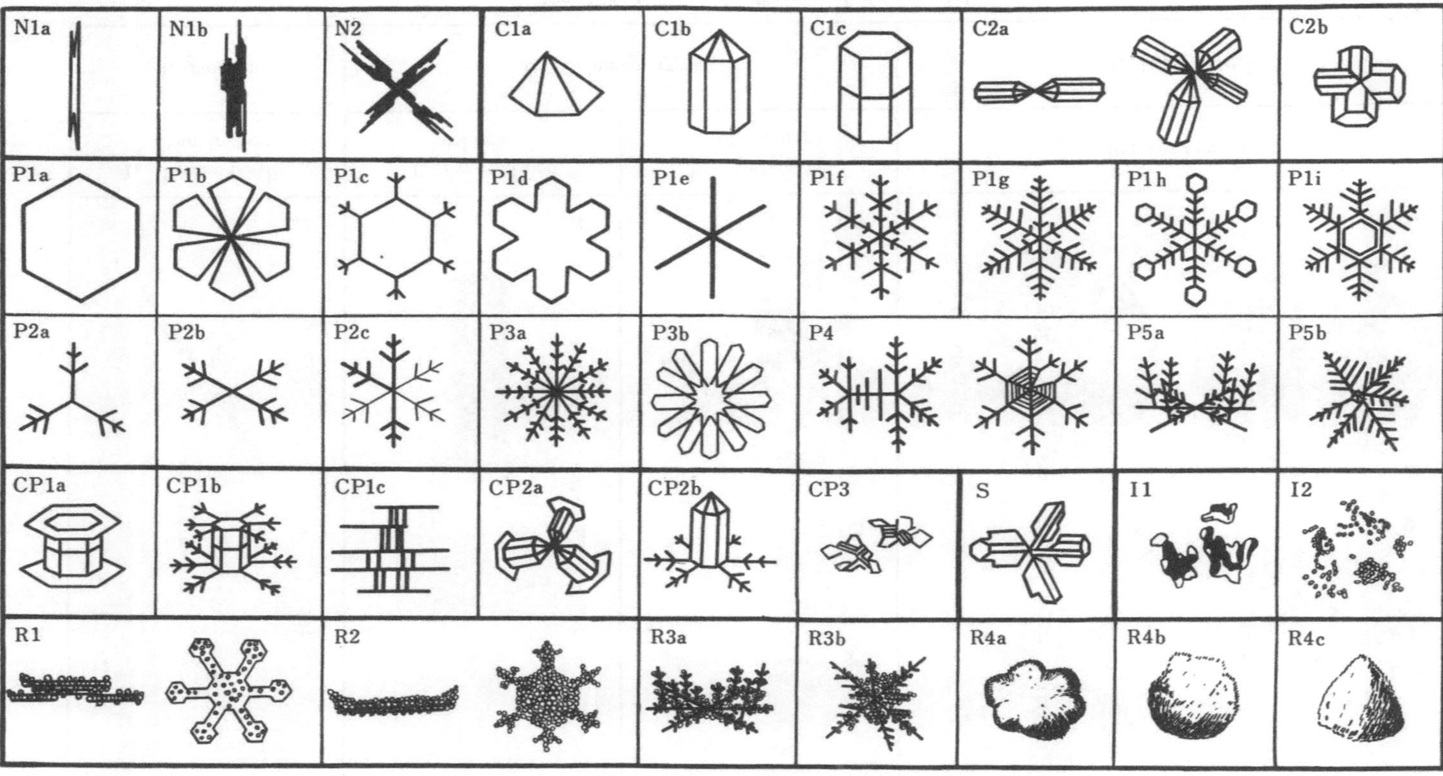

Fig. 197. General classification of snow crystals, sketches.

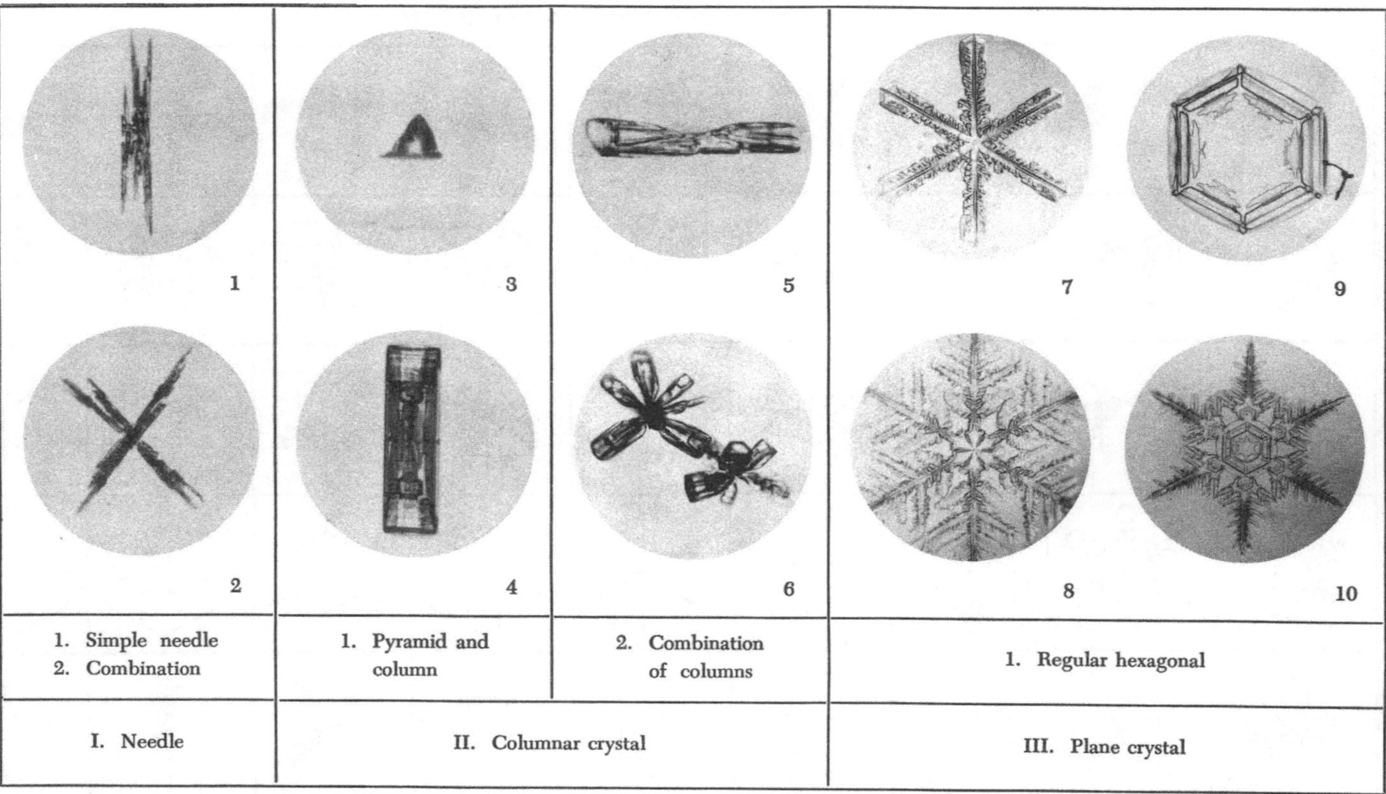

FIG. 198. General classification of snow crystals.

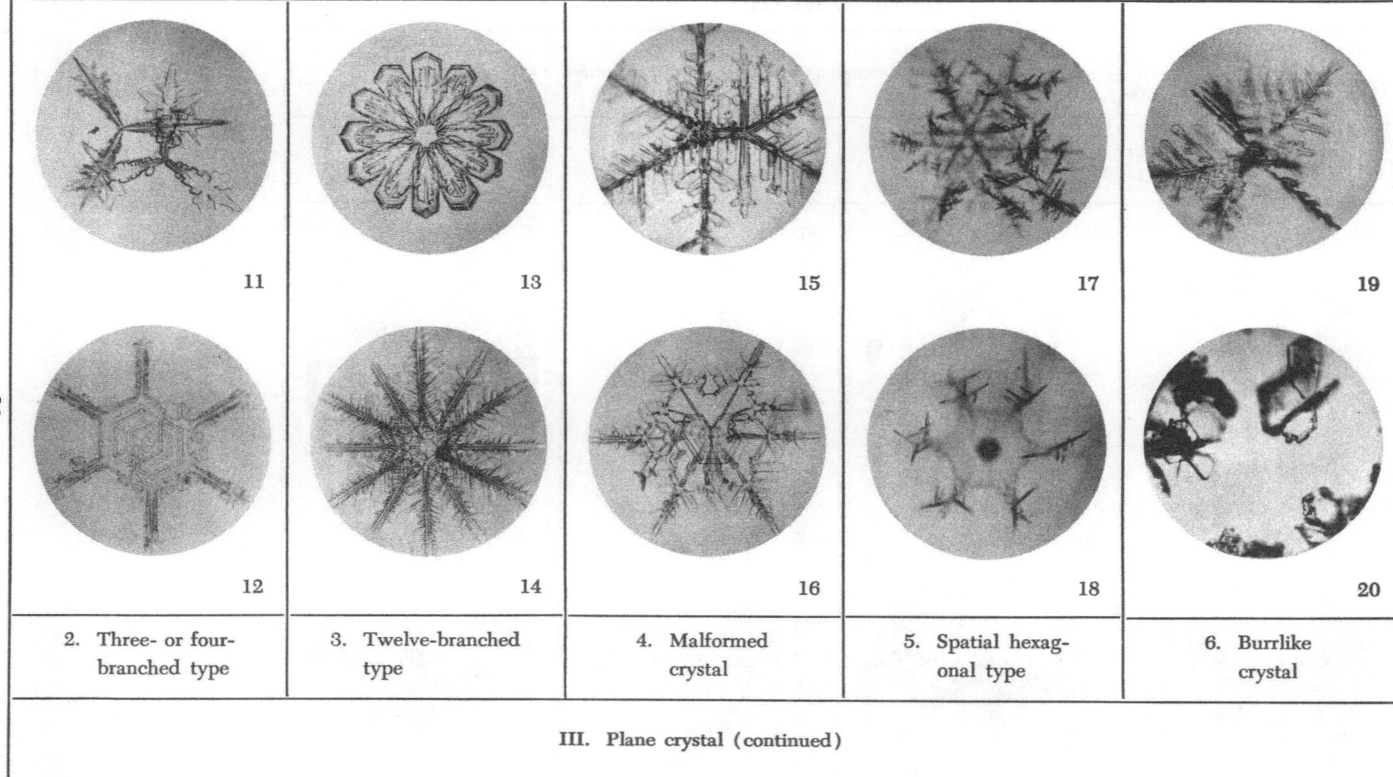

Fig. 198 (*continued*).

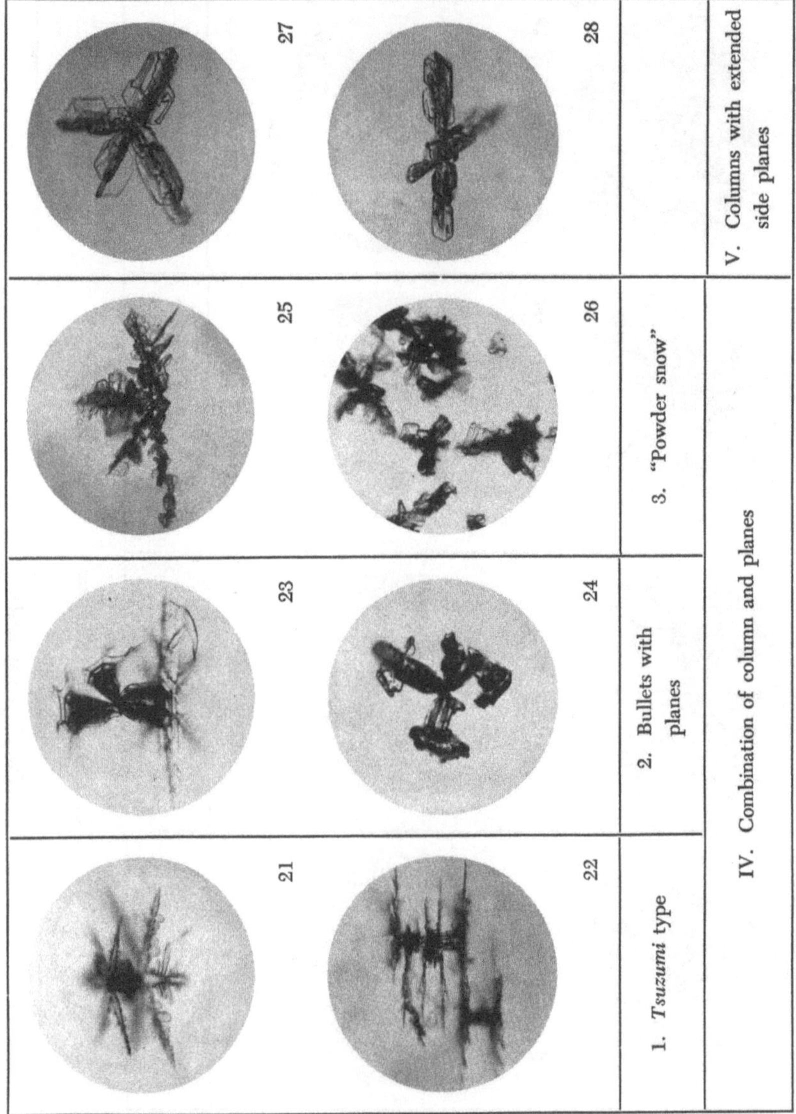

Fig. 198 (*continued*).

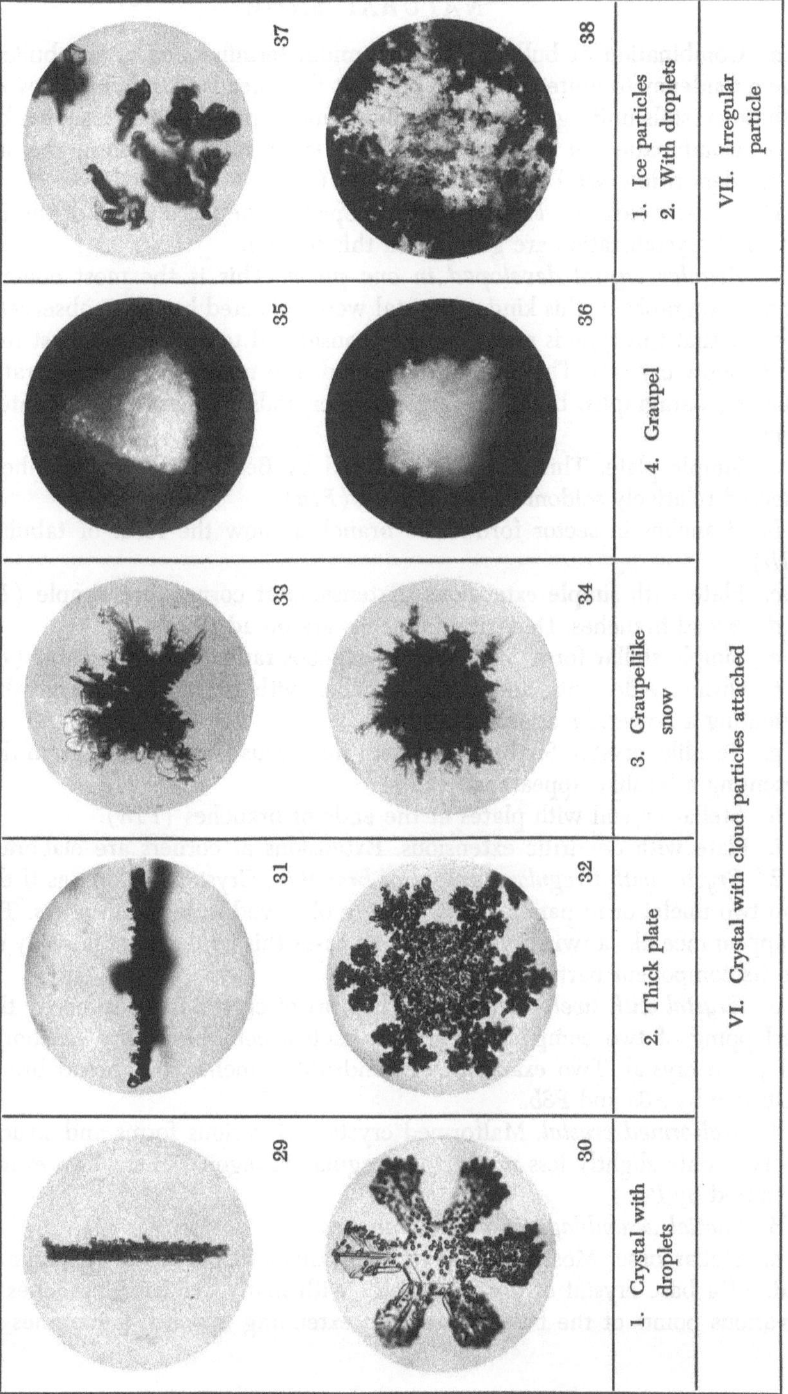

Fig. 198 (*continued*).

a. Combination of bullets. The pyramidal terminations of the bullet crystals have a tendency to unite with each other at their heads. Examples of two and four of these crystals uniting at their pyramidal ends respectively are shown by *C2a.*

b. Combination of columns. This type is considered to form the stone of a spatial dendritic crystal of radiating type (*C2b*).

III. PLANE CRYSTAL, *P.* Crystals developed in the basal plane of the hexagonal system of crystallization are grouped in this section.

1. *Regular crystal developed in one plane.* This is the most common type. Photomicrographs of this kind of crystal were collected by many observers in such numbers that this type is now generally considered to be the one most representative of snow crystals. This is again classified into nine sorts as enumerated in the following paragraphs, but there are many crystals of these sorts in intermediate states.

a. Simple plate. This sort was observed by Bentley in great numbers, but is observed relatively seldom in our climate (*P1a*).

b. Branches in sector form. The branches show the form of tabular sectors (*P1b*).

c. Plate with simple extensions. Extensions at corners are simple (*P1c*).

d. Broad branches. Dendritic branches are broad (*P1d*).

e. Simple stellar form. Six straight branches radiate from a center (*P1e*).

f. Ordinary dendritic form. Six branches with twigs radiate from the center, presenting a flowerlike appearance (*P1f*).

g. Fernlike crystal. Six branches with numerous twigs radiate from the center, presenting a fernlike appearance (*P1g*).

h. Stellar crystal with plates at the ends of branches (*P1h*).

i. Plate with dendritic extensions. Extensions at corners are elaborate (*P1i*).

2. *Crystal with irregular number of branches.* Crystals looking as if developed from two nuclei or in parallel growth were observed in large numbers. They have an appearance like a twin crystal. In many cases this twin crystal is easily separated into its component parts (*P2a, P2b, P2c*).

3. *Crystal with twelve branches.* This sort of crystal is made up by the simple overlapping of two component crystals, each a combined form of short column and plane crystal. Two examples of dendritic branches and broad branches are illustrated by *P3a* and *P3b.*

4. *Malformed crystal.* Malformed crystals of various forms and structures are observed only slightly less rarely than regular hexagonal ones. Two examples are illustrated by *P4.*

5. *Spatial assemblage of plane branches.*

a. Stellar base. Most of the crystals of this sort belong to the type which is made of a base crystal of dendritic form with many dendritic branches attached at various points of the base crystal and extending upward. Sometimes the plate crystal with extensions at the corners is observed with spatial branches attached. One example is shown by *P5a*.

b. Radiating type. This sort of crystal has dendritic branches radiating in space from the center (*P5b*). It is rarely observed that many plates are gathered together to form a spatial assemblage of plates.

IV. COMBINATION OF COLUMN AND PLANE CRYSTALS, *CP*. It is usually understood that this type means a hexagonal column with two plane crystals attached, usually one at each end. Many other crystals, however, have been observed which ought to be included in this type.

1. *Column with plane crystals at both ends; tsuzumi type*. This is called a "capped column" by Vincent J. Schaefer. The typical form of this sort of crystal is a hexagonal column with two plane crystals attached to either end. The plane crystals may be of dendritic form or plates. Two examples are shown by *CP1a* and *CP1b*. A hexagonal column is also observed with end plates and also one or more intermediate plates attached normally to the column. Sometimes a crystal is observed which is composed of many columns standing one above another and several sheets of well-developed dendritic crystals. One example is shown by *CP1c*.

2. *Combination of bullets with plane crystals*. The plane components may be of dendritic form or simple plates. Two examples are illustrated by *CP2a* and *CP2b*.

3. *Irregular assemblage of columns and plates; "powder snow."* The "powder snow" or "flour snow" is a heap of minute crystals of this type (*CP3*).

V. COLUMNAR CRYSTAL WITH EXTENDED SIDE PLANES, S. A crystal is sometimes observed that seems to be made up of an assemblage of columnar crystals with extended side planes. The structure has not yet been completely clarified, but laboratory experiments on the artificial production of frost crystals lead the author to take the plane parts, observable in the structure, as extensions of the side planes of the columns which form the skeleton of this crystal. One example is shown by S.

VI. CRYSTAL WITH CLOUD PARTICLES ATTACHED, OR RIMED CRYSTAL, *R*. Snow crystals with numerous water droplets attached are very frequently observed in our climate. These droplets are found to be cloud particles. Crystals with cloud particles and graupel are brought together in this section. It is difficult to distinguish these two types, because any sort of snow crystal is observable, the number of attached cloud particles varying continuously from snow to graupel. This type is better called a rimed crystal.

1. *Crystal with droplets*. Almost all sorts of crystals are observed with numerous droplets attached. Two examples are illustrated (*R1*), one a needle with droplets and the other a plane crystal with droplets.

2. *Thick plate*. When many water droplets become attached to a plane crystal, it turns into a thick plate, sometimes half a millimeter in thickness. Under the

microscope this plate looks like a piece of moss when observed by reflected light and appears opaque by transmitted light. The photomicrograph of the section of this crystal (*R2*, left) shows that the droplets are deposited mostly on one side of the crystal; *R2* right is the plan of this crystal.

3. *Graupellike snow.* Snow crystals that appear to be in the intermediate stages between rimed snow and graupel are described as graupellike snow. They are produced by the attachment of numerous cloud particles to the spatial dendritic crystals of ordinary snow (*R3a* and *R3b*).

4. *Graupel.* The author considers that graupel is an advanced state of graupellike snow. There are three types of the form of graupel: the lump graupel, the conelike one, and that which bears a trace of hexagonal symmetry. Examples of these three types of graupel are shown by *R4a*, *R4b*, and *R4c*.

VII. Irregular snow particle, *I*. Snow particles are sometimes observed which do not show any regular crystalline structure. They are collected under this section and are provisionally called irregular snow particles. Two examples are shown by *I1* and *I2*. The former looks like an assemblage of pieces of ice and the latter has a great many water droplets attached.

30. Frequency of occurrence

It must be one of the standard undertakings in the study of snow to investigate the frequency of occurrence of each of the various types of snow crystals as classified in the foregoing section. We have pointed out many times that the ideal crystals with hexagonal symmetry as shown in a textbook constitute only a small part of the total number of snow particles that visit our country. Former investigators in this line of research seem to have been inclined to collect mostly photomicrographs of beautiful snow crystals which had developed symmetrically to their full extent. For example, the famous collection by Bentley contains more than 2000 photographs of regular dendritic crystals and plates, while only 200 photographs of the other types are added to them. In spite of their full scientific value, these previous investigations suggest that almost all crystals of snow are of a regular plane type, with rare exceptions of a malformed or columnar sort. Our observations, however, showed that crystals other than the plane dendritic ones are not less frequent. Then the author decided to carry out the investigation of the frequency of occurrence of various types of crystals for every snowfall in the winter of 1934–35. It is rather a hard task to continue the examination of the crystal forms of snow without a break during several hours of a snowfall, the temperature being sometimes below $-10°C$. The observations were chiefly carried out by one of my assistants, Y. Sekido, now Professor at Nagoya University. Sometimes the observations were made with a magnifying glass; these are noted as "rough" in Table 5.

The "precise" observations were made with a microscope, in which case 20 to 30 crystals of snow were placed on a plane glass and the type of each crystal as well as its dimensions were examined. This was counted as one observation, and was repeated at about 5- or 10-min intervals during the period of the snowfall. Sometimes more than 50 observations were made in one snowfall, and this was found desirable in order to follow the unexpectedly rapid changes in the occurrence of certain types of crystal. The observations were carried out at Sapporo, which is situated nearly at sea level, and at a spot halfway up Mount Tokachi, 1060 m above sea level.

The number of days on which observations were made was 36 at Sapporo and 18 at Mount Tokachi, and the total number of observations was 396 at the former place and 578 at the latter. The data are given in Table 5.

The author learned, from the results of the continuous observations of snow crystals which were carried out in the manner above described, that a snowfall usually consisted of several types of snow and sometimes almost all types of crystal were found during the period of a snowfall. Occasionally only the dendritic plane crystals were observed at a given moment; it would continue thus for a short interval and then another type would begin to mix with them; after some time a third type would take its place, and so on. Thus a snowfall is found to consist of several types of snow, when we summarize the results of the continuous observations extending over a sufficiently long period, notwithstanding that occasional observations may give the impression that only a certain type of crystal was to be observed during a snowfall. For example, the snowfall on 28 December 1934 observed at Mount Tokachi was one of the most "homogeneous" snowfalls, but even in this case plane dendritic crystals, spatial dendritic crystals with a base crystal, and crystals with cloud particles attached, were found mixed with each other.

TABLE 5. Summary of observations on frequency of occurrence of various types of snow crystals.

Place	Interval of observation	No. of days on which snow was observed	"Precise" observation: No. of observations	"Precise" observation: Total time	No. of "rough" observations	Total no. of observations
Sapporo	1-22 December 1934	12	7	1 h 40 m	36	43
	5 Jan.-9 Feb. 1935	13	282	22 40	49	331
	19 Feb.-1 March 1935	4	—	—	10	10
	10-13 March 1935	2	—	—	6	6
	20-28 March 1935	5	—	—	6	6
	Total	36	289	24 20	107	396
Tokachi	23 Dec. 1934-1 Jan. 1935	10	272	27 30	28	300
	11-18 February 1935	8	255	44 00	23	278
	Total	18	527	71 30	51	578
	Total	54	816	95 h 50 m	158	974

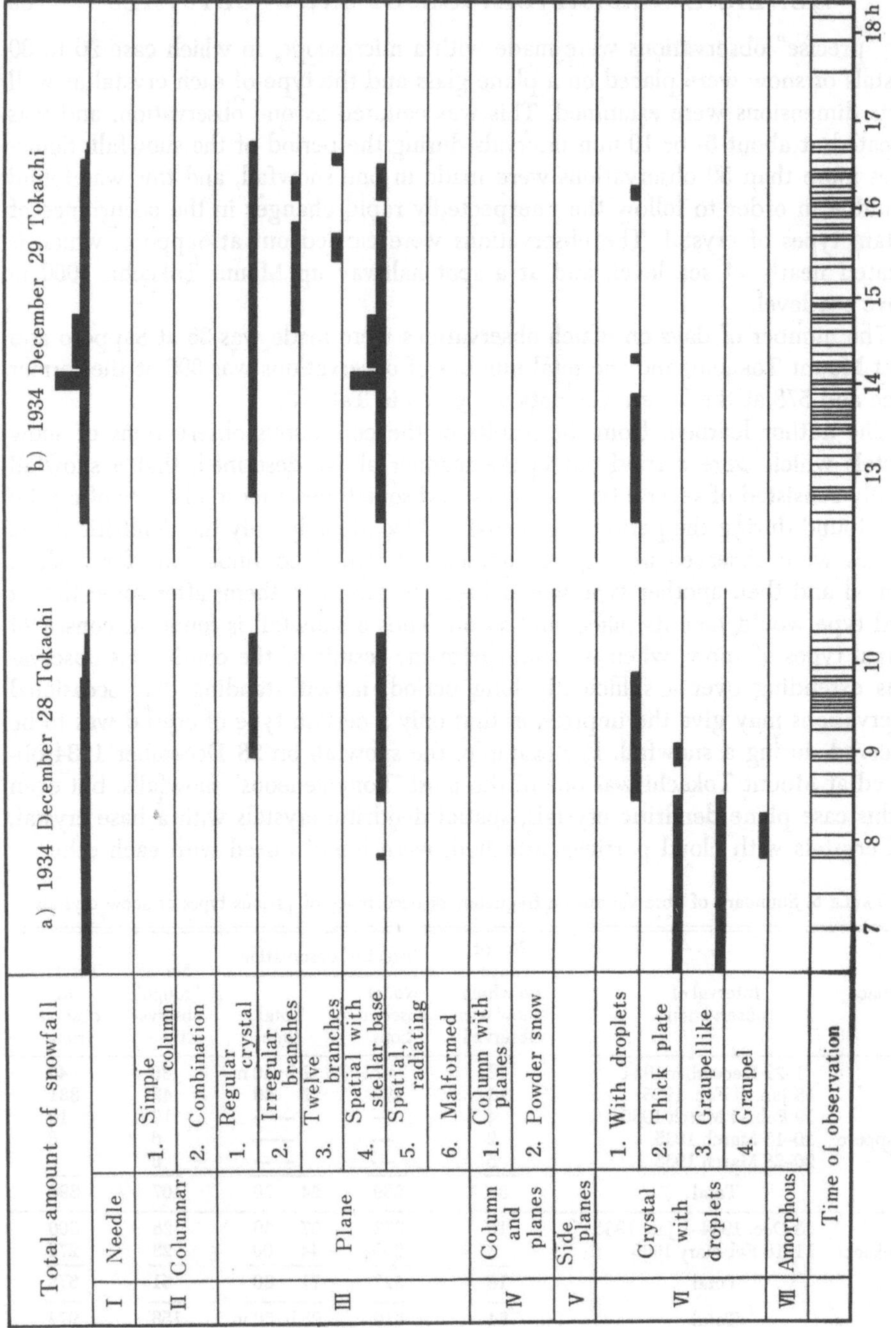

Fig. 199. Frequency diagram of "homogeneous" snowfall.

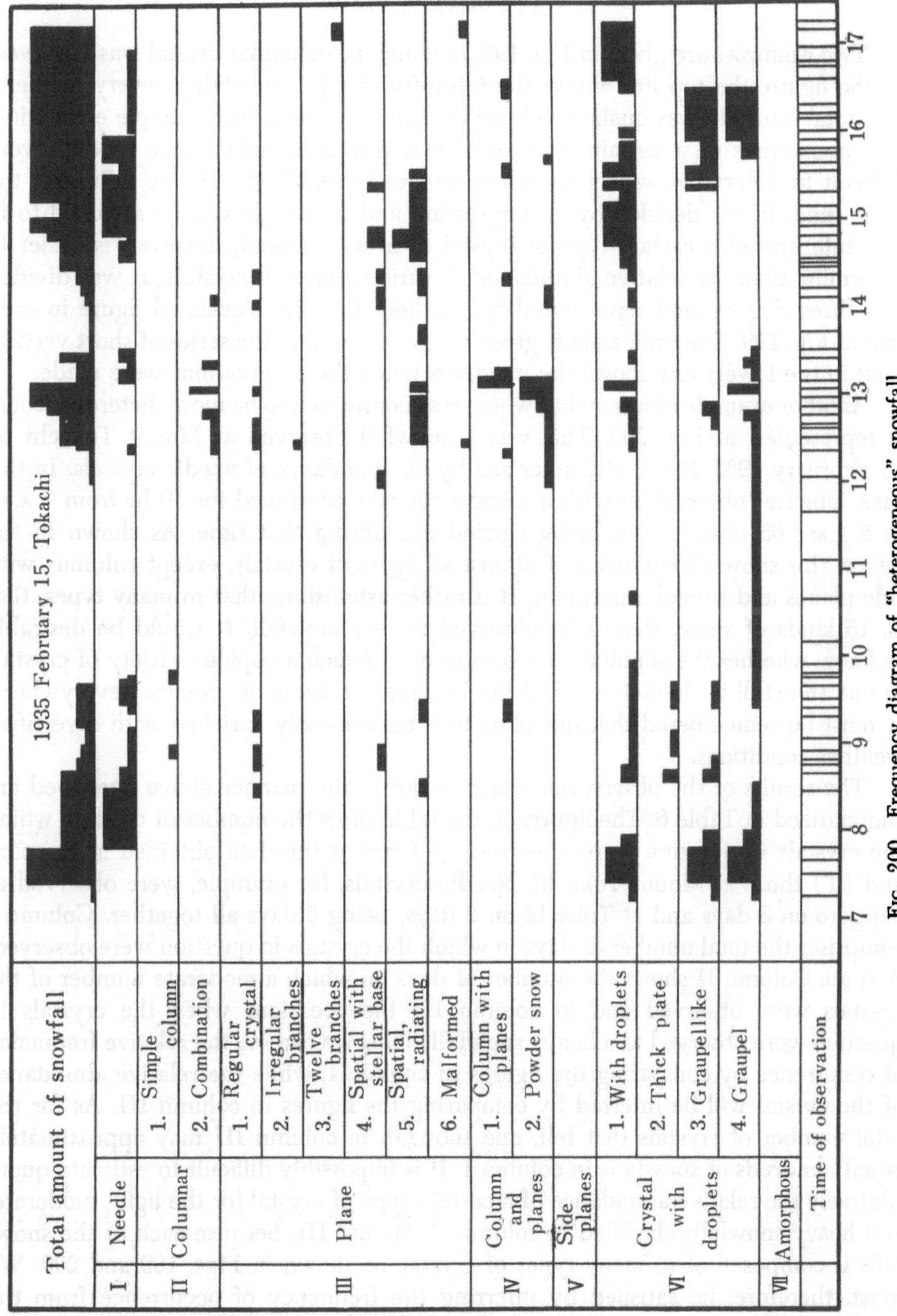

Fig. 200. Frequency diagram of "heterogeneous" snowfall.

Two examples are given in Fig. 199, in which no columnar crystal was observed. In the figure, the top line shows the intensity of total snowfall at every moment. The total intensity was qualitatively divided into five degrees by simple estimation. The abundance of a certain type of crystal compared with other types is very difficult to determine, especially when several types of crystals are falling at the same time. It was decided by an impression and no weight can be attached to it. The intensity of a certain type of crystal at every moment, however, is easier to determine than the relative abundance of various types of crystals. It was divided into three degrees and represented by the height of the blackened figure in each line of Fig. 199. The time scale is given on the x-axis, and the series of short vertical lines in the lowest row shows the moments when the observations were made.

Another example of a snowfall which was considered to be most "heterogeneous" is represented in Fig. 200. This was a snowfall observed at Mount Tokachi on 15 February 1935. It was characterized by an abundance of needle crystals. In this case "precise" observation with a microscope was continued for 10 hr from 7 A.M. to 5 P.M., 65 observations being carried out during that time. As shown in the figure, this snowfall consisted of almost all types of crystals, except columns with side planes and irregular particles. It is rather astonishing that so many types, that is, 15 kinds of snow, should be observed in one snowfall. It would be desirable to know whether the simultaneous occurrence of such a copious variety of crystals in one snowfall is characteristic of the Hokkaido climate or universal everywhere. It must be remembered that our climate is considerably enriched with diversified weather conditions.

The results of the observations carried out in the manner above described are summarized in Table 6. The figures in the table show the number of days on which the crystals in question were observed; (S) means the data obtained at Sapporo and (T) those at Mount Tokachi. Needle crystals, for example, were observed at Sapporo on 3 days and at Tokachi on 2 days, being 5 days all together. Column I comprises the total number of days in which the crystals in question were observed, if at all. Column II shows the number of days on which a moderate number of the crystals were observed and in column III the occasions when the crystals in question were observed as a heavy snowfall. Thus we can see the relative frequency of occurrence by comparing the figures in column I, while the relative abundance of the crystal will be inferred by comparing the figures in column III. As for the total number of crystals that fell, one snowfall in column III may approximately equal hundreds of snowfalls in column I. It is impossibly difficult to estimate quantitatively the relative abundance of a certain type of crystal for the light, moderate, and heavy snowfalls classified in columns I, II, and III, because each of the snowfalls is composed of so many types of crystal, as shown in Figs. 199 and 200. We must, therefore, be satisfied by inferring the frequency of occurrence from the figures in column I and the total abundance from those in columns II or III. When the figures in these three columns are all comparatively large, the occurrence of the crystal is considered to be "frequent," and in the contrary case "rare," the intermediate case being denoted as "medium." The results are summarized in the last column of Table 6, from which we can see clearly that the regular plane crystals constitute only a small portion of the total number of snow particles that actually reach the ground.

31. The dimensions of snow crystals

The problem of the size of snow crystals seems to have attracted less attention from students of snow, compared with the comprehensive study of the form and structure of the crystals. First impressions showed no definite rule on the size of crystals, but observations during the four winters revealed that the size of the crystals observable most frequently in our climate had a definite value characteristic of their types. The preceding photomicrographs are reproduced with varying magnifications, and they are not adequate to convey a clear indication of the relative dimensions of the crystals of various types.

In the course of the continuous observations of crystals described in the foregoing section, the size of the crystals was recorded at the same time. Arranging the data, an attempt was made to construct a frequency curve of the size of the crystal for each of the various types of snow crystal. Most of these frequency curves were found to represent the form of a probability curve, showing that there is a probable size for each of the various types of crystal corresponding to a certain climatic condition.

For the study of the probable size the previous classification described in Sec. 29 is not adequate. In this case the rate of growth of the crystal seems to play the most important role, and consequently attention must be paid to the habits of the branches of crystals. For example, a plane dendritic crystal of a fernlike appearance belongs to the sort called a "regular plane crystal" in the former classification, and a spatial dendritic one with a stellar base crystal belongs to another sort, the "spatial assemblage of plane branches." These two sorts were found to show the same value in regard to their probable size, as shown in Fig. 201 by curves *a* and *b*. In the case of the spatial dendritic crystal of radiating type, the probable size is about 2.5 mm (Fig. 201, curve *c*), and is decidedly less than that of the type with a stellar base. In Sec. 13 it was inferred from crystal form that the condition of the upper layers of the atmosphere required for producing one kind of spatial dendritic crystal will be quite different from that producing the other type. The present results as regards the probable size also agree with this conclusion.

In order to see whether there is any difference in the probable size between the crystals observed at Sapporo, nearly at sea level, and similar ones observed at

Table 6. Numbers of days when various types of crystal were observed at Sapporo (S) and at Mount Tokachi (T).

	I The crystal is observed	II Moderate numbers of the crystal are observed	III The crystal is observed as a heavy snowfall	
I. Needle	(S)3 (T)2 5	(S)2 (T)1 3	(S)0 (T)1 1	Simple needle Combination of needles
II. Columnar crystal	(S)3 (T)5 8	(S)3 (T)3 6	(S)2 (T)2 4	Simple column Combination of column
III. Plane crystal	(S)19 (T)17 36	(S)17 (T)16 33	(S)7 (T)12 19	Plane hexagonal Spatial assemblage, with a base crystal Spatial assemblage, radiating type Crystal with two nuclei Malformed crystal
IV. Combination of column and plane crystals	(S)5 (T)10 15	(S)4 (T)5 9	(S)2 (T)4 6	Column with plane crystals Twelve-sided crystal Combination of bullets with plane crystals
V. Column with extended side planes	(S)1 (T)1 2	(S)1 (T)0 1	(S)0 (T)0 0	
VI. Crystal with droplets and graupel	(S)20 (T)10 30	(S)18 (T)8 26	(S)12 (T)8 20	Crystal with droplets Thick plates Graupellike snow Graupel
VII. Irregular snow particle	(S)15 (T)0 15	(S)13 (T)0 13	(S)4 (T)0 4	

Table 6 (*continued*).

I	II	III		I	II	III	
(S)3 (T)2 5	(S)2 (T)1 3	(S)0 (T)1 1					rare
(S)1 (T)1 2	(S)0 (T)1 1	(S)0 (T)1 1					
(S)3 (T)4 7	(S)3 (T)3 6	(S)1 (T)1 2	cylinder	(S)3 (T)4 7	(S)3 (T)3 6	(S)1 (T)1 2	rare
			bullet	(S)0 (T)1 1	(S)0 (T)0 0	(S)0 (T)0 0	
			pyramid	(S)0 (T)1 1	(S)0 (T)0 0	(S)0 (T)0 0	
(S)3 (T)3 6	(S)1 (T)2 3	(S)1 (T)2 3	cylinder	(S)3 (T)2 5	(S)0 (T)1 1	(S)0 (T)1 1	rare
			bullet	(S)1 (T)2 3	(S)1 (T)2 3	(S)1 (T)1 2	
(S)18 (T)17 35	(S)13 (T)15 28	(S)5 (T)10 15	dendritic	(S)18 (T)17 35	(S)12 (T)14 26	(S)5 (T)10 15	frequent
			areal extension	(S)3 (T)11 14	(S)2 (T)6 8	(S)1 (T)0 1	
(S)7 (T)13 20	(S)6 (T)13 19	(S)3 (T)7 10	dendritic	(S)6 (T)13 19	(S)5 (T)12 17	(S)3 (T)6 9	frequent
			areal extension	(S)1 (T)4 5	(S)1 (T)3 4	(S)0 (T)2 2	
(S)8 (T)12 20	(S)6 (T)11 17	(S)2 (T)9 11	dendritic	(S)7 (T)12 19	(S)4 (T)11 15	(S)2 (T)9 11	frequent
			areal extension	(S)3 (T)7 10	(S)3 (T)6 9	(S)0 (T)1 1	
(S)2 (T)7 9	(S)1 (T)7 8	(S)0 (T)0 0	dendritic	(S)2 (T)7 9	(S)1 (T)7 8	(S)0 (T)1 1	medium
			areal extension	(S)0 (T)1 1	(S)0 (T)1 1	(S)0 (T)0 0	
(S)1 (T)3 4	(S)1 (T)3 4	(S)0 (T)2 2	dendritic	(S)0 (T)2 2	(S)0 (T)2 2	(S)0 (T)0 0	medium
			areal extension	(S)1 (T)3 4	(S)0 (T)3 3	(S)0 (T)2 2	
(S)4 (T)9 13	(S)3 (T)5 8	(S)2 (T)3 5	dendritic	(S)4 (T)9 13	(S)3 (T)5 8	(S)2 (T)2 4	medium
			areal extension	(S)3 (T)4 7	(S)1 (T)3 4	(S)0 (T)2 2	
(S)0 (T)4 4	(S)0 (T)0 0	(S)0 (T)0 0					rare
(S)1 (T)5 6	(S)1 (T)2 3	(S)1 (T)2 3					rare
							very rare
(S)1 (T)7 8	(S)1 (T)7 8	(S)1 (T)6 7					frequent
(S)8 (T)8 16	(S)4 (T)4 8	(S)0 (T)2 2					medium
(S)9 (T)8 17	(S)8 (T)5 13	(S)6 (T)4 10					frequent
(S)19 (T)6 25	(S)17 (T)5 22	(S)10 (T)4 14					frequent
							medium

TABLE 7. Most probable dimensions of crystals.

No.	Type of crystal	Most probable dimensions (mm)
1	Simple plate	0.25, 0.75
2	Branches in sector form	1.0
3	Simple stellar form	1.0, 2.5
4	Plate with simple extensions	1.5
5	Broad branches	2.0
6	Ordinary dendritic crystal	2.5
7	Plate with dendritic extensions	3.0
8	Fernlike crystal	4.0
9	Spatial dendritic hexagonal crystal	4.0
10	Spatial dendritic radiating crystal	2.5
11	Needle and combination of needles	1.75
12	Columnar crystal	0.5
13	Combination of bullets	0.75
14	Column with dendritic planes at both ends	1.0, 2.5
15	Column with plates at both ends	0.75
16	Spatial assemblage of plates	0.75
17	Combination of bullets with planes	1.0
18	Combination of columns with or without planes	1.0

Mount Tokachi, 1060 m above sea level, the frequency curves are compared for crystals of the same kind observed at these two places. As an example, the results obtained for the fernlike plane crystal are shown in Fig. 202, in which S shows the frequency curve for the crystals observed at Sapporo, *TI* those at Mount Tokachi in January and *TII* those at the same place in February. It will be seen that these three curves all reach a similar peak at 4 mm. Similar results were obtained for most of the other types of crystal and the author concluded that there is no essential difference in the probable size between the crystals of the same kind observable at these two places.

As the habit of the branches of the crystal is an important factor in this case, the simple plane crystal must be classified further in detail. For this purpose eight kinds were chosen from the classification of Fig. 18, and frequency curves were constructed for each of them. The stellar crystal with plates at the ends of its branches was omitted, because it is not frequently observed in our climate. The results are shown in Fig. 203. Curves 2, 5, 6, 7, 8 show more or less regularly the form of a probability curve, and the most probable sizes of the corresponding crystals can be determined from these curves. The other curves show some deviations from the regular form. One of the causes of this deviation is of a statistical nature, namely, the subclassification is not definite enough for the present purpose. From these curves it is very clearly seen that the fernlike crystal is largest and the plate is smallest among the plane crystals observed in Hokkaido. Bentley observed at Jericho, Vermont, a quite large number of relatively large plate crystals, but in our climate the plate is usually minute, measuring only 0.25 mm

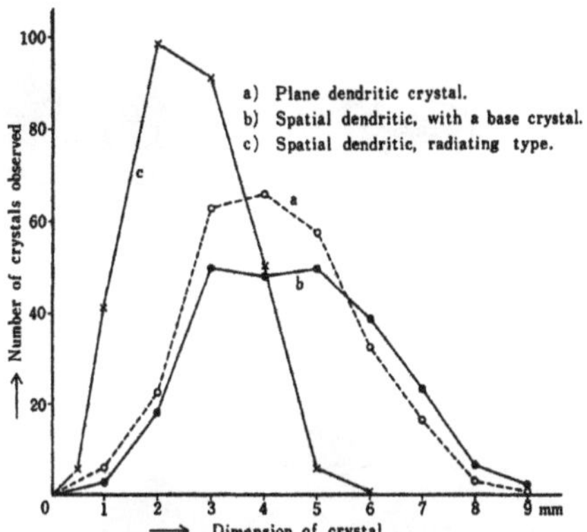

Fig. 201. The dimensions of plane and spatial dendritic crystals.

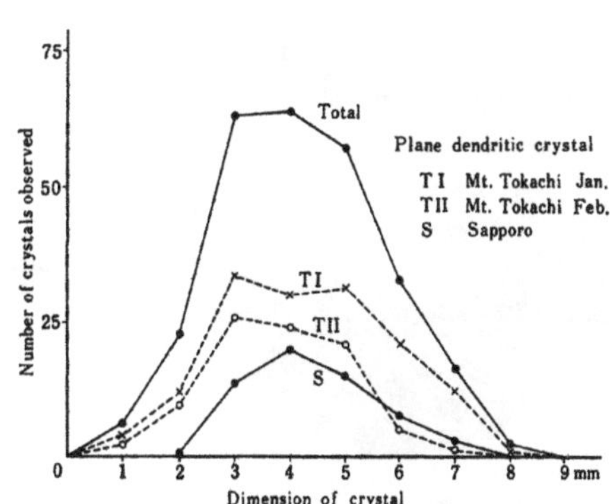

Fig. 202. The dimensions of crystals observed at Sapporo and at Mount Tokachi.

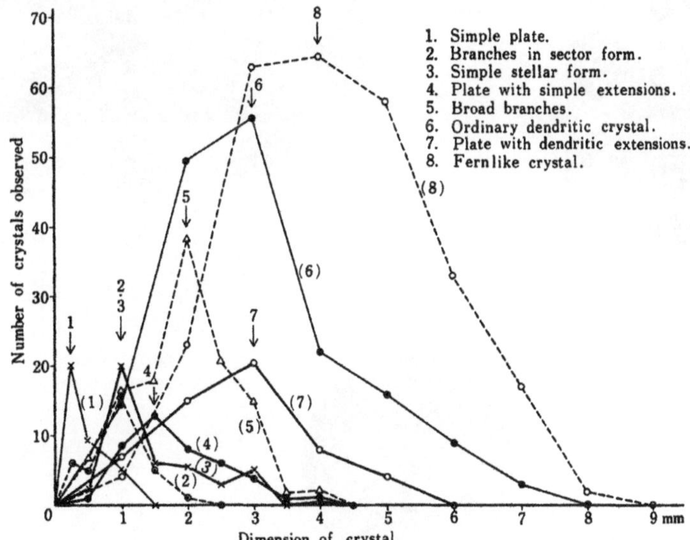

Fig. 203. The dimensions of the various types of plane crystals.

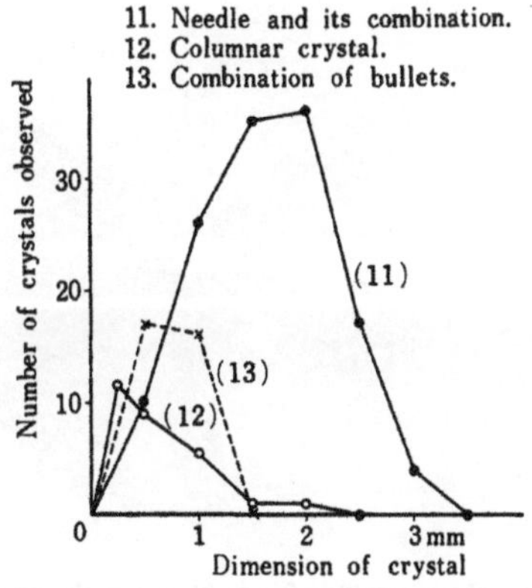

Fig. 204. The dimensions of needles, columns, and bullets.

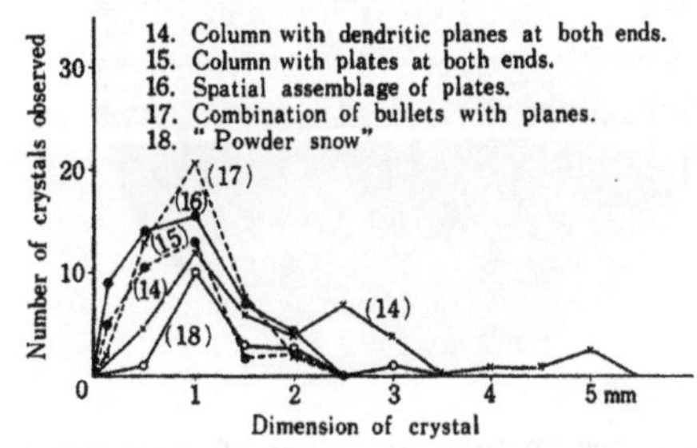

Fig. 205. The dimensions of the other types of crystals.

Fig. 206. Relative dimensions of snow crystals of various types; the magnification is 20 times the size of the average crystal of each type.

Fig. 206 (*continued*).

and at most 1.5 mm in diameter. This difference must be due to the difference in the climatic conditions.

The frequency curves of the dimensions of needles, columns, and combinations of bullets are shown in Fig. 204, and those for five kinds of crystals that are composed of columns and planes are given in Fig. 205. The most probable sizes for each of these crystals are listed in Table 7.

In order to give a clear indication of the relative dimensions of the crystals described above, typical photographs of these 18 kinds are reproduced in Fig. 206 with the dimensions proportional to their probable sizes. All crystals in Fig. 206 are shown with a magnification 20 times their probable size. Most of the hexagonal plates were observed at Mount Tokachi, at an altitude of 1030 m, and the minuteness of the crystal suggests that it is an early stage in the development of almost all kinds of plane crystals.

The large plate crystals without any dendritic extensions at the corners that are rather frequently observed by Bentley and others are very seldom observed in our climate. The difference derives chiefly from the excessive humidity of Hokkaido, which is a favorable condition for the extension of crystals in a dendritic form.

Figure 206 shows the probable size of all types of crystals at a glance. One can see clearly that in any case the dendritic branches are very easily elongated and the plate forms are slowly developed, the intermediate forms as sectors or broad branches falling between the two extremes.

32. Relation between crystal type and meteorological conditions

There are various reports concerning the relation between the forms of snow crystals and the meteorological factors controlling their occurrence. For instance, the result of observation by Heim in the antarctic was introduced in a paper by Kalb.[31] According to it, when the temperature was below −27°C the snow consisted entirely of columns, but when the temperature rose above −27°C plane crystals mingled with columns, and at last when it became −12°C only the plane crystals fell. A more elaborate study was made by Bentley. He classified snow crystals into seven types and investigated the relations among the temperature, crystal types, and the storm section on 131 snowfalls. The result of his investigation is quoted in a paper by Shedd.[32] However, we cannot expect a simple relation between the meteorological conditions observed upon the earth's surface and the types of snow crystals. Consequently, the results of observations conducted at different times in different places do not agree; occasionally they are even contrary. For instance, in Bentley's observations, fully developed dendritic crystals tend to be formed when the temperature is low, but the contrary is given in the statistics by Hellmann.[33]

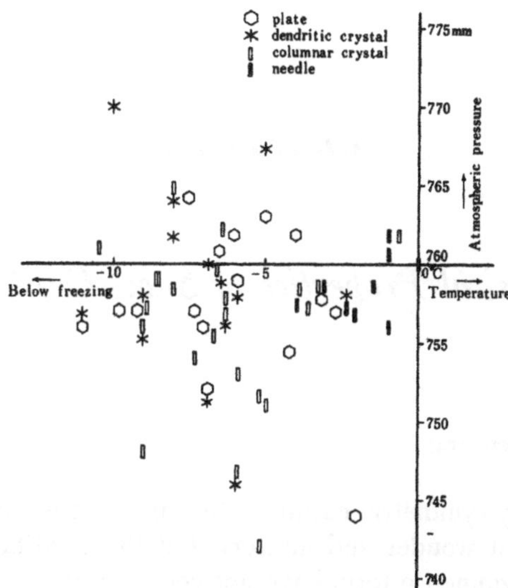

FIG. 207. Relation between crystal types and meteorological conditions.

In our observations in Hokkaido, so many kinds of crystals were observed in one snowfall in many cases, being intermixed with one another, as stated before, that we could not pick out any one type of crystal as representing the meteorological condition at the time. But despite this difficulty, we studied the relation between the type of crystal that chiefly comprised a certain snowfall and the condition of barometric pressure and temperature. The result is shown in Fig. 207, in which the temperature and the pressure are each the mean of the data at 6 h and 18 h of the day on which the crystals in question were observed. Needle and columnar crystals were recorded if any were observed. For plate and dendritic crystals, only those snowfalls were taken which were composed mostly of these particular crystals.

As is clearly seen from this diagram, no simple relation is observed in the distribution of dendritic, columnar, and plate crystals. However, the distribution of needle crystals is limited to a narrow range of comparatively warm temperature and of pressure close to the standard value, namely, near 0°C, 760 mm. It is reported that also in South Sakhalin needle crystals fall when the temperature is rather warm. Therefore, we may conclude that, at least in northern Japan, needle crystals fall when the temperature is near 0°C. This assumption was endorsed also by the study of artificial snow.

CHAPTER 3

Physical Properties of Snow Crystals

33. Symmetry of the crystal

The extraordinarily symmetric nature of the six branches of a snow crystal has been long an object of wonder and mystery, but the conditions that control the formation of such a symmetric form have not yet been discussed in detail. It need scarcely be said that the theory of the crystal lattice cannot explain the symmetry of the macroscopic form of a crystal. Figure 208(*a*) is one of the best examples showing the symmetric nature in both form and design. A schematic sketch of a part of the crystal is shown in Fig. 208(*b*). It is a remarkable phenomenon that not only the curved line marked by *A* but also the small dots *B* and even the pair of dots *C* show almost exactly the nature of hexagonal symmetry. Judging from these points, it is natural to suppose the existence of some action that leads to this symmetry.

Shedd,[34] Wegener,[35] and others consider that a dendritic crystal is first formed in a sufficiently supersaturated atmosphere and then the space between the skeleton branches is filled up with ice crystal under less supersaturated conditions, transforming the dendritic crystal into a hexagonal plate. Other dendritic branches extend from the corners of the hexagonal plate. Thus Shedd considers the crystals of the second, the third, and even the fourth growth as the stages of an evolution of snow crystals. According to this view, the symmetric nature of the design of a hexagonal plate originates in the symmetry of the dendritic branches belonging to one crystal. This similarity of the form of the branches is itself a problem that is difficult to explain. There is apparently no reason why a similar twig must grow, in the course of the growth of the crystal, from one main branch when a corresponding twig happens to extend from another main branch. Good examples are No. 20, Plate 1, and No. 99, Plate 22. In order to explain this phenomenon we must suppose the existence of some means which informs other branches of the occurrence of a twig on a point of one branch. One possible explanation is that

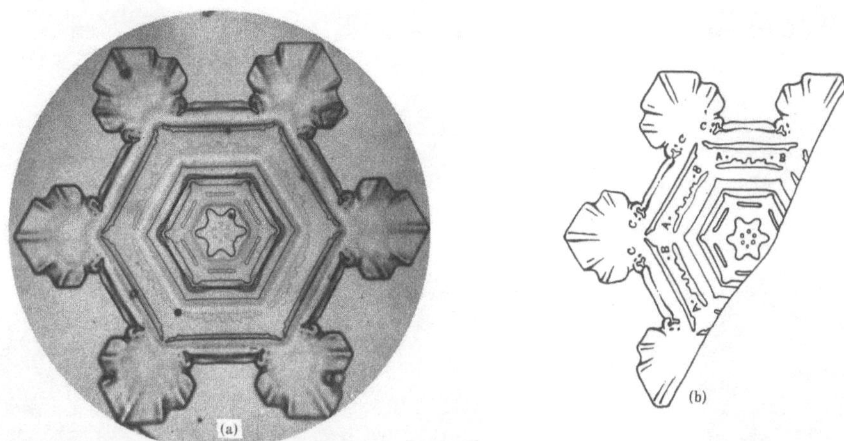

FIG. 208. Crystal showing a perfect symmetry of design: (*a*) photomicrograph; (*b*) sketch of part of the crystal.

the occurrence of a bud of the twig causes some distortion in the crystal lattice which is transmitted to the center and then to the other branches, but it is very improbable that such distortion is transmitted to the other branches through so large a number of atoms, say 10^7, without failure. The crystals that are apparently of a regular form but are made of two components, as already shown in Fig. 64, give support to this consideration; in this case the transmission of some distortion of lattice from one branch to another is most improbable.

As for the dendritic growth of crystals, Lehmann's experiment [36] on the formation of the snowlike crystal of iodoform is most instructive. Iodoform was crystallized out from a solution in a form similar to that of snow under favorable conditions. Lehmann could then observe a dilution of the solution surrounding the crystal by a lightening of the color. He named this diluted area a "halo." He explains that the concentration gradient is steeper near a pointed part of the crystal, where diffusion of the solute is quicker, resulting in a promotion of the crystal growth from that pointed part. Vogel [37] extended this idea by taking into account the heat of crystallization. We also repeated this experiment and investigated [38] the crystals obtained through evaporation of a saturated solution of iodoform dropped on a glass plate. By changing the speed of crystallization we could obtain various types of snow crystals, examples of which are shown in Fig. 209. Five photographs reproduced in (*a*) of the figure show the course of development of an iodoform crystal into the dendritic form, which is exactly similar to the process of formation of artificial snow described later. In (*b*), snow crystals are shown on the left and corresponding iodoform crystals on the right.

The forms of the two crystals are so much alike that this experiment is of

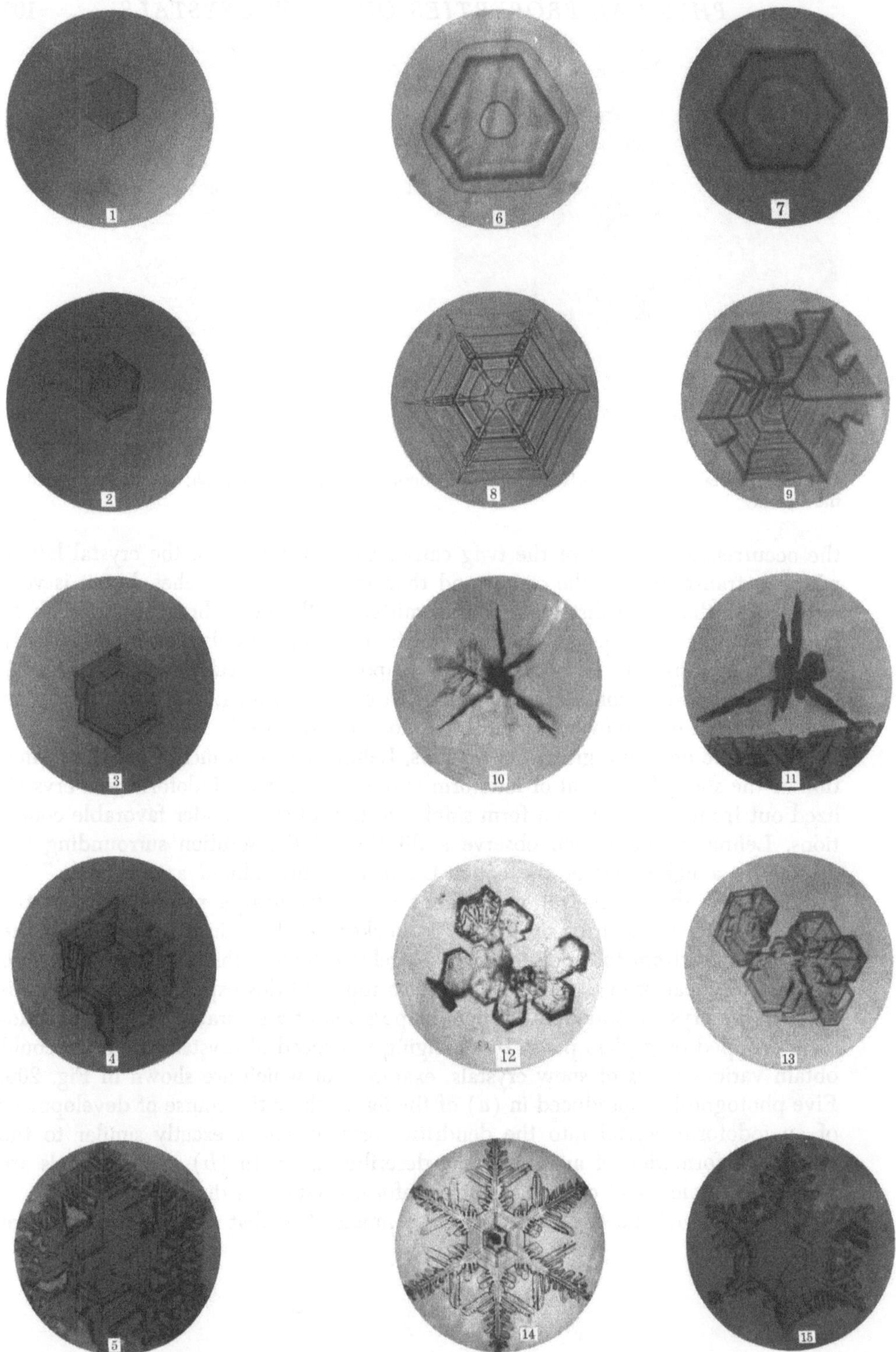

FIG. 209. (*a*) Growth of snowlike crystal of iodoform (left); (*b*) comparison of crystals of natural snow (middle) and iodoform (right).

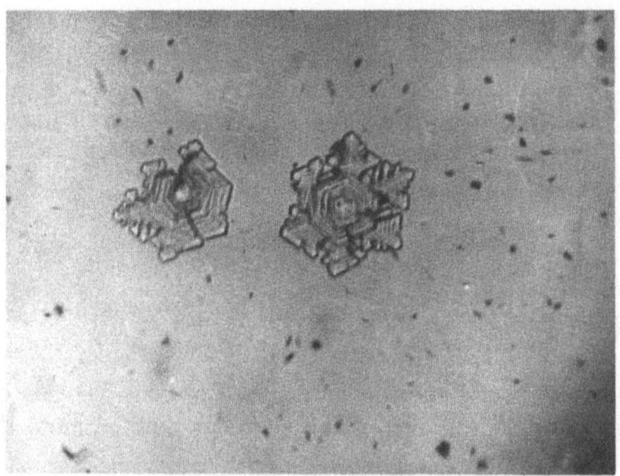

FIG. 210. Motion of the iodoform solution.

interest as a demonstration experiment. We could not observe such clear haloes as Lehmann describes. We learned that both the crystal and solution move pretty actively in the field of a microscope. As the liquid is cooled from its surface by evaporation, the upper surface of the liquid film is cool while the lower surface is warm. It is well known that in this condition the liquid film is divided into periodic columnar vortices of microscopic dimensions caused by convection. An elaborate study by T. Terada [30] revealed that this sort of microscopic columnar convection occurs whenever there exists a gradient of temperature, not only in the liquid but also in the air.

We considered that this kind of vortex has an important significance in the development of snowlike crystals of iodoform, and we examined the movement of aluminum powder in the solution by a microscope. One example of this experiment is shown in the photograph of Fig. 210, in which some of the aluminum particles appear as dots but others make short streaks because of their movement during the exposure. As seen in this picture, their movement at various parts of the solution is complicatedly varied. As we traced the movement of each particle with the naked eye, we noticed that some of them were pushed back near the crystal rim along a spiral path. This proves the existence of columnar vortices. From this experiment, we infer that convection contributes more than diffusion to the crystal formation, by carrying the solute to the crystal surface.

A snow crystal liberates a large amount of latent heat, 600 cal/gm, by condensation due to sublimation. This must make the temperature gradient surrounding the crystal very steep, causing easy occurrence of columnar vortices. Like all vortices, they have a strong tendency to occur periodically in similar form, thus contributing

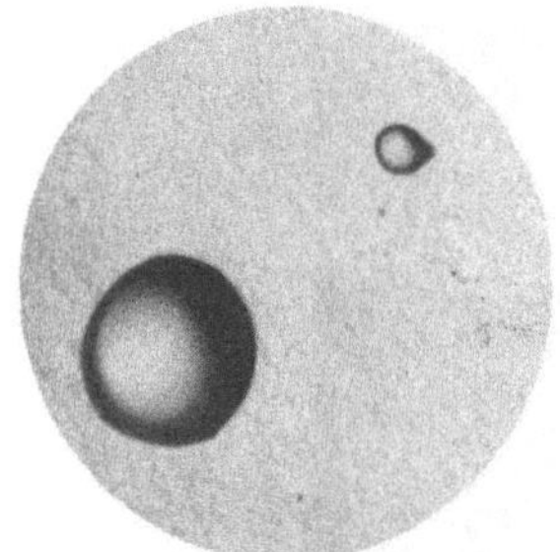

FIG. 211. Snow crystal melted into droplets.

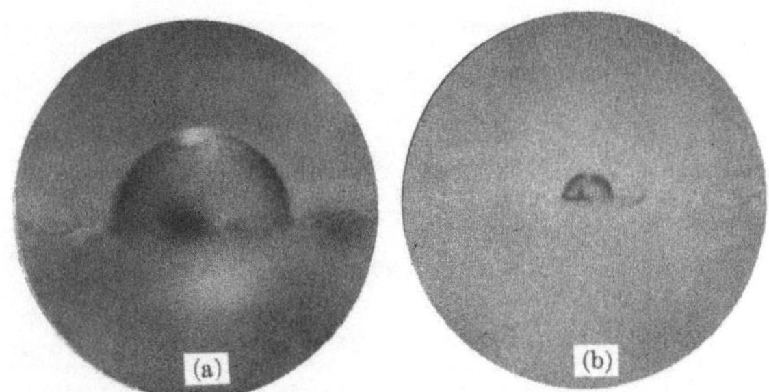

FIG. 212. Side views of droplets: (*a*) $d = 1.2$ mm; (*b*) $d = 0.32$ mm.

to the symmetric growth of dendritic branches. We consider this occurrence of columnar vortices to be at least one of the elements that promote the symmetric development of crystal forms.

These studies lead us to the conclusion that snow crystals do not always show complete hexagonal symmetry but merely have a tendency to grow in that shape. As a matter of fact, most of the natural snow crystals of the hexagonal plane type do not exhibit such complete hexagonal symmetry as is shown in Fig. 208. Such an example is a rather rare one. Minute examination of crystals reproduced in the plates of this book will show that most of the crystals, even those of the regular type, deviate more or less from complete symmetry.

In the next place, we have to consider the speed of crystal formation. In the artificial production of snow crystals, the dendritic crystals need from 30 min to 1 hr for full development. Therefore, if ten twigs develop during this time, the interval of twig growth becomes from 3 to 5 min. Since the rate of crystal growth is thus comparatively slow, we can imagine that a crystal falls through atmospheric layers of various conditions and when it comes to a supersaturated layer fit for the twig growth, twigs begin to develop simultaneously from the six main branches. This will be another important element which promotes symmetric development.

34. Measurement of the mass of a single crystal

For the purpose of measuring the mass of individual snow crystals of various types, we devised a method of measuring the mass of a single crystal. Usually the mass of these crystals is of the order of 0.1 mg to 0.01 mg. They are apt to undergo transformation by sublimation while being measured, even when the temperature is below 0°C. As the outdoor use of a chemical balance in a snowfall is impossible, we adopted the method of melting a crystal into a water droplet and measuring its diameter under the microscope.

This measurement can be safely made when the temperature is considerably below the freezing point, say below −5°C. We conducted the measurement at Mount Tokachi. First, the photograph of the crystal in question is taken. Then the crystal is moved to a glass plate, the surface of which has been previously covered by a thin film of paraffin wax of uniform thickness. The crystal is then melted by touching the under side of the glass plate beneath the crystal with the finger tip. Usually the melting begins from the tips of the branches and the melted water is attracted to the center of the crystal by capillary action, the whole crystal being finally melted into a water droplet, which freezes as soon as the finger is taken away. Of course in some cases a crystal is melted into several droplets; Fig. 211 shows an example. This water droplet is not a complete sphere but a hemisphere, as shown by the side views in Fig. 212. Therefore, the mass can be calculated from the density of ice, by computing the volume of the hemisphere from the diameter measured under a microscope. Although the shape is not strictly a hemisphere in smaller droplets, the deviation can be neglected for our present requirements. When the hemisphere was somewhat oblong, we took the mean value of the long diameter and the short one. A large dendritic crystal sometimes melted into several droplets. In such cases we took the sum of the volumes of those droplets. Three examples of the crystals and the corresponding droplets after melting are shown in Figs. 213–215. Figure 213(*a*) is a dendritic crystal of extraordinarily large size. This crystal, when melted, formed a number of water droplets as shown in (*b*), but this is rather an exceptional case. Figure 214(*a*) is an example of graupellike snow with many cloud particles attached. This type is turned into larger water droplets as shown in (*b*); in other words, it has a large mass. Figure 215 is a crystal powder snow, the mass of which is very small. Other examples are shown in Nos. 982–997, 1002, 1003, Plates 153–155.

After measuring the mass of individual crystals, the average thickness of hexagonal plane crystals can be easily computed. The areas of crystals are measured with a planimeter on the magnified photographs and the average thickness is calculated from the mass and the dimension. The result of calculation on seven crystals of fernlike and regular dendritic type observed at Mount Tokachi is shown in Table 8.

TABLE 8. Thicknesses of snow crystals.

Size of crystal (mm)	Average thickness (mm)
2.35	0.009
2.5	.012
2.8	.011
2.9	.015
3.0	.009
3.5	.011
5.0	.010
Mean	0.011 ± 0.0015

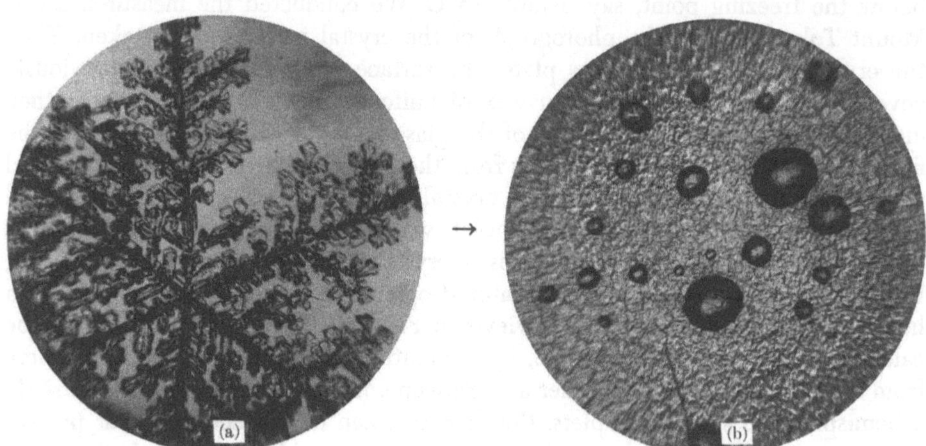

Fig. 213. Large dendritic crystal: (*a*) [980] original form, d = 8.1 mm; (*b*) [981] after melting; m = 0.072 mg, v = 49 cm/sec.

From this result we learned that the hexagonal crystals of this kind have a certain fixed thickness, irrespective of their dimensions. Most of the hexagonal plane crystals with beautiful and fine structure observed at Mount Tokachi are extremely thin, with thicknesses of about 0.01 mm. This result agrees with the knowledge obtained from the side-view photograph of the fernlike crystal, Fig. 20. Needless to say, hexagonal crystals can be thicker, depending on the meteorological conditions, and thicker hexagonal crystals are often observed in foreign countries, but their structure is usually not so fine as that of those treated in this book.

35. Measurement of the rate of fall of a single crystal

There are some observations on the rate of fall of snowflakes of various dimensions, but we are not aware of any data on the velocity of individual snow crystals. When the temperature is comparatively high, many crystals gather and form a snowflake. The rate of fall varies according to the size of snowflakes. We are familiar with the scene of a snowfall on a calm day, in which larger flakes outrun smaller flakes while falling.

In Hokkaido separate crystals are usually observed more or less mixed with snowflakes. At a spot halfway up Mount Tokachi where the altitude is 1030 m, snow falls as separate crystals almost always during the coldest three months of winter. Even at those places where the temperature is comparatively high and snow usually falls in the state of flakes, snow grows as individual crystals at a high altitude and falls through a considerable distance before forming flakes. So the

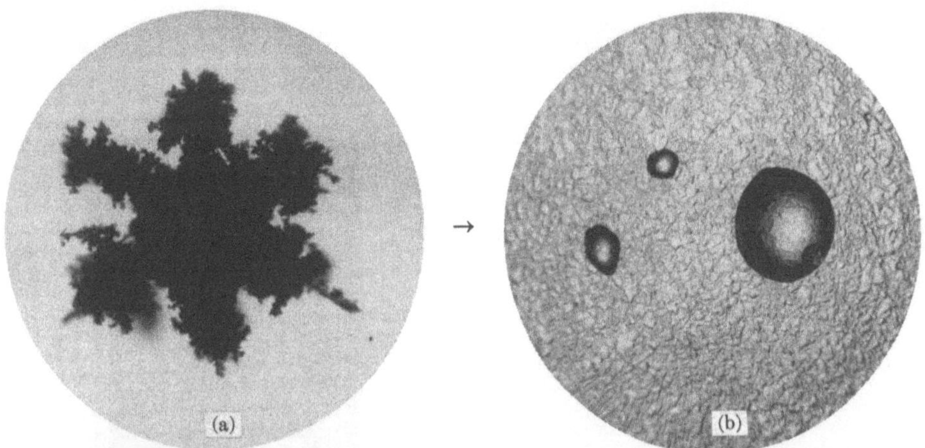

FIG. 214. Graupellike snow: (*a*) [998] original form, d = 3.4 mm; (*b*) [999] after melting; m = 0.51 mg, v = 108 cm/sec.

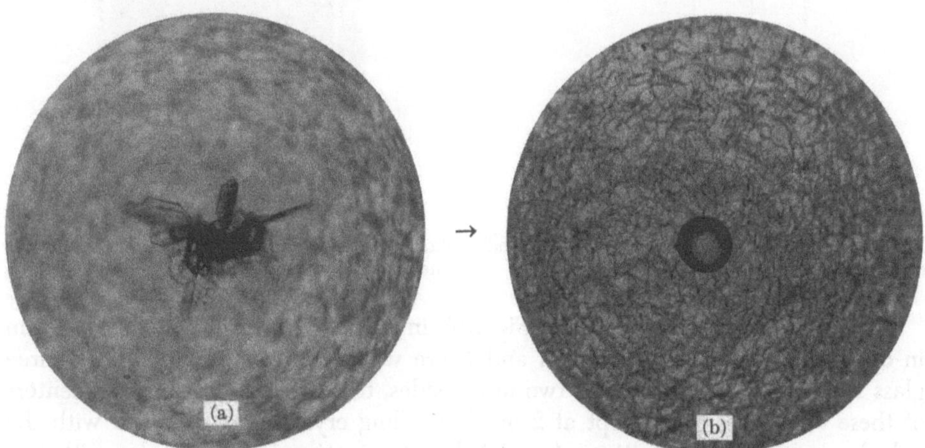

FIG. 215. Powder snow: (a) [1000] original form, d = 1.3 mm; (*b*) [1001] after melting; m = 0.027 mg.

knowledge of the velocity of fall of individual crystals is necessary for inferring the altitude at which a certain type of natural crystal begins to grow, from the knowledge of the artificial production of a similar crystal.

The rate of fall was measured directly with a stop watch by letting the crystal fall through a long closed cylinder held in a vertical position. This procedure having been safely conducted at a temperature below −5°C, all of the measurements were carried out at Mount Tokachi where the temperature was lower than −5°C.

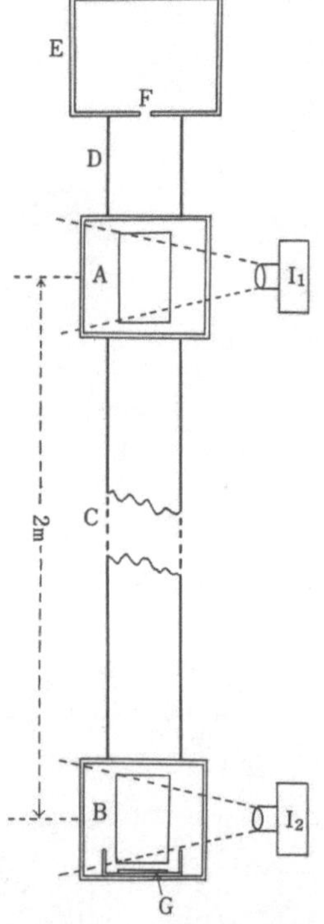

Fig. 216. Apparatus for measuring the rate of fall of single crystals.

Fig. 217. Photographic method for measuring the rate of fall of single crystals.

The apparatus used is shown in Fig. 216, in which *C* and *D* are cylinders 12 cm in diameter made of sheet iron; *A* and *B* are wooden boxes provided with three glass windows, one in front and two in the sides, the distance between the centers of these two boxes being kept at 2 m. The falling crystal was observed with the naked eye at the two positions *A* and *B* by illuminating from the sides with two portable electric lamps I_1 and I_2. The velocity was determined with fair accuracy by measuring with a stop watch the time taken for falling the distance *AB*. Another wooden box *E* with a hole *F* in the bottom covers the upper cylinder. The crystal is made to fall through *F*, which is 1 cm in diameter, into the long cylinder, and this hole is closed as quickly as possible after letting the crystal fall, in order to prevent any upward air current. At *G* is a glass plate covered with a paraffin sheet, which receives the falling crystal. The whole apparatus was set up in the open air and the experiments were carried out at Mount Tokachi when the temperature was between $-8°C$ and $-15°C$.

The point of this experiment is that the photomicrography of a crystal and the determinations of its rate of fall and mass must be carried out within the shortest possible time. The program of observations is as follows: a suitable snow crystal is chosen among many received on an ordinary glass plate and its photomicrograph is taken, the exposure being a few seconds; then the crystal is lifted with a fine splinter and brought to the hole *F* (Fig. 216), and it is made to fall into the cylinder for the determination of the rate of fall, the crystal being received on a paraffin glass and then melted into a drop for measuring its mass. After a little practice, the whole program could be carried out in about 1 or 1.5 min. Thus the author succeeded in the simultaneous observations of the mass, rate of fall, and form of the crystal for various types of snow.

The above method can be adopted only for light crystals which soon reach the terminal velocity. In the case of graupel or snow pellets, this method is not suitable and a photographic method is used. For this purpose the apparatus used for the study of the electrical nature of snow particles, to be described in Sec. 38, was employed. The track of the graupel pellet was photographed by illuminating from both sides with intense beams from arc lamps, and the velocity was determined from the time marks on the image of the track, which were made by rotating an electric fan in front of the lens. An example of those photographs is reproduced in Fig. 217.

36. Relation among the form, mass, and rate of fall of snow crystals

Figures 218–223 are examples of needle, plane dendritic, spatial hexagonal, powder snow, crystal with cloud particles, and graupel, of which we were able to take photographs and determine the mass m and rate of fall v simultaneously. Other examples are collectively shown in Nos. 1004, 1006–1019, 1021–1031, 1033, 1034, 1036–1039, 1041–1044, Plates 156–162. The values of the mass and rate of fall that we measured are given under each of the photographs (on some of the examples we failed to measure one or the other of these two quantities), together with the diameter d or length l.

Next we sought the relation among the form and size, mass, and rate of fall of each crystal, but for this purpose the general classification was too detailed and also the data obtained by our observations were not sufficient. So a simple classification was adopted as follows:

1. *Needle.* A great number of needles fell during our observation on this subject, enabling us to make sufficient study on this type.
2. *Plane dendritic.* Fernlike and ordinary dendritic crystals comprised most of the crystals of this type in our observation on this particular subject.
3. *Spatial dendritic.* Both the spatial hexagonal and the radiating type are included in this group.

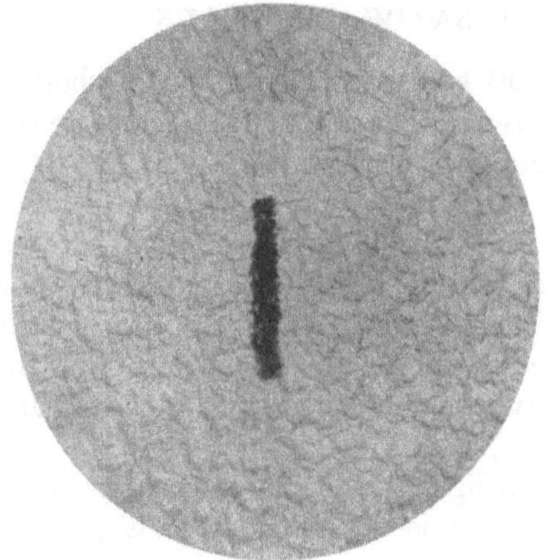

FIG. 218 [1035]. Needle; $l = 0.9$ mm; $m = 0.004$ mg; $v = 54$ cm/sec.

FIG. 219 [1005]. Plane dendritic crystal; $d = 5.0$ mm; $m = 0.072$ mg; $v = 33$ cm/sec.

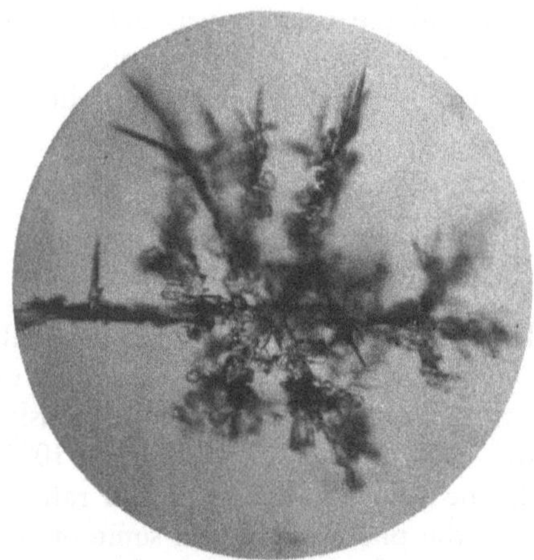

FIG. 220 [1020]. Spatial hexagonal crystal; $d = 3.0$ mm; $m = 0.064$ mg; $v = 59$ cm/sec.

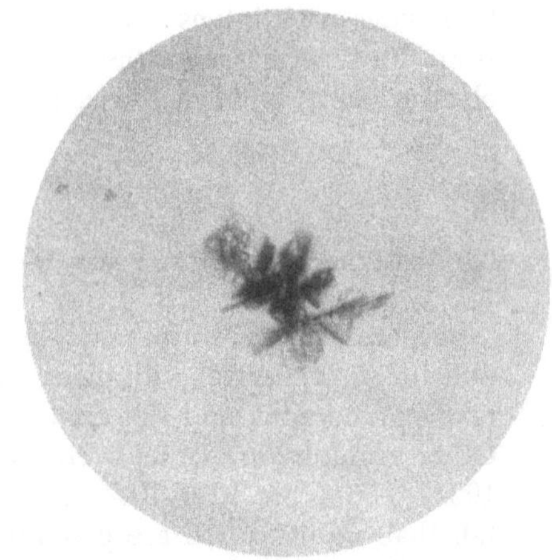

FIG. 221 [1032]. Powder snow; $d = 1.2$ mm; $v = 54$ cm/sec.

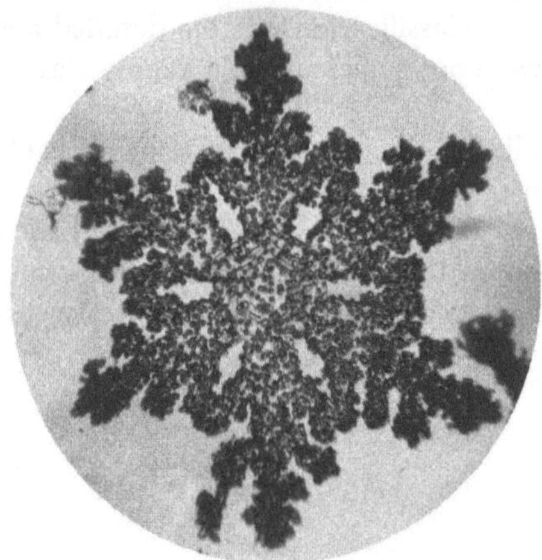

FIG. 222 [1040]. Plane crystal with droplets; $d = 3.1$ mm; $m = 0.226$ mg; $v = 65$ cm/sec.

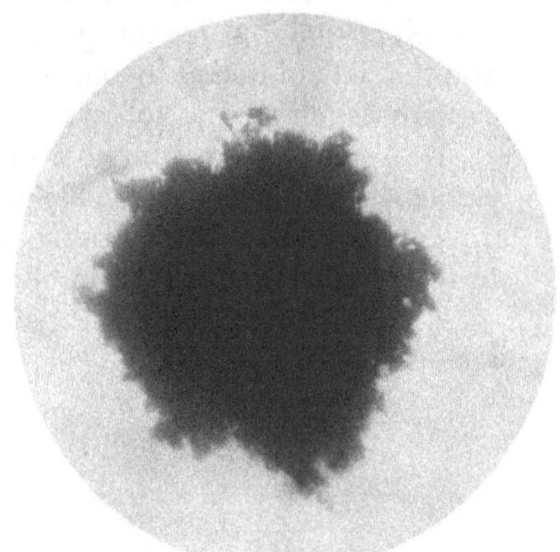

FIG. 223 [1047]. Lump graupel; $d = 2.1$ mm; $m = 0.66$ mg; $v = 139$ cm/sec.

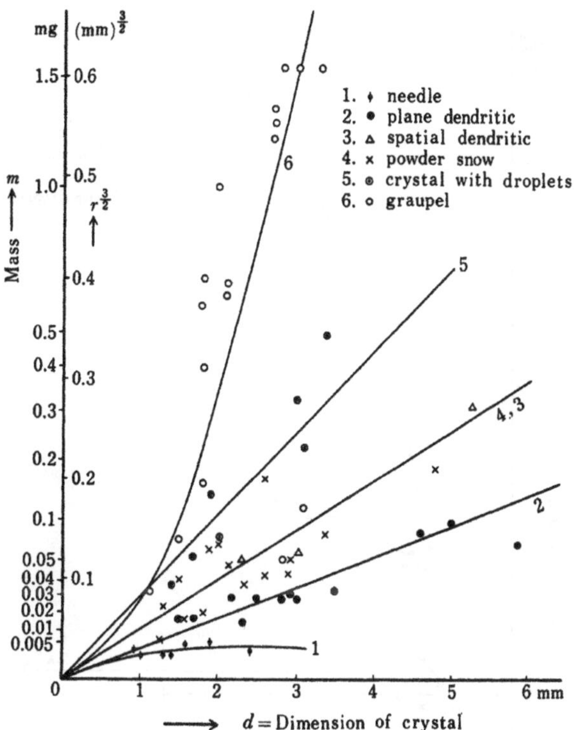

FIG. 224. Mass and dimension of various crystals.

4. *Powder snow.* Combinations of columns and plates, including minute crystals.

5. *Crystals with cloud particles.* All types except needles, including thick plane crystals and graupellike snow.

6. *Graupel.* Various kinds.

The dimension of a crystal is also measured, which means that the diameter is measured of a circle or sphere just enclosing the plane crystal or the spatial crystal, respectively, except needles, in which case the length is meant. The relation between the dimension of a crystal and its mass or rate of fall velocity was examined for each of the six types of crystals above classified.

(*a*) *The relation between the mass and the dimension of the crystal.* The radius r of the drop was first plotted with respect to the dimension d of the crystal. The diagram thus obtained showed that the curve took a form like a parabola for dendritic crystals and powder snow, while the relation was represented by a straight line for graupel. In the latter case one can interpret the result as indicating that the mass of a graupel pellet is proportional to the cube of the dimension, that

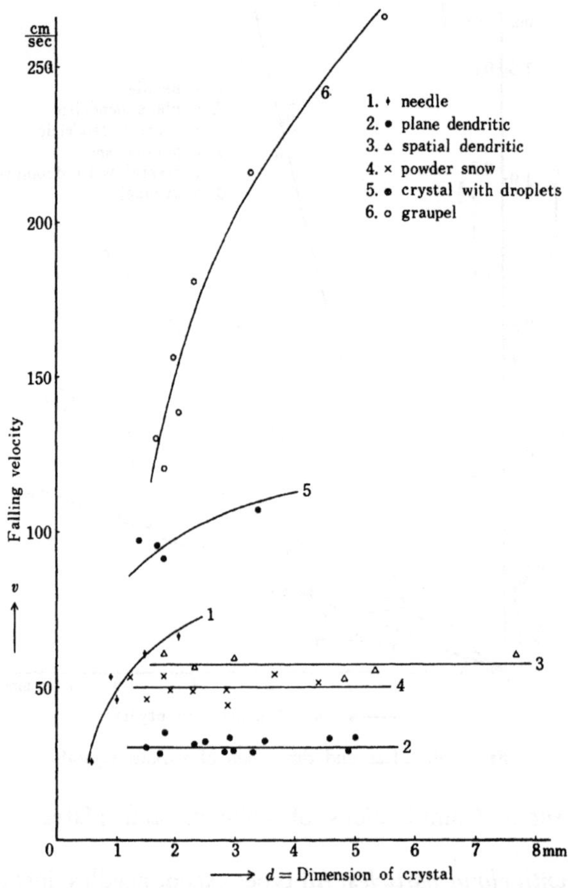

FIG. 225. Rate of fall and dimension of various crystals.

is, the density of a graupel pellet is nearly constant without regard to its dimension. As for dendritic crystals, it can be assumed that the thickness of the crystal is constant for various crystals. Then the mass is proportional to the square of the dimensions, so that $r^{3/2}$ is proportional to d. This relation holds not only for plane dendritic crystals but also for spatial ones under the assumption above mentioned. The diagram representing the relation between $r^{3/2}$ and d is shown in Fig. 224. The points are scattered to a considerable extent, but a general tendency is observed so that the points belonging to one kind of crystal are represented to a first approximation by a straight line passing through the origin, except for graupel and needles. In the figure one can see clearly the order of magnitude of the mass for various types of crystals. The powder snow and the spatial dendritic crystals have masses of the same order. The curve for graupel is concave upward, which is naturally expected from the fact that the mass is proportional to d^3 in this case. The one for the needle form is concave downward and this shows that the mass is proportional to d, that is, the thickness is nearly constant for various samples and the mass is proportional to the length. Of course needles of various sizes are observed, but at present the author deals only with the needles of ordinary size which are commonly observed in Hokkaido.

From the slopes of the straight lines, the empirical formulas representing the relation between m and d are deduced for the four types of crystals. For graupel the formula is deduced from the curve for r and d, and for needles it is obtained from that for r^3 and d. The results are as follows:

$m = 0.065\ d^3$ for graupel;
$m = 0.027\ d^2$ for crystals with water droplets;
$m = 0.010\ d^2$ for "powder" snow and spatial dendritic crystals;
$m = 0.0038\ d^2$ for plane dendritic crystals;
$m = 0.0029\ d$ for needles,

in which m is measured in milligrams and d in millimeters. Of course the variety in the forms of the crystals is so wide that these formulas hold only as a first approximation, but one can know from them the order of magnitude of the mass when one measures the dimension of the crystal, which means, as above, in the case of the ordinary snow and graupel the diameter of the circle or sphere just enclosing the crystal and in the case of the needle its length.

The density of graupel, which is constant for samples of various sizes, was calculated from the formula above described and it was found that

$$\rho\ (\text{graupel}) = 0.125\ \text{gm/cm}^3,$$

which indicates that there is considerable empty space inside a graupel pellet.

(*b*) *The relation between the rate of fall and the dimension of a crystal.* The relation between the velocity of fall and the dimension of a crystal is shown in Fig. 225. In the figure one sees the remarkable phenomenon that the velocities of plane dendritic, powder, and spatial dendritic crystals are each independent of their dimensions; that is, the velocity is nearly 30 cm/sec for all dimensions of plane dendritic crystals, 50 cm/sec for powder snows, and 57 cm/sec for spatial dendritic crystals. When water droplets are attached to the crystals, the velocity increases to the order of 100 cm/sec and in this case the velocity tends to increase with the dimensions. This tendency is most marked for graupel; the velocity of a small particle is about 100 cm/sec, while that of a large one is more than 250 cm/sec. In the case of the needles also a similar phenomenon is observed; that is, the longer needle falls more rapidly than the shorter one, the velocity ranging from 30 cm/sec to 70 cm/sec. The data are not sufficiently numerous but one will be able to estimate from this figure the order of magnitude of the rate of fall for all sorts of snow crystals.

(*c*) *The mean values of d, m, and v.* In order to get a rough estimate of the relative values of these properties for various types of crystals, the mean values of d, m, and v were compared for six kinds of crystals without regard to the distribution of frequency. The results are given in Table 9.

TABLE 9. Mean values of dimension, mass, and rate of fall for various types of crystals.

Type of crystal	$\bar{d}$ (mm)	$\bar{m}$ (mg)	$\bar{v}$ (cm/sec)	$\bar{v}/v_r$
Needle	1.53	0.004	50	1/2
Plane dendritic	3.26	0.043	31	1/6
Spatial dendritic	4.15	0.146	57	1/5
Powder snow	2.15	0.064	50	1/4
Crystal with droplets	2.45	0.176	100	1/5
Graupel	2.13	0.80	180	1/2.5

It will be understood that a crystal of snow has very small mass and an unexpectedly low rate of fall; for example, one million particles, on the average, make 40 cm^3 of water in the case of plane dendritic crystals and only 4 cm^3 of water in the case of needles, and a plane dendritic crystal will take about 1 hr to fall a distance of 1 km.

The relation between the mass and the velocity of a crystal is understood clearly by comparing the velocity of snow fall with that of a rain drop, the mass of which is equal to the mass of the crystal. For the velocity of a rain drop v_r, the value given by Schmidt [40] is used. The ratio $\bar{v}/v_r$ is also shown in Table 9. The velocity of a plane dendritic crystal is smallest, being only one-sixth of that of the corresponding rain drop, while a graupel pellet or a needle falls with a velocity one-half that of the rain drop.

37. Water droplets attached to snow crystals

It is common in our climatic conditions that snow crystals carry more or less attached water droplets. Even in our observation at Mount Tokachi, a snowfall consisting exclusively of beautiful crystals without droplets was rather rare. In order to measure the dimensions of these water droplets, photographs of crystals with water droplets were enlarged and the diameters of the droplets were measured directly with a measuring magnifier. Based on this measurement, frequency curves were constructed on two representative crystals. An example of the photographs used for this purpose is reproduced in Fig. 226. The results are shown in Fig. 227. The number of droplets used for the construction of curve I in the figure was 125,

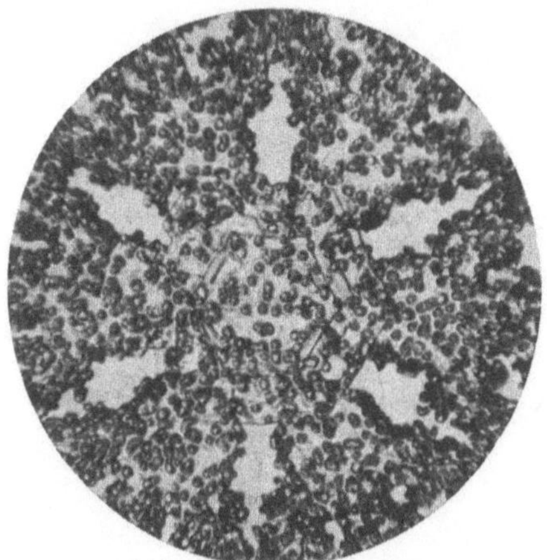

FIG. 226. Cloud particles on a crystal.

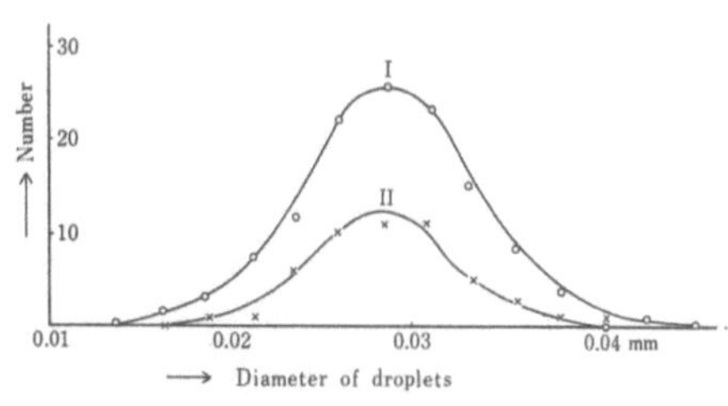

FIG. 227. Frequency curve of diameters of droplets attached to snow crystals.

and that for curve II was 50. Both curves have the typical form of a probability curve and the most probable value of the diameter is about 0.03 mm for both cases.

This value agrees with that obtained by Wagner [41] as the diameter of cloud particles. He did not measure the size of particles directly but calculated it, by the aid of a theory of optics, from the diameter of the halo produced in the cloud, and obtained the value of 0.033 mm. Pernter and others also measured the size of cloud particles by this optical method and got more or less similar results. From this result we named these crystals with water droplets "crystals with cloud particles attached."

Later a few attempts were made to measure with a microscope the diameter of cloud or fog particles received on an oil film. Hagemann [42] employed this method. In his measurements most of the particles ranged between 0.01 and 0.005 mm in diameter. Needless to say, the size of cloud or fog particles varies considerably according to the time and locality.

38. Apparatus for Measuring the Charge of Individual Snow Particles

There are not a few records of measurements on the charge of snow. The cascade effect was at one time thought to give the explanation of the origin of thundercloud electricity, and in connection with it the electrical nature of snow attracted the attention of many scientists. Simpson [43] first conducted his measurement on the charge of snow at Simla and reported that more positive snows were observed than negative, the ratio of the total positive to the total negative charges

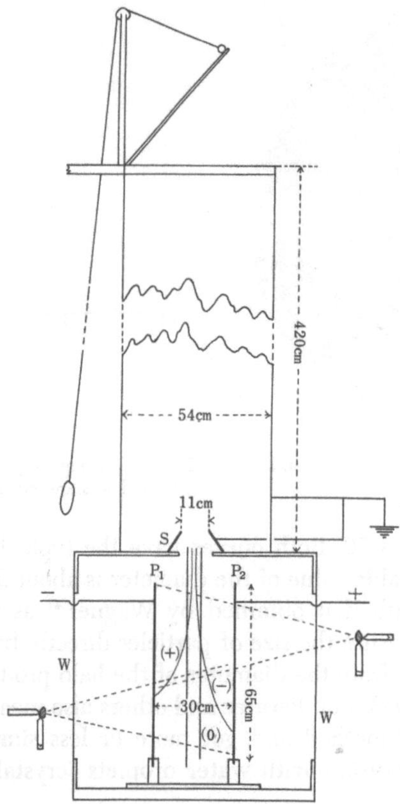

Fig. 228. Apparatus for measuring the electrical nature of snow particles.

observed having been 3.6. MacClelland and Nolan [44] made a similar observation at Dublin and they found that larger flakes tend to be positively charged and smaller particles negatively. On the other hand, Kähler's observations at Potsdam [45] showed that negative snow particles were more frequently observed than positive. All these measurements were done with a collector and an electrometer, and the charge recorded was a mean of the charges of many snow particles. This kind of measurement may therefore be called macroscopic. We tried to treat this problem microscopically, that is, to measure the charge of individual snow particles. By a "snow particle" is meant a "snowflake" as well as a snow crystal, which usually fall intermixed under the climatic conditions of Hokkaido. In this series of measurements no attempt was made to distinguish between them; they were treated as of the same weight. Some ambiguity, therefore, may be introduced in the explanation; for example, a small particle may be a small flake or a separated crystal.

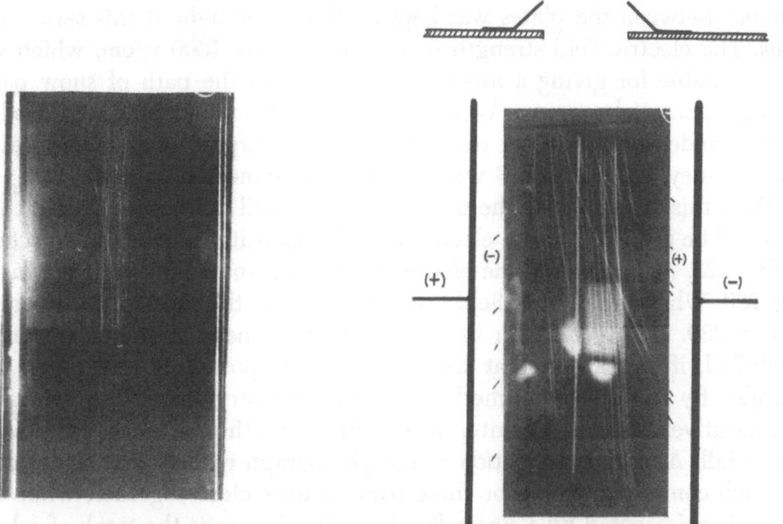

FIG. 229. Snow particles falling without electric field.

FIG. 230. Snow particles falling in electric field; $E = 1400$ v/cm; exposure, 2 sec.

Measurement of the charge of individual snow particles was done by Gschwend [46] at Freiburg. He received one or a few particles of snow in a vessel and measured the charge with a Saiten electrometer. In his observation it was noticed that larger flakes were more frequently observed to be charged positively and smaller particles negatively, in a calm-weather snowfall. Our method of observing the charge of individual particles was an electric deflection method, which was considered to be most convenient for continuous observation. It is difficult to measure the quantity of charge accurately by this method, but the sign of charge is very nicely observed. In this measurement importance was attached to the sign of charge and the quantity of electricity was only qualitatively determined.

The apparatus is illustrated in Fig. 228. It consists of a chimney and a wooden box containing electrode plates P_1 and P_2, and the whole system is set up in the open air. The chimney is made of galvanized sheet iron and is 54 cm in diameter and 420 cm in height. It is provided with a movable cover on the top and a square stop S at the base. For the deflection method it is important to let the snow particle fall in a vertical line, and the dimensions of the chimney were found to be sufficiently suitable for the purpose. The wooden box was made airtight in order to prevent an upward air current which would disturb the free fall of the particle. Three glass windows W were fitted in the walls of the box, one in the front for the purpose of observation or photography and two on the sides for illumination. The electrodes were brass plates, 65 cm × 30 cm, each having a long vertical slit 4 cm wide along the middle line of the plate, for the purpose of illumination. The distance between the plates was kept at 30 cm throughout this series of observations. The electric field strength was approximately 1300 v/cm, which was found to be suitable for giving a moderate deflection to the path of snow particles. As the high-potential source a Wommelsdorf's influence machine was used, together with a condenser and water resistances for the purpose of stabilizing the voltage. The chimney and the stop S were electrically connected to earth.

With this arrangement the neutral particles will fall vertically and the charged ones will be attracted to the electrode of the opposite sign, as shown schematically in Fig. 228. At first, while the electric field is off, snow falls vertically as shown in Fig. 229. When the electric field is on, the paths of the snow particles are as shown in Fig. 230. It may be seen in the picture that many particles are only slightly electrified, if at all, and that some of them are positively and others negatively charged. By this deflection method one may see very nicely that neutral, positive, and negative particles fall, intermixed with each other, at almost the same time in a snowfall. A close examination of the photograph reveals that some of the paths are each composed of two or three parallel lines close together, while the others are each represented by a single fine line. The former is the track of a large snowflake and the latter is that of a separate crystal or a small flake. It will be understood that the charge of a flake and also that of a crystal are both determined by this method. The photograph of Fig. 230 was taken with an exposure of 2 sec. This snowfall was a very dense one. The number of particles which are clearly charged positively is 8 and that of the negatively charged ones is 5. A number of particles are so strongly deflected that they are thrown against the electrode plates and bound back. Sometimes it happens that a large snowflake, especially of the cotton-snow type, is divided into several pieces in the field and the separated pieces are deflected in both directions. This phenomenon will be explained by the theory of separating the electric charge by induction and it is a matter of interest that is worth further investigation. In this book, however, these flakes are not considered and data regarding them are not intermixed with other data.

39. Electrical nature of snow particles

Using the apparatus mentioned in the foregoing section, we made our observations with the naked eye. We recorded the size of the snow particle and the magnitude of deflection on those particles that were obviously deflected, excluding neutral ones or those only slightly electrified. When the snowfall was fairly dense, particles dropped into the chimney incessantly and made continuous observation in the open air rather difficult. T. Terada, Jr., conducted the observations and investigated 1480 particles during six snowfalls. The magnitude of charge was estimated relatively in the following manner. The deflection and the size were each classified into three sorts, namely, large, medium, and small. As a qualitative estimate of the amount of charge, it is assumed that the charge of a small particle showing a large deflection is of the same order of magnitude as that of a large one showing a small deflection. Extending this idea, Table 10 was constructed, in which the figures show the order of magnitude of the charge. The charge of the particle

TABLE 10. Relative charges of snow particles.

Deflection \ Size	Large	Medium	Small
Large	5	4	3
Medium	4	3	2
Small	3	2	1

thus estimated was plotted with a small dot in a diagram, taking time t on the horizontal axis. This time t was chosen on an arbitrary scale, showing only the sequence of the occurrence of the particles. A typical example of the diagram is shown in Fig. 231. This is a record of a continuous observation for 32 min, numbering 67 positive particles and 136 negative ones. One can see clearly in the diagram how the positive and negative particles fall intermixed with each other. By the macroscopic method it is possible to measure only an algebraic sum of these positive and negative charges. Sometimes it may happen that the charges of the positive and negative particles received on the collector cancel each other and the electrometer shows no deflection, notwithstanding the fact that individual particles are heavily electrified.

While observing the electrical nature of snow particles, a microscopic investigation of the form of the crystal was carried out in parallel. In the case of Fig. 231, the crystal was of the thick-plate type. The frequency of occurrence of negative snow was observed to exceed that of positive snow in general. Sometimes it happened that more than 90 percent of all the snow particles observed were negatively charged, two examples of which are shown in Fig. 232(*a*), (*b*). The crystals in these cases were columnar or dendritic, both without water droplets. When water droplets are attached to the crystals, the sign of charge tends to be positive. Figure 232(*c*) is an example showing that positive particles were predominant in number when the crystal was a type of dendritic plane form with many water droplets. Even when positive particles predominate, the negative ones are usually rather numerous. It was very seldom that more than 60 percent of the total particles observed were positively charged, in the observations of one winter. Sometimes it was observed that only particles of one sign fell for a short while and then those of the opposite sign came down for some time. Figure 233(*a*) is an example of the changing of the sign from positive to negative. At that time the type of the crystal was observed to be a mixture of simple columns with end plates and dendritic

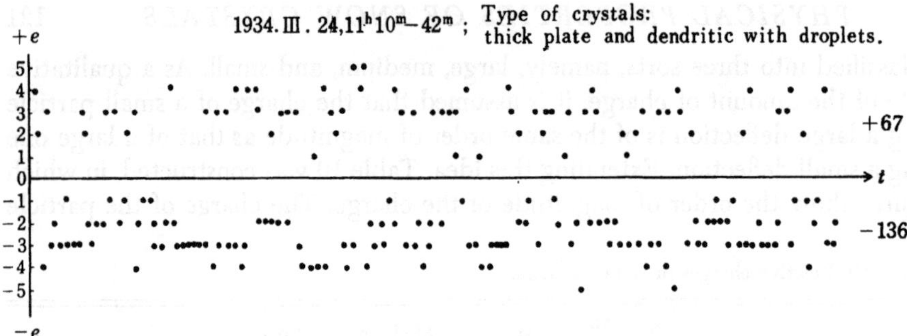

FIG. 231. Electrical nature of snow particles.

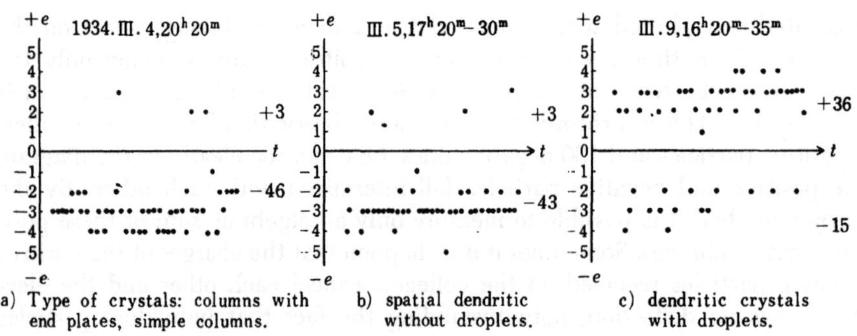

a) Type of crystals: columns with end plates, simple columns.
b) spatial dendritic without droplets.
c) dendritic crystals with droplets.

FIG. 232. Electrical nature of snow particles.

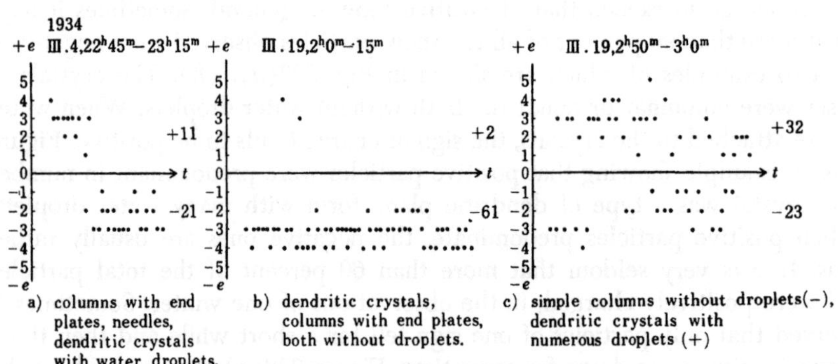

a) columns with end plates, needles, dendritic crystals with water droplets.
b) dendritic crystals, columns with end plates, both without droplets.
c) simple columns without droplets(−), dendritic crystals with numerous droplets (+)

FIG. 233. Electrical nature of snow particles.

crystals with water droplets. The relation between the crystal type and the sign of charge was not determined in this case. It was closely studied in the snowfall of Fig. 233(*b*). Between $2^h\ 0^m$ and $2^h\ 15^m$ the snowfall consisted of dendritic crystals and columns with end plates, both without water droplets; the particles were, almost all of them, negatively charged. After half an hour the observation was resumed and it was found that the character of the snow crystals had changed into a mixture of needles without droplets and dendritic crystals with numerous droplets. It was then clearly observed that the needles were negatively charged and that the dendritic crystals with droplets were positively charged, as shown in Fig. 233(*c*).

The results of the observation are summarized in Table 11. In the table it may be seen that among eighteen observations the negative particles predominated numerically in twelve cases and the positive ones in six cases. The number of the negative particles in general far exceeded that of the positive, being 63 percent of the total of 1474 particles observed. The character of the crystal is also noted in the table. When the crystal had no water droplets attached, marked by an asterisk in the table, the negative particles were more frequently observed. In seven cases out of the eight observations the negative particles always predominated. One exception was the observation at 11^h on 9 March, when the number of positive particles was 1.6 times that of the negative. When the crystals had attached water droplets, the positive particles tended to be more frequently observed. As will be seen in the table, the positive particles exceeded the negative in five out of nine observations. From these results it may be considered as a general rule that the ordinary snow crystals are more frequently charged negatively but when they have attached water droplets they tend to be positively electrified.

As for the relation between the sign of charge and the size of particle, some investigators report that larger particles are more frequently charged positively and smaller ones negatively, while others did not observe such a relation. In order to clarify this point by the microscopic method, the author constructed a table showing the ratio of the number of positive particles to that of negative ones for large, medium, and small particles, as in Table 12. If all large particles are positively and all small particles are negatively charged, the figure representing the ratio must be infinity for large particles and zero for small ones. One may see in the table that this is not the case. As a mean, the ratio is 0.71 for large particles, 0.77 for medium particles, and 0.90 for small ones; that is, the number of positive particles is smaller than that of negative ones for all sizes of particles. Taking these figures as they are, positive charges are more frequently observed on small particles than on large ones.

Kähler and Dorno [47] measured, by a macroscopic method, the charges of snow particles produced by mutual friction, and were led to the conclusion that large particles are positively and small particles are negatively charged by mutual fric-

Table 11. Number of positive and negative snow particles observed.

Time of observation (March 1934)	Number of positive snow flakes and crystals				Number of negative snow flakes and crystals				Character of snow crystals; the asterisk indicates the absence of attached water droplets
	Large	Medium	Small	Total	Large	Medium	Small	Total	
4. 20^h	0	3	0	3	21	18	7	46	* Columns with end plates, simple columns.
21	2	3	2	7	7	3	9	19	* Columns with end plates, needles, assemblage of plates.
23	7	3	1	11	9	5	7	21	Columns with end plates, needles, dendritic crystals with droplets.
5. 16^h	7	17	3	27	40	14	7	61	Dendritic crystals with droplets, without droplets.
17	2	2	1	5	9	22	24	55	* Spacial dendritic crystals without droplets, cotton snow flakes.
9. 11^h	7	45	36	88	6	23	26	55	* Assemblage of columnar crystals, spatial assemblage of plates.
15	3	7	9	19	10	17	10	37	Not observed.
16	16	14	6	36	4	9	2	15	Dendritic crystals with water droplets.
17	18	14	6	38	26	22	2	50	* Cotton snow flakes.
13. 11^h	0	2	59	61	0	1	56	57	Powder snow with droplets.
13	25	50	37	112	26	36	23	85	Cotton snow flakes without droplets, needles with droplets.
19. 2^h	1	0	1	2	26	16	19	61	* Dendritic crystals, columns with end plates, both without droplets.
3	14	13	5	32	5	9	9	23	Needles without droplets (−), dendritic crystals with droplets (+).
5	4	1	11	16	6	1	4	11	Needles without droplets, small dendritic with droplets.
6	1	0	0	1	11	6	9	26	* Needles, hexagonal plates with branches.
24. 11^h	36	14	17	67	49	43	44	136	Thick plates.
12	2	6	10	18	26	19	20	65	Columns with droplets, spatial and plane dendritic crystals without droplets.
15	1	6	7	14	7	29	64	100	* Columns with end plates, simple columns, small plates, all without droplets.
Total	146	194	211	551	288	293	342	923	
Percentage	27	35	38	100	31	32	37	100	

TABLE 12. The ratio of the number of positive to that of negative snow particles.

Time of fall (March 1934)	Large	Medium	Small	Total
4. 20h	0	0.2	0	0.1
21	0.3	0.4	0.2	0.4
23	0.8	0.6	0.1	0.5
5. 16h	0.2	1.2	0.4	0.4
17	0.2	0.1	0	0.1
9. 11h	1.2	2.0	1.4	1.6
15	0.3	0.4	0.9	0.5
16	4.0	1.6	3.0	2.4
17	0.3	0.6	3.0	0.8
13. 11h	—	2.0	1.1	1.1
13	1.0	1.4	1.6	1.3
19. 2h	0	0	0.1	0.0
3	2.8	1.5	0.6	1.4
5	0.7	1.0	2.8	1.5
6	0.1	0	0	0.0
24. 11h	0.7	0.3	0.4	0.5
12	0.1	0.3	0.5	0.3
15	0.1	0.2	0.1	0.1
Mean	0.71	0.77	0.90	0.72

tion. Relating to this problem, we made a preliminary experiment. A small window was opened in the wall of the chimney and a lump of newly fallen snow that was held on an earthed metal plate was blown into the chimney through that window by using compressed air. The particles of snow thus scattered fell into the space between the electrode plates and were deflected by the electric field. It was found that the separation of the charge was quite remarkable in this case but there were also many particles that were scarcely electrified. No such definite relation was observed between the sign of charge and the size of particles as was reported by Kähler and Dorno.

PART II

Artificial Snow

CHAPTER 4

Artificial Production of Frost Crystals

40. The correspondence of snow and frost crystals

In the vicinity of Hakugin-so cottage at Mount Tokachi the temperature is normally between $-5°$ and $-10°C$ at the beginning of the winter and between $-10°$ and $-15°C$ in midwinter, never rising above 0°C, except upon a very few occasions, during the four months of the winter. Accordingly rime or frost crystals develop in a beautiful manner on the various exposed objects surrounding the cottage. While observing them by the naked eye and by microscope the author noticed a remarkable correspondence of these crystals to those of snow.

It is well known that there are two distinct kinds of structure of rime or frost: amorphous and crystalline. As for the frost, the amorphous one is produced when the temperature is below but nearly at 0°C, while the crystalline is made at lower temperatures. In case of the rime, the amorphous sort is produced when supercooled water droplets are frozen fast to some exposed objects, and the crystalline form is obtained when water vapor is condensed by sublimation.* The difference can be seen clearly by simple eye observation, as noticed by many meteorologists. In the discussion of the crystal habits, there is no essential difference between the crystals of frost and rime. In this book, therefore, the rime crystal is included under the term frost. In any case they are both solidified in an amorphous form when they pass through a liquid phase while solidifying, and become crystalline when condensed by direct sublimation.

One example of a photomicrograph of amorphous frost is shown in Fig. 234, in which one may see clearly that it is made of numerous frozen droplets. For the present this type of frost is disregarded, and in the following only the crystalline frost is dealt with.

* The German terms are sometimes used for the classification of them; that is, *Rauhreif* for the crystalline rime and *Rauhfrost* for the amorphous one.

Fig. 234. Amorphous frost (× 33).

Fig. 235. A cavern under a tree stump.

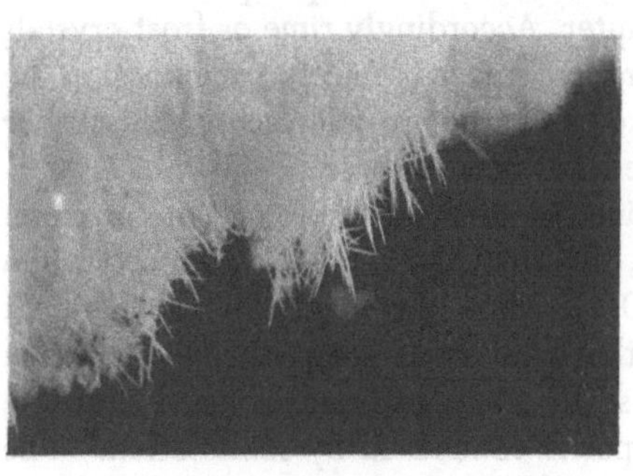

Fig. 236. Needlelike crystals of frost (× 0.85)

Fig. 237 [1052]. Photomicrograph of Fig. 236 (× 15.5).

Among the frost crystals observed at Mount Tokachi, the most beautiful are those developed on the walls of snow caverns, which are made naturally under the stumps of trees or in such places. The caverns are of various forms and are of considerable volume, say 2 or 3 m across. One example of such a cavern is reproduced in Fig. 235. Frost crystals grow on the snow walls in various forms, and especially those hanging down from the snow ceiling of the cavern show the most beautiful sight. These frost crystals can be classified into five types: (1) needle

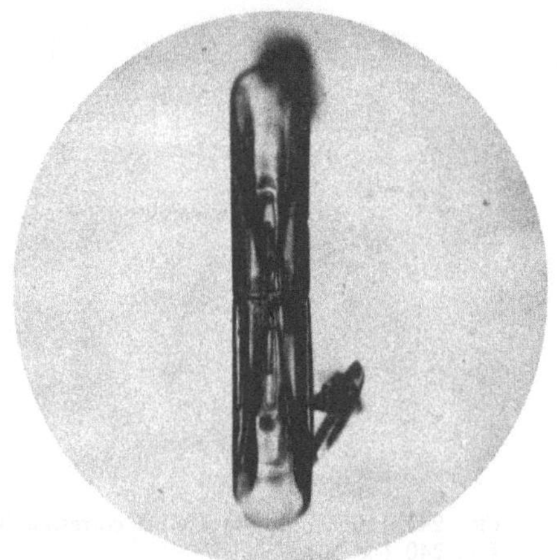

FIG. 238 [1053]. Snow crystal corresponding to Fig. 237 (× 50).

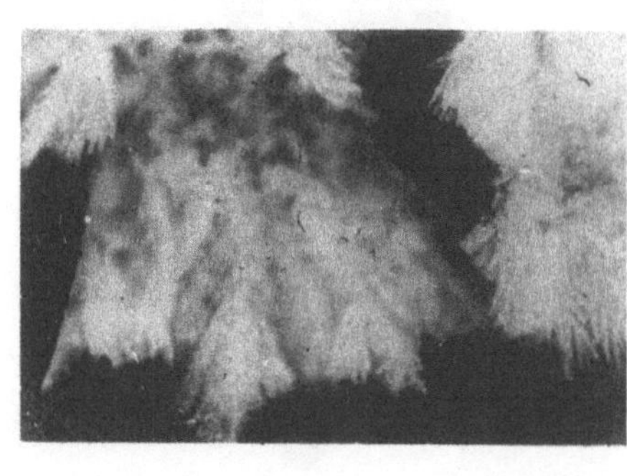

FIG. 239. Featherlike crystal of frost (× 0.7).

form, (2) feather form, (3) plate, (4) cup, and (5) dendritic crystals. They are depth hoars and surface hoars, according to Seligman.[48]

From the crystallographic point of view, each of these frost crystals seems to have a corresponding crystal of snow.

(1) *Needle crystals.* A general view of this type is shown in Fig. 236. They often grow out from a wall of snow in the form of needles 0.2–0.5 mm in diameter and about 1 cm in length. Some of the needles were detached from the wall of the cavern and a photomicrograph of them was taken by transmitted light, which is shown in Fig. 237. It is clearly seen that they are composed of an assemblage of hexagonal columns grown parallel to each other. The correspondence of this type of frost crystal to the columnar crystal of snow, as shown in Fig. 238, has already been pointed out by Stüve,[49] and it seems that there is no doubt as to this correspondence.

(2) *Featherlike crystals.* This type of frost crystal is most often observed at Mount Tokachi. Sometimes it develops to 5 or 7 cm in length. A general view is represented in Fig. 239, which shows the plumelike appearance of the crystal very nicely. A photomicrograph of a part of this rime is shown in Fig. 240. As seen in the photograph the rime is composed of an assemblage of small hexagonal columns as in the case of the needle crystals. These columns are not arranged parallel with each other as in the former case, but some columns are attached head-on to the side of the other columns, giving a branched appearance as a whole. The combination of the hexagonal columns in this manner is occasionally observed in snow crystals. An example is shown in Fig. 241, which may be considered as corresponding to a unit of the combination illustrated in Fig. 240.

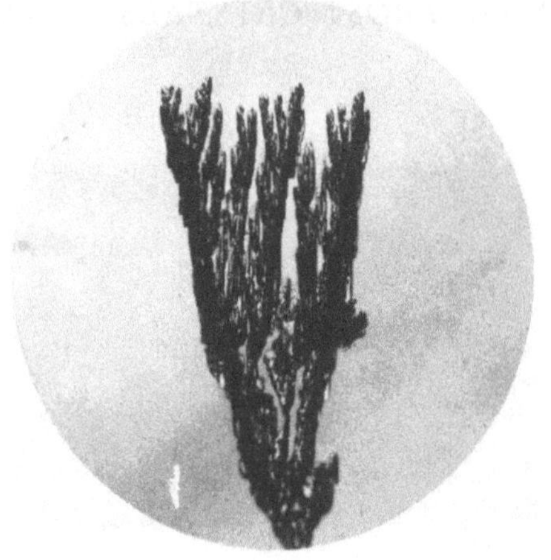

Fig. 240 [1054]. Photomicrograph of Fig. 239 (× 8).

Fig. 241 [1055]. Snow crystal corresponding to Fig. 240 (× 59).

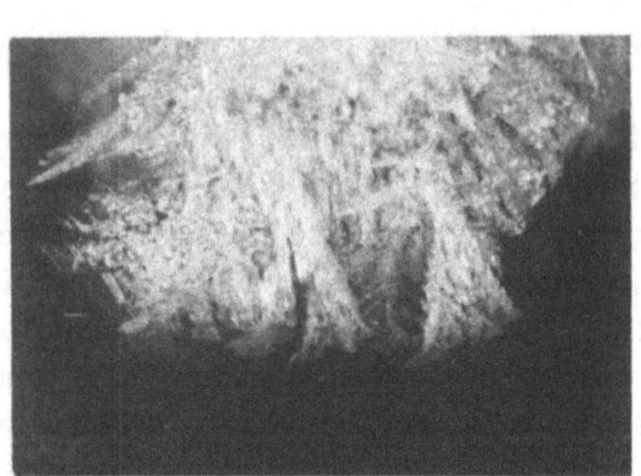

Fig. 242. Plate crystal of frost (× 0.8).

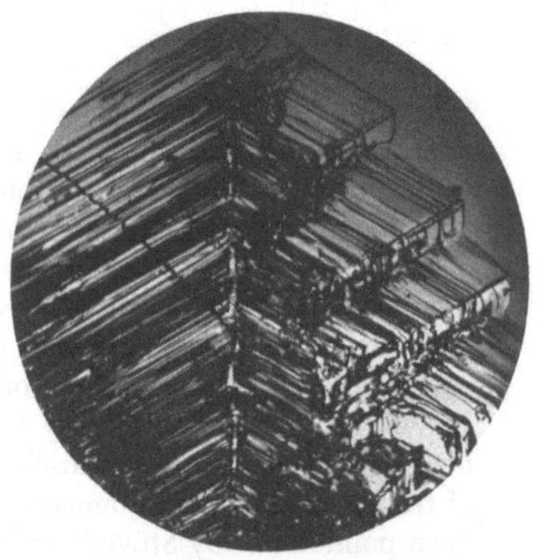

Fig. 243 [1056]. Photomicrograph of Fig. 242 (× 6.7).

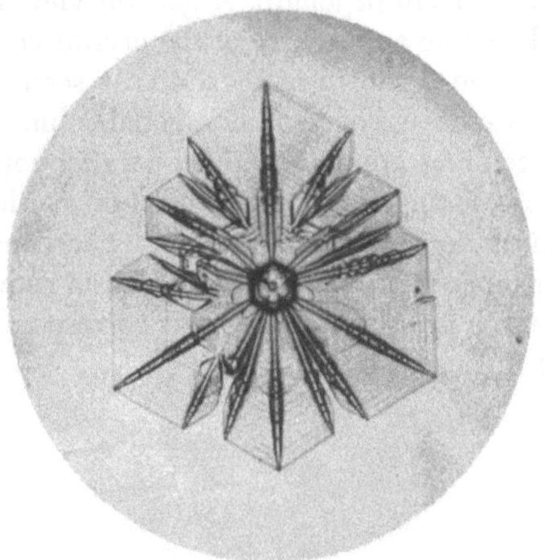

Fig. 244 [1057]. Snow crystal corresponding to Fig. 243 (× 37).

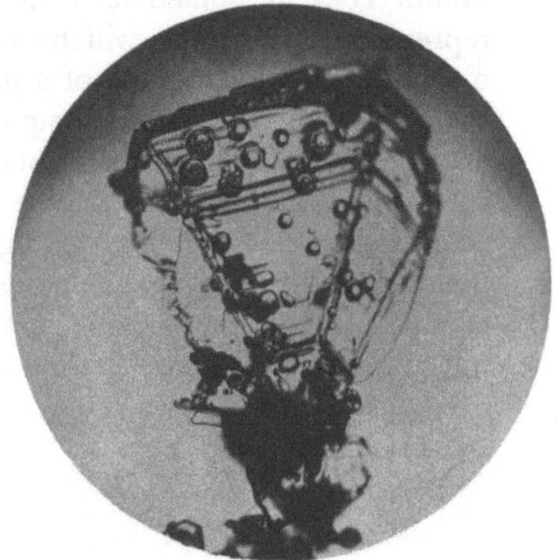

Fig. 245 [1058]. Plate crystal of frost with droplets (× 10.5).

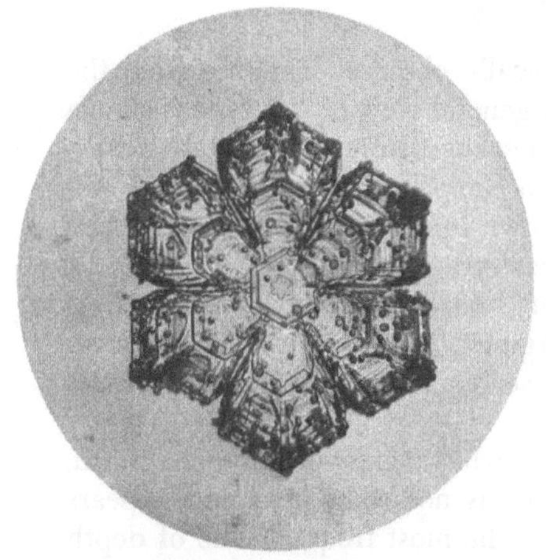

FIG. 246 [1059]. Snow crystal corresponding to Fig. 245 (× 35).

FIG. 247. Cup crystal of frost (× 11).

FIG. 248 [1060]. Crystalline frost of dendritic form (× 6.2).

FIG. 249 [1061]. Snow crystal corresponding to Fig. 248 (× 9.3).

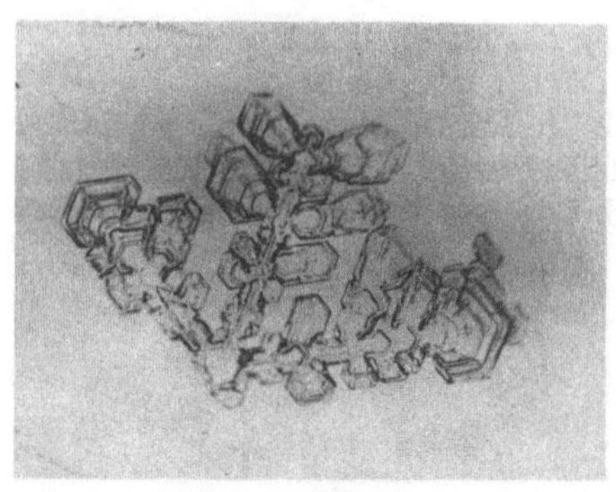

FIG. 250 [1062]. Snowlike frost observed by Hatakeyama in South Sakhalin.

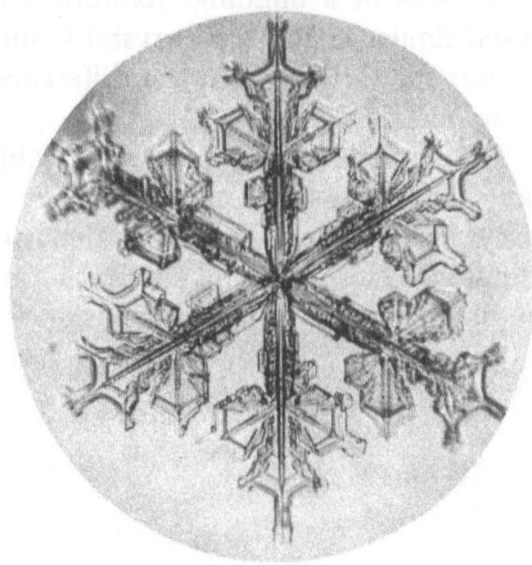

FIG. 251 [1063]. Snow crystal corresponding to Fig. 250 (× 19.5).

(3) *Plate crystals.* This type of crystal is usually observed hanging from the snow ceiling of the cavern as described above. A general view of this type is shown in Fig. 242 in its natural state. The plate was detached from the crystal and examined under the microscope; a photomicrograph is represented in Fig. 243. Comparing this plate crystal and that of snow as shown in Fig. 244, it is clearly seen that they are quite similar to each other in their crystal habits.

Plate crystals are often observed growing out from an exposed object such as the wall of a wooden box. One example with large fog particles attached is shown in Fig. 245. The corresponding snow crystal, sector form with cloud particles attached, is shown in Fig. 246.

(4) *Cup crystals.* This type takes a form like a whisky glass of hexagonal form, when developed completely. But usually one side is not completed and appears to be rolled up. Seligman considered this type as the most frequent one of depth hoars, and called it the cup crystal. One good example observed in the frozen tundra in South Sakhalin is reproduced in Fig. 247.

(5) *Dendritic crystals.* Dendritic crystals are often observed among the crystalline frost developed on the branches of standing timber when the weather is calm. This type is also frequently observed among the surface hoars. A photomicrograph of one of these crystals is shown in Fig. 248. The correspondence of this type of frost to the ordinary dendritic crystal of snow as shown in Fig. 249 needs no further explanation. The only difference is the lack of minute structure or design in the frost crystal. This structure seen in the crystal of snow is caused by the ruggedness of the surface or very narrow channels engraved on the surface of the crystal. When a snow crystal is left in an atmosphere saturated with water vapor, being kept far below 0°C throughout, it undergoes a transformation by sublimation, although it does not change its form by melting. The surface energy of the crystal acts to evaporate the molecules from pointed parts or sharp ridges and condense them onto the dented parts. As a result the ruggedness of the surface is flattened, leading to the disappearance of the design on the crystal surface.

(6) *Another snowlike frost.* A remarkable photograph of a frost crystal which is very similar to snow both in its form and structure, was taken by Hatakeyama in South Sakhalin. The photograph is shown in Fig. 250. This frost crystal was found on the wall of a building, standing almost perpendicularly to the wall. A snow crystal similar to this frost crystal is shown in Fig. 251 for comparison. These two pictures show clearly that the difference between snow and frost crystals lies only in the nuclei.

The copious variety of crystal habits of snow has attracted the attention of many physicists and meteorologists. Shedd,[50] Wegener,[51] Bentley and Humphreys,[52] and others agree that dendritic crystals are formed in an amply supersaturated atmosphere and simpler forms are obtained under a less supersaturated condition, but no further conclusion is deduced. The best way of attacking this

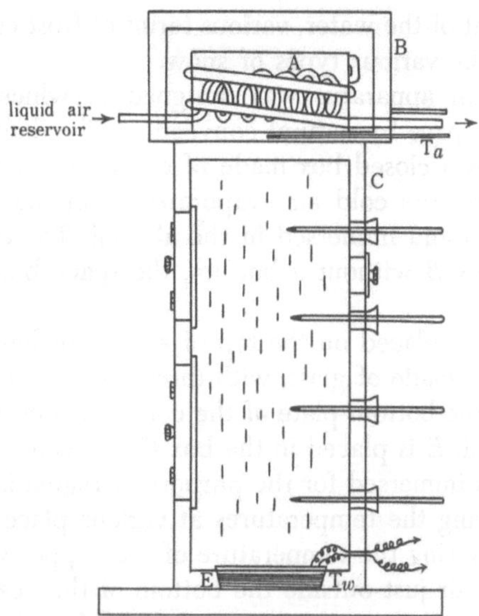

FIG. 252. Frost-making apparatus.

problem is the artificial production of the snow crystal in the laboratory, but at first there was little hope of its success. The correspondence, however, of the habits of snow and frost crystals above described suggests that the conditions for the formation of a type of snow crystal may be inferred from those for the corresponding frost crystal. We therefore began to attack this problem from the side of artificial production of frost crystals.

41. ARTIFICIAL PRODUCTION OF FROST CRYSTALS

In the artificial production of snow crystals, the difficulty exists in suspending the crystal in air for a long while. One way of attacking this problem is to consider the question of crystal development separately from that of the nucleus formation. As for the former point, Adams [53] succeeded in making columnar crystals of snow in the laboratory by mixing a cold air current with a warm and wet one. He received the nuclei thus formed on a cold glass plate and made them grow to columnar crystals of dimensions observable under a microscope, by letting a stream of wet air pass over them. He made this experiment in order to see the polarity of ice crystals, but unfortunately the conditions under which they were produced are not clearly stated. In our experiment, the water vapor, evaporated from a water surface, was brought up by natural convection and deposited as frost crystals on the bottom surface of a metal box cooled with liquid air. By varying the temperature of the box and that of the water, various forms of frost crystals were produced which correspond to the various types of snow.

After some trials an apparatus was designed in which the water vapor is brought to the cooled space by natural convection. The diagram is shown in Fig. 252. In the figure, *A* is a closed box made of copper sheet filled with alcohol. A copper tube for passing the cold air evaporated from the liquid-air reservoir is wound in a spiral form and immersed in the alcohol. The copper vessel *A* is enclosed in a wooden box *B* without a bottom, the space between *A* and *B* being filled up with cotton.

This cooling vessel is placed on the top of a tall wooden box *C*, the front and back sides of which are made of glass, with three doors in the side walls. This box *C* lacks a top, so that the bottom plate of the copper vessel *A* is exposed to the air inside the box *C*. A dish *E* is placed in the box *C* and is filled with pure water, in which a heating coil is immersed for the purpose of regulating the temperature of the water. For measuring the temperatures at various places seven thermometers are used: one for measuring the temperature of the copper vessel; another for the temperature T_a of the air just outside the bottom of the vessel, a third immersed in the water for measuring its temperature T_w, and four in the air column. The room temperature was usually between 0°C and 3°C.

For producing the frost crystals, the vessel *A* is gradually cooled to a certain temperature by passing the cold air through the coil, and after T_a reaches a desired value the current of cold air is regulated to keep it constant. Then the temperature of the water in *E* is raised to a required value. After a little while, beautiful frost crystals begin to develop on the bottom of the cooled vessel, all hanging down in the air. If this condition is maintained, fully developed crystals are detached by their own weight and fall in the box *C*. These falling frost crystals are usually very numerous and they give an aspect just like a natural snowfall. As a matter of fact this phenomenon may be called an artificial snowfall, if the question of nuclei is not taken into consideration. These falling crystals were received on a glass plate which had been previously well cooled. They were then examined under a microscope. The crystals that adhered to the bottom plate of the cooling vessel were also detached mechanically with a cold glass plate and brought under microscopic investigation. These crystals are marked "adhered" in the later descriptions.

42. Crystal habits and degree of supersaturation

In the preliminary experiments it was found that the crystal habit of frost was probably determined by T_a, the temperature of the air where the crystal is formed, and T_w, the temperature of the water in the reservoir. Attention was paid chiefly to the degree of supersaturation. The ratio of s_w, the saturation vapor pressure of

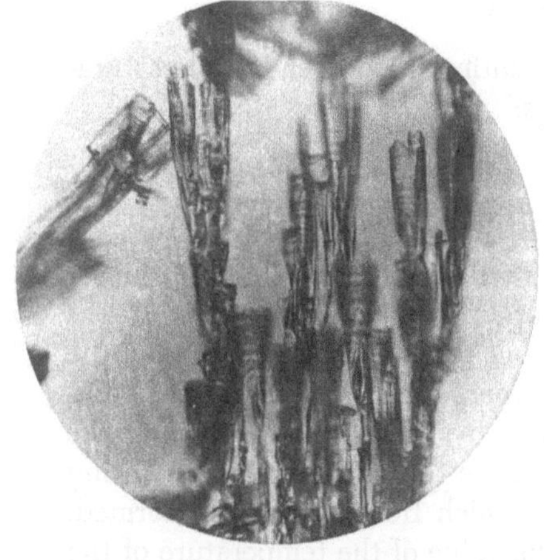

FIG. 253 [1065]. Needle crystal of frost (× 32.5).

FIG. 254. Dendritic and pseudodendritic structures.

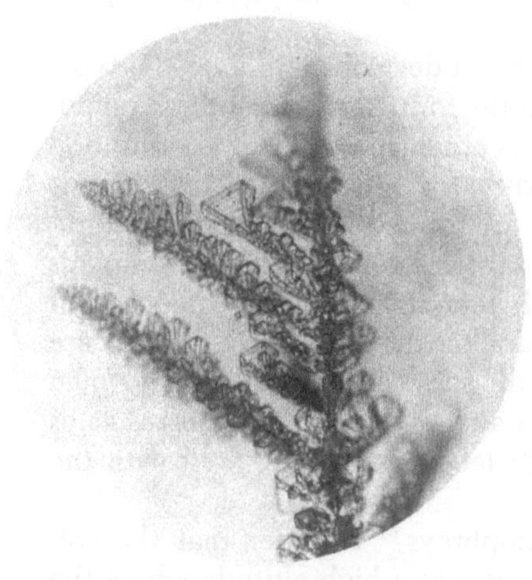

FIG. 255 [1069]. Pseudodendritic crystal of frost (× 40).

FIG. 256 [1072]. Intermediate form between pseudodendritic and plate (× 51.5).

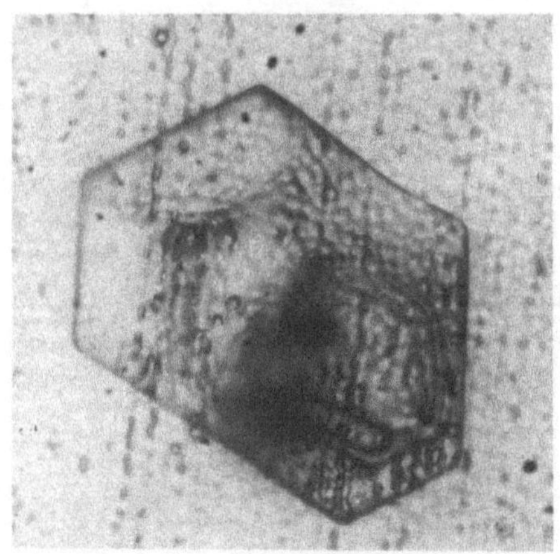

FIG. 257 [1075]. Plate crystal of frost (× 18).

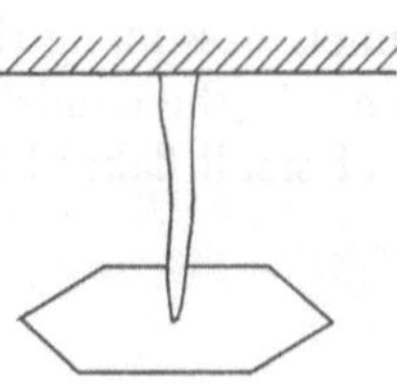

FIG. 258. Mode of development of Fig. 257.

water at T_w, to s_a, that at T_a, was chosen as representing the degree of supersaturation and it was called the supersaturation ratio s;

$$s = \frac{s_w}{s_a}.$$

The result of the experiments showed that s was suitable for the indication of supersaturation in the discussion of its bearings upon the habits of crystals. It was found that the element that determines the type of crystals that is formed is the supersaturation ratio s. From the result of the experiments on artificial snow we learned that the type of crystal is much influenced by T_a, the temperature of the air where the crystal is formed. In this experiment the thermometer for measuring T_a is just below the cooling plate of copper, on which frost crystals are formed. Then T_a measured in this apparatus is not the true value of the temperature of the air at that spot. The true value of T_a being unknown, the discussion is limited to the relation between the supersaturation ratio and the crystal habits in this experiment.

When the supersaturation ratio is large the crystal develops into a needle type, but with small supersaturation it tends to grow in the form of a prism or a pyramid. The plane type, either in dendritic or plate form, is obtained with a supersaturation ratio lying between those for needles and for prisms.

(1) *Needle type.* Crystals of the needle type are formed when the supersaturation ratio is between 6 and 8, the mean for five examples being 7.5. One example is shown in Fig. 253. It is seen in the photograph that the needle is composed of pillars growing parallel with each other. This characteristic is also observed in case of the natural snow crystal. In Hokkaido the needle snow crystals fall chiefly when the temperature is comparatively warm and the air is very humid. The result of the artificial production of this type of crystal in laboratory agrees well with the observations for natural snow.

(2) *Pseudodendritic form.* Bentley and Humphreys [54] consider that the columnar crystal of snow is produced by a slow process at a high altitude where the humidity is comparatively low, while the dendritic form is obtained in a humid region near the earth's surface. Many others also agree with this view, which is concordant with our present knowledge on crystal habits in general, but no supposition has been proposed for the degree of supersaturation that is required for the formation of the crystals. In the present experiment the dendritic form was obtained with a high supersaturation ratio, next to that for needle crystals, the mean of four examples being 6.6. The dendritic form obtained in this experiment is not the same as that of snow. In natural dendritic snow, two parallel lines due to thin channels pass along the central line of the branch, as shown schematically in Fig. 254(a). In the dendritic form of frost crystals, however, the structure is an assemblage of small flakes like fish scales, as shown in Fig. 254(b). We call this

Fig. 259 [1076]. Pyramid crystal of frost (× 127).

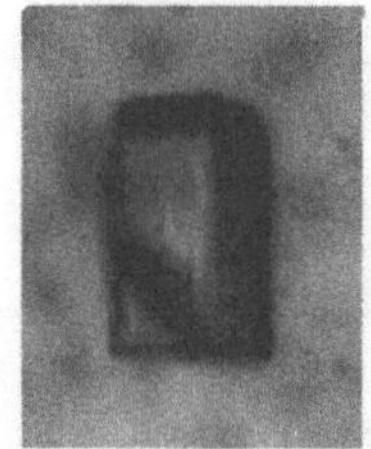

Fig. 260 [1079]. Prism crystal of frost (× 240).

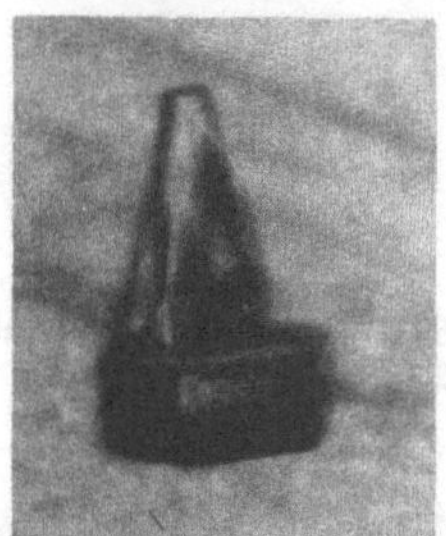

Fig. 261 [1078]. Short prism with pyramid (× 250).

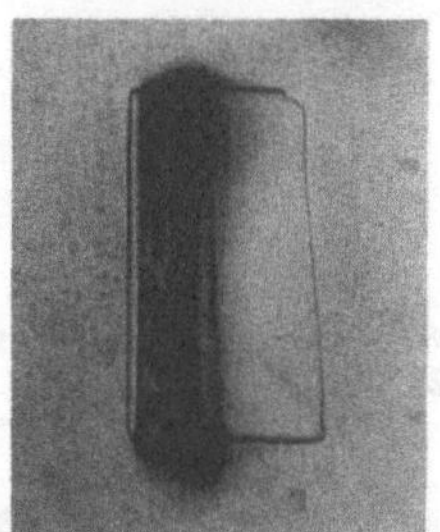

Fig. 262 [1081]. Prism with extended side plane (× 26).

apparently dendritic form "pseudodendritic." One photograph of this pseudodendritic crystal is reproduced in Fig. 255.

(3) *Intermediate form between dendritic and plate crystals.* If the supersaturation ratio is decreased, the pseudodendritic form tends to turn into a plate form, as expected from our knowledge of crystal habits in general. There are of course many intermediate states between the two extremes, one example of which is shown in Fig. 256. The range of the supersaturation ratio is fairly wide, being between 3.5 and 5.5.

(4) *Sector and plate forms.* The supersaturation ratio for the formation of sector and plate crystals ranges between about 3 and 4. The mean of six examples is 3.5. A complete form of hexagonal plate was very rarely formed and only one example was observed during this series of experiments, the photomicrograph of which is reproduced in Fig. 257. By visual observation under the microscope it was noticed that a thin strip of frost crystal was attached to nearly the center of the plate perpendicular to the plane of the crystal. Unfortunately this strip melted off before any good photograph was obtained. It is very natural to suppose that the tip of a strip of frost hanging down from the ceiling was a minute hexagonal prism or plate and that this complete hexagonal plate developed perpendicularly to the strip having this tip crystal as a nucleus. A schematic sketch is shown in Fig. 258.

(5) *Pyramidal form.* A simple pyramidal form of snow crystal is known but it is very seldom observed. In the laboratory, frost of this type is not difficult to

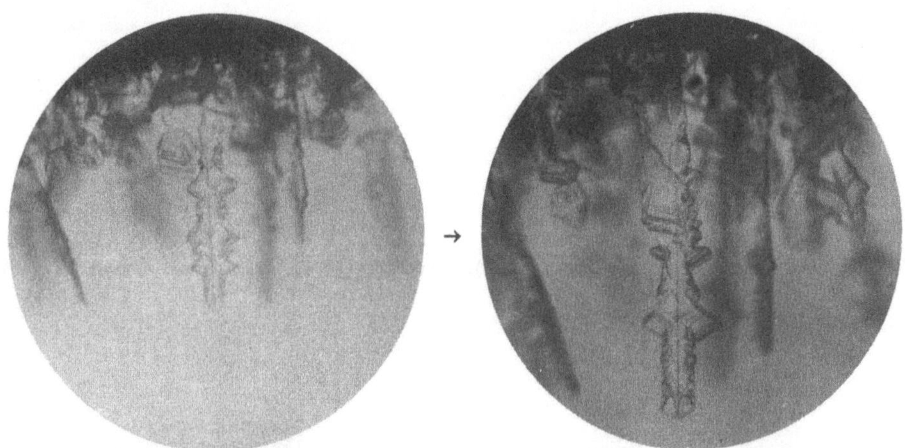

FIG. 263 [1087]. Growth of frost crystal, t = 1 hr 36 min (× 23).

FIG. 264 [1088]. Growth of frost crystal, t = 2 hr 17 min (× 23).

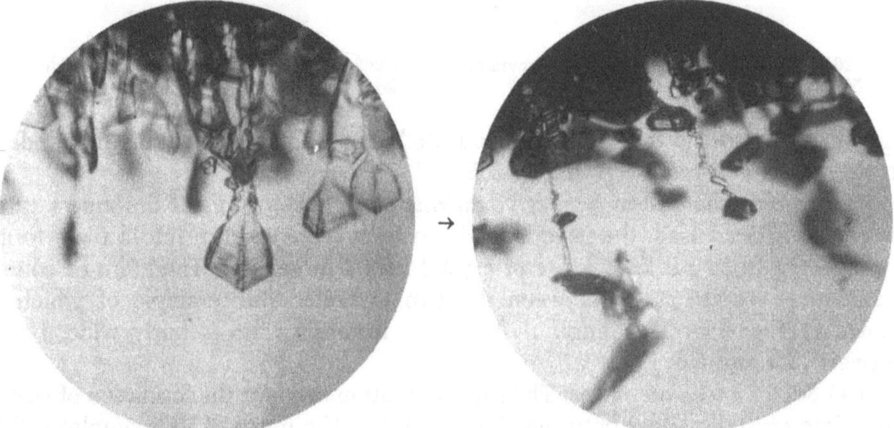

FIG. 265 [1093]. Evaporation of frost crystal, t = 1 hr 18 min (× 23).

FIG. 266 [1094]. Evaporation of frost crystal, t = 2 hr 49 min (× 23).

produce. The author observed three examples during this series of experiments, one of which is reproduced in Fig. 259. The supersaturation ratio for the production of this form of crystal was 2.2 as a mean of three examples.

(6) *Prism.* This type of frost crystal is obtained with a small supersaturation ratio, as expected. The mean of s for four examples observed is 1.7. Several photographs were taken, among which the one reproduced in Fig. 260 is a simple prism, that in Fig. 261 is a short prism with pyramid. A bullet type is seen in No. 1080, Plate 164.

(7) *Prism with extended side plane.* This type is produced in the laboratory under conditions similar to those for prisms. Two examples were observed which are reproduced in Fig. 262 and No. 1082, Plate 164. It is clearly seen in the photographs that a thin plane of crystal is attached to a prism and it appears to be an extension of the side wall of the prism. This type was found always adhering to the cold ceiling of the box. A snow crystal of this type has not yet been observed in any country of the world, so far as the literature at hand shows, but the author considers that the peculiar crystals of natural snow represented in Figs. 155 and 156 are complicated crystals of this type.

From the results above described we could infer to some extent the conditions of formation of various types of snow crystals. Although the frost crystals obtained in this experiment do not show exact coincidence with the snow crystals, the data are collectively shown in Table 13.

TABLE 13. Formation of frost crystals.

Type	No. of photograph	T_a (°C)	T_w (°C)	s	State of occurrence	Mean of s
Needle	1066	− 5.2	21.0	6.3	adhered	7.5
	1067	−14.3	13.0	8.5	falling	
	—	−14.3	12.9	8.5	falling	
	1064	−14.1	10.5	7.1	falling	
	1065	−14.0	11.0	7.3	adhered	
Pseudodendritic	1069	−22.3	0.1	7.5	adhered	6.6
	—	−21.5	0.9	7.3	adhered	
	1070	−15.8	4.3	5.4	adhered	
	—	−17.0	4.3	6.1	falling	
	1068	−17.5	6.5	—	falling	
Intermediate	—	− 8.8	11.5	4.7	falling	4.4
	—	− 7.0	11.0	3.9	adhered	
	1072	−13.2	1.0	3.4	adhered	
	1071	−14.5	4.0	4.7	falling	
	—	−16.0	4.0	5.4	adhered	
Plate	—	−12.5	1.0	3.2	adhered	3.5
	1073	−12.0	1.0	3.0	adhered	
	1074	−11.4	1.1	2.9	adhered	
	—	−12.3	4.1	3.9	adhered	
	—	−12.0	4.2	3.8	falling	
	1075	− 4.5	14.0	3.8	falling	
Pyramidal	1076	− 3.0	5.8	1.9	falling	2.2
	1077	−10.1	0.0	2.4	falling	
	—	− 9.5	0.4	2.3	falling	
Prism	1078	− 5.5	−0.1	1.6	falling	1.7
	1079	− 5.6	0.0	1.6	falling	
	—	− 5.3	0.2	1.6	falling	
	1080	− 5.6	0.7	1.7	falling	
Prism with extended side plane	1081	− 2.3	0	1.2	adhered	1.6
	—	− 2.5	5.4	1.8	adhered	
	1082	− 2.2	5.2	1.7	adhered	

43. Growth of frost crystals

In the experiment described in Sec. 42 the frost crystals detached from the cold ceiling were placed on a cooled glass plate and subjected to microscopic investigation. In the present case the crystal was intended to be observed with a microscope from outside without touching the crystal, so that the successive stages of its development were examined with ease.

In order to make frost crystals grow at a certain position, a wedge of copper was fixed to the ceiling, and the crystals developed on this wedge were observed from outside.

Examples of frost crystals developed on the wedge are shown in Figs. 263–266. Figure 263 is the state of the crystal 1 hr 36 min after the start of the experiment. In the former experiment, we could only obtain a crystal of pseudodendritic structure. In this experiment we were able for the first time to obtain a structure like that of natural snow, that is, two fine continuous channels are seen in the middle part of the ray, arranged parallel to each other and extending in the direction of the ray. These channels can be seen in the crystal of Fig. 263. At 2 hr 17 min after the start of the experiment, the crystal in Fig. 263 had developed into the form shown in Fig. 264. The crystal seen in the middle of the photograph grew 0.9 mm in length during 41 min, keeping the original form. Nos. 1089 and 1090, Plate 165 are another example, in which the crystal of sector form developed 0.3 mm in length during 25 min.

The above examples were formed when T_a was kept constant or gradually fell. When T_a rises during the experiment, evaporation takes place and the crystal undergoes transformation, becoming smaller. Figures 265 and 266 and Nos. 1091 and 1092, Plate 165 are examples.

As described in the preceding section, these frost crystals begin to fall by themselves, giving the appearance of a snowfall. Formerly it was simply supposed that they become detached from the ceiling by their own weight, and the mechanism of this process was clarified by the present experiment. While observing continuously the stages of development of the frost crystals, it was found that the part near the root of the crystal sometimes began to become attenuated. This stage can be seen in Fig. 265. After this stage the attenuated part becomes thinner and thinner, so that after some time it appears that the crystal is suspended by a fine thread of ice, and then this thread is cut so that the crystal falls down by itself. One can see this final stage in Fig. 266, which shows the transformed state after 1 hr 31 min of the crystal of Fig. 265; meanwhile, T_a increased by 2.5C deg. This phenomenon is to be explained as follows. After the crystals that grow in clusters develop to some extent, the water vapor carried upward by natural convection cannot get into the space near the roots of the crystals. On the other hand, as there exists a colder metal plate just above the frost crystals, the ice molecules of this part of the crystal sublime and condense on the minute colder crystals which adhere to the metal plate. The surface tension of the fine edge of the crystal will also accelerate this process. The structure of the thread of ice seen in Fig. 266 is not simple but appears like an aggregate of somewhat rounded granules of ice. Comparing this structure with that of irregular snow, Fig. 192, one will see a marked resemblance between the two. In this case the temperature of the air was −7.5°C and it will be understood that this structure like an aggregate of rounded granules of ice, which is often observable in natural snow, can be produced by sublimation, without melting, when the temperature is far below freezing point. Nos. 1091, 1092, Plate 165 are the similar examples. The crystal of sector form seen in No. 1091 transformed into No. 1092 in 38 min, T_a having risen by 1C deg during that time.

From these results we were lead to conclude that in order to get a well-developed crystal of ice that should be very like that of snow we must remove the cooling box and carry on the experiment by setting up the whole apparatus in a large cold chamber.

44. Low-temperature laboratory

In the spring of 1935 the authorities of the Hokkaido University began to build a low-temperature laboratory, which was completed at the beginning of 1936. This laboratory is provided with a cold chamber 4 × 4 × 4 m in dimensions, the whole chamber being designed to be cooled down to −50°C by a refrigerating mechanism. It is also provided with an accessory chamber 2 × 4 × 4 m in dimensions, which can be cooled down to −30°C. The former experiments on artificial frost crystals were carried on in this cold chamber and further the author succeeded in producing a snow crystal having six dendritic branches extending outward almost symmetrically from a nuclear center. The nucleus of this artificial snow crystal was suspended by being attached to a thin rabbit hair or cotton filament.

Usually we worked at the temperature of −30°C, and occasionally the chamber was cooled down to −45°C, which was sufficient for our purpose. We had to wear a flying suit while carrying on the experiments. The cold chamber is shown in Fig. 267.

45. Artificial production of frost crystals in the cold chamber

Two sets of apparatus were used for this experiment. One is almost the same as that used in the former experiment, Fig. 252. The copper box for cooling was replaced by a copper sheet and the temperature of the cold chamber was between −20°C and −30°C.

Another apparatus, as shown in Fig. 268, was designed for experiments when very pure water or heavy water is used. The whole vessel was made of glass, and

Fig. 267. The author and the snow-manufacturing apparatus in the cold chamber.

the principle is similar to the former apparatus except for the method of heating the water. Slender alcohol thermometers, specially designed for the purpose, were used for measuring T_a and T_w, the temperature of the air where the crystal is made and that of the water in the reservoir. Two kinds of wedge *E* were used, one of wood and the other of metal.

Artificial frost could be developed to its full extent, as expected, when the experiment was carried on in the cold chamber.

In the case of any crystal it is well known that a dendritic form is quickly produced in a sufficiently supersaturated condition, whereas a crystal in solid form is developed slowly in a state of less supersaturation. In the case of snow also our crystallographic knowledge suggests that the dendritic form is produced most quickly, the plate form requiring more time, and the column or prism even more. The difference in the modes of formation of the dendritic ray and the plate form is clearly shown in Figs. 269 and 270. The plate form represented in Fig. 269 was made with the room temperature at $-25°C$, the temperature of the water reservoir

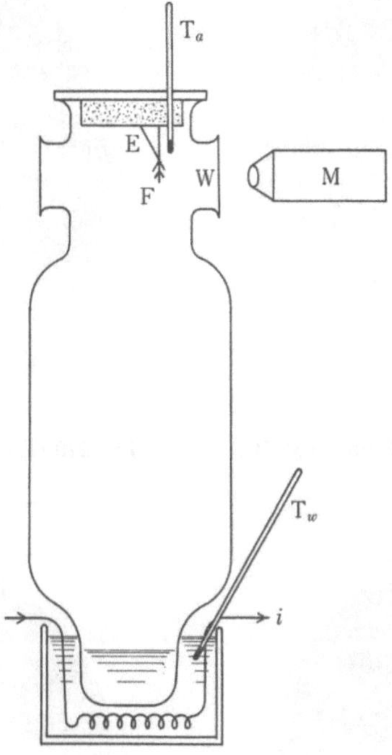

FIG. 268. Apparatus used for frost making in the cold chamber.

at nearly 0°C, the apparatus being left overnight. The rate of growth is about one-tenth that of the dendritic one. The dendritic crystal, shown in Fig. 270, was made with T_a at -16°C and T_w at $+10$°C, and the photograph shows the state nearly 2 hr after the beginning of the experiment. In this case the minute structure of the ray is like that of an ordinary snow crystal, and one can see the continuous channel along the middle line of the ray.

Similar experiments were repeated with heavy water, which was supplied by the Norsk Hydro-Elektrisk Kvelstofaktieselskaf of Norway; the content of D_2O was 99.6 percent. The apparatus of Fig. 268 was used throughout. Heavy water weighing altogether 5 gm was used and it was adequate in bulk for many beautiful crystals to be obtained. No essential difference was observed between the characteristics of crystals made from heavy water and those made from ordinary water. As the melting point of heavy-water ice is $+4$°C, the temperature of the water reservoir had to be raised a few degrees higher than is necessary in the case of ordinary water, in order to obtain a similar form of crystal. When the heavy water was used, the inner structure of the dendritic branch was liable to take the form of an assemblage of scales. Figure 271, made when T_a was -17°C and T_w $+25$°C, is a typical example showing the inner structure like an assemblage of scales. A

FIG. 269 [1098]. Plate crystal of frost (× 19).

FIG. 270 [1100]. Dendritic crystal of frost (× 19).

FIG. 271 [1113]. Pseudodendritic frost, heavy water (× 18).

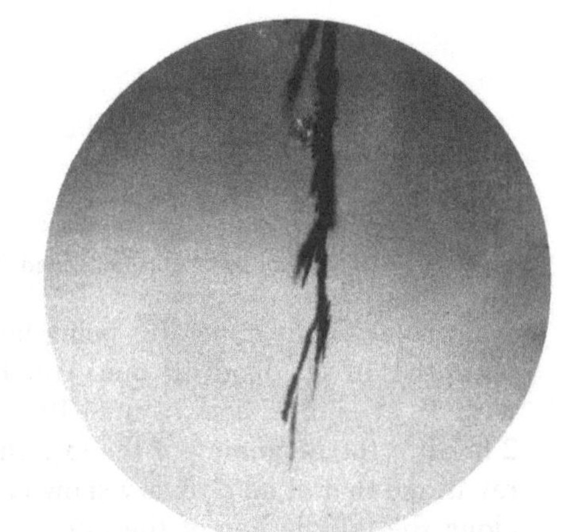

FIG. 272 [1114]. Side view of Fig. 271 (× 18).

FIG. 273 [1101]. Natural snow crystal (× 17).

TABLE 14. Formation of frost crystals in the cold chamber.

Photo. No.	Water used	Wedge	T_a (°C)	T_w (°C)	t* (hr)	(min)	Notes
1095	distilled	metal	−16	+18		50	Slender plate
1096	distilled	metal	−13	+14	1	56	Slender plate
1097	city	wood	−18	+11	2	15	Sector form
1098	city	wood	—	~ 0	~12		Left overnight, large sector
1099	distilled	wood	−15	+17	2	20	Pseudodendritic
1100	city	metal	−16	+19	1	55	Complete dendritic
1111	heavy	wood	−16	+28		25	1112, side view
1113	heavy	wood	−17	+25	1	10	1114, side view

* t = time after the start of the experiment when the photograph was taken.

close examination of a crystal of this type revealed that it was made by the overlapping of one piece of flake on another. The side view of the crystal shown in Fig. 271 is reproduced in Fig. 272, which shows nicely the mode of overlapping of component flakes. In another type, which has fine continuous channels along the middle line of the ray, the whole crystal, or at least the main ray, grows in one plane.

The conditions for formation of the crystals above described are tabulated in Table 14.

The chief object of this experiment was to observe the successive stages in the development of a crystal. One example of a series of photographs showing the manner in which a dendritic crystal develops is given in Fig. 274. Figure 274(a) shows the crystal about 30 min after the beginning of the experiment. This time interval varies considerably, depending on the initial condition of the apparatus. The rate of growth of the crystal is shown in the figure, taking the stage (a) as the initial state, at time $t = 0$. The rate of growth of the extremity of the crystal is 0.047 mm/min between stages (a) and (b), 0.044 mm/min between (b) and (c), and 0.045 mm/min between (c) and (d). In this case the minute structure of the ray of the crystal belongs to the type of the assemblage of scalelike flakes. The rate of growth of the ray of a crystal in dendritic form varies considerably with the condition of formation, that is, not only with the temperatures of the water and atmosphere, but also with the mode of convection of water vapor, but usually it can be taken as nearly 0.05 mm/min. Comparing the frost crystal of Fig. 274(d) with one branch of the natural snow crystal shown in Fig. 273, we can infer to some extent the mode of growth of the natural snow crystal.

Hexagonal plates with dendritic extensions at the corners are often observed in the Hokkaido climate. According to the view above presented, these crystals are made when the upper layer of the atmosphere is relatively dry and the layer near the ground is sufficiently humid. One example is shown in Fig. 275. Artificial pro-

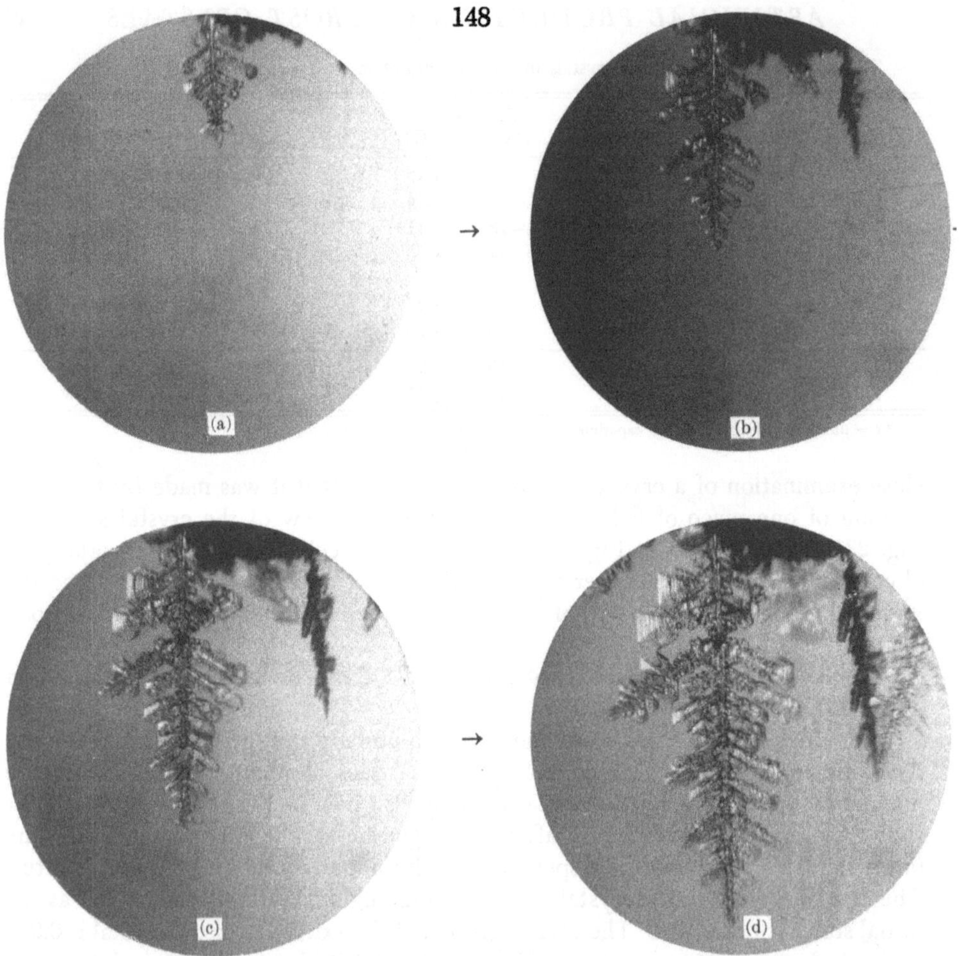

FIG. 274. Development of dendritic frost crystal: (*a*) [1102] $t = 0$; (*b*) [1103] $t = 15$ min; (*c*) [1104] $t = 30$ min; (*d*) [1105] $t = 45$ min ($\times$ 17.5).

FIG. 275 [1106]. Natural snow crystal ($\times$ 20.5).

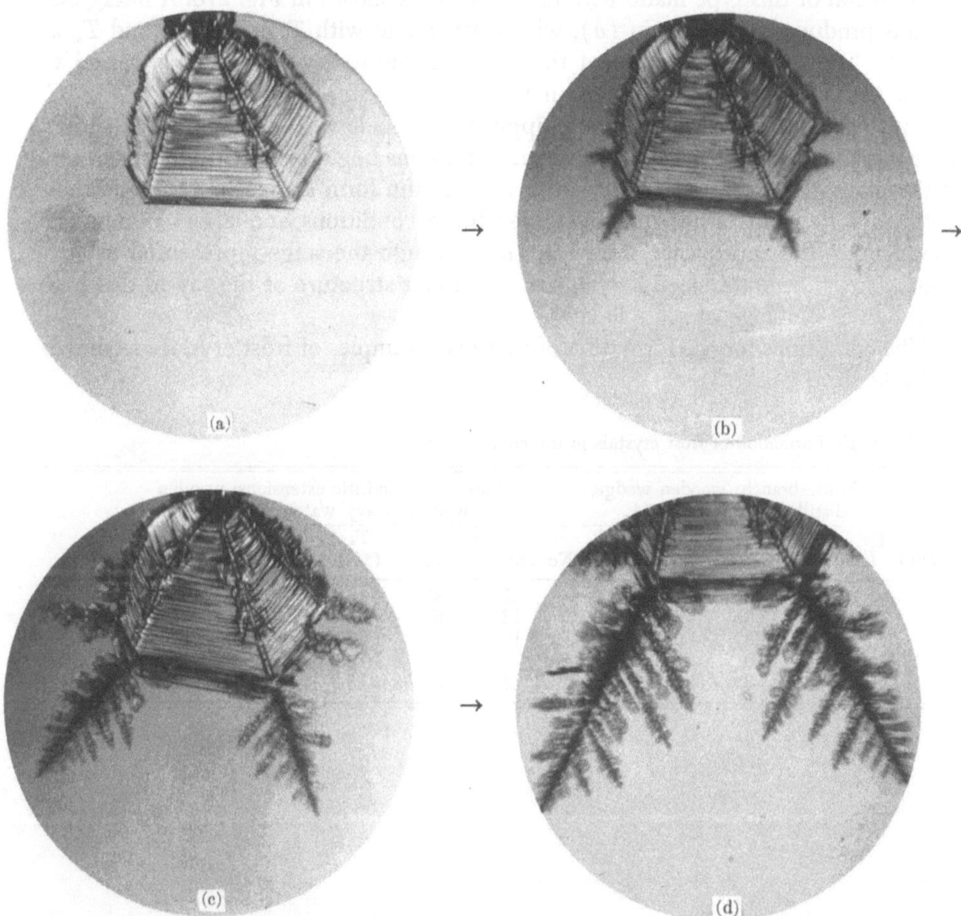

Fig. 276. Extension of dendritic branches from a plate: (*a*) [1107] $t = 0$; (*b*) [1108] $t = 15$ min; (*c*) [1109] $t = 30$ min; (*d*) [1110] $t = 45$ min ($\times$ 18).

duction of this type of frost crystal was attempted in the following way. The temperature of the water reservoir was kept low and the hexagonal plate was obtained; then the temperature of the water was raised suddenly by increasing the heating current, so that we could expect dendritic extensions at the corners. In this case it was found that the excessive humidity must be supplied quite suddenly, otherwise the crystal develops in a quite different form from the so called "hexagonal plate with dendritic extensions at the corners" observable in the case of natural snow. From this result it will be supposed that the boundary between the dry and humid layers in the atmosphere is fairly sharp when this type of snow is falling.

A crystal of this type made with heavy water is shown in Fig. 276. A hexagonal plate is produced as shown in (a), which was made with T_a at -19°C and T_w at $+9.7$°C. Then the temperature of the water in the reservoir is quickly raised to $+22.5$°C. The temperature of the air where the crystal is made rises to -18°C, owing to the excessive warm vapor supplied from below. Taking the stage shown in (a) as the initial state, $t = 0$, dendritic extensions begin to start from the corners of the plate and after 15 min the latter assume the form shown in (b). The dendritic rays develop rather quickly under these conditions and after 45 min the crystal takes the appearance shown in (d), through the stage represented in (c), which shows the state when $t = 30$ min. The inner structure of the ray in this case is also an assemblage of scale flakes.

The conditions for the formation of these two samples of frost crystal are shown in Table 15.

TABLE 15. Formation of frost crystals in the cold chamber.

Dendritic branch; wooden wedge, distilled water; Fig. 274					Plate with dendritic extensions; wooden wedge, heavy water; Fig. 276				
t (hr)	t (min)	T_a (°C)	T_w (°C)	Photo. No.	t (hr)	t (min)	T_a (°C)	T_w (°C)	Photo. No.
	12	−16.3	+20.5	—	11	30	−19.4	+ 9.7	—
	30	−15.3	+20.5	1102	12	45	−19.0	+19.8	1107
	45	−15.3	+20.5	1103	13	0	−17.2	+22.5	1108
1	0	−16.0	+20.5	1104	13	15	−18.0	+22.5	1109
1	15	−15.3	+20.5	1105	13	30	−18.3	+22.3	1110

CHAPTER 5

Artificial Production of Snow Crystals

46. The first crystal of artificial snow

Now we are led to conclude that, for the artificial production of snow crystals, we are to make frost crystals in the cold chamber in a state suspended freely in the air. From the rate of growth of frost crystals, it is supposed that it takes at least 1 or 2 hr for the complete development of a snow crystal. In order to keep the crystal suspended in air for such a long time, we tried to produce it by hanging it from a thin filament.

In the apparatus for making the frost crystals, Fig. 252, many filaments of cotton were hung from the top of the box, and many tentative experiments were repeated to make ice crystals develop on some point of these filaments. It always occurred, however, that many frost crystals grew all over the filament, making it look like a hairy caterpillar. One example is shown in Fig. 277. In such a case, the mutual disturbance due to neighboring crystals prevents the free development of an ice crystal as if it were suspended in the atmosphere.

In the course of the preliminary experiments, repeated under varied conditions, by chance a crystal was obtained on 12 March 1936, which can be called an artificial snow crystal. This first crystal of artificial snow is reproduced in Fig. 278. The conditions of its formation were $T_a = -16°C$, $T_w = +9°C$.

47. Earlier experiments on artificial snow

After the first chance success of the artificial production of a snow crystal, described in the preceding section, we repeated the experiment many times, but it very rarely occurred that an isolated crystal was obtained on a spot of the filament. Usually the filament was covered all over with the frost crystals. In order to

Fig. 277 [1115]. Condensation of numerous frost crystals (× 18).

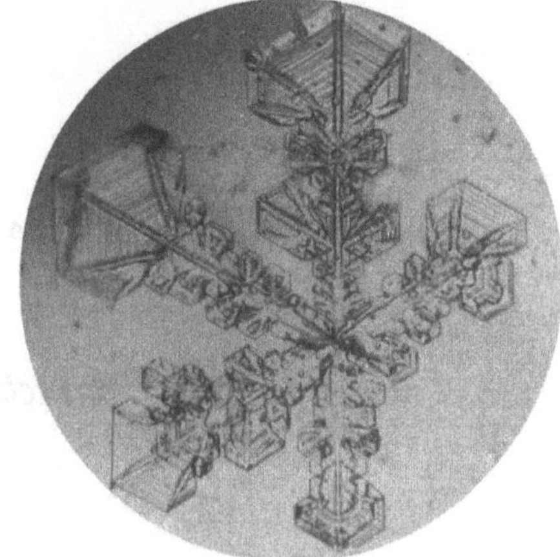
Fig. 278 [1116]. The first snow crystal made in the laboratory (× 22.5).

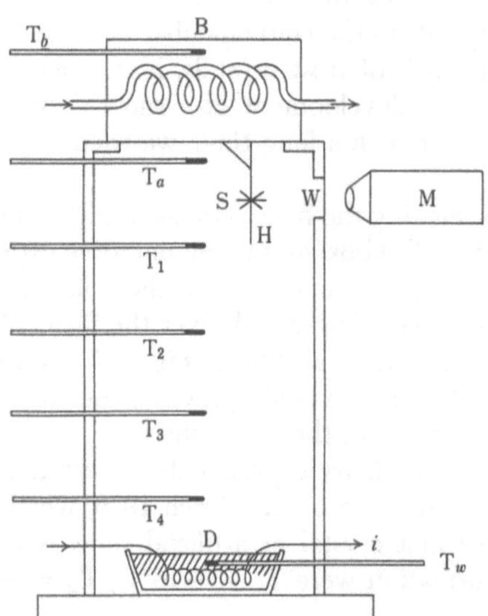

Fig. 279. Apparatus used in the earlier experiments.

avoid this we found after some trials that a fine rabbit hair which was thoroughly dried in a desiccator was most suitable for getting isolated ice crystals on the filament. We call in this book the very early stage of a snow crystal a "germ." Using this filament we could make a few germs attach themselves to the filament and develop artificial snow from these isolated germs.

TABLE 16. Formation of snow crystals in the cold chamber.

t (hr)	t (min)	T_a (°C)	T_w (°C)	Photo. No.	Notes
	20	−15.8	+25.2	1117	heavy water
	50	−18.0	+24.8	1118	
1	30	−19.2	+21.0	1119	temperature lowered
4	0	−20.5	+ 7.0	1120	
	0	−16.0	+17.8	—	heavy water
	30	−16.7	+19.0	1121	
	50	−17.0	+18.2	1122	

The apparatus used in our earlier experimentation on artificial snow is shown in Fig. 279. Similar experiments were carried out also by using the apparatus of Fig. 268.

Figure 280 shows two sorts of artificial snow crystal, the upper crystal belonging to a plane dendritic type and the lower being an assemblage of dendritic branches. These crystals were made with heavy water. At 30 min after the state shown in Fig. 280, the crystals developed into No. 1118, Plate 167. The lower crystal was detached after taking the photograph, and the upper one was made to grow under the same conditions. The upper crystal developed through the stage of No. 1119 into the form shown in No. 1120 (Plate 167). The temperature was lowered after No. 1118, and the spatial attachment of small plates observable in No. 1120 was proved by later experiments to have been due to the lowering of the temperature. Another example is shown in Nos. 1121 and 1122, Plate 167. The conditions for the formation of these crystals are shown in Table 16.

48. Production of various types of crystals

The experiments on artificial snow made some progress, and various types of snow crystals were made. The elements controlling the variety in form of the crystal were found, as a first approximation, to have been T_a and T_w. Changing T_a and T_w in various manners, almost all sorts of snow crystals were obtained. They are shown in Figs. 281–285 and Nos. 1123, 1124, 1126–1129, 1133, Plates 167 and 168. Figure 281 is a typical dendritic crystal. This crystal was made in 2 hr, during which interval T_a was nearly −16°C and T_w varied between +6°C and +8°C, the supersaturation ratio being nearly 7.5. The inner structure of the ray is exactly the same as that of the ordinary dendritic snow crystal of a fernlike type. A plane crystal of snow with three branches is sometimes observed in natural snow. A similar crystal is also obtained in the case of artificial snow. One example is shown in No. 1126, Plate 168, which was made with T_a at nearly −17°C and T_w at +6.5°C. Nos. 1127 and 1128, Plate 168, are two crystals produced on different hairs

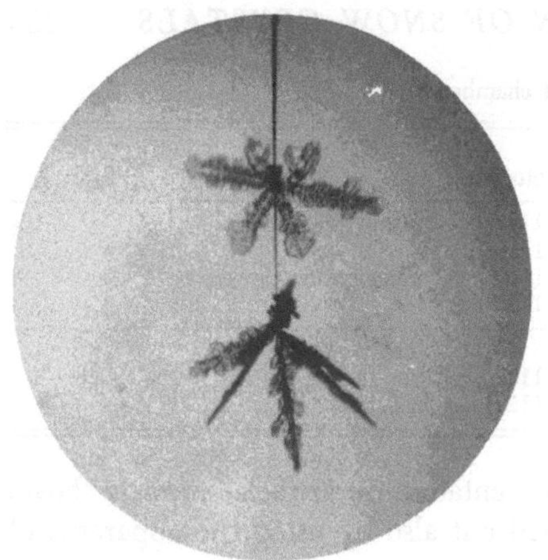

Fig. 280 [1117]. Plane and spatial dendritic types of artificial snow crystals (× 18).

Fig. 281 [1125]. Artificial snow, fernlike crystal (× 37).

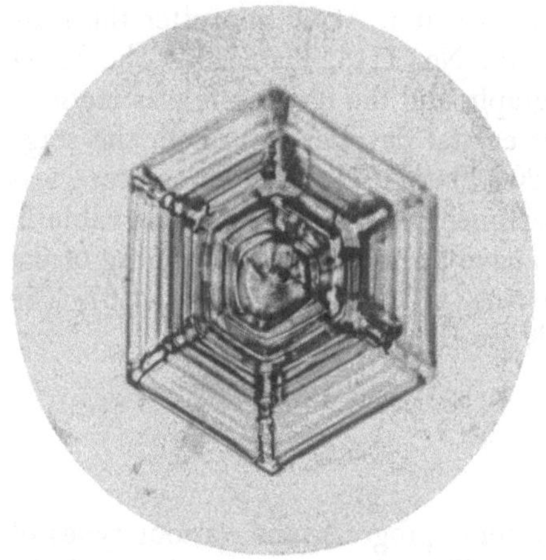

↔

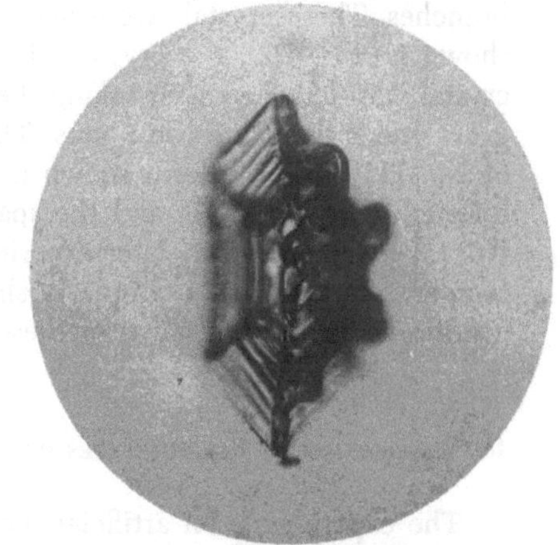

Fig. 282 [1131]. Front view of a cup crystal (× 31).

Fig. 283 [1132]. Side view of Fig. 282 (× 31).

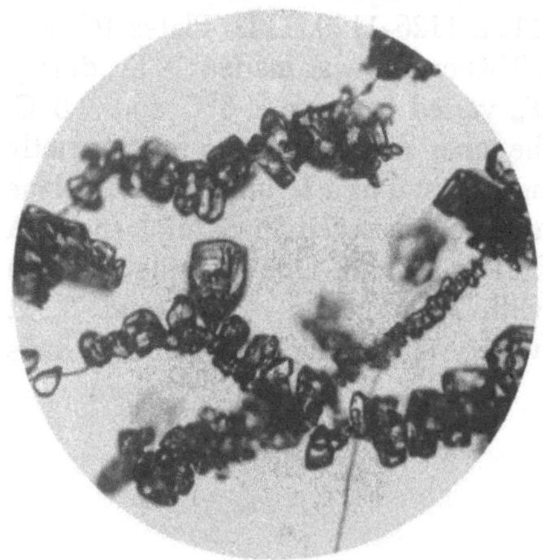

Fig. 284 [1130]. Bullets of artificial snow (× 27).

Fig. 285 [1134]. Artificial snow, thick plate (× 12.5).

in one experiment. When similar crystals are obtained in one experiment at different positions in the apparatus, the convection of water vapor in the apparatus can be considered fairly uniform, but that is not always the case.

When the degree of supersaturation of the water vapor was decreased, the dendritic crystals above described were transformed into a plate type. Under the same condition as the plate crystal a cup crystal was sometimes obtained, which is the type so often observed in the crystals of depth hoar, that is, a sublimation product growing beneath the surface of fallen snow. Photomicrographs of depth hoar can be seen in Seligman's book.[55] One example of this cup crystal is reproduced in Figs. 282 and 283, the former being the front and the latter the side view. This crystal takes the form of a shallow hexagonal dish. In this case T_a was nearly $-15°C$ and T_w was $0°C$, the surface of the water in the reservoir having been covered with a thin sheet of ice. The supersaturation ratio was nearly 3.3. This cup was found on the wedge on the top of the box, the bottom of the cup adhering to the surface of the wedge. These photographs show the state of the crystal about 7 hr after the beginning of the experiment. In the same experiment it was noticed that the cotton filaments hanging down from the wedge were coated with minute crystals of ice. The photomicrograph, Fig. 284, shows that those minute crystals are an assemblage of columns and bullets. In this case the amount of water vapor that was supplied to the neighborhood of the filament was distributed among so many crystals that the rate of growth was particularly slow. After 7 hr they attained an average length of only 0.3 mm. This type of snow crystal observed in nature is also very small.

The thick-plate type of snow crystal is frequently observed in Hokkaido. It is produced when many water droplets – cloud particles – attach themselves to a plane crystal of snow. A similar type is also obtained in the case of artificial snow. One example is shown in Fig. 285. In this case the temperature of the water was very high and some of the water vapor brought up by convection condensed as fog and attached itself to the crystal. The process is almost the same as the supposed mechanism of formation of this type of natural snow. The crystal shown in Fig. 285 was made with T_a at $-16.3°C$ and T_w at $+24°C$. It took 2 hr for the crystal to reach the state shown in the photograph.

Summarizing the results of the experiments above described it may safely be stated that almost every sort of snow crystal can be made in the laboratory. The chief factor governing the formation of various types of snow is the rate of crystal growth and the latter is strongly influenced by the supersaturation ratio. The temperature of the air where the crystal is produced was found in the present experiment to be also of great importance in determining the crystal habit of snow. Besides, the mode of convection of the water vapor also has some noticeable influence. The argument given in Sec. 42, to the effect that the form of the crystal is chiefly determined by the supersaturation ratio, was found to be insufficient.

49. Improvement of the snow-making apparatus

By the use of the apparatus of Fig. 279 it was difficult to get the artificial snow crystals in a reproducible manner. The difficulty lies chiefly in the indefiniteness of the mode of convection of water vapor in the apparatus. So the apparatus was modified to give different modes of possibly steady convection of water vapor. In order to get various manners of circulation of air, four sets of apparatus were constructed. Apparatus No. 1 is shown schematically in Fig. 286. Two concentric glass cylinders are held vertically, so that warm vapor is driven upward inside the inner tube while the cooled air comes down through the space between the two cylinders. The water in the reservoir *R* is warmed electrically as in the previous experiment. With this apparatus the mode of convection of water vapor seemed to be fairly steady and a certain type of snow crystal could be obtained in an almost reproducible manner under given conditions. The cover *C* is a metal sheet and *D* is a plate of cork or wood. The latter was introduced in order to interpose something between the cold metal plate *C* and the crystal *S*. This plate *D* was removed in some cases. The wedge *W′* may be of metal or wood. The temperature T_a of the air where the snow crystal is made varies considerably with the thermal nature of the materials surrounding the crystal. Suitable materials were chosen by trial according to the form of crystal to be manufactured.

By changing the form of the upper part of the apparatus above the wooden ring *W*, two modifications of apparatus No. 1 were constructed. They are shown in Figs. 287 and 288. In apparatus No. 2, the form of the upper cylinder is the same as that of No. 1, but its length is 10 cm less. The supply of warm water vapor is in this case more abundant than in apparatus No. 1 and T_a is accordingly higher, although the temperatures of the room and of the water in the reservoir remain the same. With apparatus No. 3 it was intended to reduce the rate of supply of water vapor to the place where the crystal is made.

Apparatus No. 4 was constructed on a different principle, as shown schematically in Fig. 289. With this apparatus it was intended that a circulation of air should be caused by using two vertical cylinders kept at different temperatures. The ascending air current goes up in the glass cylinder, which is kept warmer than room temperature by an electric heater carrying a small current *i′*. The outside of the cylinder is wrapped with a piece of felt *F*. The side tube represented in the figure by a heavy line is made of copper. This tube is always kept cold at nearly the same temperature as that of the room. It is attached for the purpose of accelerating the downward convection of the air in the apparatus. By adjusting the heating current *i′*, which is of the order of few tenths of an ampere, the rate of convection of air could be made fairly high so that the hair suspending the crystal was swayed about violently by the stream of air. Large fernlike crystals of snow were rather quickly made in this apparatus.

50. The convection of air in the apparatus

The mode of air convection in apparatus No. 1 was first studied. For this purpose the room temperature T_r and the temperature of water in the reservoir T_w were kept constant and the isothermal lines were constructed by repeatedly measuring the temperatures at various points in the cylinder by means of a slender alcohol thermometer of diameter 3.3 mm. As a matter of fact, the room temperature could not be kept constant; it varied within a range of 2 or 3 C deg. The effect of this variation upon T_a was determined experimentally and corrections were made to the observed value of the isothermals so that the room temperature could be always considered constant at $-23°C$ in this series of experiments. Measurements were made with four different values of T_w, namely, $+5°C$, $+10°C$, $+15°C$, and $+20°C$. The isothermals with T_w at $+5°C$ are shown in Fig. 290(*a*). In that figure one may clearly see that the warm air containing excess water vapor goes up through the inner cylinder and is cooled gradually as it ascends from the outlet. The rate of effusion is smaller than in other cases when T_w is higher, which will be clearly seen from the more gentle gradient of the temperature distribution in this case. The accumulation of warm air in the upper part of the apparatus is shown by the existence of a high-temperature region *H*, which is always observable at any temperature of the water. The isothermals with T_w at $+15°C$ and $+20°C$ are shown in Figs. 290(*b*) and (*c*), respectively. When T_w is $+15°C$ the trend of the isothermals is similar to those when T_w is $+5°C$. When the temperature of the water is raised to $+20°C$, the accumulation of warm air in the upper part of the apparatus becomes more conspicuous and there appears a low-temperature region just below the high-temperature one, as shown by *L* in the figure. This phenomenon can be explained by assuming that under these conditions the convection of air becomes so violent that some eddies occur in this part of the apparatus. The eddies are shown roughly by the broken-line arrows in Fig. 290(*c*).

From the isothermal lines represented in Fig. 290, it is apparent that the temperature of the air where the crystal is made, T_a, is an increasing function of T_w when T_r is kept constant. The relation between T_a and T_w is shown graphically in Fig. 291, T_r being constant at $-23°C$.

It may not be needful to say that the temperature measured by a thermometer represents the mean value of temperature at that position. The isothermals drawn in Fig. 290 show the mean value of temperature with respect to both space and time. The dimensions of the snow crystal to be manufactured are microscopic and accordingly the mode of its growth must be influenced strongly by spatial fluctuations of the conditions. The fluctuations in conditions with respect to time were found by later experiments also to have an important effect upon the form of the crystal. These fluctuations are more notable when the convection of air becomes more violent, that is, when the isothermals run more closely together.

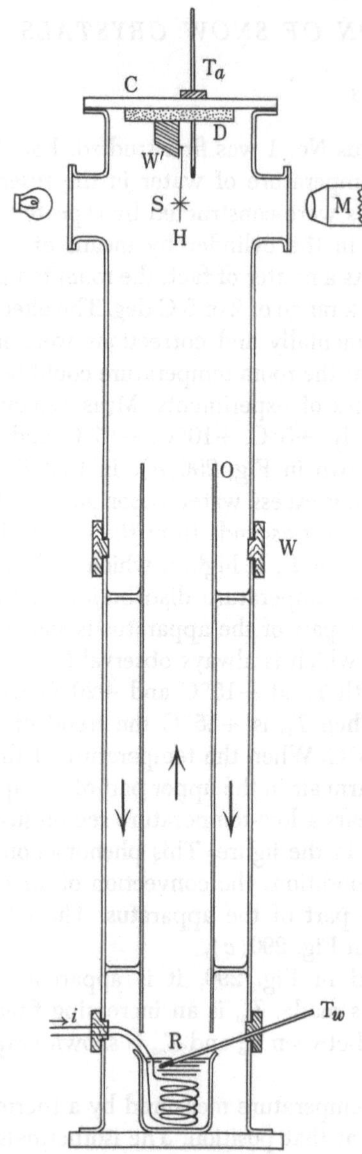

Fig. 286. Apparatus No. 1.

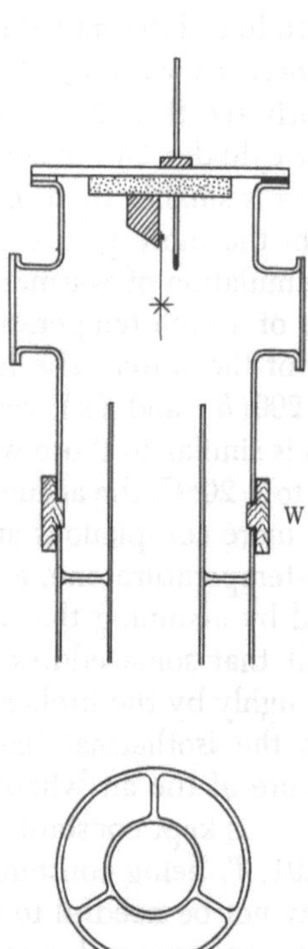

Fig. 287. Apparatus No. 2.

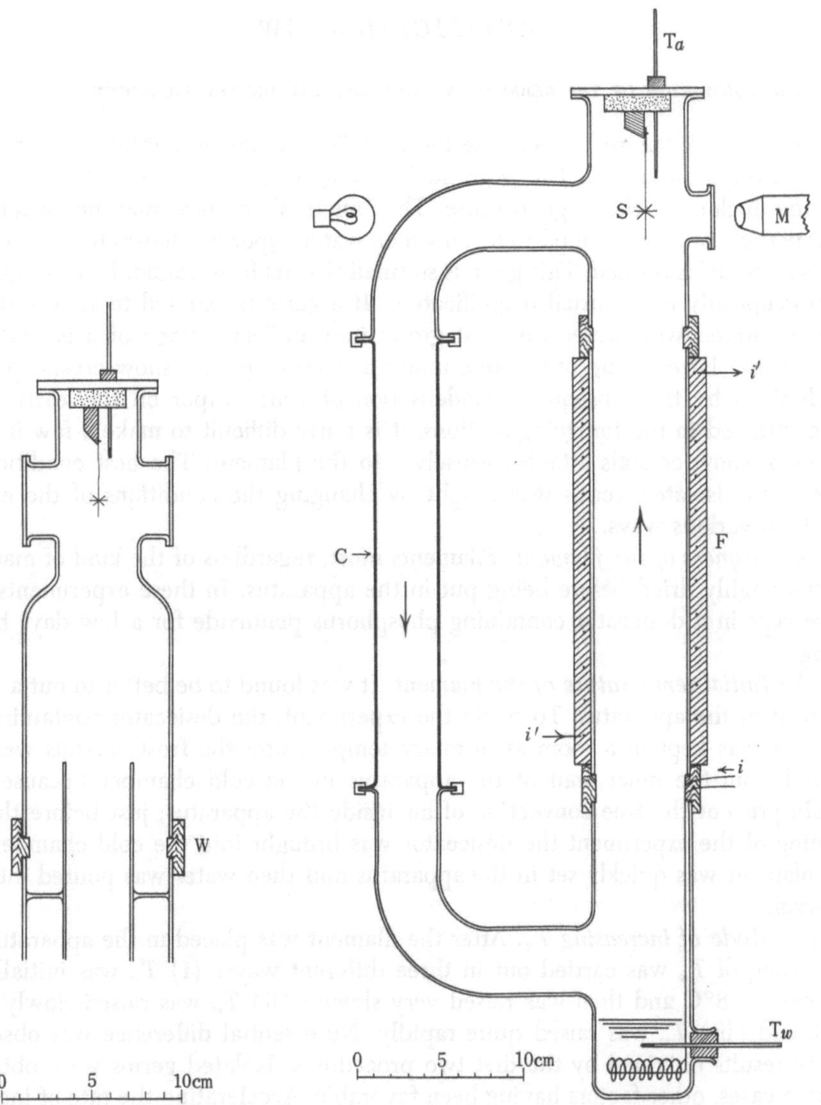

FIG. 288. Apparatus No. 3.

FIG. 289. Apparatus No. 4.

51. The formation of the germ of a snow crystal on the filament

In this book the author will use the term "germ" for describing the very early stage of the snow crystal. The word nucleus would better be reserved for designating the nucleus in the original sense. The "nucleus" of snow may be an ion or a dust particle or an aerosol particle, on which water vapor condenses by sublimation. Thus a "germ" is formed. This germ is so small that its form cannot be distinguished microscopically under usual magnification. If a germ is exposed to an atmosphere supersaturated with water vapor, it grows into an "early stage of a crystal," the form of the latter being observable under a microscope. A "snow crystal proper" is obtained by the subsequent condensation of water vapor on this early stage. As mentioned in the foregoing sections, it is fairly difficult to make a few isolated germs of snow crystals attach themselves to the filament. The best condition for getting the isolated germs was sought by changing the conditions of the experiments in various ways.

(*a*) *Dryness of the filament.* Filaments must, regardless of the kind of material, be thoroughly dried before being put in the apparatus. In these experiments they were kept in a desiccator containing phosphorus pentoxide for a few days before using.

(*b*) *Initial temperature of the filament.* It was found to be better to put a warm filament in the apparatus. To begin the experiment, the desiccator containing the filament was kept in a room at ordinary temperature; the frost crystals were removed from the inner wall of the apparatus in the cold chamber because they would prevent the free convection of air inside the apparatus; just before the beginning of the experiment the desiccator was brought into the cold chamber and the filament was quickly set in the apparatus and then water was poured into the reservoir.

(*c*) *Mode of increasing T_w.* After the filament was placed in the apparatus the increasing of T_w was carried out in three different ways: (i) T_w was initially set at nearly +8°C and then was raised very slowly; (ii) T_w was raised slowly from 0°C; and (iii) T_w was raised quite rapidly. No essential difference was observed in the results obtained by the first two procedures. Isolated germs were obtained in both cases, other factors having been favorable. Accelerating the rate of increase of T_w, more germs appear on the filament, in 1 cm of which six or seven germs are usually obtained. When T_w is increased very rapidly, the filament is always covered with numerous germs, and no artificial snow crystals could be made to develop. In the light of these results, we chose the process for executing the experiments as follows: after setting the filament, the temperature of water T_w, which was initially between 0°C and +7°C, is raised very slowly; waiting till a few germs appear on the filament, T_w is then raised rather quickly to the desired value corresponding to the type of crystal to be produced.

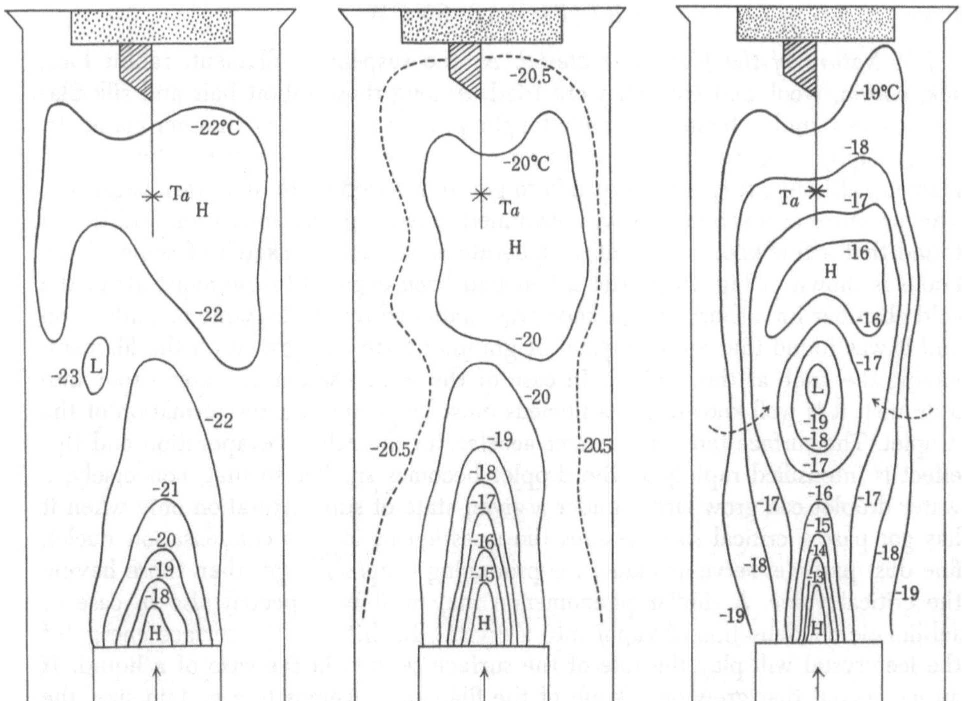

Fig. 290. Convection of vapor in the apparatus, $T_r = -23$°C: (a) $T_w = +5$°C; (b) $T_w = +15$°C; (c) $T_w = +20$°C.

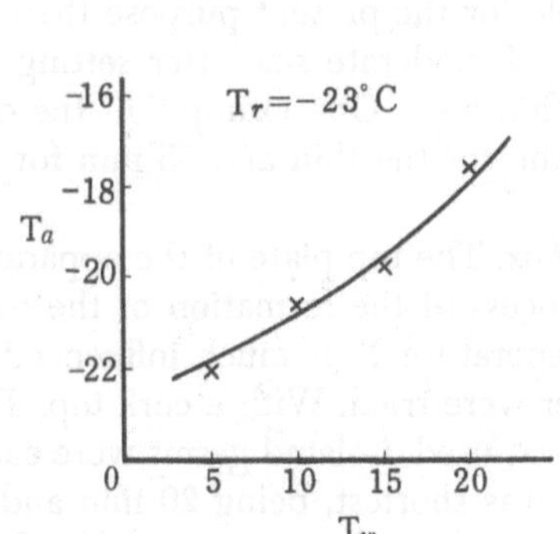

Fig. 291. Relation between T_a and T_w.

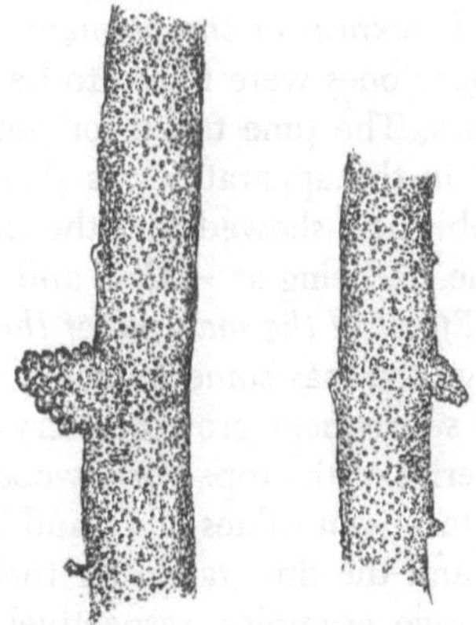

Fig. 292. Structure of rabbit hair.

(*d*) *Nature of the filament material.* As the suspension filament, rabbit hair, silk, cotton, wool, and cobweb were tried. Among these rabbit hair and silk filament were found to be most suitable for the present purpose. With other filaments, the generally observed tendency was for many germs to attach themselves on the filament, although it is not impossible to get an isolated germ upon rare occasions. The structure of a rabbit hair was examined under high magnification and it was found that a few knobs occur at a moderate separation. A sketch of some of the knobs is shown in Fig. 292. After a hair had been exposed to the moist air in the cold chamber for a short time, a knob was examined under the same magnification and it was found that an ice crystal of granular form had grown on the filament, having the knob as the nucleus. In case of the condensation of water vapor into a droplet, it is well known that a nucleus must be present for the formation of the droplet. The surface tension of water accelerates the rate of evaporation and this effect is intensified rapidly as the droplet becomes smaller so that, conversely, a water droplet can grow larger under a given state of supersaturation only when it has got past a critical size. Besides the question of ions as condensation nuclei, fine dust particles serve as nuclei for producing droplets larger than those having the critical value. A similar phenomenon may well be expected also in case of sublimatic condensation of vapor into a crystal. In this case the surface energy of the ice crystal will play the role of the surface tension in the case of a liquid. If an ice crystal that grew on a knob of the filament develops to a certain size, the greater part of the vapor afterward conveyed from the reservoir will condense onto this crystal, the rest of the filament being kept free of ice crystals. Later on, it was found that a silk filament which had previously been exposed to the vapor of boiling paraffin for a short time was suitable for getting isolated germs. The silk filament is convenient to handle under the microscope, since it is very flexible.

(*e*) *Thickness of the filament.* Among various filaments of the same material, the thinner ones were found to be more suitable for the present purpose than the thick ones. The time taken for getting a germ of moderate size after setting the filament in the apparatus was shorter for the thin ones. One example in the case of a rabbit hair showed that the time was 35 min for the thin and 55 min for the thick one, T_a being at $-20°C$ and T_w at $+13°C$.

(*f*) *Effect of the material of the top of the box.* The top plate of the apparatus, *D* in Fig. 286, has some influence upon the process of the formation of the germ and the subsequent growth of crystal. The temperature T_a is much influenced by the material of the top. Cork, wood, and copper were tried. With a cork top, T_a is highest for given values of T_w and T_r. When it was used, isolated germs were easily formed and the time taken for their formation was shortest, being 20 min and 35 min for two examples, respectively. In promoting the subsequent growth of the snow crystal the cork ceiling is not always favorable, since T_a sometimes becomes too high for many sorts of crystals except a fernlike one, for the production of which type the cork top is the most suitable. With a copper top T_a is lowest and the time taken in the formation of nuclei is longest, lying between 1 and 2 hr. The effect of a wooden top is intermediate between the two extremes.

52. The early stage of artificial snow crystals

As described in the foregoing section, the hair is set in the apparatus and the water temperature is raised gradually; after half an hour or so, a germ of snow is seen to appear on the filament. At first the form of this germ is not clearly seen, but before long it develops into a defined form, the early stage of a snow crystal, observable under a microscope with the usual magnification. The form of the early stage is not simple. Almost all types of snow crystals are seen among their early stages, and moreover some rare types were found which are not often to be seen in natural snow. Twelve types of them are shown schematically in Fig. 293, and photographs of them are reproduced in Fig. 294. The conditions under which the early stage of the crystals are produced are summarized in Table 17.

The twelve types of these early stages can be roughly classified into three groups.

I. *Crystals that are rather quickly formed.* The small dendritic plane form (1, Fig. 293), the irregular form (2), and frozen droplets (3) are formed rather quickly. As a usual thing they develop to the stages shown in the photographs of Fig. 294 within half an hour. It was often observed that under exactly similar conditions the crystal sometimes develops into a dendritic form and in other cases grows into a spatial assemblage of sectors. In the former case the crystal develops quickly and in the latter slowly. The difference may be caused by the difference in the form of the germ, which itself cannot be observed microscopically under usual magnification.

II. *Crystals that are formed by an intermediate process.* The spatial assemblage of sectors (4) and the thin hexagonal plate (5) are usually formed within 2 hr. The small hexagonal plate seen in Fig. 294(5) is a crystal of type 5 and the larger flake in the same photograph seems to be a crystal developed from this early stage. The subsequent growth of the snow crystal is influenced by the form of the crystal in its early stage. This point will be discussed in detail in Sec. 54.

III. *Crystals that are slowly formed.* The cylinder with end plates (6), bullets (7), thick hexagonal plates (8 and 9), the cup crystal (10), the skeleton form of prism (11), and the solid needle (12) belong to this group. It takes more than 5 hr for the development to the stages shown in the photographs of Fig. 294. The time t in Table 17 the interval between the moment of setting the hair in the apparatus and that of taking the photograph. Most of the values of t exceed 15 hr, but this value has no essential importance, because in much less time the crystal

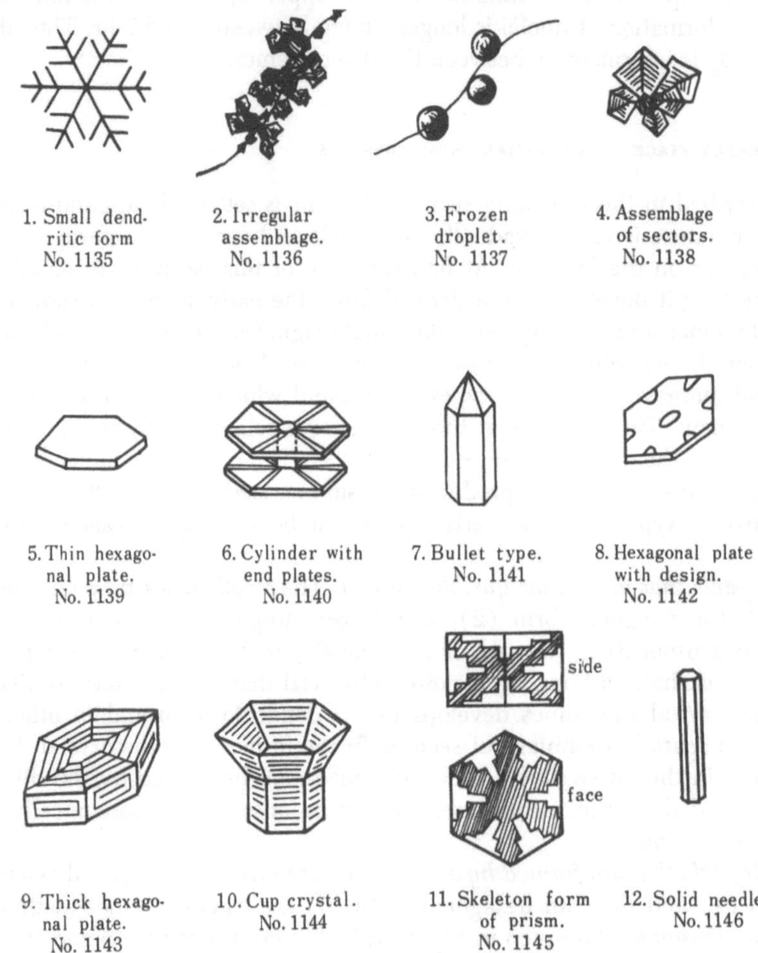

Fig. 293. Schematic diagrams of initial stage.

develops into a similar form only slightly smaller in size than that shown in photograph.

The crystals numbered 8, 9, 10, and 11 in Fig. 293 belong to the same class. The first three are all modifications of the skeleton form of hexagonal prism shown in 11, in which the shaded parts represent the space filled up with ice. The design in the hexagonal plate (8) is due to the cavities, which correspond to the stepped hollows in case of the thick plate (9). These stepped hollows develop to their full extent in 11, causing the crystal to assume a typical skeleton form. When the parts filled with ice become very thin the crystal grows in the form of a cup (10).

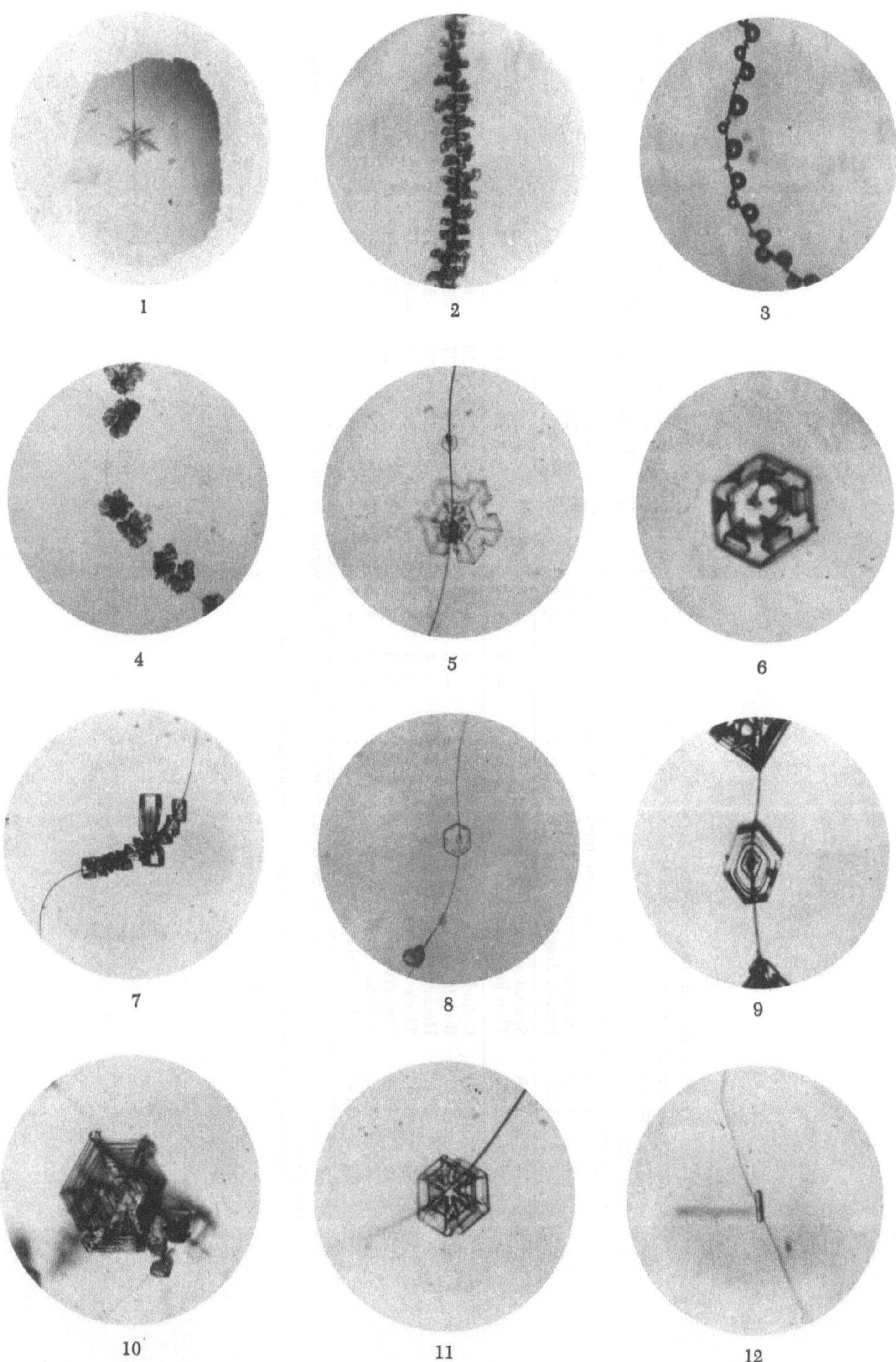

FIG. 294. Initial stages of artificial snow crystals: (1) [1135]; (2) [1136] (× 20); (3) [1137] (× 16); (4) [1138] (× 7.5); (5) [1139] (× 18); (6) [1140] (× 27); (7) [1141] (× 24); (8) [1142] (× 21); (9) [1143] (× 40.5); (10) [1144]; (11) [1145] (× 34); (12) [1146] (× 25.5).

TABLE 17. Formation of early stages of snow crystals.

No. in Fig. 293	Photo. No.	Form of early stage	App. No.	T_w (°C)			T_a (°C)			t *	Rate of growth
				Initial	Middle	Last	Initial	Middle	Last		
1	1135	Small dendritic plane form	Fig. 252	+5	+10	+15	−19	−17	−18	30 min	high
2	1136	Irregular assemblage of plates	1	+3	+35	+20	−20	−13	−15	30 min	high
3	1137	Frozen droplets	1	+6	+46	+25	−22	−10	−12	30 min	high
4	1138	Spatial assemblage of sectors	1	+10	+16	+19	−15	−13	−12	1.5 hr	medium
5	1139	Thin hexagonal plate	1	+6	+10	+12	−24	−22	−16	2 hr	medium
6	1140	Cylinder with end plates	1	0	+8	+20	−20	−20	−14	5.5 hr	low
7	1141	Bullets	4	+3	+5	+10	−29	−27	−25	7 hr	low
8	1142	Hexagonal plate with design	4	+4	+4.5	+4.5	−26	−27	−28	15 hr	low
9	1143	Thick hexagonal plate	4	+2	+4	+1	−38	−36	−39	18 hr	low
10	1144	Cup crystal	1	+5	+4.5	+4	−21	−21	−19	17 hr	low
11	1145	Skeleton form of prism	4	+6	+8	+10	−22	−22	−20	20 hr	low
12	1146	Solid needle without structure	4 †	+4	+4	+4	−30	−30	−28	25 hr	low

* The time taken for the development of the crystal to the stage shown in the photograph, measured from the moment the hair was exposed to the supersaturated atmosphere.

† Apparatus No. 4 with a stop diaphragm for the circulation of air current.

The cylinder with end plates is usually composed of a hexagonal cylinder with two plates attached one to each end, but sometimes a curious form is obtained in case of artificial snow. One example is shown in Fig. 294(6) and similar ones in Figs. 377 and 405. In these cases the cylinder is a solid needle of ice without marked structure. This sort of needle can be obtained by itself as shown in Fig. 294(12). It is formed at a lower temperature with a low rate of supply of water vapor.

53. Experimental procedure in making artificial snow

The procedure in making artificial snow is as follows. The apparatus is set in the cold chamber and left till it cools down nearly to the temperature of the room. Then a few rabbit-hair or silk filaments, which have previously been thoroughly dried, are mounted in the apparatus and at the same time some water is poured into the reservoir. The temperature of the water T_w is raised slowly. One must wait till a few germs appear on the filament, then T_w is adjusted to the required value corresponding to the form of crystal to be manufactured. Three values of temperature, namely, T_r, temperature in the cold chamber, T_a, and T_w, are recorded from time to time. They are plotted in a graph as functions of time and this graph is taken to represent the history of the development of a snow crystal. Three examples of these graphs are shown in Figs. 295–297.

Figure 295(*a*) shows the course of the process by which a fernlike plane crystal is formed. The photograph of the crystal produced is represented in Fig. 295(*b*). When the rabbit hair was set in the apparatus, T_w stood at +8°C. It was raised gradually to +10°C within about 1 hr, and a crystal germ was observed on the hair. Then the heating current was increased so that T_w was raised to +15°C in 35 min. The room temperature T_r was kept at nearly −30°C till the appearance of the germ, after which it was raised to −27°C. The example graphed in Fig. 295(*a*) shows that a fernlike plane crystal grows very rapidly, compared with the other sorts of snow. In this case the rate of growth amounted to 11 mm/hour. In this book the rate of growth of a crystal means the increase in 1 hr of the main dimension, which is taken to be the diameter of a circle or a sphere enclosing the crystal.

Figure 296(*a*) shows the course of the formation of a plane dendritic crystal of broad-branched type. The condition for the formation of germs is similar to the former case. After the germs were observed on the filament, T_w was raised from +8°C to +16°C in 80 min. The mean rate of growth was about 1.2 mm/hour in this case. The crystal thus obtained is shown in Fig. 296(*b*).

Figure 297(*a*) shows the course of the formation of a crystal consisting of a spatial assemblage of sectors. In this case T_w was increased at a higher rate than in the previous two cases and consequently, as described in Sec. 51(*c*), several

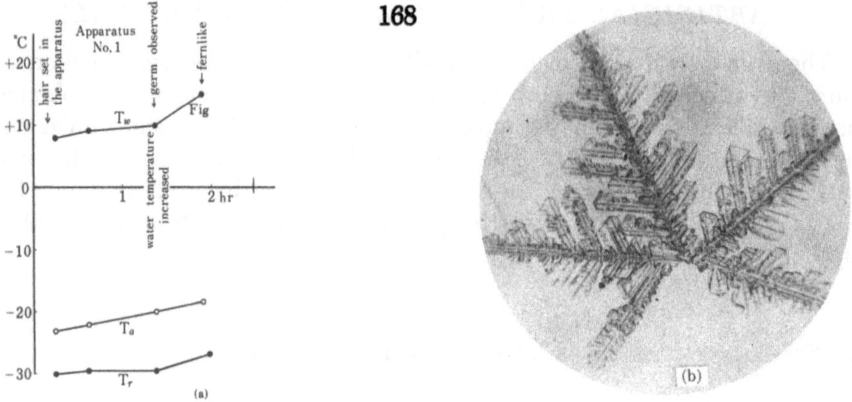

FIG. 295. (*a*) Conditions of formation (I) and (*b*) crystal form.

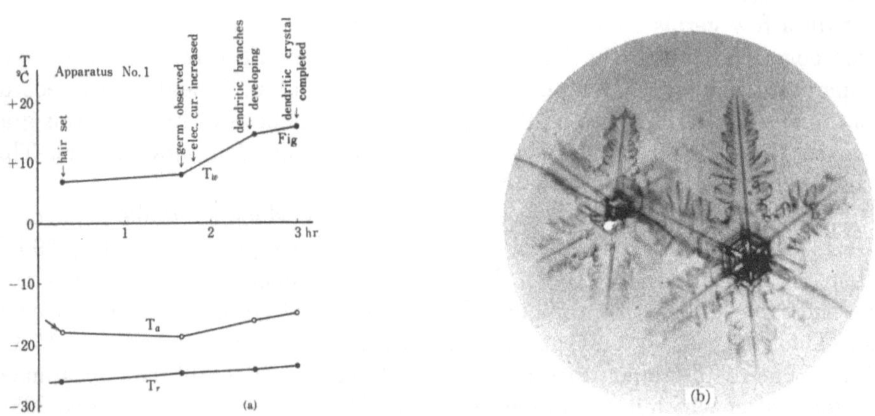

FIG. 296. (*a*) Conditions of formation (II) and (*b*) crystal form.

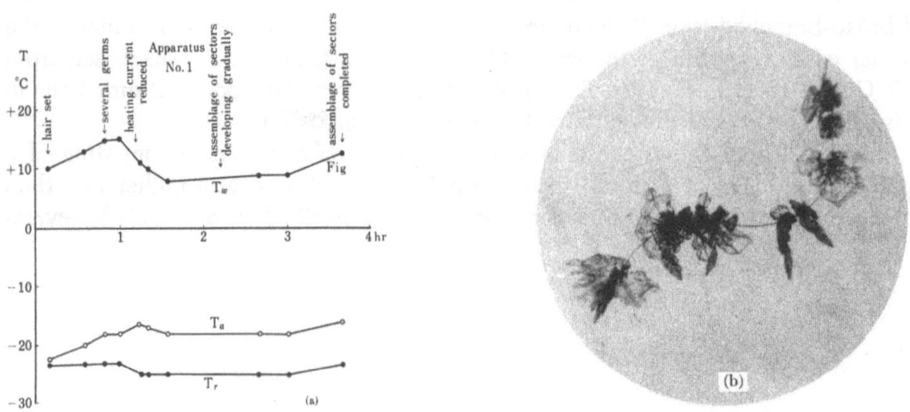

FIG. 297. (*a*) Conditions of formation (III) and (*b*) crystal form.

germs formed on the filament. Then T_w was reduced to nearly +9°C and kept there for 2 hr. The snow crystals thus manufactured developed into a spatial assemblage of sectors, the photograph of which is shown in Fig. 297(*b*). The rate of growth is less in this case than that for the broad branches, being on the average 0.6 mm/hr. In the above three experiments apparatus No. 1 was used throughout.

It is generally accepted and it is true that the form of a snow crystal is determined chiefly by the degree of supersaturation of the surrounding atmosphere during the course of its formation. From the physical point of view the degree of supersaturation is one of the most important factors governing the rate of growth of a crystal. This rate of growth in turn is itself one of the factors controlling the variation of crystal form. Besides the degree of supersaturation there is another important factor, namely, the temperature of the air where the crystal is formed. It is better to take the rate of growth directly, rather than the degree of supersaturation, as the parameter for describing the variation of crystal form.

54. Form of the early stage and its influence upon the subsequent growth

It can reasonably be expected that the form of snow is strongly influenced by the form of the crystal in its early stage. In natural snow three sorts of dendritic crystals are observed with similar frequency in Hokkaido. They are the plane dendritic one, the spatial assemblage of dendritic branches with a stellar base, and the spatial assemblage of radiating type. It is quite natural to consider that the first sort develops from a small dendritic crystal, when the subsequent conditions are favorable for the dendritic growth. This was verified experimentally. The plane dendritic crystal shown in Fig. 299 developed from the small dendritic crystal seen in Fig. 298. Another example is shown in Nos. 1149 and 1150, Plate 169. Other similar experiments led us to conclude that a plane dendritic crystal must originate from the early stage of a small dendritic one.

The second sort, the spatial hexagonal type, is probably produced by the attachment of other nuclei to the various points of the branches of the plane dendritic one, as described before.

The third sort, the radiating type, has two kinds, one having a stone at the central region and the other lacking such a stone. The stone observable at the center consists of an assemblage of prisms, as has been described in detail in Sec. 13. The latter type was found by the present experiment to have been developed from a spatial assemblage of sectors. One example of the course of formation of this type is shown in Figs. 300 and 301. In this case T_w was kept rather low and germs were obtained two hours after the beginning of the experiment. The early stage of the crystal developed from this germ took the form of a spatial assemblage of sectors, as reproduced in Fig. 300. After the photograph was taken the hair

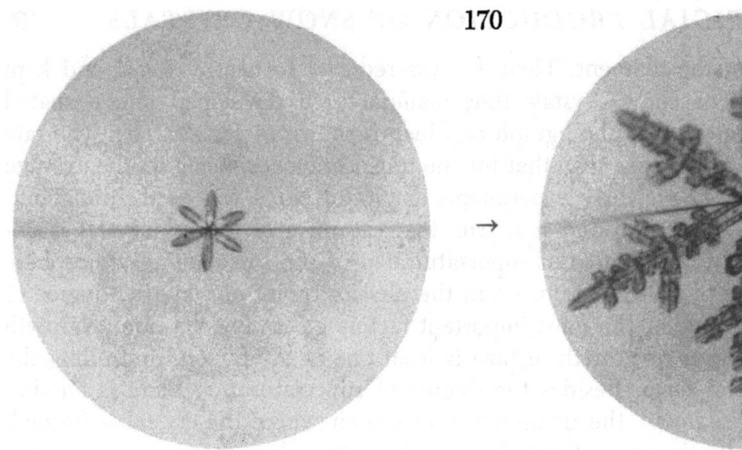

→

FIG. 298 [1147]. Initial stage, small plane dendritic crystal (× 18).

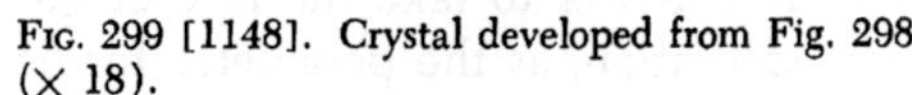

FIG. 299 [1148]. Crystal developed from Fig. 298 (× 18).

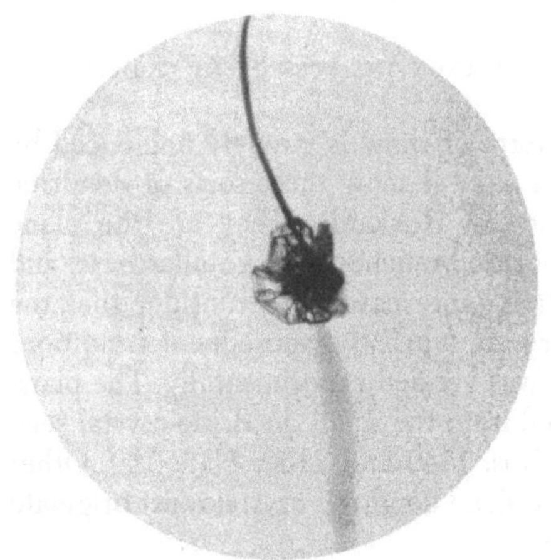

→

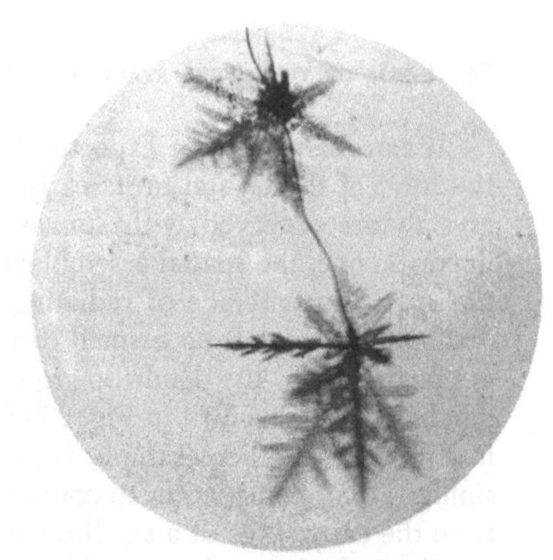

FIG. 300 [1151]. Initial stage, assemblage of sectors (× 25).

FIG. 301 [1152]. Crystal developed from Fig. 300 (× 9.1).

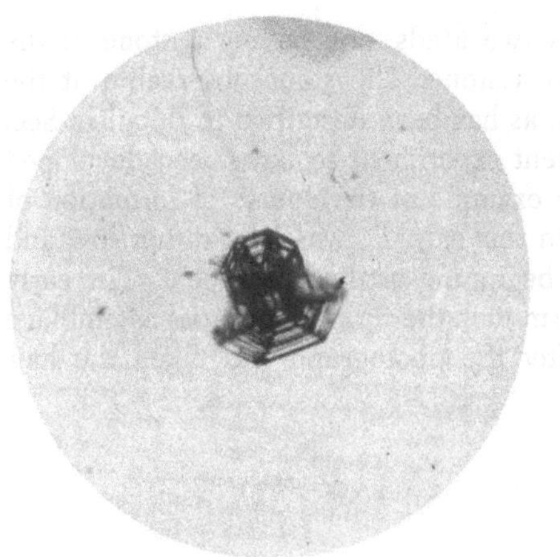

→

FIG. 302 [1157]. Initial stage, cup (× 18.5).

FIG. 303 [1158]. Crystal developed from Fig. 302 (× 18.5).

was again set in the apparatus and the experiment was continued by increasing T_w rapidly so that dendritic extensions were expected. After 1 hr the crystals developed into a spatial assemblage of radiating type, as shown in Fig. 301. The crystal developed from the germ reproduced in Fig. 300 is the upper one in Fig. 301.

Numbers 1153 and 1154, Plate 169, show irregular crystals developed from frozen droplets. In this case T_w was suddenly raised by pouring hot water into the reservoir. Many frozen droplets were obtained on the filament. The crystals developed from these frozen droplets assumed the form of irregular assemblages of sectors, as shown in No. 1154. Some irregular particles observed in natural snow may have developed from frozen droplets; such a phenomenon is likely to occur when a marked temperature inversion takes place in the atmosphere.

As an example of the crystal made by a slow process, thick hexagonal plates are shown in No. 1155, Plate 169. The crystal of apparently cylindrical form that is seen in the photograph next above the hexagonal plate is the side view of a similar plate. They were obtained in the old apparatus of Fig. 279 by keeping T_w between +18°C and +11°C for 16 hr. Then T_w was raised to nearly +30°C. Plane branches began to extend out from the edges of the plate and after 2 hr the crystal developed into the state shown in No. 1156. The side view shows that two plane crystals attach themselves to each face of the plate.

Figure 302 is an example of a cup crystal. This was made in apparatus No. 1 by a slow process. The temperature T_w was kept at nearly +5°C for 17 hr. After the photograph was taken, the hair was reset in the apparatus and T_w was gradually raised to +21°C within 2 hr. Broad plane branches extended from the edges of the hexagonal cup and the crystal developed into the form shown in Fig. 303.

From the examples above described one may clearly see the influence of the form of the early stage upon the subsequent growth of a crystal.

55. Convection of Air and Form of Crystal

In the foregoing discussions, the degree of supersaturation was represented by the supersaturation ratio s. This quantity can be taken as a measure for determining the rate of supply of water vapor, but it will not represent strictly the degree of supersaturation at the spot where the crystal is made, because the entire amount of water vapor evaporated from the reservoir is not conveyed to the place of crystal formation. Some portion of the vapor will condense as frost crystals on the inside of the wall. The form of the upper part of the apparatus will modify the manner of convection and consequently modify the rate of supply of water vapor to the crystal being produced. Experiments were carried out for the purpose of clarifying these points.

(*a*) *Effect of frost crystals adhering to the inside of the wall.* If a single apparatus is used continually for a few series of experiments, the lower part of the inside of the glass wall becomes covered with frost crystals. It was found that these frost crystals growing in clusters prevent, to a considerable extent, the free circulation of an upward air current. For getting a certain type of crystal it was necessary to increase T_w much higher than the usual value when no frost was observed on the wall.

In one example when many frost crystals grew in clusters on the inside of the wall, a sector form was obtained in apparatus No. 1 by keeping T_w at nearly +30°C for 6 hr, which is extraordinarily high. The mean rate of growth of the crystal amounted only to 0.3 mm/hr. In another experiment, using the same apparatus after it had been thoroughly defrosted, a dendritic crystal as shown in Fig. 429(*a*) was obtained. In this case it took only 1 hr to obtain this form, T_w being kept at nearly +7°C. The rate of growth was 3.5 mm/hr.

From these experiments the influence of frost crystals in hindering the convection of water vapor was clearly understood. In the later experiments, therefore, the cylinders were entirely defrosted each time just before a series of experiments was started.

(*b*) *Form of the space where the crystal is made.* The form of the upper part of the apparatus where the artificial snow is made has a considerable influence upon the form of the crystal produced. In apparatus No. 2, which is similar in construction to No. 1 but shorter, T_a is higher than that in apparatus No. 1 for the same T_r and T_w, because the hair is much nearer the air outlet (cf. Figs. 286 and 287). This effect will be clearly understood from the distribution of the isothermal lines of T_a, which were measured in apparatus No. 1 and are shown in Fig. 290.

In apparatus No. 3 the convection of water vapor in the top portion seems to be weakened, this part having been made slender as shown in Fig. 288. In order to attain the same rate of supply of water vapor in No. 3 as in No. 1, T_w must be raised. The crystal shown in Fig. 305 was made in apparatus No. 3. The course of the formation is shown by solid lines in Fig. 304. After the appearance of the germ, T_w was raised from +6°C to +18°C within about 2 hr. With this value of T_w, the crystal would be developed into a dendritic form if apparatus No. 1 were used. In the present case, however, the crystal grew into a hexagonal plate (Fig. 305) with extensions of broad branches. The mean rate of growth was 0.9 mm/hr. Figure 306 is shown for comparison. In this case apparatus No. 1 was used and the crystal developed into a beautiful fernlike type. The course of the formation is shown by broken lines in Fig. 304. The values of T_r and T_a were more or less the same as the former case, and T_w was kept at nearly +12°C after the formation of the germ, which is decidedly lower than the former case. The rate of growth was about 3.5 mm/hr.

This difference must be assigned to the difference in form of the apparatus where the crystals are made, which hinders or facilitates the convection of water

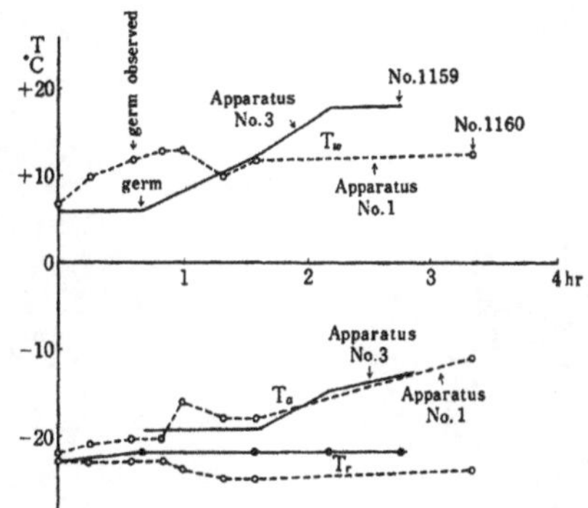

Fig. 304. Conditions of formation of two crystals.

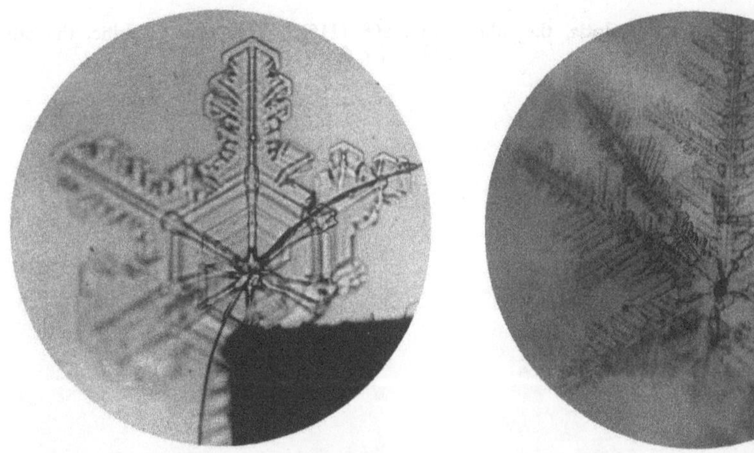

Fig. 305 [1159]. Crystal made in apparatus No. 3 (× 30).

Fig. 306 [1160]. Crystal made in apparatus No. 1 (× 11).

vapor. In later discussion, therefore, the results obtained with the same apparatus must be studied as one group and those with the other apparatus as another group.

(*c*) *Nonhomogeneity of convection current.* The nonhomogeneity of the convection current will be seen from the isothermals shown in Fig. 290. When several filaments are set at different positions in an apparatus, the crystals attached to the different filaments are usually not the same in form. It is quite natural that this sort of nonhomogeneity should take place in the apparatus.

During the course of the experiments it was found that the nonhomogeneity of the convection current can occur also on a microscopic scale, which is seen from the form of some peculiar crystals. Two examples of such crystals are reproduced

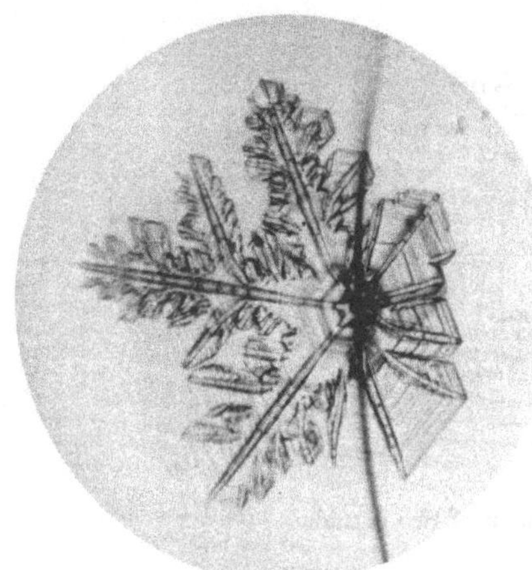

FIG. 307 [1162]. One side dendritic, the other sector.

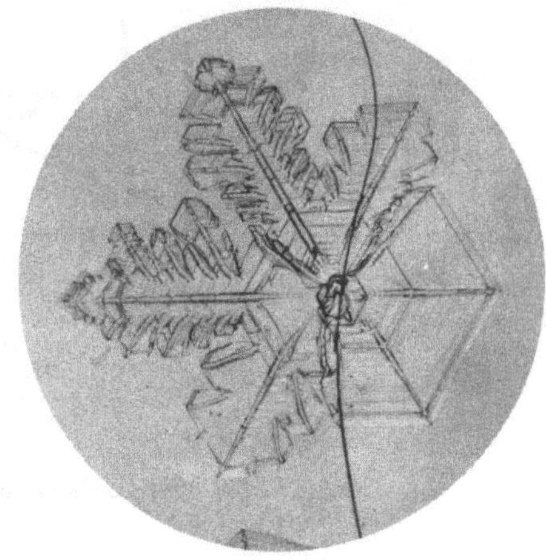

FIG. 308 [1161]. One side dendritic, the other plate (× 21.5).

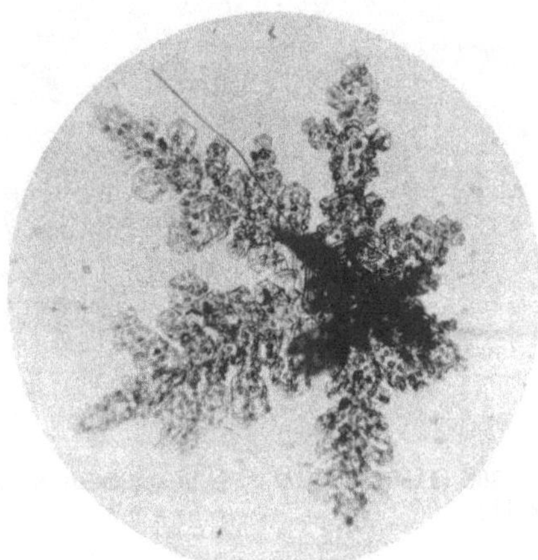

FIG. 309 [1169]. Crystal made with ample supply of vapor (× 21).

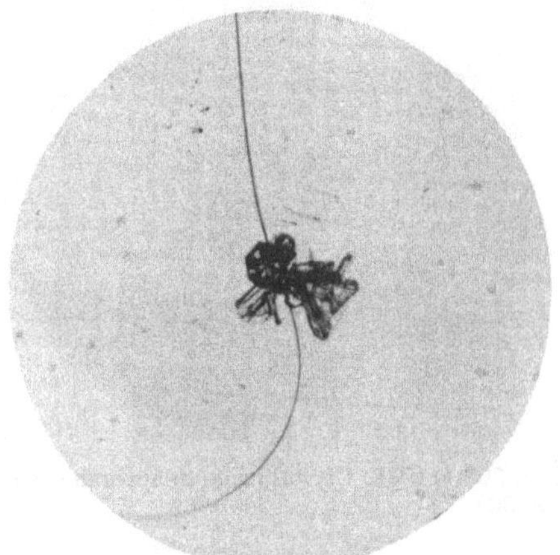

FIG. 310 [1170]. Crystal made with scanty supply of vapor (× 20.5).

in Figs. 307 and 308. The crystal shown in Fig. 307 is composed of dendritic branches on one side and sectors on the other. This crystal was made in apparatus No. 3 under the conditions that T_a was nearly $-22°C$ and T_w increasing from $+5°C$ to $+12°C$ within 3 hr. The other example, Fig. 308, is composed of dendritic branches and a half portion of a hexagonal plate. This crystal was made in apparatus No. 4 with T_a at $-16°C$ and T_w gradually increasing from $+11°C$ to $+13°C$ within 4 hr. Other examples are shown in Nos. 1163–1168, Plate 170. In nature,

TABLE 18. Formation of crystals when stop diaphragms were used.

Diameter of aperture (cm)	Time for germ formation (hr)	Temperatures when germ is formed		Mean of temperature during subsequent growth		Time needed for subsequent growth (hr)	Form of crystal produced	Photo. No.
		T_w (°C)	T_a (°C)	T_w (°C)	T_a (°C)			
3	1	+ 6	−22	+ 8	−21	3	spatial dendritic with droplets	
2	0.5	+12	−22	+10	−20	3	spatial dendritic with droplets	
1	1	+16	−24	+30	−22	5	spatial dendritic with droplets	
—	3	+14	−22	+21	−20	2.5	spatial assemblage of plates	
—	22	+18	−23	+13	−23	20	irregular assemblage of small plates	
0.6	1.5	+21	−23	+25	−22	3.5	plane dendritic with droplets (Fig. 309)	1169
—	24	+18	−24	+20	−24	17	irregular assemblage of small plates (Fig. 310)	1170

this sort of crystal is not formed because of the rotation of the crystal while it is falling.

(*d*) *Effect of introducing a stop diaphragm upon the convection of water vapor.* In order to clarify the effect of the rate of supply of water vapor to the spot where the crystal is made, stop diaphragms of various apertures were introduced in apparatus No. 1. The diaphragm was made of a brass plate having a circular hole in it. It was placed on the top of the inner glass cylinder, at *O* in Fig. 286. Experiments were carried out with four sorts of diaphragms, the diameter of the circular apertures being 3 cm, 2 cm, 1 cm, and 0.6 cm, respectively. The results are summarized in Table 18.

From the data in the table one will see that the smaller the diameter of the aperture, the longer the time for the germ formation becomes. Additional interesting phenomena were observed in the subsequent growth of the crystal following the formation of the germ. When the stop diaphragm was used, T_w was, as expected, much higher than when the diaphragm was not used, for obtaining a similar type of crystal.

When the aperture is 1 cm in diameter, T_w must be raised to +30°C in order to get a dendritic crystal. It was found that dendritic crystals thus produced had many water droplets attached. These droplets may be explained as fog particles that condensed from the supersaturated air current when the latter was cooled down by passing the cold diaphragm. This sort of crystal, the spatial assemblage of dendritic branches with droplets, was obtained with T_w at +10°C when the aperture was 2 cm in diameter, though it had to be as high as +30°C when the diameter was 1 cm. With the aperture 3 cm in diameter, the corresponding value of T_w was +8°C.

Upon decreasing T_w to +21°C when the aperture was 1 cm, the crystal became a spatial assemblage of plates. If T_w was still further reduced, to +13°C, an irregular assemblage of small plates was obtained. A similar effect was observed when the aperture was 0.6 cm. Two examples are shown in Figs. 309 and 310. Figure 309 is a plane dendritic crystal with water droplets and Fig. 310 an irregular assemblage of small plates. The former was obtained with T_w at +25°C and T_a at −22°C, and the latter with T_w at +20°C and T_a at −24°C.

It will be clearly seen from the results of the experiments above described that the increase of the aperture of the diaphragm has a similar effect to the increase in T_w.

CHAPTER 6

Investigations on Artificial Snow

56. Fernlike and dendritic crystals

The development of detailed technique for experimentation as described in the preceding chapter made it possible for us to produce nearly all types of snow crystals almost at will. To restate it briefly, the procedures are as follows:

First, a thoroughly dried apparatus is put in a thermostat placed in the cold chamber and left until it is cooled down to about −20°C. Meanwhile, the reservoir, which has been removed from the apparatus, is left outside the thermostat, and the temperature of the water in it is maintained at about +12°C. When the above temperatures are obtained, a rabbit hair that has been thoroughly dried in a desiccator containing phosphorus pentoxide placed in the laboratory room of normal temperature, is carried into the cold chamber in a covered glass dish. As soon as the reservoir is fixed to the bottom of the apparatus, the hair is set on the inside of the lid of apparatus. When T_a and T_w are near the above-mentioned values, germs begin to develop in many cases sporadically on the hair with 15 to 30 min after the hair is set. Under this condition, the crystal tends to become the plane hexagonal type in its early stage of development.

To obtain the fernlike type or the ordinary dendritic crystal which is regarded as the representative of snow crystals, the temperatures of the thermostat and of the water in the reservoir are controlled so that T_a becomes −15°C or −16°C and T_w rises above +15°C. In this case the crystal develops into a beautiful dendritic type. Examples of the fernlike and ordinary dendritic crystals thus obtained are shown in Figs. 312–323. Some of them were obtained without using the thermostat.

In our experiments thus far, for getting a beautiful dendritic crystal it was technically necessary, as explained in Sec. 53, to raise T_w gradually after the appearance of germ.

The beautiful fernlike crystal reproduced in Fig. 295(*b*) was formed, as under-

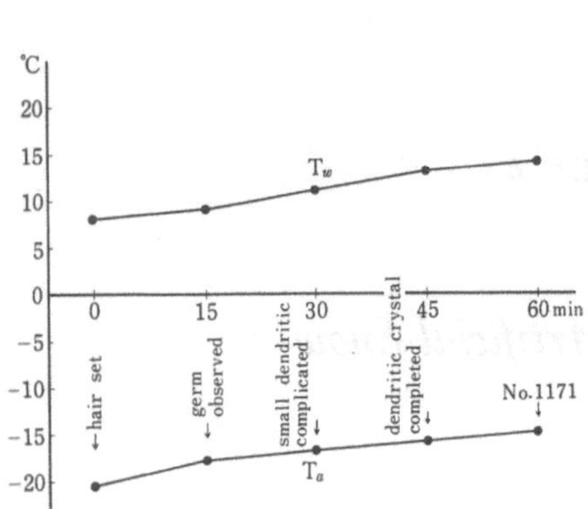

FIG. 311. The course of formation of a fernlike crystal.

FIG. 312 [1171]. Fernlike crystal produced by the course shown in Fig. 311.

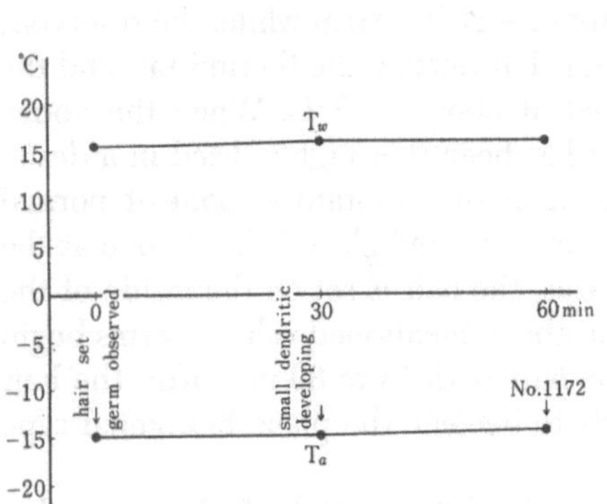

FIG. 313. The course of formation of the crystal of Fig. 314.

FIG. 314 [1172]. Fernlike crystal produced by keeping T_a and T_w constant (× 24).

stood from the course curve shown in Fig. 295(*a*), when the water temperature was gradually raised. Another example of such a crystal, reproduced in Fig. 312, also required the gradual raising of T_w, as shown in Fig. 311. It is naturally expected that a larger supply of water vapor will be necessary when the crystal grows larger. But, as stated later, T_a being a more important factor than T_w in deciding the crystal type, the dendritic crystal can be obtained by keeping T_a and T_w constant, if their values satisfy the necessary condition.

The crystal shown in Fig. 314 was obtained with T_a and T_w constant within the range of 1C deg. As seen in Fig. 313, both temperatures were constant throughout the course of formation. A gradual increase in the water temperature, however, is nevertheless advantageous for getting beautiful dendritic crystals. This experiment merely proves that it is not the indispensable factor. When the water temperature is raised, the temperature of the air where the crystal is formed goes up with it. Therefore care must be taken so that T_a may not rise beyond the range for the dendritic development.

Conditions for the formation of these crystals will be stated in detail in Sec. 79, but the data for representative ones are listed in Table 19. The conditions for the dendritic formation are: T_w, in general above $+12°C$ and T_a, between $-14°C$ and $-17°C$. In this case T_w differs for different types of apparatus, and shows nothing but the relative degree of supersaturation. The value of T_a, however, has physical significance.

In the artificial snow crystals, especially in the dendritic crystal, the branches seldom grow symmetrically. The reason is that, the crystal being suspended by a hair and the supply of water vapor coming chiefly from beneath, the lower branches are apt to grow faster. It is assumed that in natural snow symmetrical development is easier because crystals in the air keep an almost horizontal position and rotate around a vertical axis while they fall. In our experiment, as the phenomena are clarified by investigating a few completely developed branches, we were satisfied with this extent of success. Other examples of dendritic crystals obtained artificially are shown in Plates 171 and 172.

Figure 323 shows plan and side views of the same crystal. Occasionally the spatial hexagonal crystal, as in this example, is obtained, but the conditions are not clarified yet.

57. Dark-field illumination

For the purpose of showing the crystal as a white image against a dark background, we made various attempts to take photographs with dark-field illumination, using artificial snow crystals. For this purpose, the Zeiss Ultropak microscope was used. A weak light was projected from below in addition to the reflected light. Controlling the intensities of these two lights in various ways, we took photographs as shown in Figs. 324 and 325. This method having proved to be unsuitable for clarifying the minute inner structure of crystals, we decided not to adopt it for the present investigation, although this kind of illumination can effectively be used with higher aesthetic value in taking microscopic motion pictures.

Fig. 315 [1173]. Artificial snow, fernlike crystal. (× 18).

Fig. 316 [1175]. Artificial snow, fernlike crystal (× 20).

Fig. 317 [1176]. Artificial snow, fernlike crystal (× 39).

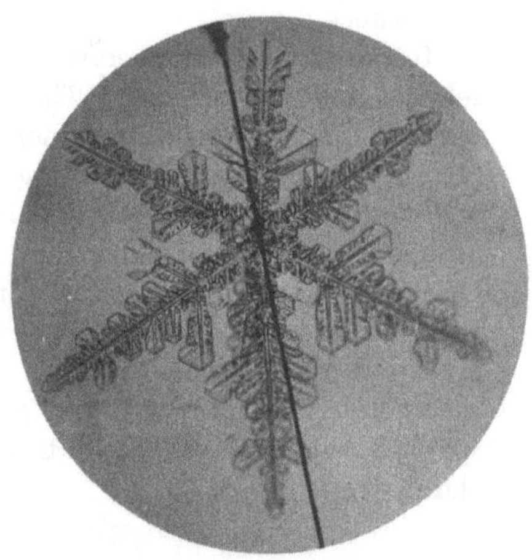

Fig. 318 [1177]. Artificial snow, fernlike crystal (× 16.5).

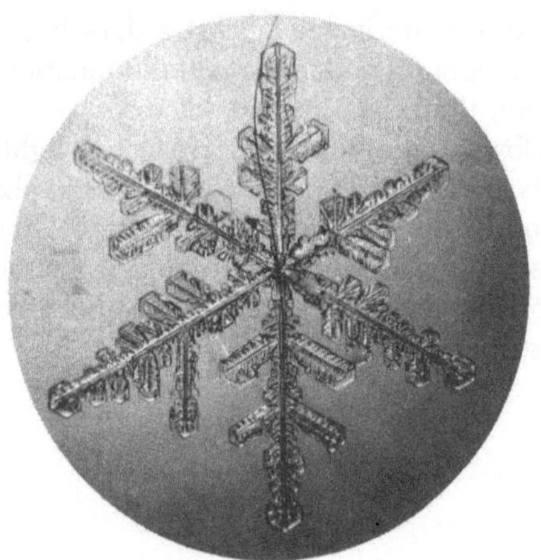

Fig. 319 [1183]. Artificial snow, dendritic crystal (× 15).

Fig. 320 [1184]. Artificial snow, dendritic crystal (× 33).

Table 19. Formation of fernlike and dendritic crystals.

Fig. No.	Photo. No.	Temp. in thermostat T_t (°C)	T_w (°C) Initial	T_w (°C) Middle	T_w (°C) Final	T_a (°C) Initial	T_a (°C) Middle	T_a (°C) Final	Diameter * (mm)	t † (min)	Rate of growth (mm/hr)
315	1173	−21	+17	+18	+18	−20	−16	−15	3.1	30	6.2
	1174	−20	+ 9	+13	+13	−16	−16	−16	3.3	45	4.4
316	1175	−20	+10	+12	+13	−16	−16	−16	3.1	30	6.2
317	1176	−22	+12	—	+17	−17	—	−16	—	15	—
318	1177	−28	+13	+15	+15	−19	−17	−17	3.5	60	3.5
	1178	−25	+11	+14	+18	−17	−16	−18	6.6	120	3.3
319	1183	−22	+16	+17	+18	−16	−16	−14	3.8	40	5.6
320	1184	−20	+ 9	+12	+14	−18	−16	−15	1.9	45	2.5
321	1189	−21	+13	+17	+10	−17	−16	−15	1.0	30	2.0
322	1190	−20	+12	+14	+16	−17	−16	−15	1.6	30	3.2

* Diameter of a circle enclosing the crystal, or twice the mean of the lengths of six branches.
† Time interval between germ formation and completion of crystal.

Table 20. Formation of radiating-type crystals.

Fig. No.	Photo. No.	Initial stage, when spatial assemblage of sectors is formed: T_a (°C)	T_w (°C)	Time (min)	Subsequent conditions for dendritic development: T_a (°C)	T_w (°C)	Time completed (min)	Mean of the lengths of branches (mm)
326	1201	−20 — −19	+12 — +14	30	−16 — −15	+14 — +23	20	2.9
327	1202	−21 — −19	+ 6 — +13	50	−19 — −16	+13 — +19	30	3.3
328	1203	−20 — −18	+ 7 — +10	15	−17 — −15	+13 — +18	30	2.6

Table 21. Formation of plate or sector with dendritic extensions.

Fig. No.	Photo. No.	Earlier stage, plate or sector formation: T_a (°C)	T_w (°C)	Dimension * (mm)	Time for growth (min)	Rate of growth (mm/hr)	Later stage, dendritic development: T_a (°C)	T_w (°C)	Dimension † (mm)	Time for growth (min)	Rate of growth (mm/hr)
	1207	−18	+ 6	0.9	30	1.8	−16	+15	1.95	20	5.8
330	1208	−18	+ 9	.8	40	1.2	−15	+16	1.27	20	3.8
	1209	−18	+ 3	.77	40	1.5	−15	+17	1.73	15	6.9
	1210	−17	+11	.83	30	1.7	−15	+14	1.52	20	4.6

* Diameter of a circle enclosing the plate or the sector.
† Twice the mean length of dendritic portions.

58. Spatial assemblage of dendritic branches, radiating type

As mentioned in Sec. 54, the radiating-type crystals, which have dendritic branches radiating from the center, grow from the early stage of the assemblage of sectors or a similar stage of spatial structure. The plane hexagonal crystal develops, as described in the preceding section, in a certain region on the T_a–T_w diagram. In the early stage of crystal formation, if the condition is outside that region, especially when T_a is low, a small spatial assemblage of sectors is formed. Later if T_a goes up and the conditions become suited for the dendritic development, dendritic branches grow out of each projecting point of the spatial assemblage and the crystal of radiating type is obtained. Four examples of such crystals are reproduced in Figs. 326–329. Taking three of the above examples, we will describe the conditions for their formation in Table 20. As is readily seen in the table, the central part is assumedly formed in the upper air where the temperature is low or the supersaturation is insufficient. As it falls it passes through an atmospheric layer fitted for dendritic development, and this radiating type of crystal is formed.

59. Plate or sector with dendritic extensions

In a case similar to that mentioned in the preceding section, if a considerable length of time is given for the first stage, a plate-type or sector-type crystal is obtained. Then the subsequent conditions are changed to those for dendritic development. We can thus get the plate or the sector with dendritic extensions. It is very difficult to get a complete central plate. The best one we could obtain with our technique is the semiperfect plate shown in Fig. 330. Other crystals of this kind are reproduced in Nos. 1207, 1209, 1210, Plate 174.

Upon examining the conditions for formation of three crystals of Nos. 1207–1209, it is found that T_a is $-18°C$ for all of them and T_w is $+6°C$, $+9°C$, and $+3°C$, respectively, during the earlier stage. In other words, when T_a is lower than $-17°C$ and the water temperature is too low, the central part grows in a sector form or in a shape close to a plate. In the later stage the condition was changed to that suitable for dendritic growth and then the dendritic branches extended from the points of the plate or the sector. In the case of No. 1210, T_a was $-17°C$ and T_w was $+11°C$ in the earlier stage. Although the value of T_a was suitable for dendritic development, the supersaturation of water vapor was yet a little insufficient. The above conditions for formation are tabulated in Table 21.

In the case of Fig. 331, first a small hexagonal crystal of broad-branched type was formed, with T_a at $-16°C$ and T_w at $+14°C$, which later developed into a somewhat sectorlike form, because the water temperature was kept constant for some time. In the later stage, the water temperature was raised to the value suitable

FIG. 321 [1189]. Artificial snow, dendritic crystal (× 53).

FIG. 322 [1190]. Artificial snow, dendritic crystal (× 30).

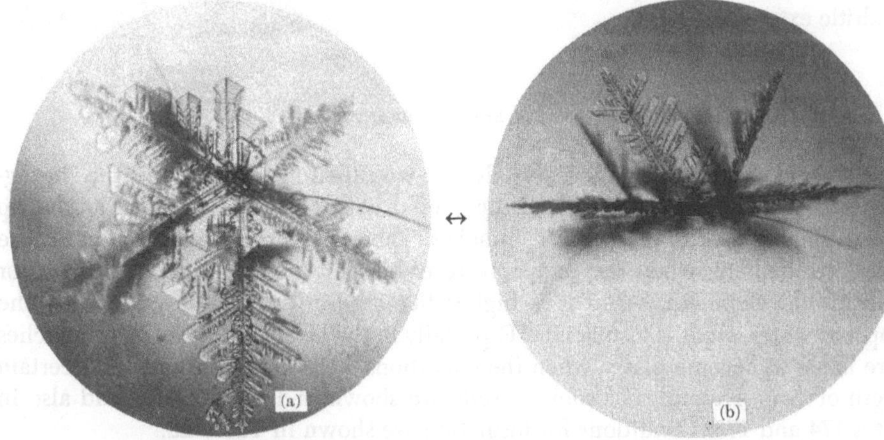

FIG. 323. Artificial snow, spatial hexagonal type: (*a*) [1193] plan view: (*b*) [1194] side view (× 17.5).

for dendritic development, and the dendritic branches, as expected, extended out from it.

The conditions in the earlier period of Fig. 332 were T_a at $-12°C$ and T_w at $+18°C$. In this case the dendritic development did not occur, owing to the high temperature of the air. As will be stated later, when the temperature of the air is high, the plate becomes thick and the double-sheet structure is obtained. So in this case the sector part of this crystal assumed a special pattern, as seen in the photograph. Its branches extended out later when T_a and T_w were lowered to

TABLE 22. Formation of stellar crystals with plates at ends of branches.

Fig. No.	Photo. No.	Earlier stage, dendritic development T_a (°C)	T_w (°C)	Time for growth (min)	Later stage, tips of branches widened T_a (°C)	T_w (°C)	Time for growth (min)	Remarks
333	1213	−15	+14	15	−14	+14	20	Temp. of air increased
	1214	−16	+16	20	−13	+20	20	T_a and T_w increased, double-sheet structure
	1215	−14	+16	15	−13.5	+15	15	
	1216	−16	+13	15	−14	+15	15	Temp. of air increased
	1219	−17	+15	30	−19.5	+13.5	30	Temp. of air decreased
334	1220	−17	+15	30	−19.5	+13.5	30	Temp. of air decreased
335	1221	−15	+13	30	−13	+15	30	Temp. of air increased
	1222	−15	+13	30	−13	+15	30	Temp. of air increased
	1223	−16	+15	15	−13.5	+15.5	15	Temp. of air increased
	1224	−15	+14	20	−16.5	+12	40	Temp. of water decreased, kept constant

−16°C and +15°C, respectively, so that the conditions would be suitable for dendritic extension.

60. Stellar crystals with plates at ends of branches

We could easily produce the crystals that we called in natural snow the "hexagonal with plates," that is, the dendritic crystal with branches widened at the tip or with small plates at the ends of branches. There are three conditions that cause these tips, namely, when the temperature of the air is lower than the range for dendritic development, when it is higher than the critical value, and when the supply of water vapor is insufficient. Especially in the last case, the ends of branches were liable to become plates when the conditions were kept constant for a certain length of time. Examples of such crystals are shown in Figs. 333–335 and also in Plates 174 and 175. Conditions for formation are shown in Table 22.

When the tips of branches are widened, the crystal assumes a shape close to a plate or sector if the conditions for the later period are given from the beginning. Good examples are seen in Figs. 337 and 338, the course of formation being shown in Fig. 336. These two crystals were obtained in the same experiment. Conditions for dendritic growth were given first and then in the later period the temperature of the air was raised. In the crystal of Fig. 337, a dendritic one with plates at the ends of branches, the germ was formed earlier and the crystal passed through conditions for both dendritic and sector formation. In the crystal of Fig. 338, a sector type, the formation of the germ was late and the crystal did not pass through the stage of stellar formation, having directly assumed the shape of sectors.

TABLE 23. Formation of broad branches and plates.

Fig. No.	Photo. No.	T_a (mean) (°C)	T_w (mean) (°C)	Dia.* (mm)	Time for growth (hr)	Rate of growth (mm/hr)	Remarks
	1225	−13.5	+13	2.7	3	0.9	Water temperature constant
	1226	−14	+15	1.2	0.5	2.4	
339	1227	−14	+17	0.58	0.75	0.77	Conditions constant from the beginning
	1228	−17.5	+15	1.05	0.5	2.1	
340	1229	−17	+10	1.95	1.5	1.30	$T_w = +8° \rightarrow +12.5°$
	1230	−23	+ 8	4.5	16	0.29	Left overnight
	1231	−17	+12	1.27	1	1.27	Conditions kept constant
	1238	−18	+10	0.84	1.5	0.56	T_a and T_w kept constant
341	1239	−14.5	+12.5	2.4	1	2.4	
	1240	−18	+13	2.5	19	sublime	Left overnight, conditions constant
342	1241	−17	+12.5	1.17	1	1.17	

* Diameter of a circle enclosing the crystal, or twice the mean of lengths of six branches.

61. Broad branches and plate type

As stated in the preceding section, if the conditions for widening the branch tips are given from the early stage of crystal formation, a broad-branched hexagonal crystal with a feature of plate type is obtained. If the conditions are carried still further, the shape of the crystal approaches the sector or plate form. A plate is apt to develop when the supply of water vapor is small and the conditions are kept constant. Its rate of growth is so low that in some cases it barely reaches a dimension of a few millimeters, even when left overnight. It is difficult to determine the difference in conditions for the formation of sector-form crystals and of plates. The detailed description will be given in Sec. 79. Figures 339–342 and Plates 176 and 177 show crystals of these types, and the conditions for the formation of representative ones are listed in Table 23.

62. Needle crystals

Needle crystals of natural snow are said to be very rarely observed in foreign countries but they occur fairly frequently in Hokkaido. As has already been mentioned, in Sec. 32, when crystals of this type fall, the atmospheric temperature is moderately high, approaching 0°C in many cases.

The conditions for artificial production of needles coincide very well with those of natural ones. In other words, when T_a is increased to approximately −6°C, needles are obtained. In this case it is necessary to make T_w sufficiently high that

FIG. 324 [1195]. Dark-field illumination (× 20).

FIG. 325 [1198]. Dark-field illumination (× 20).

FIG. 326 [1201]. Artificial snow, burrlike type (× 18.5).

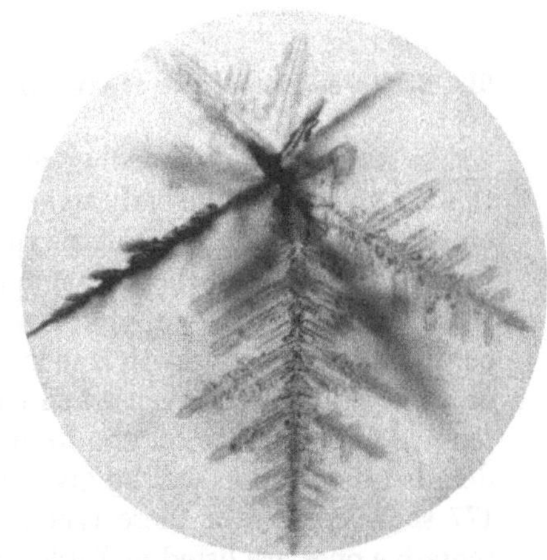

FIG. 327 [1202]. Artificial snow, burrlike type (× 17.5).

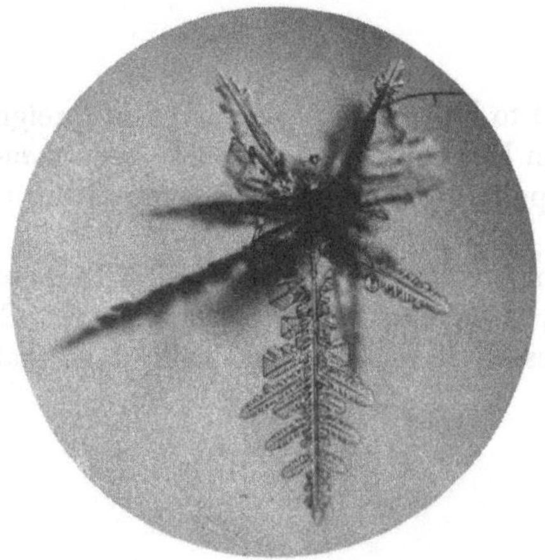

FIG. 328 [1203]. Artificial snow, burrlike type (× 18.5).

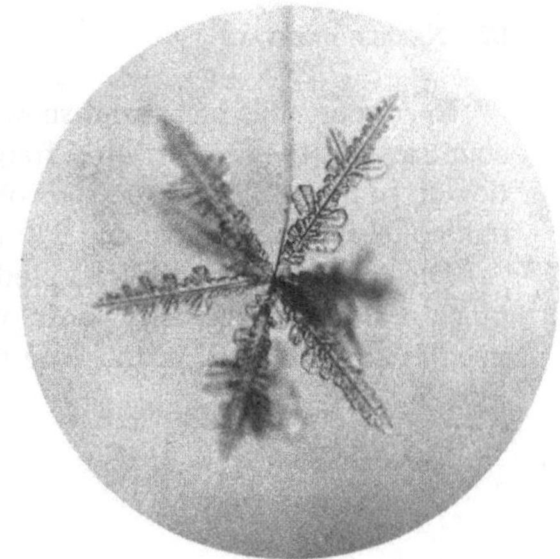

FIG. 329 [1205]. Artificial snow, burrlike type (× 20).

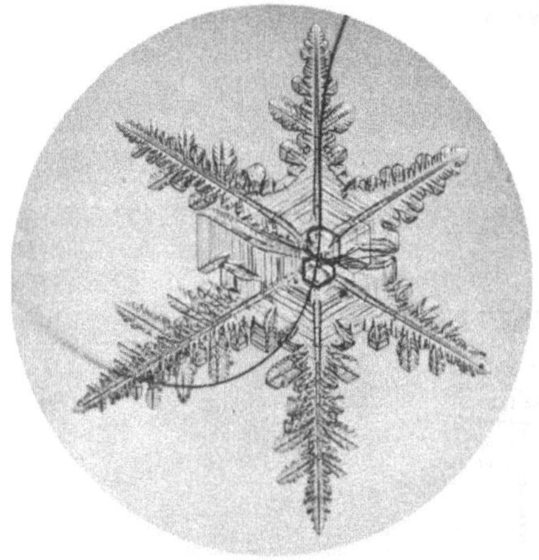

FIG. 330 [1208]. Sectors with dendritic extensions (× 20).

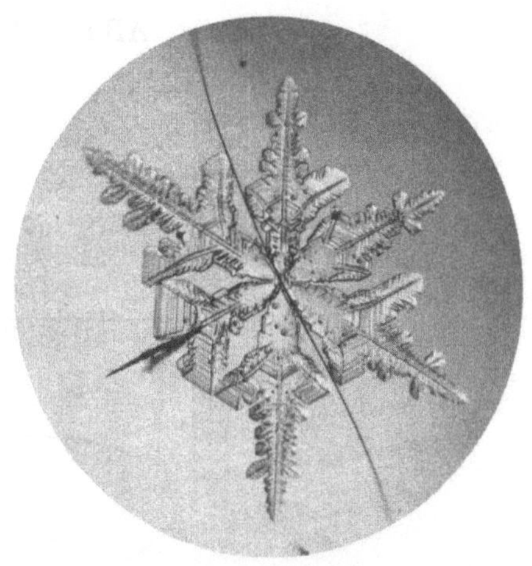

FIG. 331 [1211]. Sectors with dendritic extensions (× 20.5).

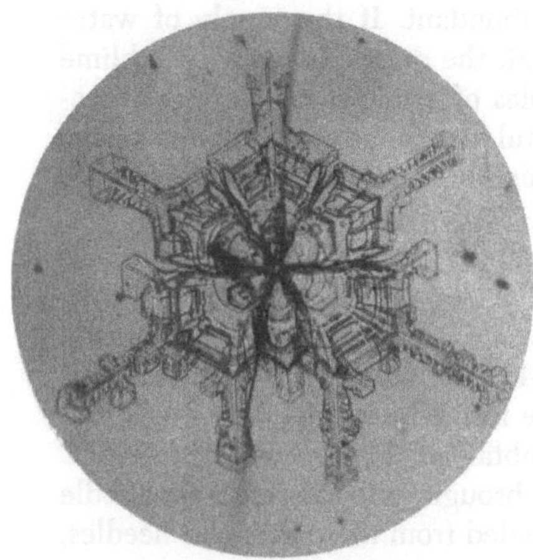

FIG. 332 [1212]. Sectors of double-sheet structure with simple extensions (× 23).

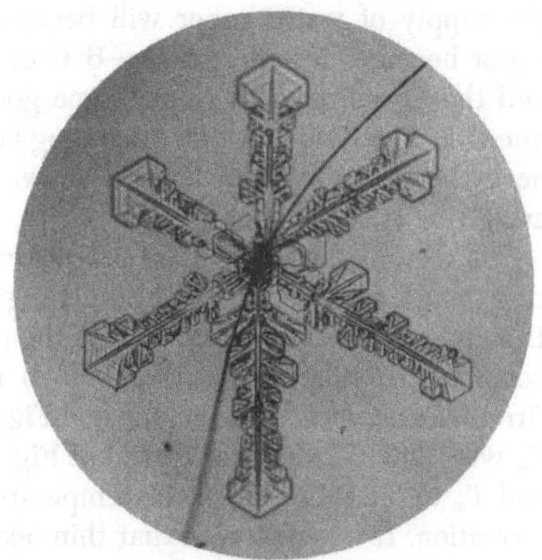

FIG. 333 [1213]. Stellar crystal with plates at ends of branches (× 22).

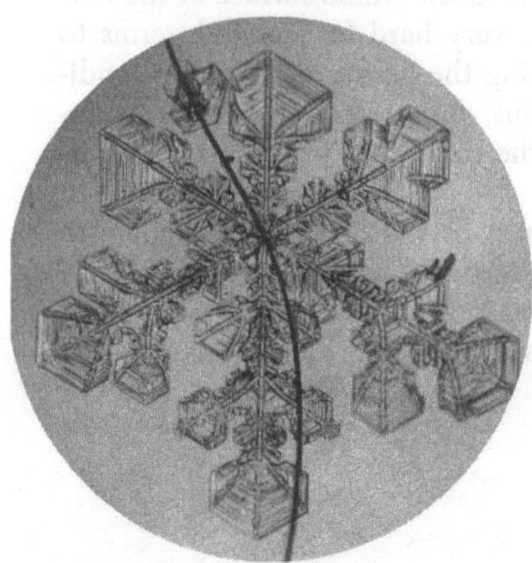

FIG. 334 [1220]. Stellar crystal with plates at ends of branches (× 16.5).

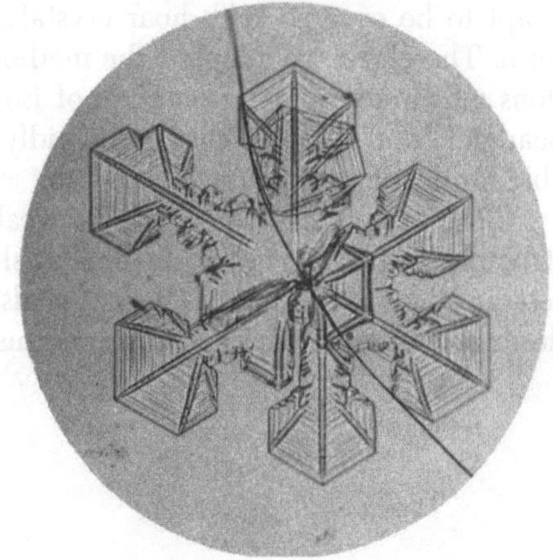

FIG. 335 [1221]. Stellar crystal with plates at ends of branches (× 23).

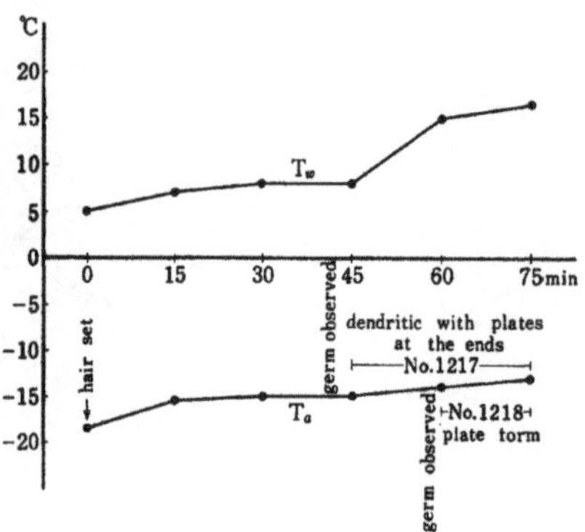

Fig. 336. The course of formation of Figs. 337 and 338.

the supply of water vapor will become pretty abundant. If the supply of water vapor becomes insufficient at −6°C or thereabout, the crystal is likely to sublime and the crystal growth stops. Some good examples of artificial needles are reproduced in Figs. 343 and 344. Such long and beautiful needles as these examples have never been observed in natural snow. Conditions for formation of these needles are given in Table 24.

As seen in the table, there are distinct limits, upper and lower, in the temperature range for needle formation, ranging about 2 deg above and below −6°C. If T_a is above this limit and falls in the range between −4°C and −1°C, the crystal becomes irregular in form, as shown in Fig. 345. We call this type of crystals "irregular needles." The crystals of Fig. 345 were formed when T_a was −4°C and T_w was +40°C. After the crystal of Fig. 345 was obtained, T_a was lowered to −6°C and T_w to +31°C, that is, the temperatures were brought into the range for needle formation. It was observed that thin needles extended from these irregular needles, as expected. The final state is seen in Fig. 346.

To make needles develop on a hair is quite difficult. As the temperature of the air is high and the supply of water vapor is abundant, the whole surface of the hair is apt to be covered with hoar crystals, and it is very hard for isolated germs to form. Therefore, we adopted the method of making the germs under other conditions which allow easy formation of isolated germs, and as soon as the germs appeared T_a and T_w were changed rapidly so that the conditions would become suitable for needle formation.

Close examination of crystals reveals that most of the needles are hollow in structure and have a cut at the end, as shown in Fig. 347. These features are clearly seen in Figs. 348 and 349. This cut is also observed in the scroll-form crystals to be described in Sec. 64. The origin of this cut was supposed to be that during the crystal growth water vapor is supplied only from beneath. An attempt was made to rotate the crystal during the course of its formation, so that the supply of water vapor upon the crystal surface would be uniform. The result was contrary to expectation and the crystals thus produced still showed the cuts. Moreover, it was found that the directions of these cuts on several needles growing on one hair were not always the same. These attempts led to the supposition that these cuts are due to the unstable nature of the crystal growth, so that once such a cut is accidentally formed by a minute fluctuation in the conditions for formation, it grows on as the crystal develops.

63. Combination of hexagonal plane crystal and needle

The experiment of producing the combination of a plane hexagonal crystal and needles was attempted. The result showed most clearly that the needle is a crystal that extended along the principal axis of crystallization. When a hexagonal plane crystal was first produced and then needles were made to grow from it, needles extended in the direction perpendicular to the plane of the base crystal, and vice versa.

The crystal shown in Fig. 350 is one in which needles are made to extend from a fernlike base crystal. In this example, T_a was about $-9°C$ and a little too low for the development of regular needles, so that they are piled up askew one upon another. This type is often observed in frost crystals.

No. 1258, Plate 179 is a contrary case. First needles were formed, with T_a at $-6°C$ and T_w at $+30°C$, and then the conditions were changed to those favorable for dendritic growth, that is, T_a at $-15°C$ and T_w at $+18°C$. Hexagonal plane crystals developed at the tips of the needles, the plane being perpendicular to the axis of the needle.

Figure 352 and Nos. 1259–1261, Plate 179, are side views of needles extending from complete hexagonal crystals, among which Figs. 352(*a*) and (*b*) are photographs of the same crystal taken by adjusting the focus of the microscope to different portions of the crystal. The conditions for formation of this crystal are shown in Fig. 351. A sketch of one of its branches is drawn on the right side in the figure, in which needles are shown not to grow directly from the dendritic base but from a scroll state lying between them. This scroll is formed, as stated in the following section, under an intermediate condition between those for needle and for dendritic crystals. As seen in Fig. 351, during the course of transition from the dendritic to the needle form, the crystal passes through the intermediate range for the formation of a scroll. Consequently, the complicated combination of dendritic, scroll, and needle forms was obtained. The scroll part is clearly observed also in the photograph of Fig. 352(*b*).

Fig. 337 [1217]. Crystal produced by the course of Fig. 336 (× 18.5).

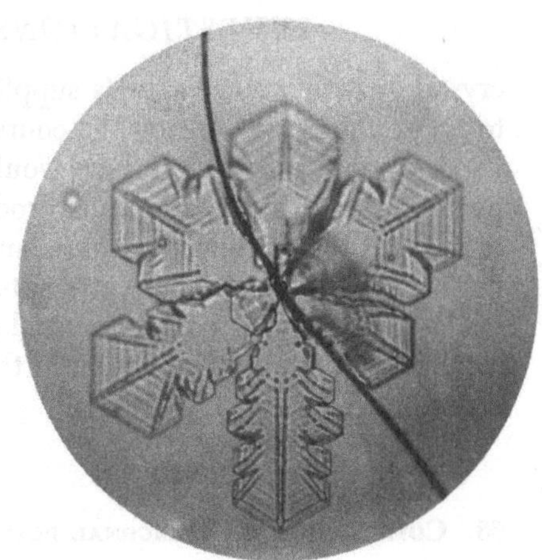

Fig. 338 [1218]. Crystal produced by the course of Fig. 336 (× 26).

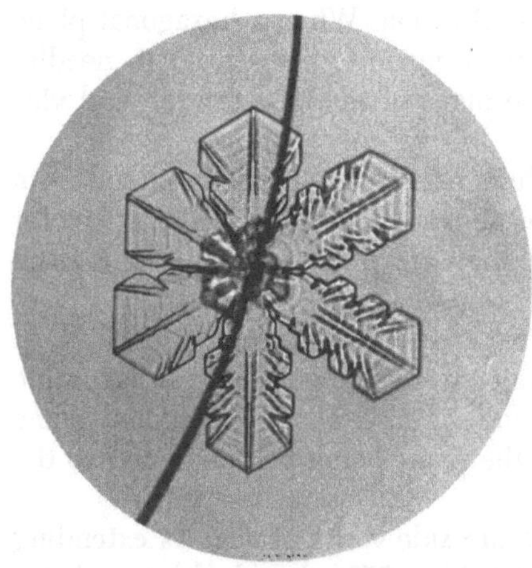

Fig. 339 [1227]. Broad branches (× 70).

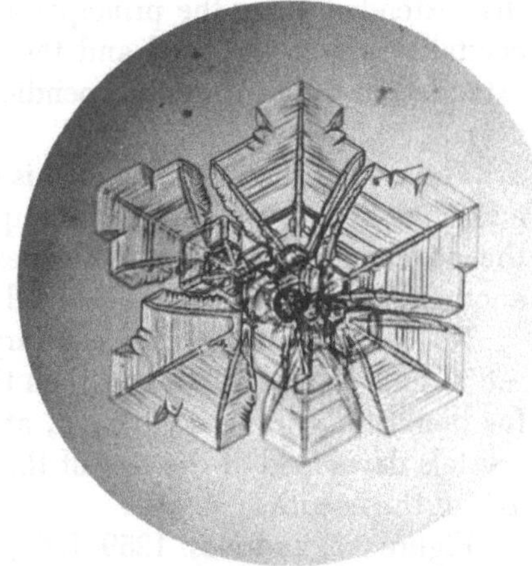

Fig. 340 [1229]. Sector form (× 24).

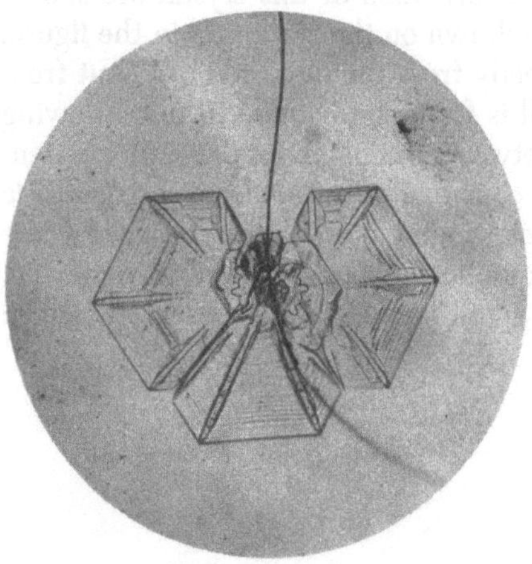

Fig. 341 [1239]. Sector form (× 16).

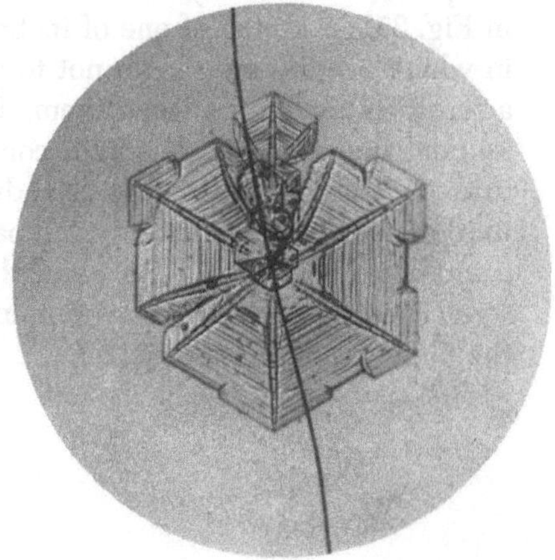

Fig. 342 [1241]. Crystal of nearly plate form (× 32.5).

Table 24. Formation of needle crystals.

Fig. No.	Photo. No.	T_a (°C) Initial	T_a (°C) Middle	T_a (°C) Final	T_w (°C) Initial	T_w (°C) Middle	T_w (°C) Final	Length when completed (mm)	Time for growth (hr)	Rate of growth (mm/hr)	Remarks
343	1243	− 9	− 6	− 6	+27	+30	+29	16	3.5	4.6	No. 1249 obtained in the same experiment
344	1244	− 9	− 6	− 6	+27	+30	+29	6.5	—	—	Start of crystal unknown
348	1246	− 4	− 5	− 6	+34	+33	+32	—	—	—	
	1247	− 4	− 5	− 6	+34	+33	+32	—	—	—	
349	1248	− 6	− 7	− 8	+29	+28	+28	—	ca 1	—	
	1251	− 7	− 5	− 4	+25	+26	+27	2.3	2	1.1	Same crystal as No. 1250
	1252	− 6	− 5	− 5	+28	+29	+31	4.2	2	2.1	

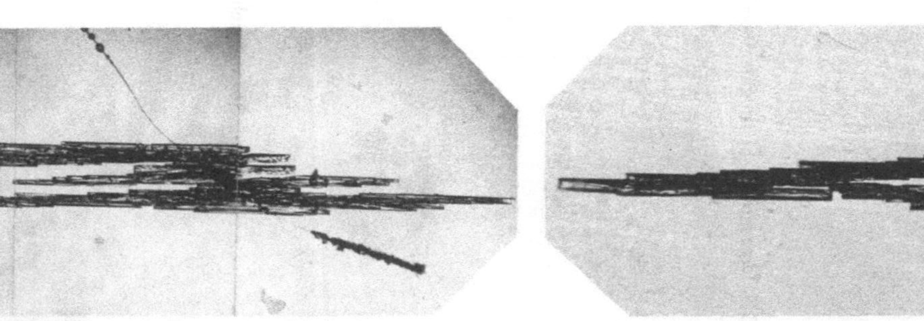

Fig. 343 [1243]. Long needle crystals (× 6.1).

Fig. 344 [1244]. Long needle crystals (× 5.5).

Fig. 345 [1254]. Assemblage of irregular needles (× 22).

Fig. 346 [1255]. Needles extending from the assemblage of irregular needles (× 22).

Fig. 347. Sketch of the ends of needle crystals.

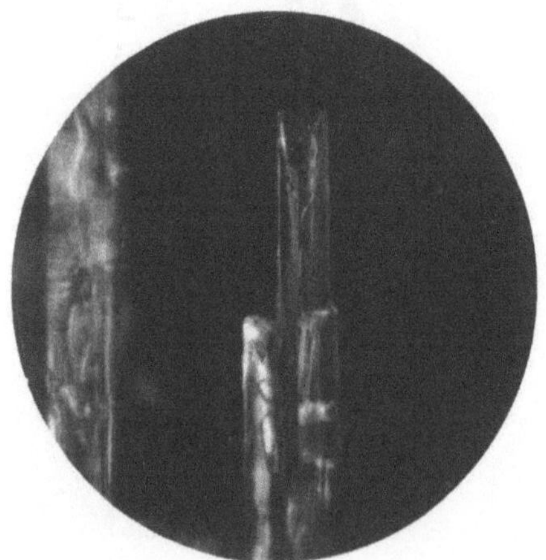

Fig. 348 [1246]. The cut at the end of a needle crystal.

Fig. 349 [1248]. The cut at the end of a needle crystal (× 39).

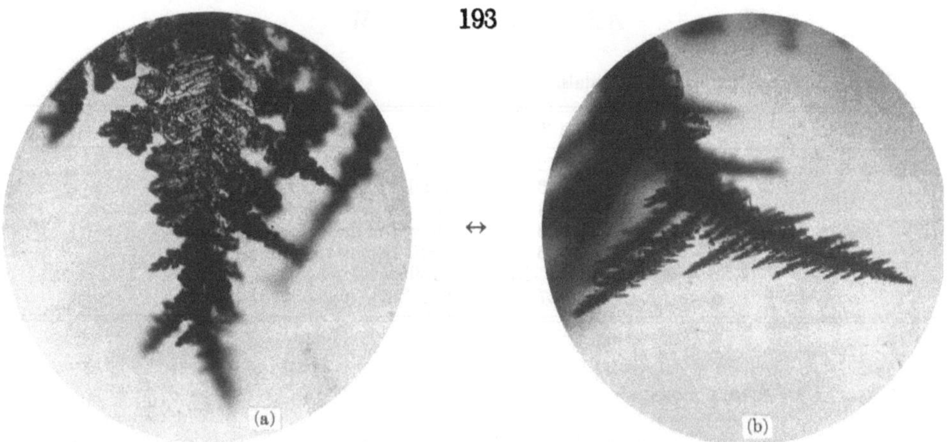

Fig. 350. Irregular needles extending from a hexagonal crystal: (*a*) [1256] plan view; (*b*) [1257] side view.

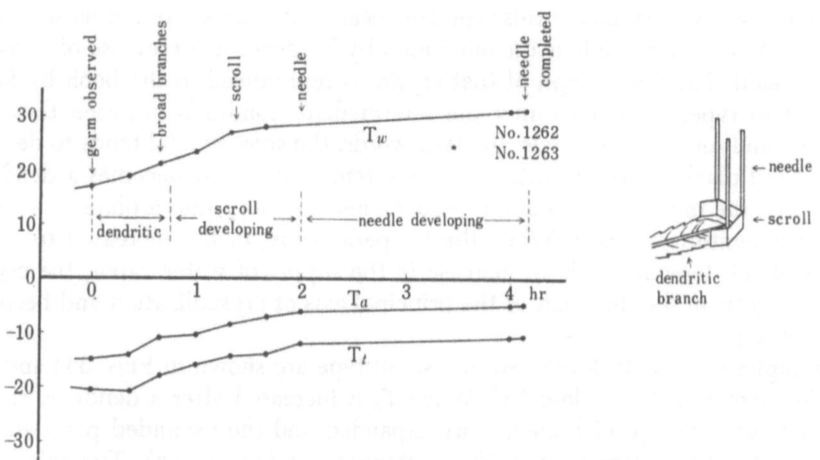

Fig. 351. The course of formation of a stellar crystal with needles.

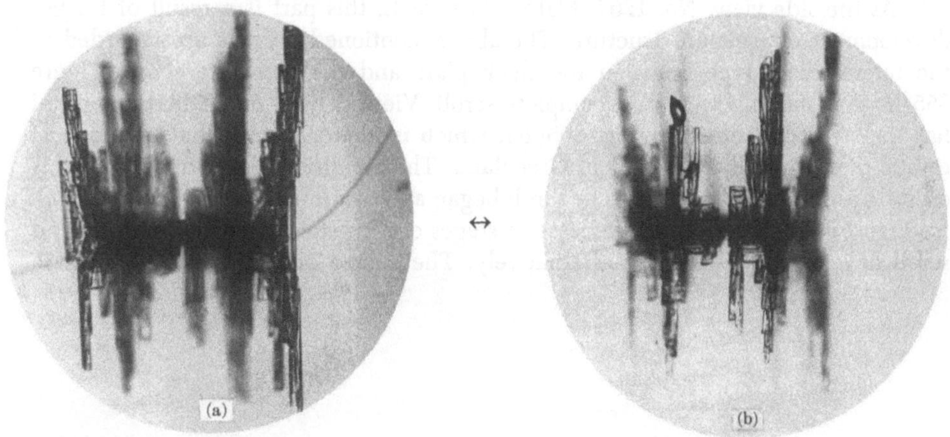

Fig. 352 [1262, 1263]. The crystal produced by the course of Fig. 351; two views taken with the microscope focused on different portions of the crystal (× 16.5).

TABLE 25. Formation of scroll crystals.

Time (hr)	T_a (°C)	T_w (°C)	Remarks
0	−22	+13	germ observed
0.3	−13	+16	small hexagonal, broad branches
1	−10	+23	hexagonal, sector type
1.5	− 9	+26	growth of hexagonal stops
2	− 8	+26	scroll begins to grow
2–4	− 8	+26 − +28	scroll developing
4.5	− 7.5	+27	No. 1268, Fig. 355(*a*)

64. Crystal in scroll form

As an extreme case of skeleton structure, there is a type shaped like a column with only its side planes developed. We call it a scroll type, because a completely developed one assumes the form of a scroll folded hexagonally, like the model shown in Fig. 353. Crystals of this type are occasionally discovered in natural frost. The most beautiful example is the one found by Wegener in a crevasse of a glacier in Greenland. The photograph of that crystal is reproduced in the book by Seligman.[56] This type is formed under the intermediate condition between those for dendritic and for needle crystals. In other words, the snow crystal tends to develop in the basal plane of crystallization at lower temperatures. It becomes a dendritic form when the temperature is about −15°C, and develops into a plate or a sector if the temperature is raised. When the temperature is further increased to −8°C or thereabout, together with an increase in the supply of water vapor, the crystal begins to grow in the direction of the principal axis of crystallization and becomes this scroll type.

Examples of crystals developed in a scroll type are shown in Figs. 354 and 355 and Nos. 1266 and 1267, Plate 180. When T_a is increased after a dendritic crystal is formed, first the tips of branches are expanded and the expanded parts assume the characteristic pattern seen in the photograph of Fig. 354(*a*). The side view, Fig. 354(*b*), clearly shows that this pattern is caused by the skeleton development. Further developed, the sectorlike parts become thicker, as seen in No. 1266, Plate 180. As the side view, No. 1267, Plate 180, reveals, this part is a result of further development of skeleton structure. The above-mentioned crystals are regarded as the intermediate type between the thick plate and the complete scroll. Figure 355(*a*) is a slanting view of a complete scroll. Viewed from one side, this crystal has the structure shown in Fig. 355(*b*), which is identical with that of the frost crystal discovered by Wegener in Greenland. The conditions for formation of this crystal are shown in Table 25. The scroll began and continued to grow at T_a −8°C.

Figures 357–362 represent successive stages of a crystal on which dendritic and scroll developments occurred alternatively. The course of formation is shown in

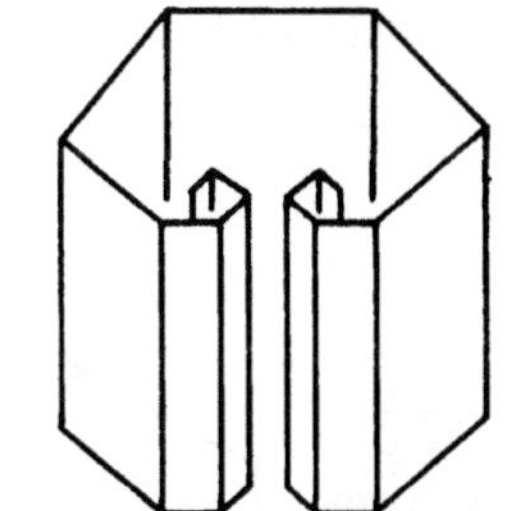

Fig. 353. Model of a scroll crystal.

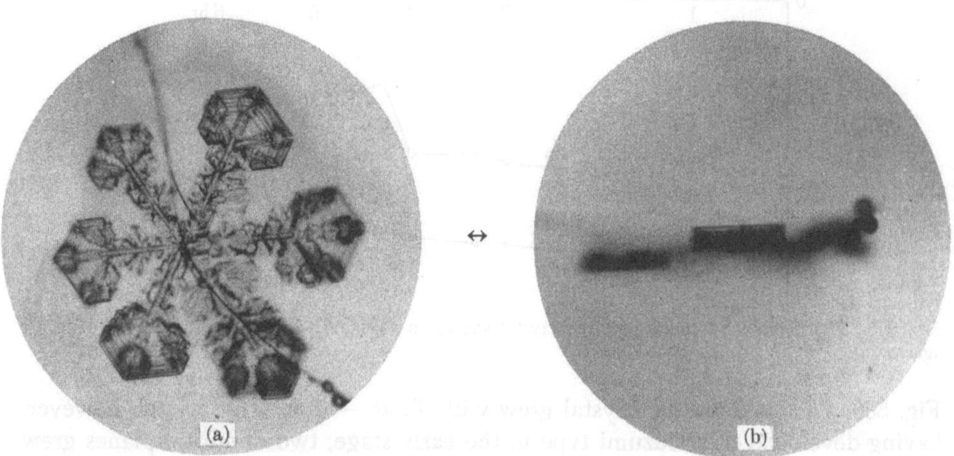

Fig. 354. Ends of branches developing into scroll form: (*a*) [1264] plan view; (*b*) [1265] side view (× 14).

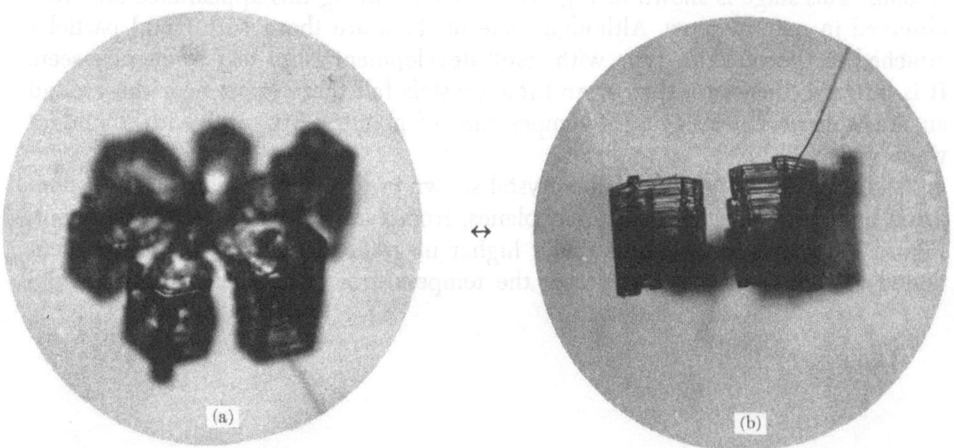

Fig. 355. Scrolls developing from a stellar crystal: (*a*) [1268] slanting view; (*b*) [1269] side view (× 22).

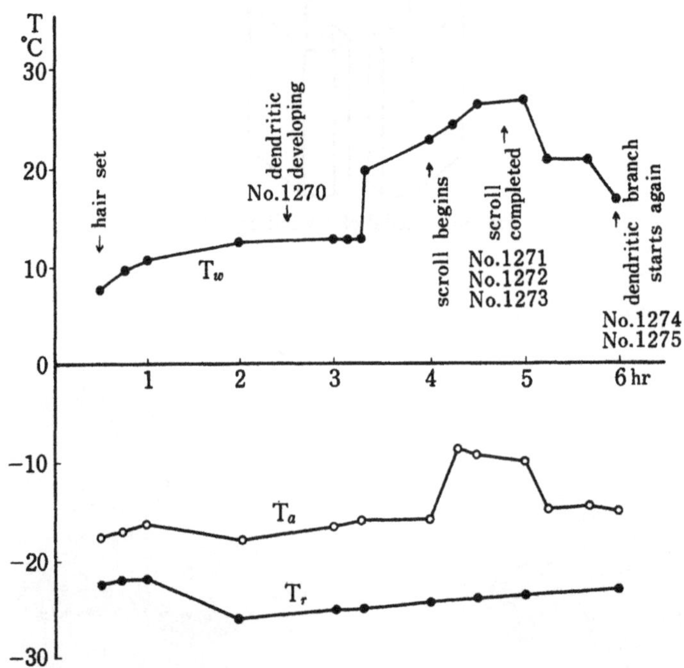

FIG. 356. The course of formation of the crystal shown in Figs. 357–362: dendritic, scroll, dendritic again.

Fig. 356. First, a dendritic crystal grew with T_a at −17°C. The crystal, however, having developed in a tsuzumi type in the early stage, two dendritic planes grew one upon the other, as represented in Fig. 357. After this crystal had formed, the water temperature was raised to make the supply of water vapor abundant, which automatically increased the temperature of the air in the apparatus until it became −9°C, just favorable for the growth of scroll. Then the rim of the crystal began to be thickened and it appeared under a microscope to have been hemmed with black ribbons. This stage is shown in Fig. 358. Crystals having this appearance are often observed in natural snow. Although some of them are those with cloud particles attached to the rim, this type with scroll development is no less frequently seen. It is inferred, therefore, that when these crystals fall there exists near the ground an atmospheric layer having a temperature of about −8°C and with abundant water vapor.

The side view, Fig. 359, of the crystal shown in Fig. 358, reveals that two dendritic branches develop in different planes, from each of which scrolls grow out. Figure 360 shows these scrolls under higher magnification. After the above-mentioned development was completed, the temperatures of air and water were de-

FIG. 357 [1270]. The original dendritic crystal (× 14).

FIG. 358 [1271]. Scroll develops from the crystal of Fig. 357 (× 12).

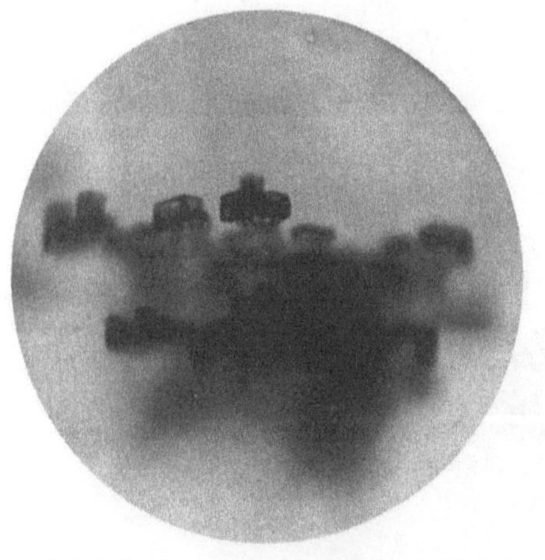

FIG. 359 [1272]. Side view of the crystal of Fig. 358 (× 20.5).

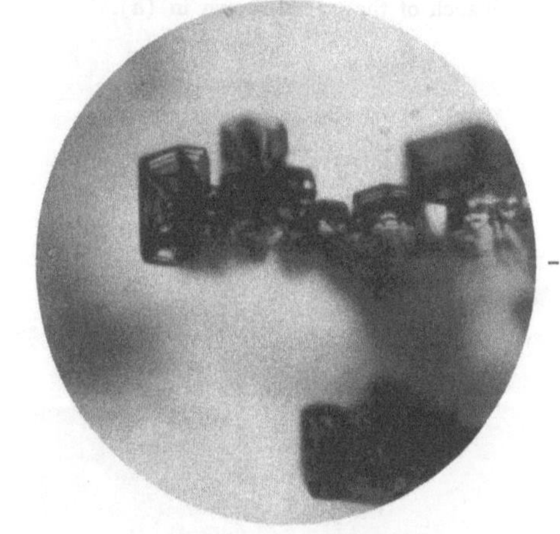

FIG. 360 [1273]. Enlargement of part of Fig. 359 (× 51).

FIG. 361 [1274]. Dendritic branches develop again from the crystal of Fig. 358 (× 8.4).

FIG. 362 [1275]. Side view of the crystal of Fig. 361 (× 8.4).

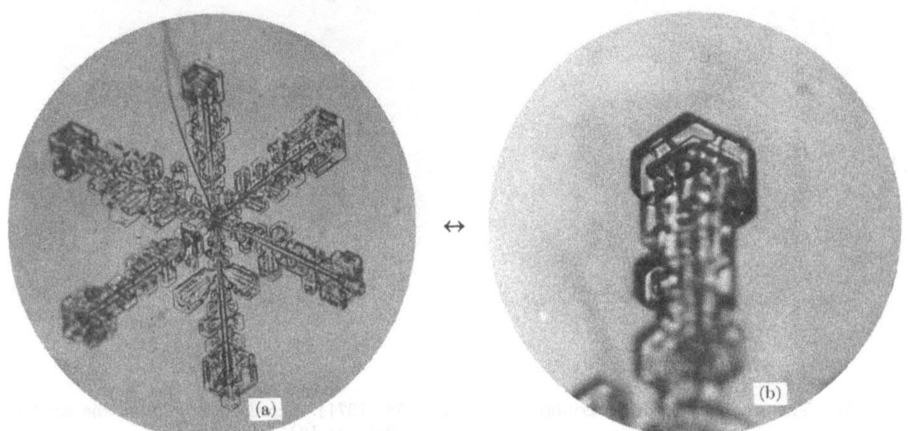

FIG. 363. Branches growing thick: (*a*) [1276] plan view (× 17; (*b*) [1277] enlarged photograph of one branch of the crystal shown in (*a*).

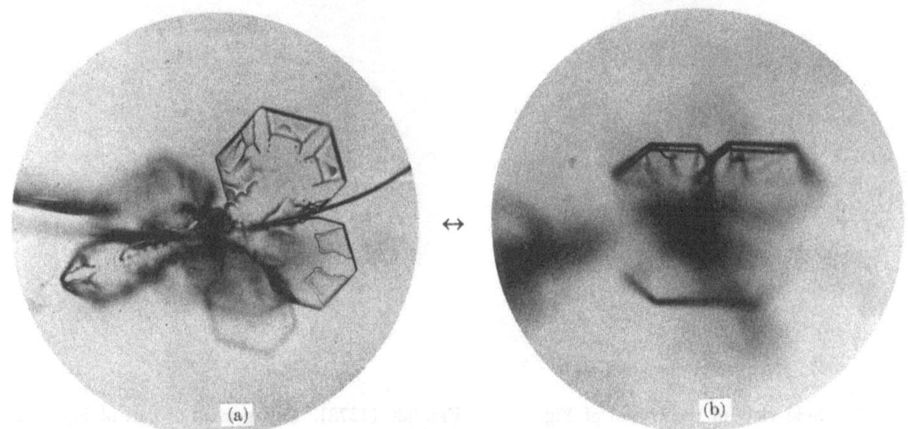

FIG. 364. Sectors with double-sheet structure: (*a*) [1278] general view; (*b*) [1279] slanting view of branches (× 54).

creased again so that the condition would be favorable for dendritic development. Under this condition, dendritic branches began to extend out from both ends of scrolls and the whole crystal became the complicated one shown in Fig. 361. The inner black part is the former crystal of Fig. 358. Figure 362 is a side view of Fig. 361. The course of formation of this crystal with such a complicated structure can be clearly understood from the above explanation.

Through the above series of experiments, we consider that the conditions for the formation of scroll-type crystals are ascertained.

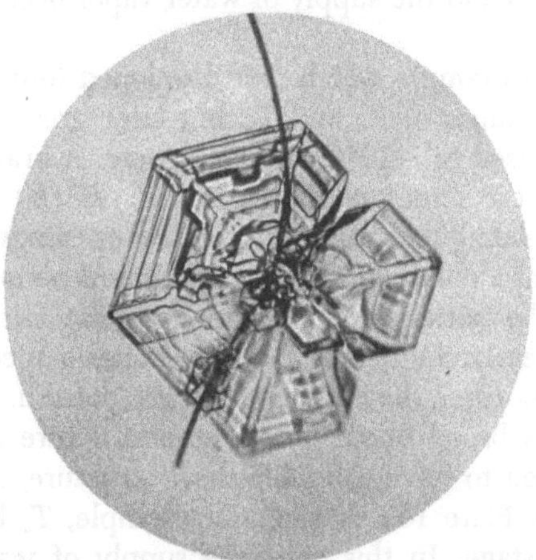

Fig. 365 [1283]. Skeleton prism produced with a scanty supply of water vapor (× 46).

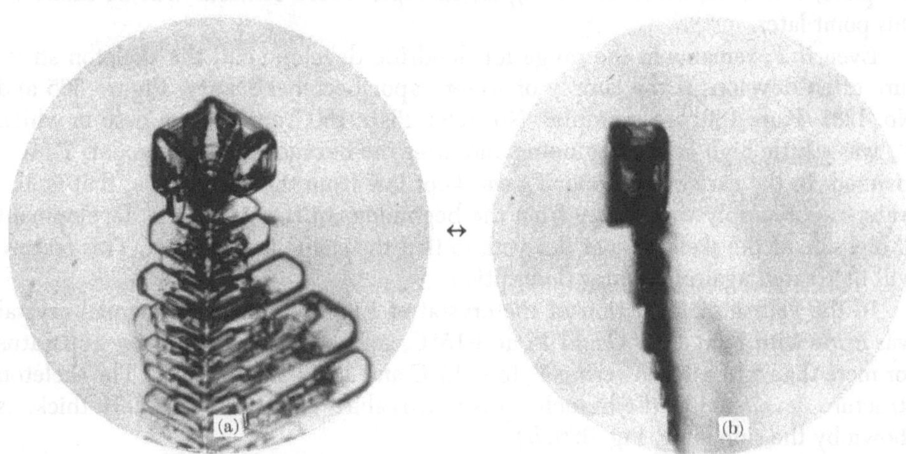

Fig. 366. Branch showing marked skeleton structure (*a*) [1284] plan view; (*b*) [1285] side view (× 31).

65. Structure of the branch

When the conditions for crystal formation go a little beyond the range for the dendritic development, the tips of branches are either widened or thickened. For this phenomenon, both T_a and T_w are important. For instance, if T_a increases or decreases beyond the critical values for the dendritic development, the tips of branches become thick and grow into a form like a skeleton structure of a prism.

Likewise, if T_w decreases and the supply of water vapor becomes less, the branches take a similar structure.

Figure 363(*a*) is an example which first developed into a beautiful hexagonal crystal under suitable values of T_a and T_w but later grew in the thick-branched type because T_a was raised to $-12°C$ and kept there. A branch tip of this crystal, viewed aslant, reveals the structure shown in Fig. 363(*b*). This structure is the same as that of a thick hexagonal plate with hollow openings on the sides, namely, the skeleton structure of a columnar crystal, which will be mentioned later.

If the period for the initial hexagonal development is short and T_a is raised immediately after the early stage, the crystal becomes a special type of sector. In this case also, sector branches show a characteristic pattern, as reproduced in Fig. 364(*a*). This pattern is based upon the skeleton structure of the sector. Viewed aslant, the sector is seen to have a double-sheet structure, as is clearly shown in Fig. 364(*b*). No. 1280, Plate 180, is a similar example, T_a having been raised to $-12°C$ in the earlier stage. In this case, the supply of water vapor was rather abundant. If it had not been, the plate would have become nearer to a transparent ice plate, No. 1281, Plate 180 being an example. More remarks will be made on this point later.

Even if T_a remains in the range for dendritic development, the skeleton structure often develops if the supply of water vapor becomes scanty. Figure 365 and No. 1282, Plate 180, are examples. No. 1282, Plate 180, represents a case in which T_w was a little high at the beginning and after the hexagonal development, T_w was lowered. In the case of Fig. 365, T_w was kept low from the early stage; that is, the water-vapor supply was scanty from the beginning. In the latter, the development of one side of the skeleton was delayed, so that the shape is imperfect. This feature will be treated again in a later description.

In the course of formation of the crystal of Fig. 366, first a hexagonal crystal was made with T_a at $-15°C$ and T_w at $+15°C$, and then it was left in the apparatus for more than 2 hr after lowering T_a to $-22°C$ and T_w to about $+9°C$. The skeleton structure developed at the branch tip is remarkable. That portion is very thick, as shown by the side view, Fig. 366(*b*).

The course of formation of the above-described crystals is given in Table 26.

The skeleton structure at the branch tips can be classified roughly into four kinds, as shown in Fig. 367. In this figure, (*a*) is a complete skeleton with hollow openings on the sides of the hexagonal plate, one good example having already been shown in Fig. 363 (*b*); (*b*) lacks one of the two bases, Fig. 365 showing a photograph of this kind of crystal taken from above. A clearer example is Fig. 375, for which explanation is given in the following section. If the characteristics of (*b*) are further intensified, the skeleton comes to lack one of the bases completely. In this case, the double-sheet structure is no longer seen, but the crystal looks like an etched plate with only the ridges remaining. Figure 367(*c*) is the model, an example of which is

TABLE 26. Formation of thick branches, double-sheet structure, and skeleton prisms.

Fig. No.	Photo. No.	T_a (°C) Initial	T_a (°C) Middle	T_a (°C) Final	T_w (°C) Initial	T_w (°C) Middle	T_w (°C) Final	Time for growth (hr)	Time for growth (min)	Remarks
363	1276–7	−17	−15	−12	+13	+16	+18	1		At first hexagonal dendritic, T_a increased later, vapor supply abundant
364	1278–9	−14.5	−13	−12	+12	+18	+21		45	T_a increased in earlier stage, vapor supply abundant, small shape
	1280	−15	−12	−11	+12	+17	+19		45	
	1281	−17.5	−14.5	−14.5	+ 3.5	+ 5	+ 6	1	30	Vapor supply scanty, intermediate state of transparent ice plate and thin plate of skeleton structure
	1282	−17	−14	−16	+10	+14	+ 7	2	30	After formation of hexagonal crystal, T_w decreased
365	1283	−17	−17	−17	+ 8	+ 7	+ 7	2	15	Water vapor scanty from the beginning, small shape
366	1284–5	−15	−22	−22	+15	+ 8	+10	3	30	After formation of hexagonal crystal, T_a greatly decreased

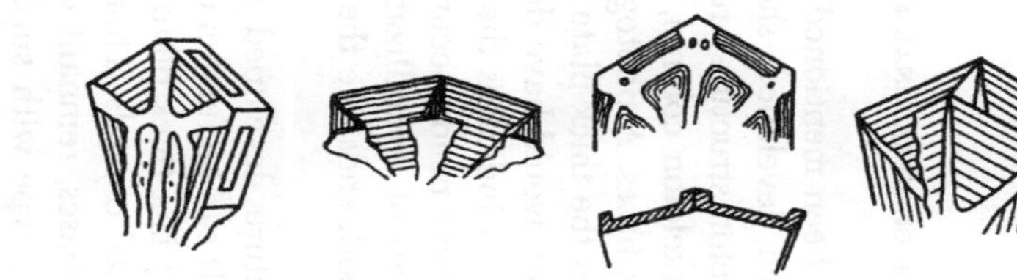

a) No. 1277 b) No. 1283 c) No. 1293 d)

FIG. 367. Classification of skeleton structure observable at the ends of branches.

seen in Fig. 379(*b*). Figure 367(*d*) is an intermediate type between (*a*) and (*b*). This is the type that partially retains the double structure. Detailed explanations will be given in the following section about this skeleton structure and its variations.

The photograph of No. 1281, Plate 180 already shows a tendency of the branch to become a transparent ice plate. Similar examples are shown in Fig. 368, and also in No. 1287, Plate 180. Figure 368 is a small crystal formed in about 30 min with T_a at -15.5 to -16.5°C and T_w at $+14$ to $+12$°C. The plate portions of crystal No. 1287, Plate 180, were formed with T_a at -19 to -16°C and T_w at $+10$ to $+20$°C. The columnar parts had been formed in the apparatus left overnight. Conditions for the formation of this transparent ice plate are not clarified as yet.

66. Skeleton form of hexagonal prism and its variations

It has already been mentioned in Sec. 52 that the skeleton structure of hexagonal column, when developed, shows a considerable variation of shape. Various types of such skeleton structure are described in Fig. 369(1)–(4). In (1) is shown a representative skeleton column, with staircaselike hollows on each of its sides and on both of its bases. A photograph is shown in Fig. 370. When the column is shorter, it becomes the thick plate shown by Fig. 369(2) and Fig. 371, which, had it been still thinner, would have developed into the thin plate with pattern shown by (3). This pattern indicates the hollow spaces formed by the skeleton development. A similar phenomenon occurs when the crystal grows chiefly in the direction of the principal axis of crystallization. In this case, the crystal becomes a needle, with hollows on both ends like the kicks of glass bottles. This type is shown in (4) of the figure.

Skeleton structure developed on the tips of branches was mentioned in the foregoing article. It can be further classified into seven types, as shown in (5)–(11), Fig. 369. Here (5) is a thick plate almost without hollow, which is often observed in the slowly formed crystals. Though no hollow is seen on its sides, it has a slight depression on its bases, remnant characteristics of the skeleton structure. In (6) is shown a similar type with small cavities opening on its sides, some of which often form a notch as shown in the sketch.

The structure in (7) is similar to that of the thick plate (2), a photograph of one example being reproduced in Fig. 372. With larger hollows and less ice, this type of crystal becomes the one shown in (8). Fig. 373 shows a crystal of this type with two dendritic branches extended. Viewed from the side, this type of crystal appears to be a rectangular box. Rectangular crystals of ice are found on rare occasions both in natural frost and in artificial snow, as will be mentioned later, but it seems that most of such rectangular crystals belong to this type.

When the supply of water vapor is scanty, a part of the base planes of (8) is

Fig. 368 [1286]. Branch of ice plate without design (× 50).

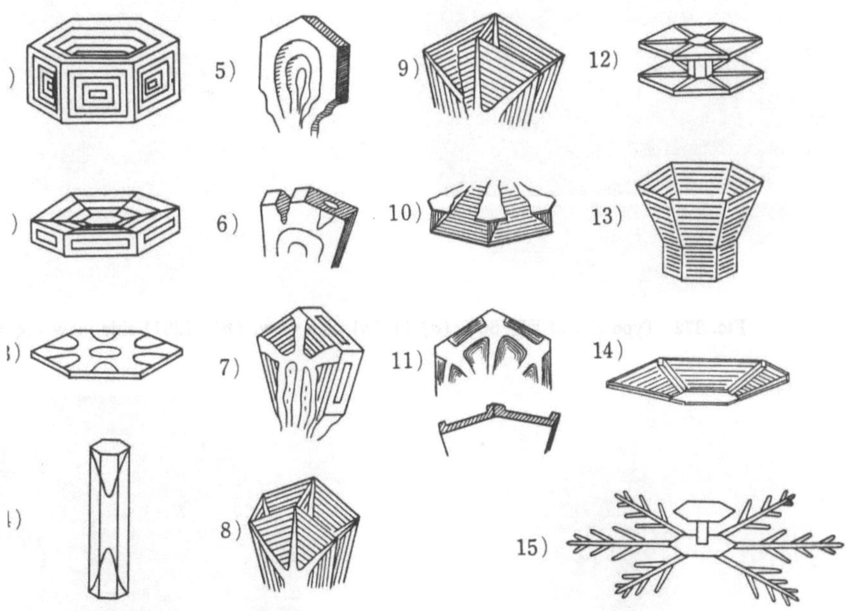

Fig. 369. Variations of skeleton structure of hexagonal prism.

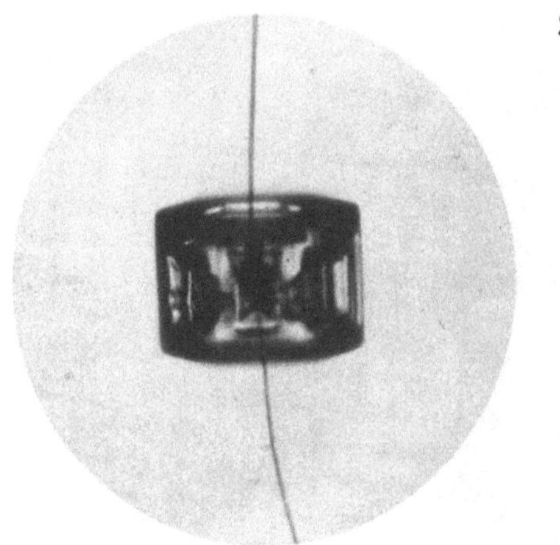

FIG. 370 [1288]. Type (1) of Fig. 369 (× 46.5).

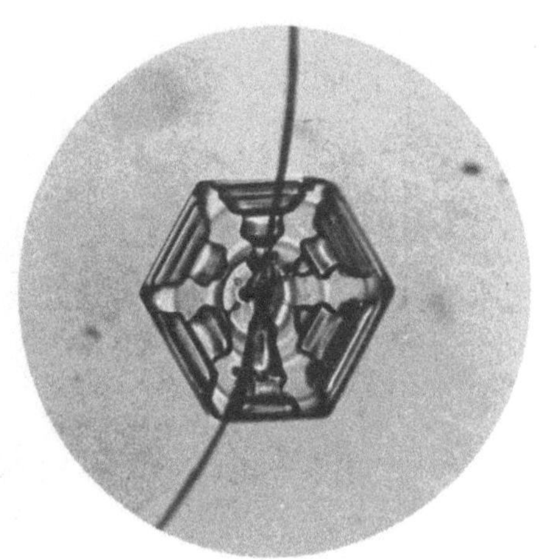

FIG. 371 [1289]. Type (2) of Fig. 369 (× 81).

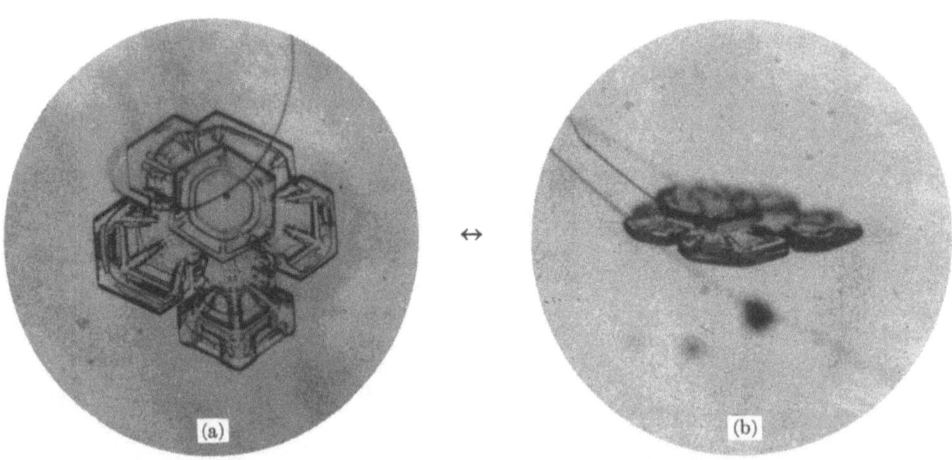

FIG. 372. Type (7) of Fig. 369: (*a*) [1296] plan view; (*b*) [1297] side view (× 30).

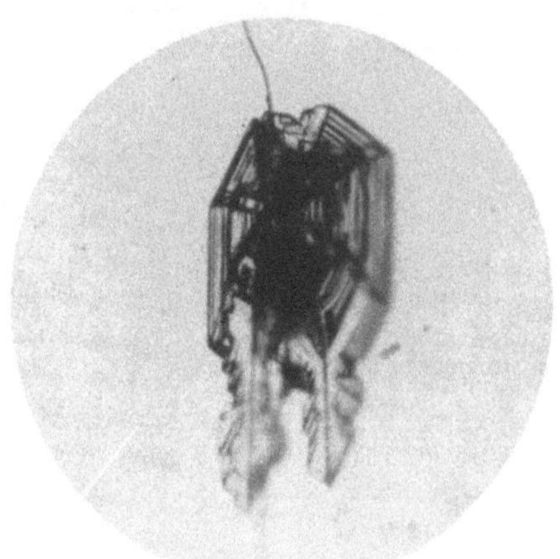

FIG. 373 [1295]. Type (8) of Fig. 369, with dendritic extensions (× 43.5).

FIG. 374 [1352]. Type (9) of Fig. 369 (× 63).

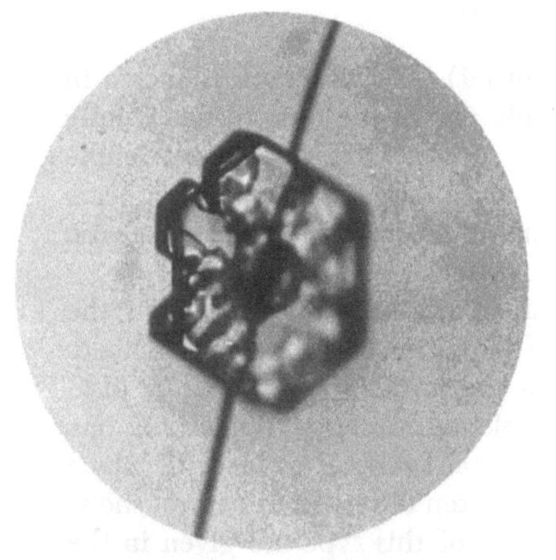

FIG. 375 [1294]. Type (10) of Fig. 369 (× 90).

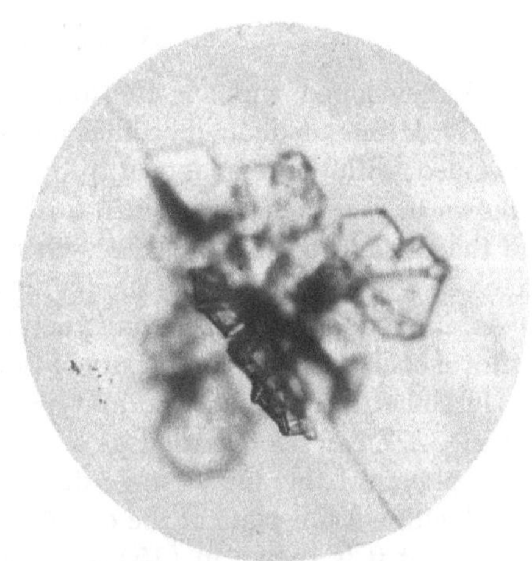

FIG. 376 [1342]. Type (11) of Fig. 369 (× 25).

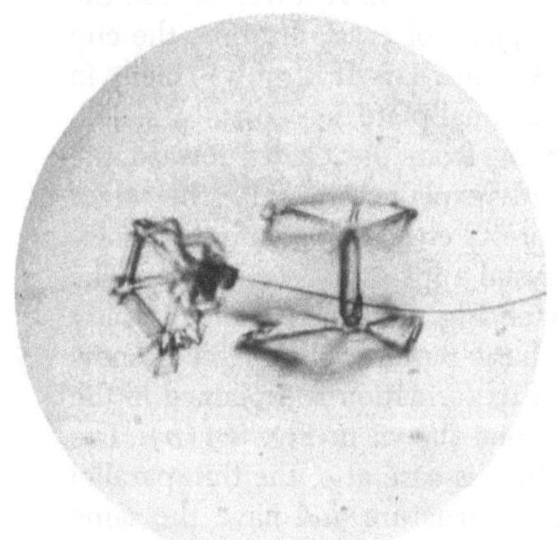

FIG. 377 [1354]. Type (12) of Fig. 369 (× 20).

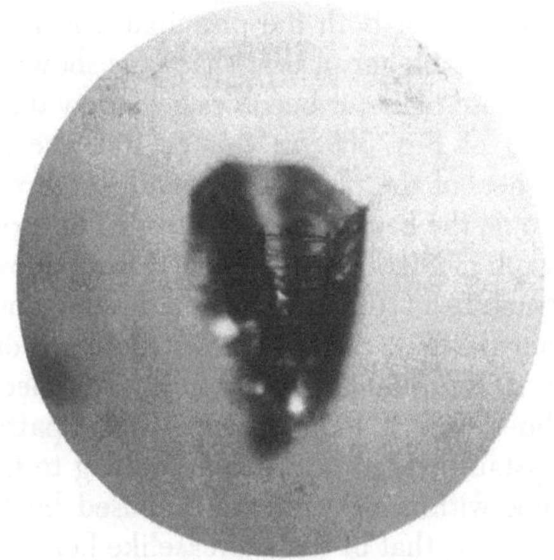

FIG. 378 [1379]. Type (13) of Fig. 369.

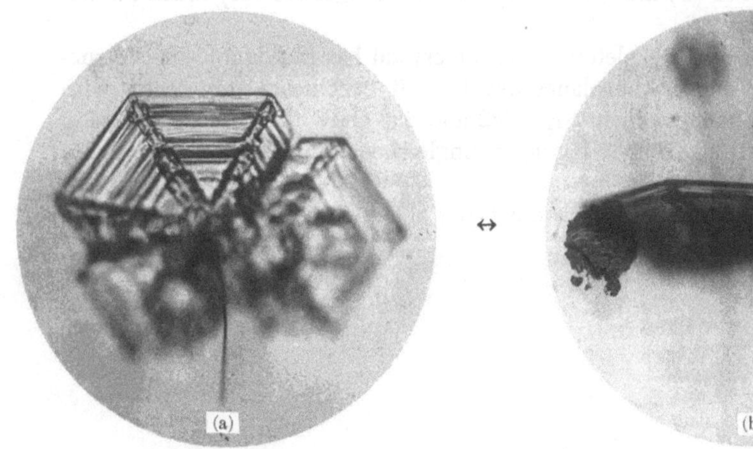

FIG. 379. Type (14) of Fig. 369: (*a*) [1292] plan view; (*b*) [1293] side view (× 34.5).

often undeveloped, showing the form sketched in (9). A photograph is shown in Fig. 374. If this tendency becomes intensified, only one of the base planes is fully developed, with the remnants of the other plane as seen in (10). One good example is shown in Fig. 375. Carried still further, the second plane completely disappears and the crystal becomes a plane with ridges running on it; (11) is the schematic sketch of it, and Fig. 376 is a photograph of an example.

As stated above, the skeleton structure has many varieties, and by clarifying the nature of each type, the complexity of the types of snow crystals is largely explained. For instance, the special tsuzumi-type crystal in the initial stage shown in (12) and Fig. 377 can be regarded as a specially developed skeleton. When only one of the planes of this initial stage develops, the crystal becomes a dendritic plane crystal with a small plate at the center; this type is often observed in natural snow. A sketch of it is shown in (15) and good examples of this type are given in the photographs of Fig. 380.

Type (13) of Fig. 369 has hitherto been called a "cup." This cup is observed already in the initial stage and evidently is a kind of skeleton. If it develops in the directions of both the principal axis and the base plane of crystallization, the cup becomes bigger in that shape, as shown in Figs. 378 and 419. If it grows chiefly in the base plane, it becomes a shallow dish or a hexagonal plate, as shown by sketch (14) of Fig. 369. In this crystal, the lines running from the center toward the corners of the plate correspond to the ridges already explained in (11); the ridges run on the back side of the dish. The structure is pretty clearly shown in the photograph of Fig. 379(*b*). Viewed from above, the crystal appears as Fig. 379(*a*). The structure of the sector seen in the photograph of Fig. 379(*a*) is schematically shown in Fig. 381(*a*). This structure is one of the basic patterns observable in snow crystals of plate or sector form. The mechanism of its formation is explained in the above way. A more frequent basic pattern is the one shown in Fig. 381(*b*). The crystals of Figs. 333–335 all belong to this type. In this case also, the thin parallel lines within the sector are caused by the skeleton structure and have the same nature as that of the staircaselike hollows in the skeleton column.

When the sector becomes still thinner, the crystal becomes like the one shown in Fig. 382. The structure is schematically shown in Fig. 381(*c*), and it is substantially the same as that of (*a*) and (*b*), except that the ridges are very much thinner and unrecognizably low.

When both planes of the skeleton grow, the crystal has the double-sheet structure, and when only one of the planes develops, it becomes a thin sector with ridges upon it. The photographs in Figs. 382 and 383 show the comparison using similarly shaped crystals, Fig. 382 having a single-sheet structure and Fig. 383 a double-sheet structure.

Conditions for the formation of various types of skeleton and its variations are shown collectively in Table 27.

Table 27. Formation of various types of skeleton.

Fig. No.	Photo. No.	T_a (°C)			T_w (°C)			Time for growth (hr)	Remarks
		Initial	Middle	Final	Initial	Middle	Final		
370	1288	−22	−22	−20	+ 6	+ 8	+10	20	Typical skeleton column made by very slow process at low temperature and with scanty supply of vapor
371	1289	−18	−18	−18	+ 8	+ 9	+ 9	2.5	Temperature a little higher than for No. 1288, but vapor supply still scanty
	1290–1	−23	−20	−20	+ 4	+ 4	+ 6	17	Plan and side views of the same crystal, thick plate of skeleton structure
379	1292–3	−17	−17	−18	+ 7	+ 8	+ 9	3	Plan and side views of the same crystal, one plane with ridges on the back side
375	1294	−17	−17	−18	+ 7	+ 8	+ 9	—	One base plane developed, the other plane showing remnants
373	1295	−21	−20	−19	+11	+11	+12	4.5	Dendritic branches extending out from a skeleton crystal
372	1296–7	−12	—	−20	—	+ 2	—	—	Plan and slanting views of the same crystal, both two sheets showing skeleton structure
382	1298	−22	−20	−18	+ 7	+ 8	+ 9	3	One base plane only developed, thin sector
383	1299	−15	−13	−11	+13	+17	+19	1.7	Two base planes developed, both showing double-sheet structure

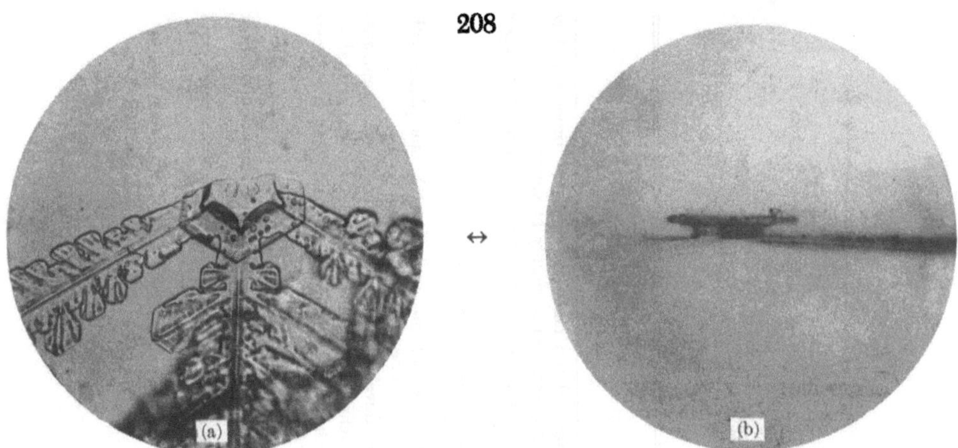

FIG. 380. Type (15) of Fig. 369: (*a*) [1316] plan view; (*b*) [1317] side view (× 33.5).

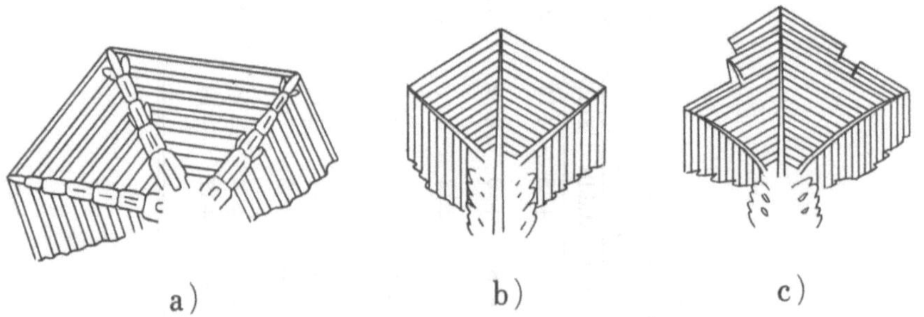

FIG. 381. Three kinds of sector form.

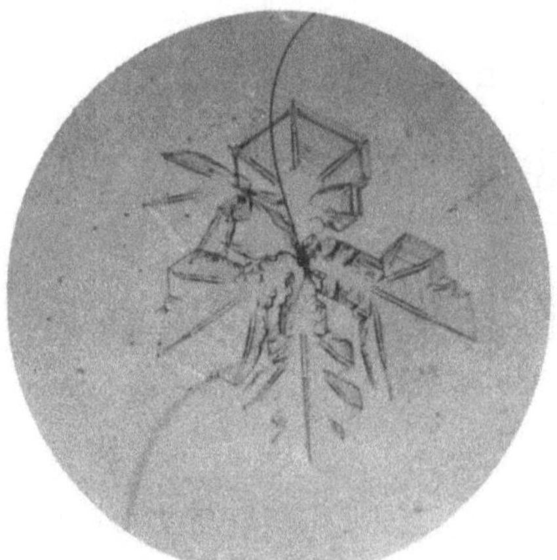

FIG. 382 [1298]. Crystal having thin sectors (× 14).

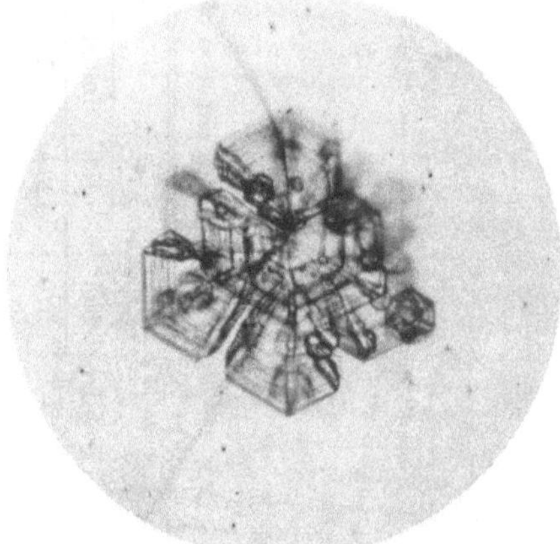

FIG. 383 [1299]. Crystal similar to Fig. 382, but with two sheets of sectors (× 29).

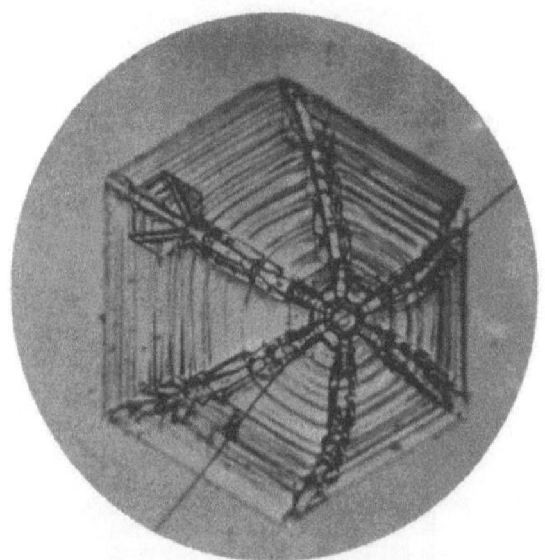

Fig. 384 [1300]. Dish-form crystal (× 27).

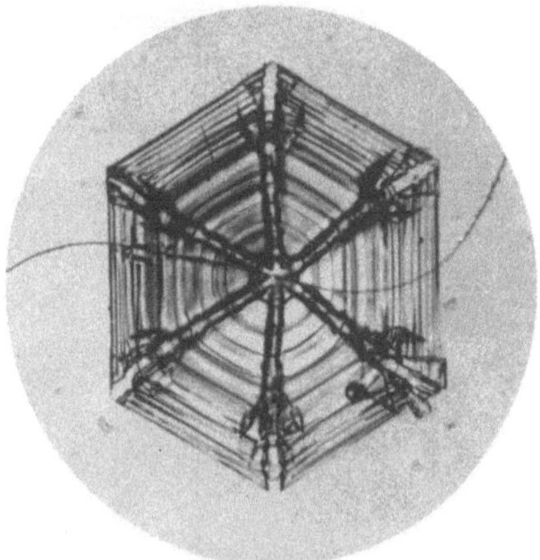

Fig. 385 [1301]. Dish-form crystal (× 27).

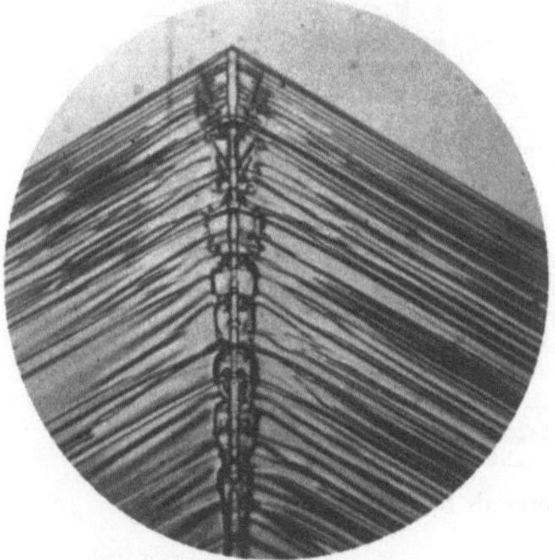

Fig. 386 [1302]. Ridge consisting of several skeleton crystals (× 25).

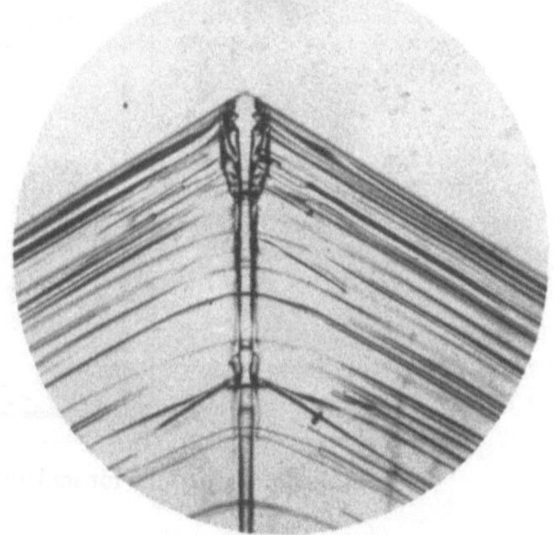

Fig. 387 [1305]. Narrow smooth ridge (× 25).

67. Structure of plate and sector

The plate and sector crystals are, according to the previous discussion in Sec. 66, special cases of the column of skeleton development. The dish crystal exists as an intermediate type between them and the skeleton column. Figures 384 and 385 are good examples of the dish type. In this type, the lines radiating from the center toward the corners of the hexagon are ridges attached to the back side of the crystal. In Fig. 384 one of these ridges shows near the end a small sector in skeleton

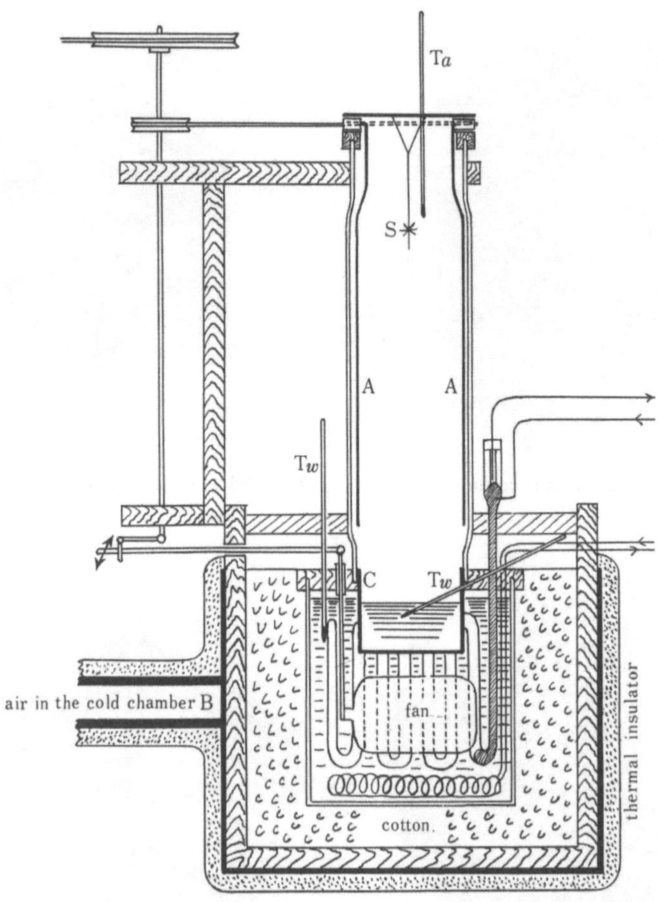

FIG. 388. Apparatus for making snow crystals by diffusion of vapor.

development. In the crystal of Fig. 385, all six of the ridges show the skeleton development at their end. In so far as the dish type is a special case of the skeleton column, such structure is naturally to be expected, and these two examples serve to confirm the aforementioned inference.

A similar structure is sometimes observed also in the case of hexagonal plates. For example, in the crystal of Fig. 386, many small skeletons are arranged like scales along the ridge. In the ordinary plates, however, these scalelike skeletons disappear and each ridge appears as two parallel lines, as shown in Fig. 387. But sometimes crystals of this type show a remnant of the skeleton development. In Fig. 387 also, a half-developed skeleton is observed toward the end of the ridge.

Upon close examination of these plates, it is revealed that many of the numerous thin parallel lines that originate from the staircaselike structure of the plane are continued to the adjacent ones, crossing the ridges. This phenomenon is clearly seen in the crystal of Fig. 387. It means that the crystals of this type have grown so as to make growth rings like the annual rings in tree trunks, although the form is hexagonal. Probably these "growth rings" were formed by some fluctuation in conditions. According to this inference, formation of dendritic crystals ought to be based on still larger fluctuation in condition.

We therefore attempted to make a crystal under conditions as constant as possible. This experiment was mainly carried out by Y. Sumi. First the apparatus shown in Fig. 388 was made. It was constructed so that the supersaturated air current would be brought to the place of crystal formation in a condition near to diffusion. The inner glass cylinder that was used in the former apparatus was removed, and a simple glass cylinder was employed. A copper case *C* was fixed to its bottom and the temperature of the water in it was controlled by the temperature of the water in the outside reservoir. A sensitive toluol regulator was set in the reservoir and the water temperature was kept constant to within 0.01C deg. As the frost crystals formed on the inside wall of the glass cylinder affect the mode of convection considerably, three thin bamboo sticks, shown in the figure at *A*, were fitted alongside the wall so that they could be rotated slowly, constantly sweeping off the growing frost crystals. The whole apparatus was set in a thermostat. As heat is generated in the reservoir, it was covered by an outer copper case *B*, one part of which was exposed to the air in the cold chamber by a copper tube.

Using this apparatus, we learned that plates and sectors were very easily formed, and also much larger crystals than before could be obtained by this contrivance. The crystals shown in Figs. 384–387, 389, 390 were formed in this apparatus. The six crystals shown in Plate 181 also belong to this class. Conditions for formation of these crystals are listed in Table 28. In this table, T_a has a physical significance, as before, but there is not much meaning in comparing T_w here with values used with former apparatus, because the supply of water vapor and therefore the amount of supersaturation at the place of crystal formation are very different in this case even if T_w is the same.

As seen in the table, though T_a was varied from -13°C to -21°C and T_w was

TABLE 28. Formation of plates and sectors.

Fig. No.	Photo. No.	T_a (°C)	T_w (°C)	Fig. No.	Photo. No.	T_a (°C)	T_w (°C)
384	1300	−21	+13		1306	−17	+15
385	1301	−20	+15		1307	−13	+14
386	1302	−20	+14		1308	−15	+15
	1303	−14	+14		1309	−15	+15
	1304	−20	+16	389	1310	−17	+ 3
387	1305	−17	+14	390	1311	−19	+18

raised from +3°C to +18°C, we always obtained the plates. From the values of T_a and T_w, Nos. 1308 and 1309, Plate 181, were expected to grow in the dendritic form, but in actuality they developed into plates of large size. In these cases the crystal was very thin and quite different from the ordinary plates. From the results of these experiments we infer that although the conditions are suited to dendritic growth, the crystal grows into a plate if the conditions are kept constant and the fluctuation is made as small as possible. This is, however, a very significant problem and no further argument would be possible without conducting many careful experiments in future.

68. Section of crystal

In discussing the crystal structure, knowledge of the minute structure that causes the interior pattern is indispensable. As mentioned in Sec. 1, we can examine the structure as a whole, by photographing the surface ruggedness with skillfully controlled illumination. But this method is unsatisfactory for determining, for example, whether the parallel lines running along the middle of a dendritic branch indicate a thin ditch upon the crystal surface or a vein with a capillary tube inside.

The most direct method would be to cut a crystal at the required point and examine the section with a microscope. However, this method is accompanied, as naturally would be expected, by immense technical difficulties. Crystals are apt to melt or sublime during the operation when the temperature is near 0°C. At the low temperature of −15°C or −20°C, they become so fragile that they are broken to pieces under the cutting blade. But these difficulties were at last surmounted by the efforts and perseverance of Y. Miyazaki.

In the method finally developed, a sheet of rubber is spread on a glass plate and the crystal is put upon it; then the crystal is cut with a sharp safety-razor blade with a somewhat sawlike motion. We tried various methods, but this one proved to be the most successful. Needless to say, we had to try this experiment on many crystals and get into the knack of carrying out this difficult technique. One more difficulty added to the rest was that the experiment had to be conducted in the cold chamber.

Of the 39 examples in which cutting and photography were successful, 12 representative examples are reproduced in Figs. 392–396 and Plate 182. All of the left-hand photographs show the crystals after cutting seen from above and the right-hand photographs show their cross sections.

Figures 392(*b*) and 393(*b*) are photographs of sections of a branch of a dendritic crystal cut perpendicularly to the direction of the channel. As seen in Figs. 392(*a*) and 393(*a*), a dendritic branch has always two parallel lines running along its center. As shown by the photographs of the cross sections, these lines are caused

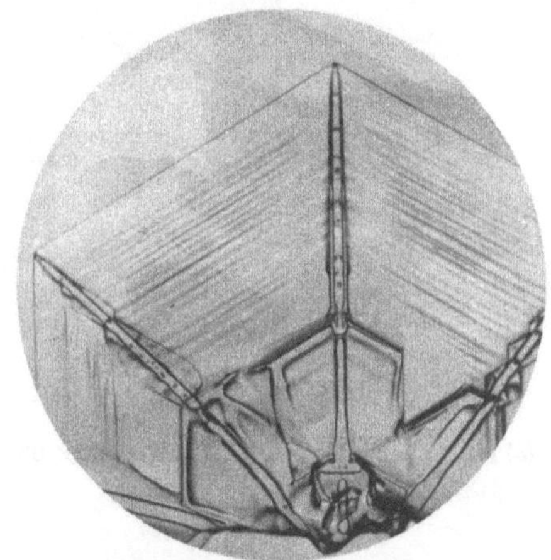
FIG. 389 [1310]. Thin-plate crystal produced by diffusion of vapor (× 26).

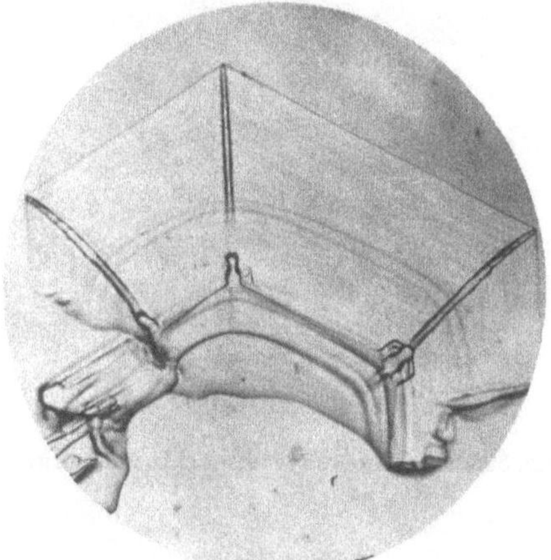
FIG. 390 [1311]. Thin-plate crystal produced by diffusion of vapor (× 26).

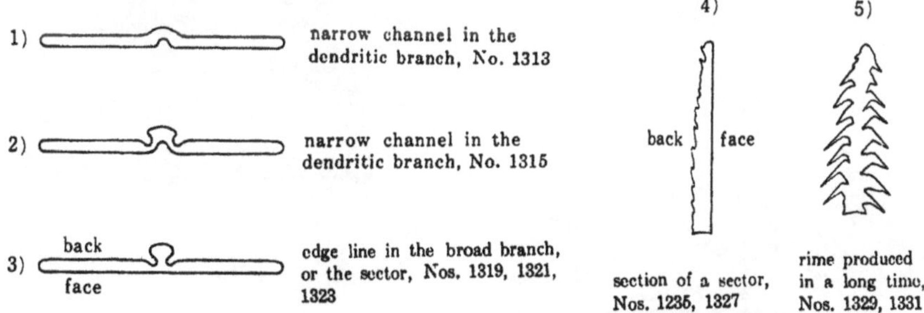

FIG. 391. Sketches of cross sections of crystals.

by a narrow ditch on one side and a corresponding ridge on the other side. Its structure differs in each crystal, but this narrow ditch occurs almost invariably, as shown in Fig. 391(1) and (2). The raised part of (2) would be the same object as the ridge in a sector or plate for which detailed explanation was given in Sec. 66.

This ridge is very clearly observed in broad-branched and sector crystals. The crystal of Fig. 394(*a*) is a representative sector, and its section shown in Fig. 394(*b*) indicates that it has the same structure as that of Fig. 369(14). The same thing can be said about the sector of No. 1320, Plate 182. No. 1318, Plate 182, is one of the branches of a broad-branched type, in whose section a similar ridge is recognized. According to the hypothesis proposed in Sec. 66, that the sector-form crystal is one of the special cases of skeleton structure, the above-mentioned fact is naturally to be expected. Therefore, as seen in Fig. 391(3), the side that has the ridge is the "back" side.

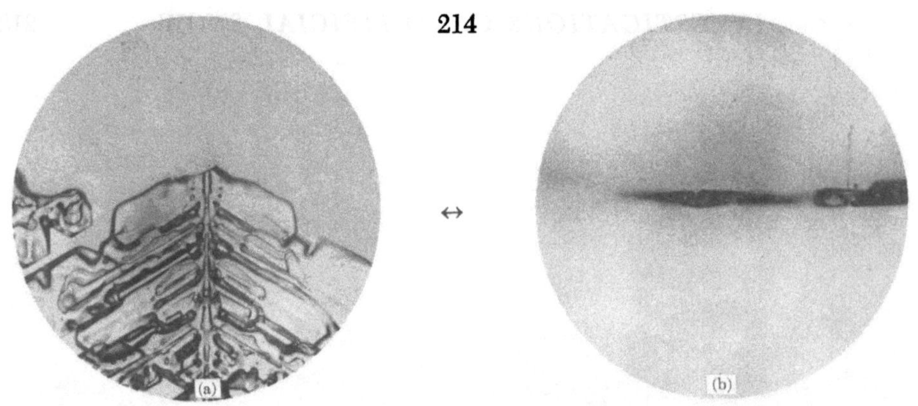

Fig. 392. Structure of the channel in a dendritic branch: (*a*) [1312] plan view; (*b*) [1313] section (× 38).

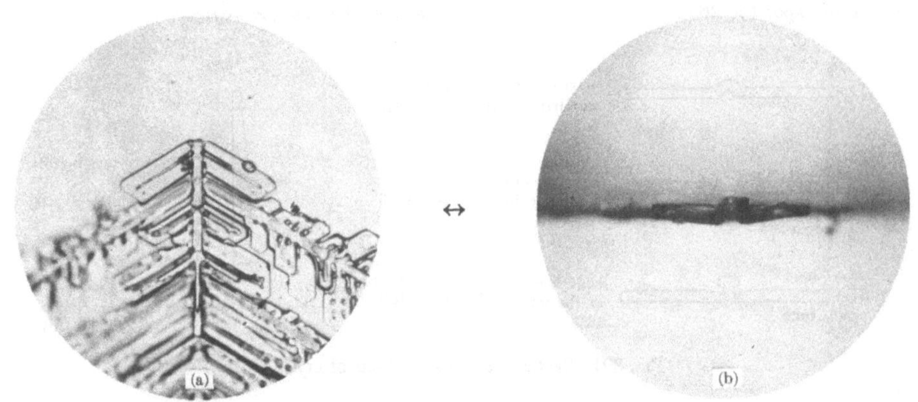

Fig. 393. Structure of the channel in a dendritic branch: (*a*) [1314] plan view; (*b*) [1315] section (× 42).

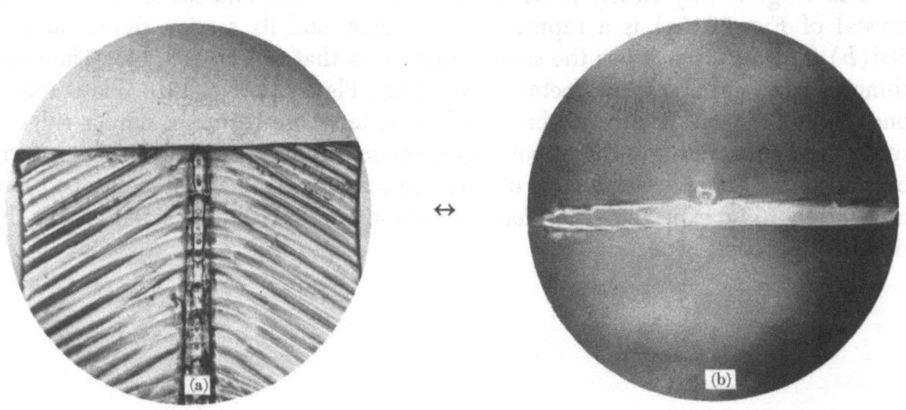

Fig. 394. Structure of the channel in a sector crystal: (*a*) [1322] plan view; (*b*) [1323] section (× 38).

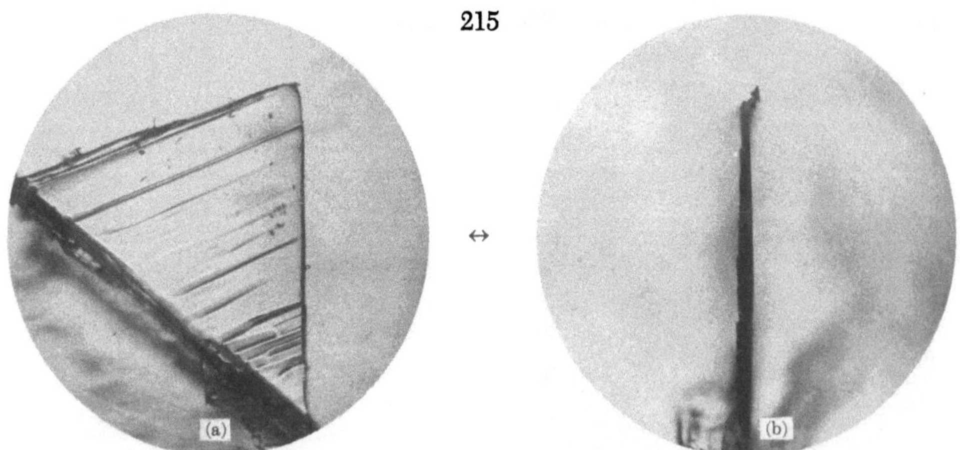

FIG. 395. Minute structure of a sector: (*a*) [1324] plan view; (*b*) [1325] section (× 40).

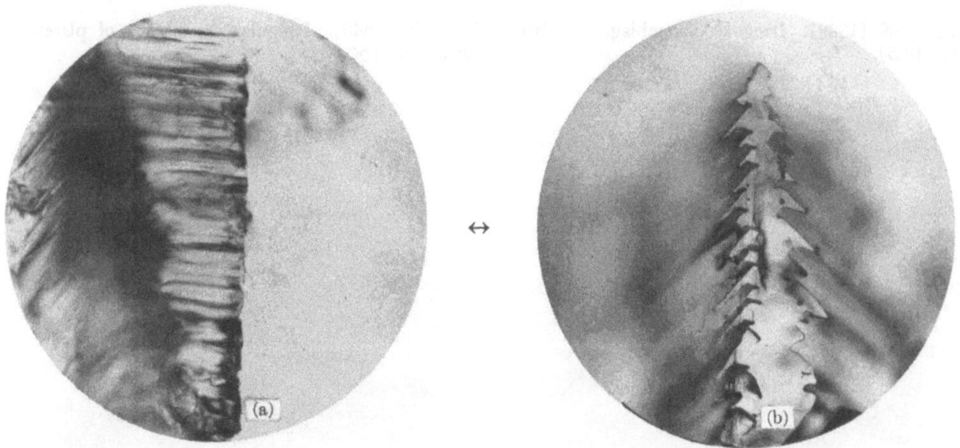

FIG. 396. Frost crystal made by slow process: (*a*) [1330] plan view; (*b*) [1331] section (× 22).

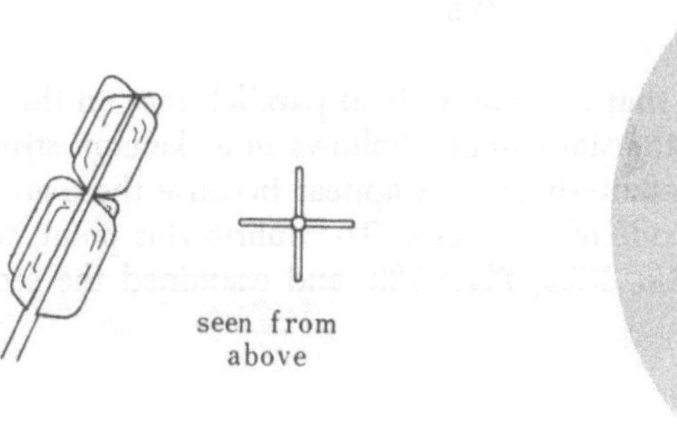

FIG. 397. Cross structure of assemblage of plates.

FIG. 398 [1337]. Stellar crystal with spatial small plates (× 18.5).

Fig. 399 [1339]. Irregular assemblage of plates (× 18.5).

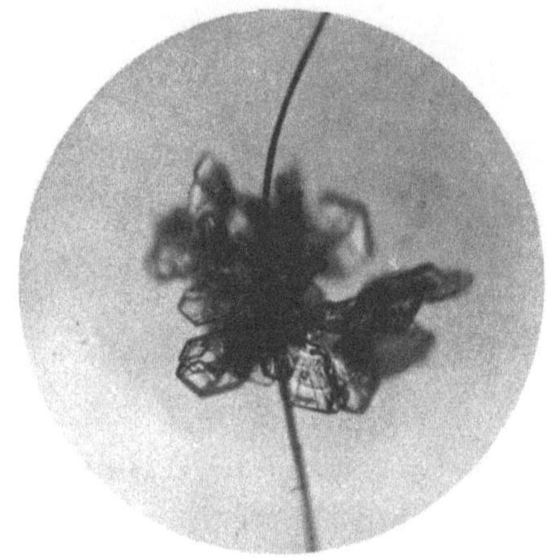

Fig. 400 [1340]. Irregular assemblage of plates, early stage (× 59).

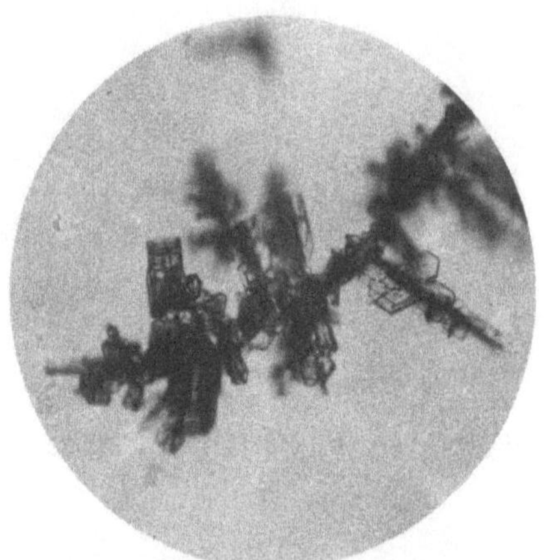

Fig. 401 [1344]. Irregular assemblage of plates, with scanty suppply of vapor (× 17).

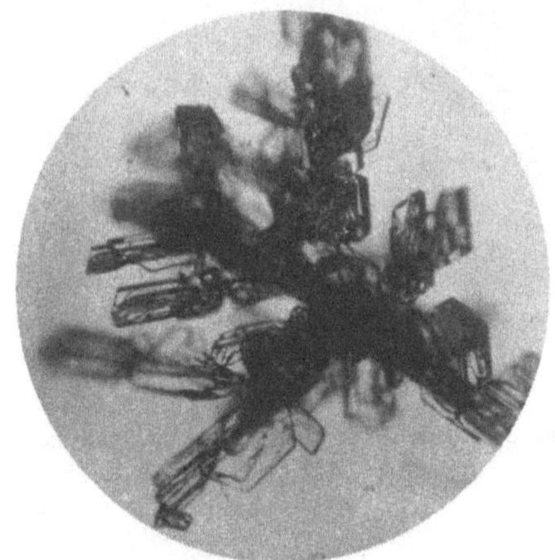

Fig. 402 [1346]. Crystal with extended side planes (× 17.5).

It has already been mentioned that the infinitesimal parallel lines on the sector have the same nature as those of the staircaselike hollows in a skeleton structure. We therefore inferred that these infinitesimal lines appear because there are staircaselike depression on the "back" side of the sector. To confirm this point, we cut the sectors like Fig. 395(*a*) and No. 1326, Plate 182, and examined the sections.

The results were just as we had expected; the "face" side was even, while the "back" side had the staircaselike depression as seen in Fig. 395(*b*) and No. 1327, Plate 182, as well as in Fig. 391(4). Although we said "even," there will be such slight staircaselike depressions as are invisible under this order of magnification on the "face" side also, if the sector is a special case of the skeleton. This is inferable from the structure of frost crystals made by a slow process.

No. 1328, Plate 183, is a frost crystal belonging to the cup type, which was formed by a slow process on the wooden part of the lid of the apparatus, with T_a at $-22°C$ and T_w at $+10°C$. Its cross section, as shown in No. 1329, Plate 183, reveals small projections on both sides like a Venetian blind. Figure 396 shows frost crystals growing on the cooling pipe of the cold chamber, and they have a similar structure.

69. Spatial assemblage of plates

When the temperature goes down to about $-20°C$ or lower, irregular forms of spatial assemblage of small plates are apt to be produced. In general it is difficult to find definite rules for the types of spatial assemblage, but the one shown in Fig. 397 has a comparatively plain form, namely, plates developed crosswise from a needle or a thin column. For convenience, we call it a cross structure. Various types of crystals which are obtained at low temperature are reproduced in Figs. 398–402 and Plates 183 and 184. Conditions for formation of these crystals are listed in Table 29 with explanations.

In the case of Fig. 398, the early conditions having been suitable for dendritic development, with T_a nearly $-15°C$ and T_w nearly $+14°C$, a plane hexagonal crystal was formed. Later, when T_a was decreased to $-20°C$ or lower, small plates grew spatially out of the hexagonal crystal and finally developed to the crystal shown in the photograph. Crystals of this kind are often observed in natural snow. No. 1336, Plate 183, is a similar crystal. In this case the supply of water vapor was abundant, while it was scanty in the case of Fig. 398. The cross structure is clearly seen on one of the branches of Fig. 398. In Figs. 399 and 401 also one may see this cross structure, which is very often formed when the temperature is about $-20°C$ or lower and the supply of water vapor is not abundant.

If the conditions are suitable for a spatial assemblage of plates from the early stage, the whole object becomes an irregular spatial assemblage, as will be expected in view of the above results. Two examples are reproduced in Fig. 399 and No. 1338, Plate 183. These photographs were taken 1.5 hr after the formation of the germ. The crystals in the early stage had shapes similar to these final ones. Figure 400 and No. 1341, Plate 183, are crystals formed under similar conditions, but they show a comparatively early stage, 30 min after the formation of the germ.

Table 29. Formation of crystals obtained at low temperature.

Fig. No.	Photo. No.	T_a (°C)			T_w (°C)			Time for growth		Remarks
		Initial	Middle	Final	Initial	Middle	Final	(hr)	(min)	
	1336	−15	−19	−20	+13	+14	+15	1	30	Early stage plane hexagonal; spatial plates when T_a decreased
398	1337	−16	−20	−24	+15	+12	+ 9	2	30	Similar to No. 1336; T_a still lower, cross structure observed
	1338	−20	−24	−23	+14	+13	+12	1	20	T_a low from beginning; irregular spatial assemblage
399	1339	−20	−19	−19	+12	—	+10.5	1	40	Similar to No. 1338, cross structure observed
400	1340	−19.5	—	−19.5	+11	—	+10.5		30	Relatively early stage of spatial assemblage of plates produced at low T_a
	1341	−19.5	—	−19.5	+11	—	+10.5		30	
376	1342	−17.5	−17.5	−17.5	+13	+12	+12	2	30	Ridges running on one side of sector; special case of skeleton
	1343	−23	−23	−23	+ 8	+10	+12	1		Irregular assemblage obtained at T_a = −23°C; a cup is seen
401	1344	−38	−37	−33	0	− 6	+ 2	2	50	Temperature very low, scanty supply of vapor; cross structure is seen
	1345	−29	−32	−30	+ 6	− 3	+ 5	4		Temperature very low, relatively ample supply of vapor
402	1346	−26	−24	−22	+ 7	+ 6	+ 5	20		Assemblage of columns with extended side planes, similar to this type of natural snow
	1347	−22	−23	−24	+ 7	+ 8	+ 8	2	30	Similar to No. 1346; relatively early stage

Figure 376 was formed when T_a was a little lower than for dendritic development. The ridges running on the back side of the sector are clearly observed. As a whole, this is a spatial assemblage of plates.

When the air temperature goes still lower, say in the range between $-23°C$ and $-33°C$, the crystal becomes an extremely irregular form, as shown in Fig. 401 and No. 1345, Plate 184. In Fig. 401, T_w was near 0°C and the supply of water vapor was extremely scanty. The cross structure is very well developed in this example. In No. 1345, Plate 184, the vapor supply was comparatively abundant.

Figure 402 and No. 1347, Plate 184, belong to the new type that was named "columnar crystals with extended side planes" in natural snow. We therefore infer that this type of natural snow is formed when the temperature is lower than $-20°C$ and the supply of water vapor is scanty.

70. Solid needle

In the early stage of crystal development, we often obtain solid needles or columns. They are formed when T_a is low and the supply of water vapor is scanty. Examples are shown in Figs. 403 and 404. In Fig. 403, a minute solid needle is observed on the upper part of the hair, with two small columns, nearly solid, hanging on the lower part. These minute crystals were formed by keeping T_a at $-25°C$ and T_w at $+4°C$ for 20 hr. That is to say, when the growth is extremely slow, the crystal becomes solid, as common knowledge of crystallography tells us. The solid needle of Fig. 404 was obtained under similar conditions, but growing on a glass surface.

This solid needle is often observed in the tsuzumi-type crystal in its early stage; Figs. 377, 405, and 406 are examples. Figures 377 and 405 are good examples in which photography was successful. Hexagonal plates are attached to both ends of a solid needle perpendicularly to the axis. In the case of Fig. 377, T_a was $-18°C$, while in Fig. 405, T_a was $-20°C$. In both cases, the supply of water vapor was scanty. In these cases, if T_a and T_w are raised toward the conditions for dendritic development, the crystals become tsuzumi-type ones with a solid thin needle as a pillar. One example is shown in Fig. 406. Another example is shown in Nos. 1356 and 1357, Plate 184. In No. 1356, the hair was set when T_a was $-20°C$, and then T_a was gradually raised. When T_a was $-18°C$, T_w being $+13°C$, a primitive tsuzumi was formed. Later, T_a was kept at $-16°C$ and T_w at $+15°C$ for 2 hr and the sector-form tsuzumi as shown in the photograph was obtained. The tsuzumi crystal with dendritic extension of Fig. 406 was obtained by keeping the primitive tsuzumi under conditions suitable for dendritic growth, namely, $T_a = -15°C$, $T_w = +18°C$ for 1 hr 15 min. These conditions for formation and the forms of crystals to be completed can be expected from our knowledge already obtained.

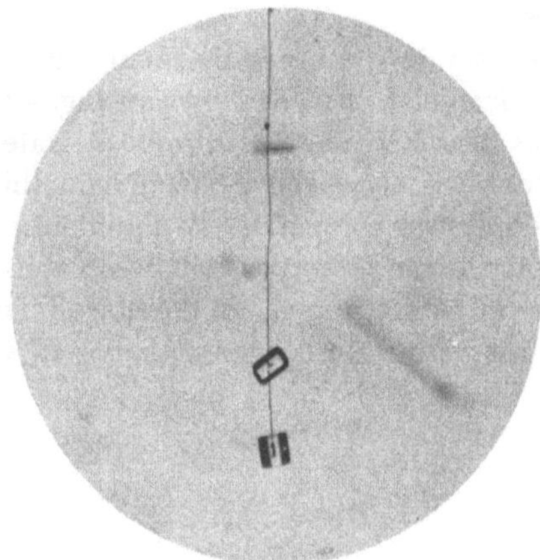

FIG. 403 [1348]. Minute solid needle (× 19).

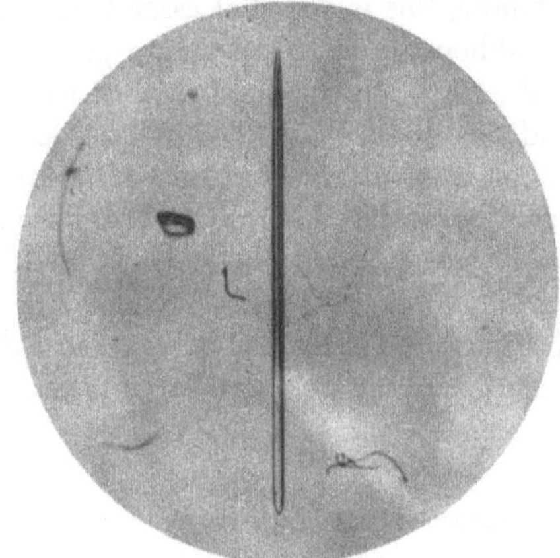

FIG. 404 [1349]. Solid needle (× 36).

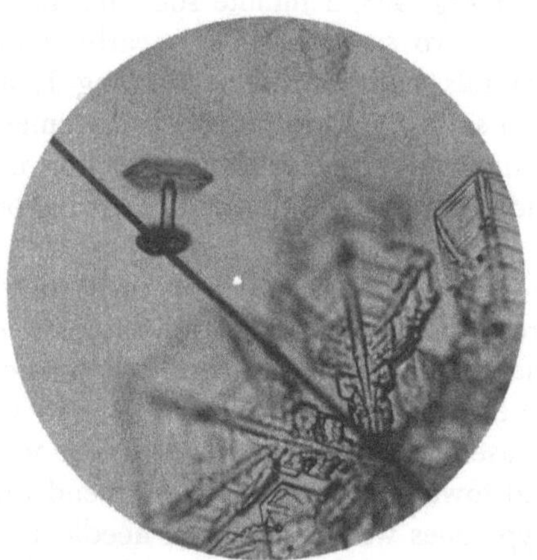

FIG. 405 [1355]. Tsuzumi with solid needle, early stage (× 36).

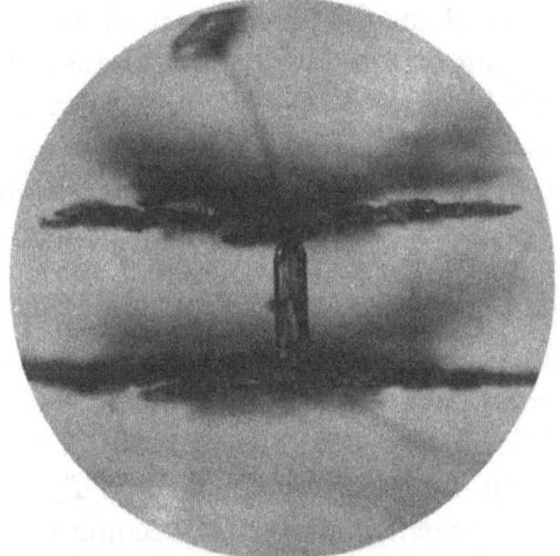

FIG. 406 [1358]. Tsuzumi developed from an early stage similar to Fig. 405 (× 30).

If a primitive tsuzumi such as is shown in Fig. 377 is kept under the initial condition for a long time after its formation, the upper and lower plates become gradually thickened and the central solid needle becomes thicker. In one experiment, a primitive tsuzumi like the one in Fig. 377 was formed with T_a at $-20°C$ and T_w at $+4°C$. It was left overnight, and when it was examined, after 13 hr, it had developed into the form shown in No. 1359, Plate 185; the space between the plates was nearly filled with ice. In another experiment, a primitive tsuzumi was obtained with T_a at $-22°C$ and T_w at $+9°C$, which, after being left in the same condition for 19 hr, changed into a skeleton column similar to Fig. 370. At least one of the mechanisms of formation of a skeleton column was thus clarified.

When a thin column is formed in the early stage, it does not grow much longer but becomes thicker if the conditions are maintained for a long while. This phenomenon was also observed in the development of proper tsuzumi-type crystals, as will be described in the following section.

71. Tsuzumi-type crystal

Crystals of the tsuzumi type are also easily produced from what we have hitherto learned in making artificial snow. First, the conditions suitable for formation of a column are established and maintained until the column grows to a sufficient size, and then the conditions are changed suddenly into those suitable for the dendritic development. Plane crystals begin to extend from both ends of the column and a tsuzumi-type crystal is formed. Examples of tsuzumi-type crystals thus formed are given in Figs. 407–412 and Plates 185 and 186.

Figure 407 is a column formed with T_a at $-19°C$ to $-17°C$ and T_w at $+3.5°C$, maintained for 2 hr 15 min. Columns are formed in this way when T_a is low and the supply of water vapor is scanty, as mentioned before. Generally, columns are thin in the early stage. Kept under the same conditions for a sufficient time, these thin columns become thicker without growing much longer, and columns of ordinary size are obtained. When a column was nearly completed, as shown in Fig. 407, both T_a and T_w were raised to the level suited to dendritic development, and a tsuzumi-type crystal, Fig. 408, was obtained. The time interval between Figs. 407 and 408 was 45 min. In this case it takes a certain length of time before the dendritic crystals begin to extend from the column. During this period, for about 30 min, the ridges of the column show a strong tendency to extend. This period is named the "transition period" and is shown in Table 30. During this transition period the bases of the column also begin to grow out, and then the dendritic branches start. After this start, the dendritic development is rather rapid. The course of these changes is schematically shown in Fig. 409, in which (3) represents the transition period.

Figures 410–412 are photographs showing the plan view and the side view of each of three examples, the right-hand photographs being side views of the crystals on the left. The crystal of No. 1365, Plate 186, was formed by a very slow process, so that the whole crystal has a nearly solid form. But other crystals were formed under more or less similar conditions. Those conditions are tabulated in Table 30.

The conditions for formation of tsuzumi-type crystals having been thus clarified,

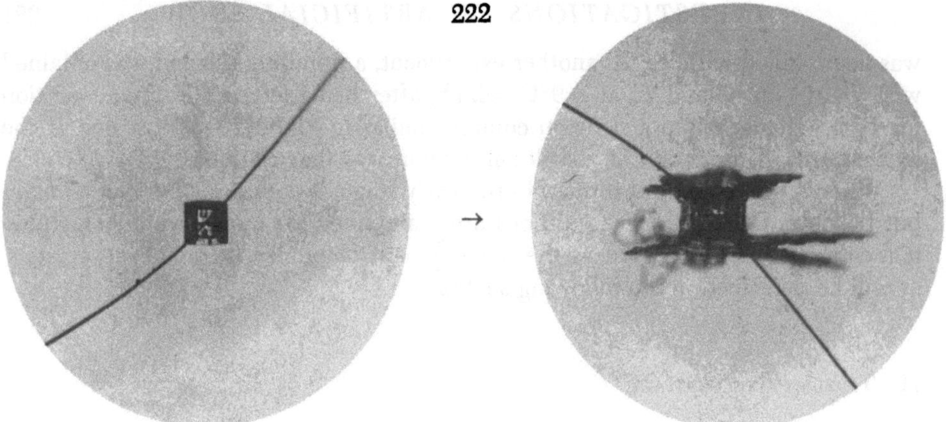

FIG. 407 [1360]. Hexagonal prism.

FIG. 408 [1361]. Tsuzumi developed from Fig. 407.

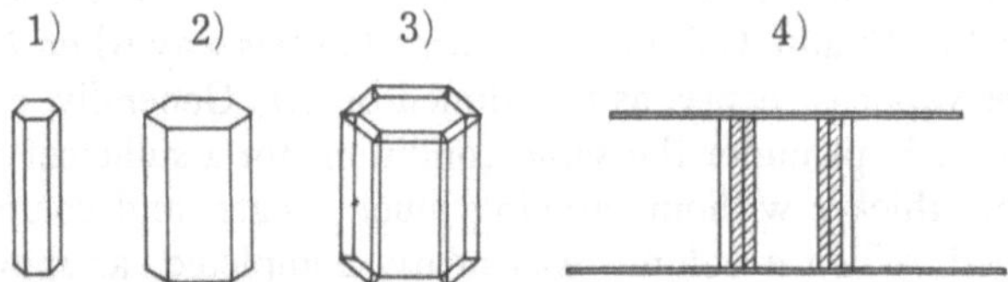

FIG. 409. Stages in development of tsuzumi crystal.

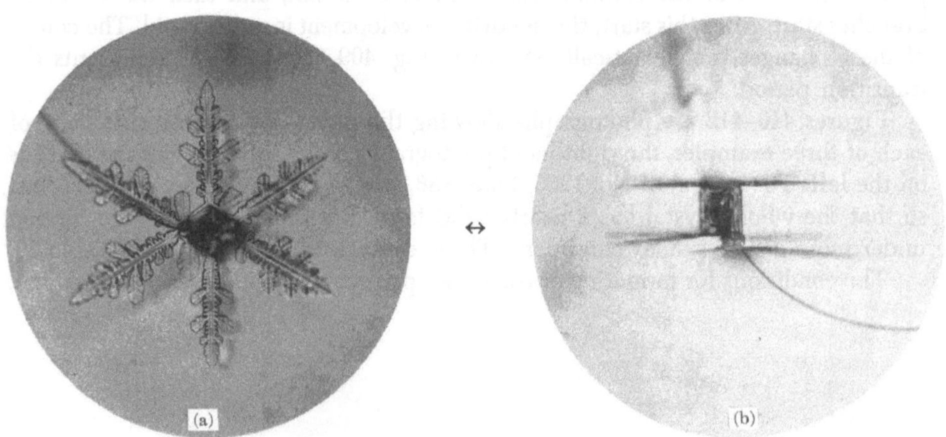

FIG. 410. Ordinary tsuzumi crystal made artificially: (*a*) [1366] plan view; (*b*) [1367] side view (× 34).

TABLE 30. Formation of tsuzumi crystals.

Fig. No.	Photo. No.	Columnar development				Transition period (min)	Dendritic development				Remarks
		T_a (°C)	T_w (°C)	Time for growth (hr)	Time for growth (min)		T_a (°C)	T_w (°C)	Time for growth (hr)	Time for growth (min)	
407–8	1360–1	−19 – −17	+ 7.0 – + 3.5	2	15	30	−17 – −16	+12.5 – +12.0		15	Fig 407 shows the last stage of columnar development
	1362	−21 – −20	+ 4.0 – + 8	2	45	30	−16 – −15	+16 – +16		30	One side, spatial hexagonal
	1363	−22 – −20	+ 5 – +10	1		30	−18 – −15	+10 – +15	1	30	The same experiment as No. 1368
	1364	−20 – −19	+ 1.0 – + 4.0	2	30	30	−14.5 – −15.5	+17.5 – +18		15	Few water droplets attached
	1365	−22 – −20	+ 4.5 – + 6.0	22		—	−20.5 – −19.5	+ 5.0 – + 5.0	27		Made by a slow process
410	1366–7	−23 – −19	+10 – +14		45	—	−18 – −15	+14 – +15		15	Good example of dendritic tsuzumi
411	1368–9	−22 – −20	+ 5 – +10	1		30	−18 – −15	+10 – +15	1	30	Sector-form tsuzumi
412	1370–1	−22 – −17.5	+ 3.0 – +10.0	5		30	−16 – −14.5	+14.5 – +16		15	Tsuzumi of broad-branched type

any required type of tsuzumi can now be artificially obtained. The problem lies in the inside column. When the time for column formation is short, it becomes thin, as shown in Fig. 411(*b*), and when the time is long, it becomes thick, as shown in Fig. 412(*b*). This point is the secret for making the required type of tsuzumi crystals.

72. Sheath-type crystal

According to Seligman,[57] a crystal composed of only the side planes of the hexagonal column is observed, though rarely, among ice crystals in nature. We [58] observed a crystal which is a very good example of this type in a cavity of frozen tundra in South Sakhalin. The photograph of this crystal is reproduced in Fig. 413. This is of course a frost crystal, but in natural snow also we observed crystals of this rare type at Mount Tokachi. Although the crystal of Fig. 413 is not complete in all of the six side planes, this photograph may be regarded as representing a typical crystal of this kind.

In artificial snow, we learned that there are three kinds of variation in this type, as schematically sketched in Fig. 414; (1) is a hexagonal cylinder with thin walls, the frost crystal of Fig. 413 being a typical example of this kind; (2) has a shape like a hexagonal cylinder but one of its side planes is cut and the flaps are folded like a scroll. The frost crystal discovered by Wegener in a glacier crevasse in Greenland, mentioned in Sec. 64, belongs to this scroll-form type. The crystal of type (3) has a special structure whose plan and elevation are shown in the sketch. We group the crystals belonging to these three kinds under the general term of "sheath-type crystals." *

Photographs of these crystals are reproduced in Figs. 415–418 and Plate 186. The crystal of Fig. 415(*a*) has the shape shown in its end view, Fig. 415(*b*), so that this crystal belongs to (1). In general, crystals of this type are slowly formed when the temperature is low and the supply of water vapor is not so scanty. Figure 415 is a crystal which was formed overnight, with T_a at $-25°C \pm 2°C$ and T_w at $+6°C \sim +12°C$, about 15 hr being required for its formation.

Figures 416 and 417 show crystals made by a slow process with T_a at approximately $-25°C$, which is much lower than that for dendritic development. Data are given in Table 31. The peculiar structure of the sheathlike crystal of type (3) may be very well seen in Fig. 418. Further careful investigations will be worth carrying out with respect to this peculiar type.

* It seems to be more reasonable to call (1) and (2) of Fig. 414 the "scroll type" and treat (3) separately as the "sheath type," but our investigation so far is not sufficient to assure this point.

TABLE 31. Formation of sheathlike crystals.

Fig. No.	Photo. No.	T_a (°C) Initial	T_a (°C) Middle	T_a (°C) Final	T_w (°C) Initial	T_w (°C) Middle	T_w (°C) Final	Time for growth (hr)	Time for growth (min)	Remarks
416	1374	−26	−26	−25	+4.7	+ 7.9	+ 9.2	4	10	The flaps of a scroll are rolled in, (2)
417	1375	—	−28	—	—	+11	—		—	Sheathlike crystal of peculiar form, (3)
	1376	−24	−26	−25	+4	+ 8	+12	5	30	
	1377	−24	−24	−22	+9	+11	+12	5	30	Sheathlike crystal of peculiar form, (3)

73. CUP CRYSTAL

It is widely known that crystals in the shape of a hexagonal whisky glass are occasionally observed in natural frost. We found that this cuplike crystal also exists in snow as a special case of the skeleton column. The cups do not grow large when they are formed in the air. But we can obtain cups of considerable size attached to the lid or glass wall within the apparatus for artificial snow. Examples of such cups are reproduced in Fig. 419 and Plate 187.

The fact that these crystals are likely to grow on the surface of some solid object suggests that one of the requirements for their formation is the rapid removal of heat of crystallization by conduction. In most cases, cups were formed when the apparatus was left overnight in the cold chamber. So another favorable condition seems to be that the convection of air and water vapor is weak. Generally, a low temperature was suitable for their growth, but there was an exceptional example, shown in No. 1382, Pl. 187, in which T_a was −10°C. Conditions for their formation are shown in Table 32.

TABLE 32. Formation of cup crystals.

Photo. No.	T_a (°C)	T_w* (°C)	Remarks
1379	−20 − −25	− 2	Grew on the inside of cork lid of the apparatus, about 24 hr; water in the reservoir was frozen
1381	−21	+2 − +10	Grew on the inside of glass window of the apparatus, left overnight
1382	−10	+ 9	Grew on the inside of the cork lid, left overnight
1383	−20	+19	Grew on the inside of the cork lid

* Since the mode of convection of water vapor is unknown, this quantity has not much physical meaning.

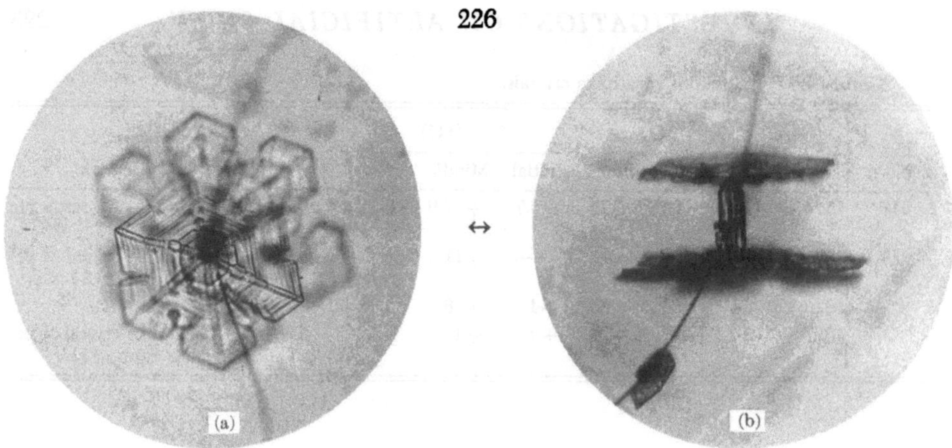

FIG. 411. Tsuzumi with thin column: (*a*) [1368] plan view; (*b*) [1369] side view (× 35.5).

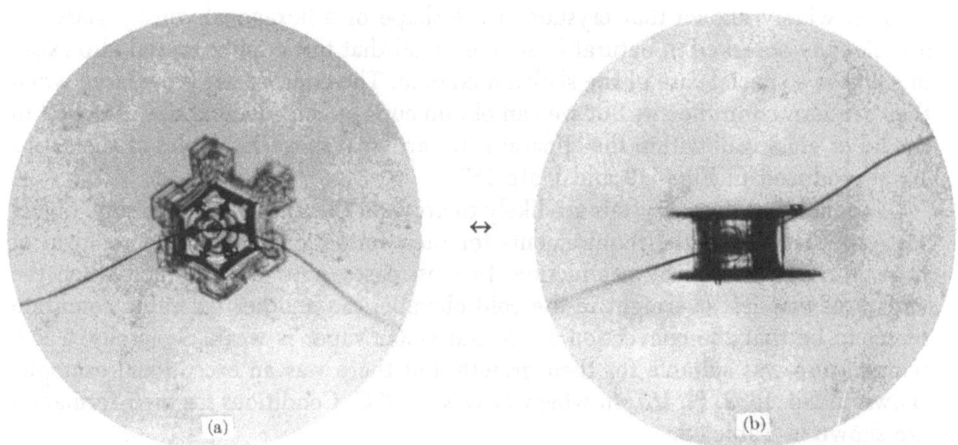

FIG. 412. Tsuzumi with thick column: (*a*) [1370] plan view; (*b*) [1371] side view (× 31).

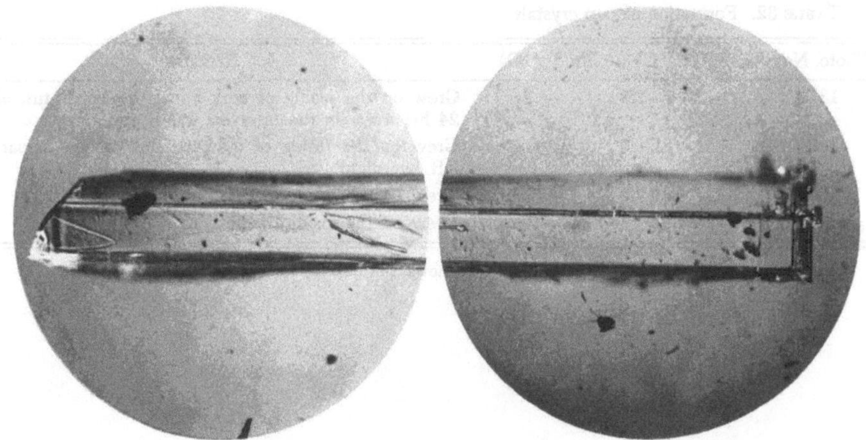

FIG. 413. Sheathlike crystal of frost found in frozen tundra of South Sakhalin.

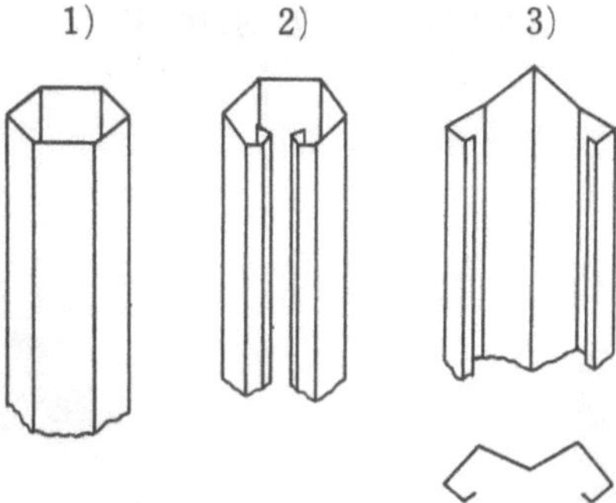

Fig. 414. Three kinds of sheathlike crystal.

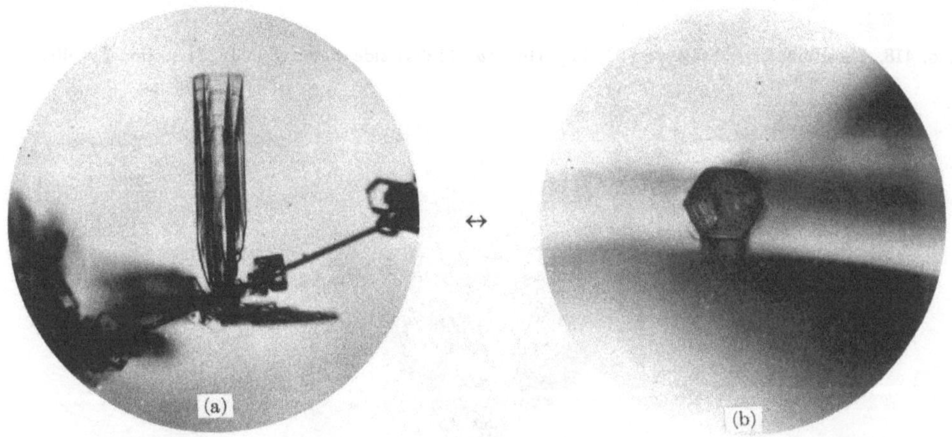

Fig. 415. Sheathlike crystal of type (1), Fig. 414: (*a*) [1372] side view; (*b*) [1373] end view (× 14).

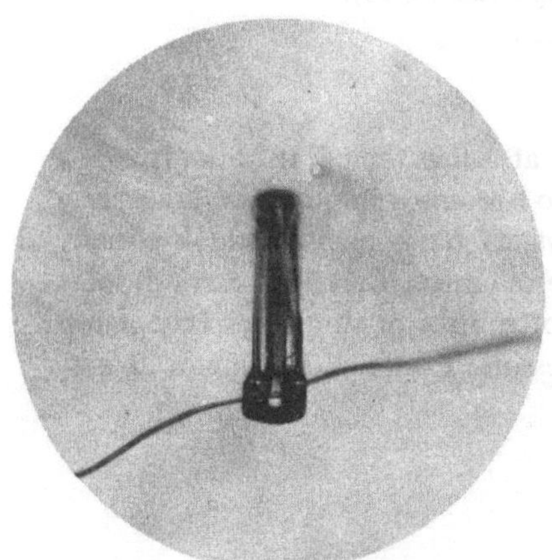

Fig. 416 [1374]. Sheathlike crystal of type (2), Fig. 414 (× 46).

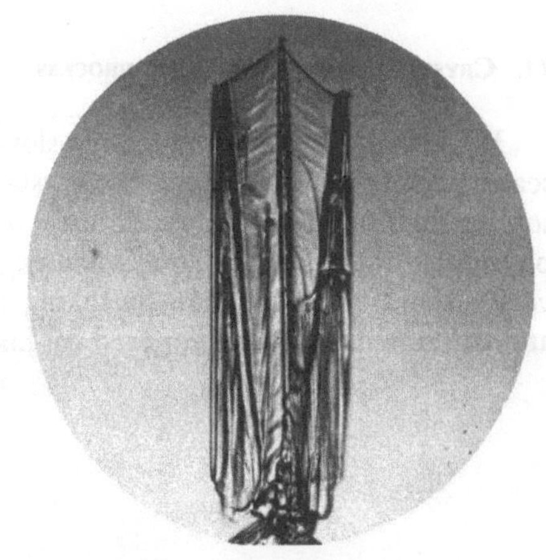

Fig. 417 [1375]. Sheathlike crystal of type (3), Fig. 414 (× 23).

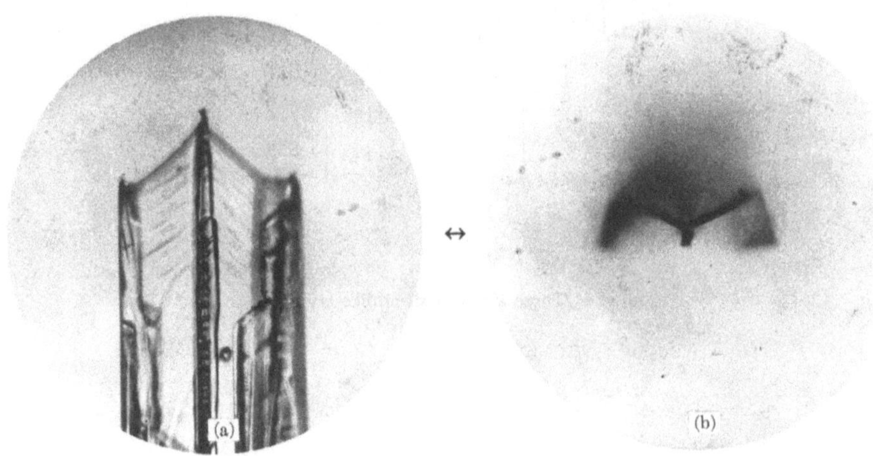

FIG. 418. Sheathlike crystal of type (3), Fig. 414: (*a*) [1334] side view; (*b*) [1335] section (× 40).

FIG. 419 [1378]. Well-developed cup crystal (× 26).

74. CRYSTALS MADE BY A SLOW PROCESS

If we make a crystal by a very slow process at a low temperature and with a scanty supply of water vapor, the crystal ought to become a single crystal of nearly solid structure. X-ray investigations of ice crystals have been made by many physicists, but a definite conclusion is yet to be obtained. Barnes [59] carried out a very elaborate investigation by taking Laue photographs of single crystals of ice, and confirmed the view proposed by his predecessors, Tamman, Bridgman, Bragg, and others that the ordinary ice crystal belongs to the hexagonal system. He went a step further and says that there seems to be every reason to believe that the ice crystal is D^4_{6h} (dihexagonalbipyramidal). But Brandenburger [60] opposes Barnes's view, saying that the photographs taken by the latter provide little basis for such conclusion.

Mügge [61] is opposed to the hexagonal-system theory and considers that ice is to be assigned to the rhombohedral system. Seljakov [62] conducted x-ray studies on ice crystals obtained by freezing water in various ways, and concluded that two modifications, α and β, exist in ordinary ice, α being hexagonal and β rhombohedral. Also Cohen and van der Horst [63] insist upon the existence of the ice crystal of the cubic system on the basis of their experiments. There are as yet various unclarified points with regard to the crystallization system of ice. It might be more complicated than is usually assumed.

The hexagonal prism is the easiest one to be obtained by a slow process of formation. This crystal is often formed at the tips of the branches of hexagonal plane crystals. Figure 420(*a*) shows one example. In this case, after an ordinary hexagonal crystal was formed, the temperature and the water-vapor supply were reduced to $T_a = -22°C$ to $-21°C$, $T_w = +8°C$ to $+12°C$, and kept there for about 4 hr. Typical skeletons are developed on the branch tips, some of which show the form of a single crystal of nearly solid structure. In (*b*) is shown a side view of the branch tips of this crystal. It is clearly observed in this photograph that the thick one shows a skeleton structure but the thin one is nearly solid. No. 1386, Plate 187, is another crystal, obtained in the same experiment with Fig. 420.

If a skeleton prism is formed upon a hair from the beginning, it develops into a complete hexagonal prism. Some of these crystals have already been shown, but Fig. 421 is a beautiful one. This crystal took 6 hr for its formation, with T_a at $-20°C$ to $-25°C$. In this experiment T_w was lowered and the water surface was partially covered with ice, that is, T_w was kept at approximately 0°C.

We could obtain a beautiful example of a complete column when we made it with T_a at nearly $-30°C$ and T_w nearly 0°C by a very slow process, taking 70 hr for the formation. Figure 501(*b*) is the crystal thus obtained and No. 1388, Plate 187, is the same crystal viewed aslant. In this case, we fitted a stop diaphragm in the apparatus to control the rising current of air and made the supply of water vapor extremely scanty. Its shape indicates without doubt that this crystal belongs to the hexagonal system. Figure 422 is a thick plate of solid structure, which was obtained on a different hair in the same experiment when Fig. 501(*b*) was formed.

While repeating similar experiments, we came across crystals apparently belonging to the rhombohedral system. Figures 424 and 425 are such crystals. In this case, the room temperature was the same as the former case, about $-30°C$, but the supply of water vapor was a little more abundant, T_w being about $+9°C$. Therefore, the crystals developed into a skeleton structure to some extent. The growth

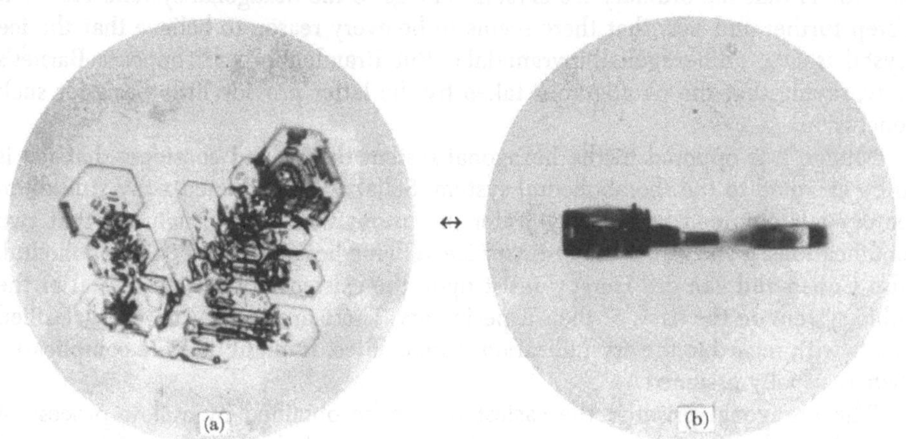

FIG. 420. Dendritic crystal of nearly solid structure: (*a*) [1384] general view (× 34.5); (*b*) [1385] end view of branches (× 33).

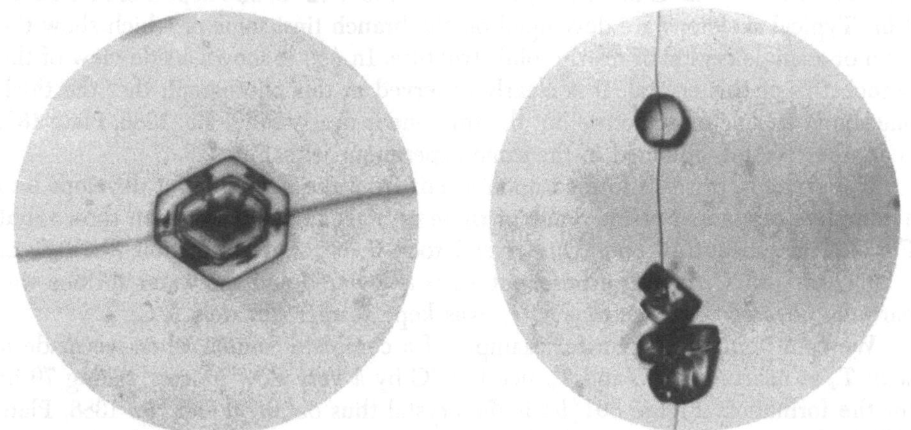

FIG. 421 [1387]. Solid hexagonal plate with trace of skeleton structure (× 73).

FIG. 422 [1389]. Solid hexagonal plate (× 32).

was also a little more rapid. The photographs show the stage at 31 hr after the beginning of the experiment. These crystals are shown schematically in (1) and (2) of Fig. 423. Judging from their shapes, they may be classified as the β-ice belonging to the rhombohedral system.

In the next place, we obtained a rectangular crystal that seemingly belonged to the cubic system. Figure 426 is an example, which is formed in 20 hr with the

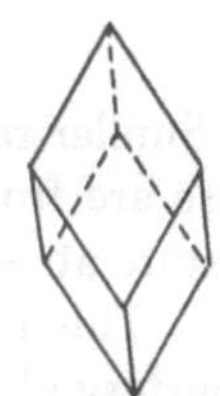
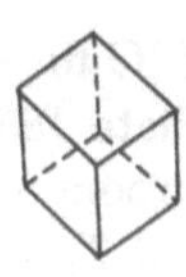
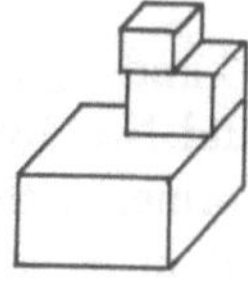
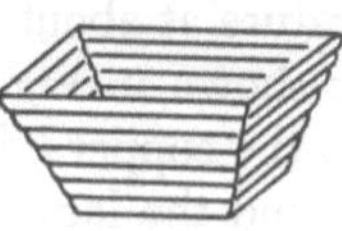

1) No.1390 2) No.1391 3) No.1392 4) No.1393

Fig. 423. Sketches of apparently rhombohedral and cubic crystals.

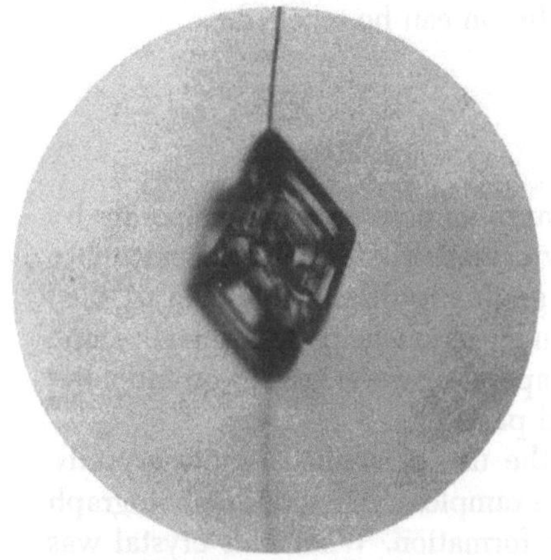

Fig. 424 [1390]. Apparently rhombohedral crystal (× 42).

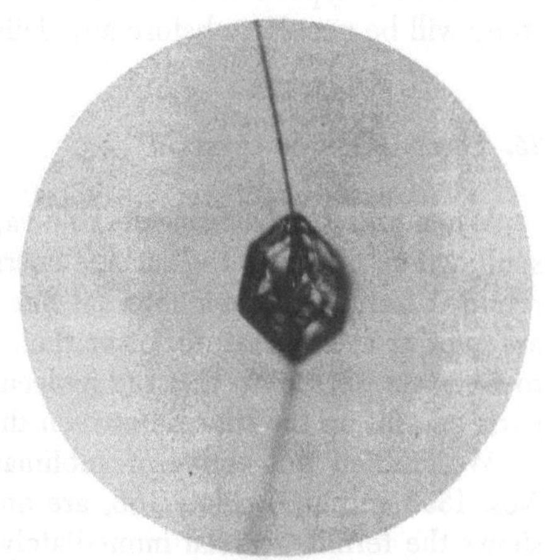

Fig. 425 [1391]. Apparently rhombohedral crystal (× 45).

Fig. 426 [1392]. Apparently cubic crystal (× 20).

Fig. 427 [1393]. Apparently cubic crystal (× 26).

room temperature at about −28°C and T_w at −6°C to +7°C. Similar rectangular skeletons are shown in Fig. 427 and No. 1395, Plate 188. These are frost crystals developed on the copper lid, the former with the room temperature at −30°C and T_w at +9°C, and the latter with T_a at −24°C and T_w at +8°C. This rectangular skeleton is often observed also in natural frost. Through the courtesy of Dr. Hatakeyama, we are able to reproduce the photograph of a natural frost of this kind taken by him at Toyohara, South Sakhalin, in No. 1394, Plate 188. Although crystals of this type may be regarded as ice of the cubic system, following Cohen, we cannot decide so by these photographs alone, because the hexagonal skeleton can grow to this rectangular shape if only one of its sides is developed. To show the shapes of Figs. 426 and 427 more clearly, they are schematically sketched in (3) and (4) of Fig. 423. The difference in the conditions for formation of such extraordinary types cannot be decided yet. Further investigations along with x-ray study will be necessary before any definite conclusion can be reached.

75. Sublimation of crystals

When a crystal has considerable vapor pressure, its pointed parts evaporate by sublimation, owing to the surface energy. Snow crystals therefore degenerate into rounded forms and their internal fine pattern gradually disappears, even if they are kept at a temperature lower than 0°C. Condensation helps this phenomenon to a certain degree, so that the molecules of evaporated water gather on indented parts and fill up the spaces between the pointed parts.

We studied this action of sublimation by the use of artificial snow crystals. Nos. 1396 and 1397, Plate 188, are one of the examples. The former photograph shows the fernlike crystal immediately after its formation. When this crystal was left in the cold chamber at −19°C, the internal fine pattern soon began to disappear and its outline came to be gradually rounded, until after 4 hr it was so degenerated that its original form was hardly imaginable, as shown in No. 1397. Walls and cooling pipes within the cold chamber are all covered with frost crystals, so that the air in the room must be regarded as nearly saturated with water vapor. Therefore, the sublimation is due to the surface energy of the crystal. Needless to say, sublimation is remarkably accelerated by wind. So the rate of evaporation becomes smaller if the crystal is kept inside a closed bottle.

Figure 428(a) is a fernlike crystal just after its formation. Preserved within a liter bottle at −16°C, suspended by a filament, this crystal does not show much transformation after 1 hr, as shown in (b). But evaporation gradually goes on even in the bottle. After 2 hr the crystal becomes (c); after 20 hr at −20°C, (d). In other words, sublimation is gradually continued even at the low temperature of −20°C and with the crystal kept in the closed bottle saturated with water vapor.

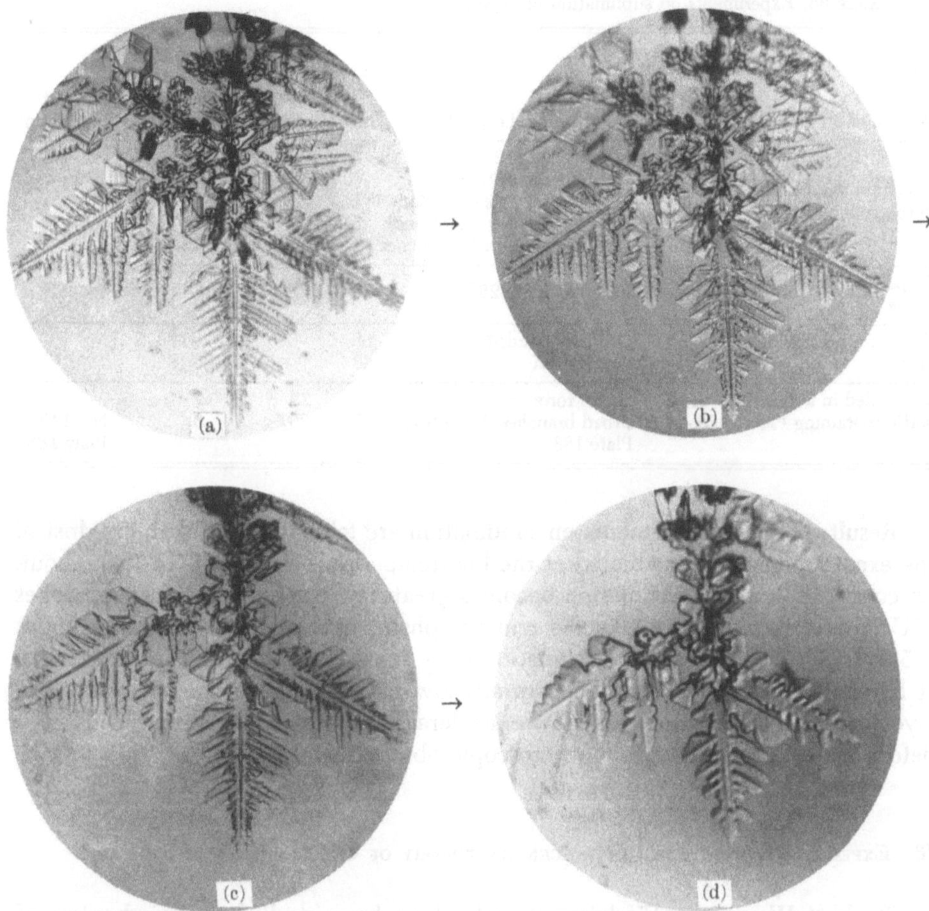

Fig. 428. Evaporation of crystal by sublimation: (*a*) [1398] immediately after formation; (*b*) [1399] after 1 hr; (*c*) [1400] after 2 hr; (*d*) [1401] after 20 hr (× 12).

To minimize the transformation, a glass bottle with a stopper was chosen and frost crystals were made to grow in clusters, covering the whole interior of the bottle. When a snow crystal is suspended therein, it is exposed to the atmosphere which is a little supersaturated with water vapor. This vessel was called a "preserving bottle." When a crystal was kept within this bottle, the diminution in dimension was as small as 10 percent after 40 hr. On the contrary, the rate of sublimation was very great in a bottle dried with P_2O_5, just as anticipated. Even at the low temperature of −26°C, the broad-branched crystal of No. 1402, Plate 188, became as small as No. 1403 after 1 hr and most of the internal pattern disappeared.

TABLE 33. Experiments on sublimation of crystals.

Method of preservation	Room temp. (°C)	Original crystal form	Preserved time (hr)	(min)	Dimension after sublimation, as fraction of original value	Form after sublimation
Left on a glass plate in the cold chamber	−19	Fernlike, No. 1396, Plate 188	4		0.48	No. 1397, Plate 188
	−20	Ordinary dendritic		50	.66	
	−20	Plate with simple extensions	1		.83	
	−26	Plate with sector extensions	5		.35	
	−28	Plate with dendritic branches	1		.75	
Suspended in a closed bottle	−16 – −20	Fernlike, Fig. 428(*a*)	20		.84	Fig. 428(*d*)
Kept in a preserving bottle	−35	Ordinary dendritic	68		.83	
	−25 – −30	Sector form	43		.9	
Suspended in a closed bottle containing P_2O_5	−16	Sector form	1	25	.2	
	−26 – −30	Broad branches, No. 1402, Plate 188	1	20	.5	No. 1403, Plate 188

Results of these experiments on sublimation are tabulated in Table 33. Most of the experiments were conducted at the low temperature of −20°C or thereabout. Of course the rate of sublimation becomes greater as the temperature approaches 0°C. Therefore, when we take the photographs of natural snow crystals at about −5°C to −10°C, the rate of sublimation is by far greater than those in this table. It is readily assumed that when natural snow falls through a comparatively dry layer before reaching the ground, a considerable part of it has already sublimed before the crystal is brought to microscopic observation.

76. Experimental criticism of Wegener's theory of crystal growth

Shedd,[64] Wegener,[65] Findeisen,[66] and others have studied the mechanism of snow-crystal formation by arranging many photographs of snow crystals. In any case, the data regarding the place where snow crystals grow, in other words, the meteorological conditions in the upper atmospheric strata during the snowfall, are either insufficient or absent. So it cannot be helped that the argument goes no further than the inference from the crystal forms. Of these, Wegener's theory was selected as a representative one, and his theory was examined experimentally by the use of artificial snow crystals.

According to Wegener, a primitive hexagonal crystal is first formed in a sufficiently supersaturated zone of the atmosphere and then the space between the branches is filled up with ice while the crystal is falling through a less supersaturated layer, transforming the dendritic crystal into plate form. Next, strata of sufficient supersaturation will add other dendritic appendages. When this crystal comes down through the next layer, of less supersaturation, it becomes a larger plate. By the repetition of processes similar to these, copious variations occur in the form of snow crystals. Although this theory deals only with the plane crystals, variations in the form of plane crystals, especially the hexagonal symmetric patterns inside the plates, are nicely explained by this theory. It is understood by common sense that the plates with dendritic extensions are formed in this way, and this is also confirmed by our experiments in artificial snow.

The point that has a physical significance in Wegener's theory is that a dendritic crystal is transformed into a plate in a less supersaturated atmosphere. This point has to be confirmed by experiment.

We carried out the experiment on this point using the dendritic crystal shown in Fig. 429(*a*). This crystal was left in the apparatus with T_w reduced to +4°C, that is, under the condition of less supersaturation than before; T_a was kept at about −20°C. The result was that, although the tendency for the space between branches to be filled up with ice was observed to some extent, the main change was that the branches became thicker. After 36 hr, the crystal was transformed into the shape shown in Fig 429(*b*). This is the development into skeleton structure mentioned in Sec. 66, and the tendency of transformation from a dendritic crystal to a plate was not observed at all.

A similar experiment was carried out with a broad-branched type of crystal, shown in Fig. 430(*a*). When this crystal was formed, the reservoir was removed so that the crystal was exposed to the atmosphere without supersaturation. Although left for 42 hr in this condition at −23°C, the crystal showed no change but the sublimation described in the preceding section. The reservoir was introduced again, and T_w was raised from +2°C to +9°C to make the branches grow once more. The crystal thus formed is the one shown in Fig. 430(*b*). Besides the above, we conducted various experiments to decrease the supersaturation in different degrees, but could not observe any such phenomenon as expected from Wegener's theory. We therefore cannot support his theory, for we learned in our investigations on artificial snow that plates and dendritic crystals grow in different conditions from the beginning, and this will also hold true in the case of natural snow crystals.

77. Crystal with cloud particles attached

Crystals with cloud particles attached are often observed in natural snow. Especially in the level land of Hokkaido, crystals with cloud particles are even more frequently observed than those without droplets.

In the experiment on artificial snow, if T_w, the temperature of water in the reservoir, is sufficiently high, white vapor is observed within the apparatus and

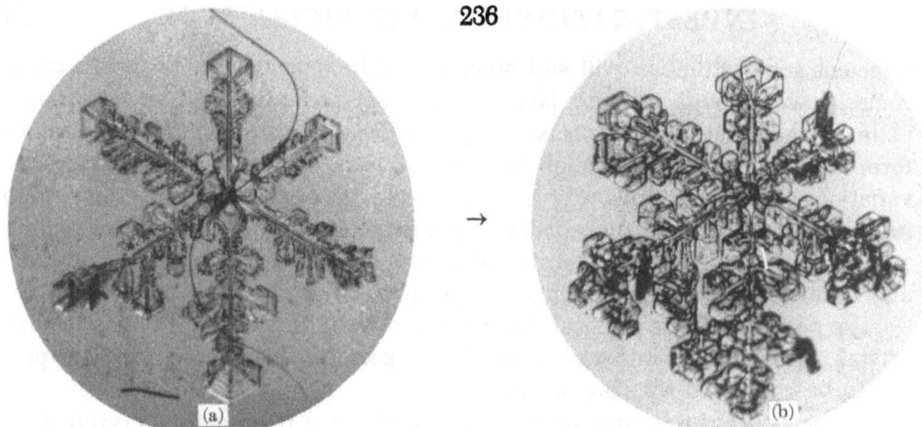

FIG. 429. Experimental criticism of Wegener's theory: (*a*) [1404] original crystal; (*b*) [1405] after 39 hr in less supersaturated atmosphere (× 14.5).

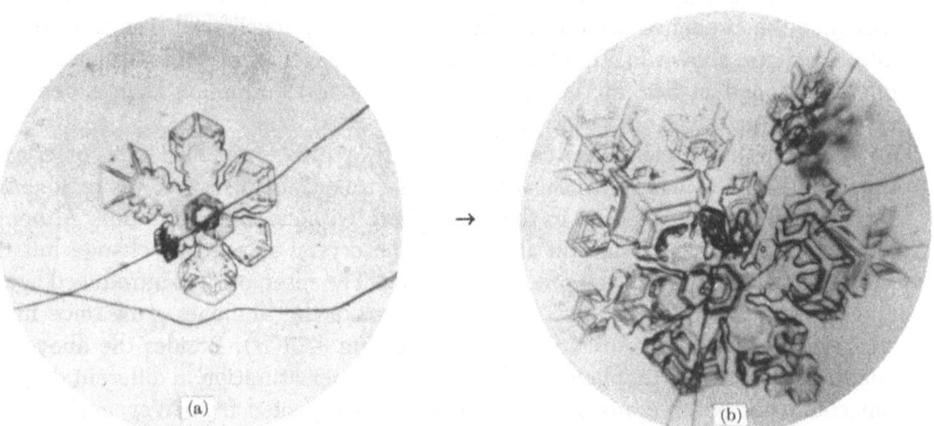

FIG. 430. Experimental criticism of Wegener's theory: (*a*) [1406] original crystal (× 25); (*b*) [1407] after transformation (× 23).

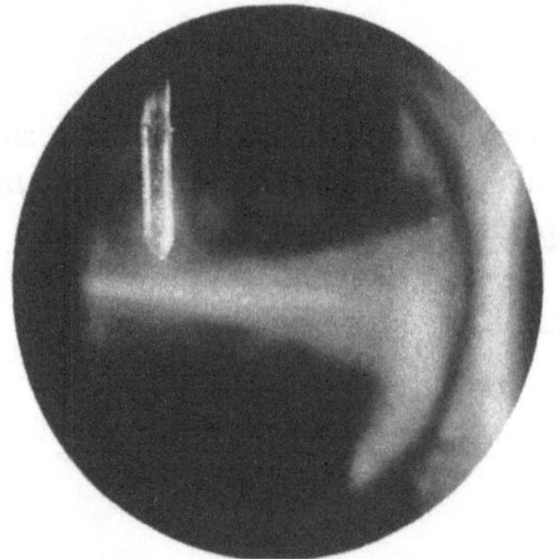

FIG. 431 [1408]. Intense beam of light projected into the apparatus (× 1.2).

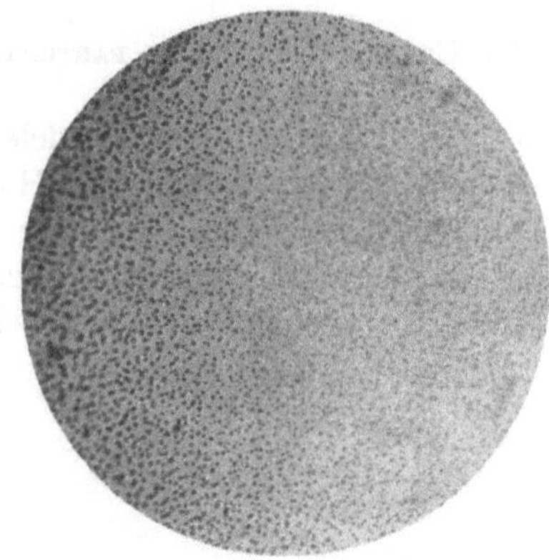

FIG. 432 [1409]. Minute fog particles (× 170).

water droplets are attached to the growing crystal. But it was found that these cloud particles or fog particles exist abundantly in the air current, even when T_w is not so high and the crystals formed are without droplets. We found that in the rising air current, which looked transparent by ordinary illumination, the beam of light appeared white in the dark field, if a strong illumination was given from one side. Figure 431 shows this condition inside the apparatus, photographed through its window. The thermometer whose end is seen in the photograph is the alcohol thermometer for measuring T_a.

When the photograph of Fig. 431 was taken T_a was $-17.0°C$ and T_w was $+16.5°C$; these were the conditions for obtaining dendritic crystals without droplets. In fact, a beautiful dendritic crystal without any droplet was obtained in this case, despite the presence of fog particles, which was proved by the white beam of light.

In order to examine the dimensions of the minute fog particles in the air current, a glass plate covered with a film of unfrozen oil was exposed to the rising current within the apparatus for a while. When this plate was examined under a microscope with high magnification, a great many minute droplets were observed in the oil film. Figure 432 shows an example. In this case the exposure to the rising current was continued for 10 min, and many of the droplets were about 2 μ in diameter, the smaller ones being about 1 μ. We therefore inferred that when water droplets with a diameter of 1–2 μ are attached to a crystal, they are not frozen to it in the form of droplets but spread on the crystal surface and help the growth of the crystal, acting just as if they are in the gaseous state.

Next, T_a was raised to the value suitable for obtaining crystals with droplets attached. The particles within the rising current were examined by the same method. In this case, the droplets were larger in comparison with the former ones, and many of them were as big as those observable in the "crystals with cloud particles," that is, 20–30 μ in diameter. An example of the latter case is reproduced in Fig. 433. This photograph shows the water droplets caught in the oil film, which was exposed to the rising air current for 5 min when T_a was $-17.0°C$ and T_w was $+29°C$. If these bigger droplets attach themselves to a crystal, they are frozen to it as they are, so that the crystal becomes the one with cloud particles attached.

Examples of artificial crystals with cloud particles are reproduced in Figs. 434–436. Conditions for their formation and some explanations are shown in Table 34.

One of the problems concerning these droplets is whether or not a small droplet attached to a crystal grows into a bigger one. While we were carrying on many continued observations of crystal formation with a microscope, we could not observe any such tendency. Sometimes the larger droplets, having diameters of the order of 30 μ, were observed to jump into the field of the microscope, and freeze to the crystal. We also confirmed on many examples that a droplet once frozen to a crystal retains its size for a considerable length of time. These observations led us

TABLE 34. Formation of crystals with cloud particles.

Fig. No.	Photo. No.	T_a (°C) Initial	Middle	Final	T_w (°C) Initial	Middle	Final	Remarks
434	1411	−18.0	−17.0	−18.0	+13.0	+13.5	+13.5	After the final stage, T_w increased to +26°C, crystal was exposed to steam for 5 min
435	1412	−17.5	−15.5	−15.0	+13.0	+17.5	+24.5	Droplets attached in 30 min of middle and final stages; largest diameter, 40 μ; smallest diameter, 7.5 μ
436	1413	−15.0	−14.5	−15.5	+14.0	+16.0	+17.0	A few droplets sometimes attached under the condition for ordinary dendritic crystals

to the conclusion that droplets do not grow after attaching to the crystal. Some of the frozen droplets show the trend of becoming hexagonal. This tendency is clearly observed in Fig. 435.

The fact that the minute droplets, of diameter 1 μ or so, take the same action as the water vapor in the process of condensation by sublimation, while the larger droplets, of diameter 20–30 μ, attach frozen to the crystal in that form, is both interesting and suggestive, and seems to be worth further investigations.

78. INITIAL STAGE OF SNOW CRYSTAL[67]

What we have hitherto called crystals in the initial stage were those whose existence can be clearly recognized with the naked eye and whose structure can be thoroughly examined under a microscope with a magnification of about 50 or so. We had no means to examine the shape of the crystal at the moment when the germ formation was recognized as a bright dot in the dark field. At this stage, crystals, especially those of the plane hexagonal type, are ready to sublime and disappear by the time they are brought under a microscope.

To investigate this most primitive stage, we adopted a method of catching the tiny crystals in an oil film to prevent sublimation and examining them under a microscope with high magnification. We tried six kinds of oil and found that ordinary buffer oil of low solidifying temperature was the most suitable. Any buffer oil that has sufficient fluidity at −15°C or so can be used. First, T_a and T_w are controlled and a required condition is given to the apparatus. Then a hair is exposed in it. In most cases, a germ is formed on the hair in a few minutes. As soon as the germ is recognized, it is quickly dipped in an oil film on a glass plate. The crystal caught in oil is scarcely transformed for about 30 min, but the part outside the oil

TABLE 35. Formation of initial stages of crystals.

Fig. No.	T_a (°C)	T_w (°C)	Crystal form	Size (mm)	Remarks
439	−15	+15	Small hexagonal tsuzumi	0.07	Formed immediately after hair is set
	−15	+15	Broad branches, hexagonal	.35	Similar form to Fig. 437
440	−21	+12	Minute column	.06	Formed immediately after hair is set
	−18	+12	Minute column	.04	Formed 2–3 min after hair is set
	−14	+15	Sector form	.27	
	−27	+14	Minute column	.07	Formed immediately after hair is set
	−25	+14	Minute column, side planes	.10	
438	−15	+15	Small hexagonal, dendritic	.34	Formed 20 min after hair is set
441	−16	+14	Plate with skeleton design	.10	Formed immediately after hair is set
442	−16	+14	Side view	.06	Side view of similar one to Fig. 441
437	−16	+15	Small hexagonal, dendritic	.38	
	−15	+14.5	Broad branches	.30	Formed 10 min after hair is set
	−15	+14.5	Sector form	.15	
	−16	+ 8	Minute column	.04	

evaporates very rapidly. For instance, in Fig. 437 one of the branches has lost its tip, because that part was outside the oil. Crystals in the initial stage formed under various conditions are listed in Table 35.

As shown in the table, crystals already have various structures – plane hexagonal, column, side planes, skeleton plate, etc. – even in this most primitive stage. And the conditions for formation are mostly the same as those discovered for the larger snow crystals proper. We therefore learned that the conditions for formation of snow crystals which we had investigated can be applied from the very first stage of crystallization. A good example is a plate with skeleton structure shown in Fig. 442. It shows already a double-sheet structure at the size of d = 0.06 mm, which is only twice the diameter of an ordinary cloud particle, for which d = 0.03 mm. This fact shows that our condition for snow-crystal formation is effective even in such a primitive stage of crystallization.

79. External conditions controlling the form of the crystal [68]

From what we have hitherto mentioned, most of the conditions for formation of snow crystals are made clear. We discovered that the factors deciding the types of snow crystals are the air temperature and the degree of supersaturation of water vapor at the spot of crystal formation. Of these factors, the air temperature T_a was measured with a specially made thin alcohol thermometer, but as regards the degree of supersaturation we could only determine the relative values by the temperature of water in the reservoir, T_w. Even if T_w is fixed, the supersaturation at the spot

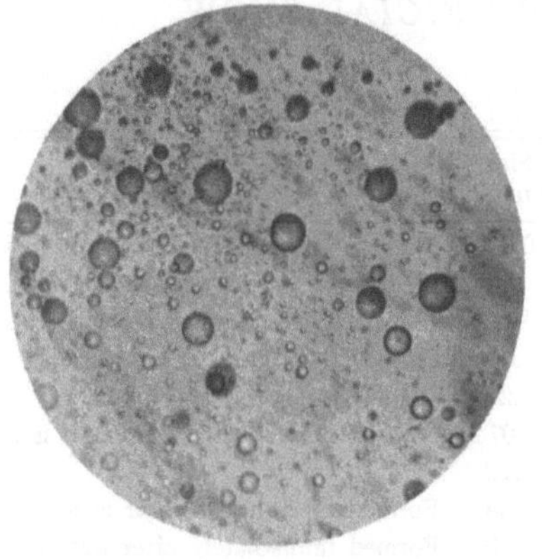

FIG. 433 [1410]. Larger droplets mixed with minute fog particles (× 170).

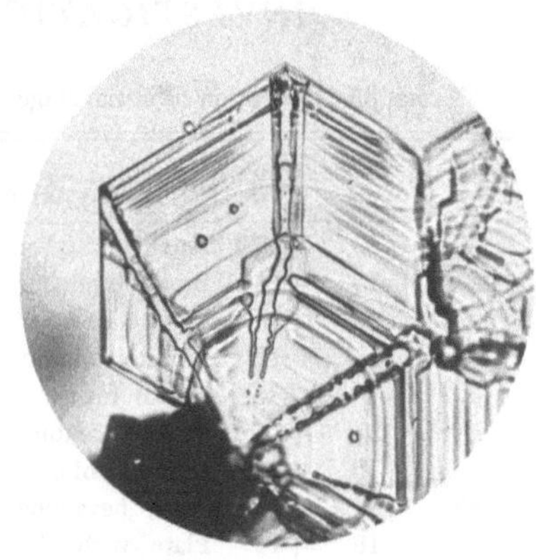

FIG. 434 [1411]. Few droplets on sector-form crystal (× 41.5).

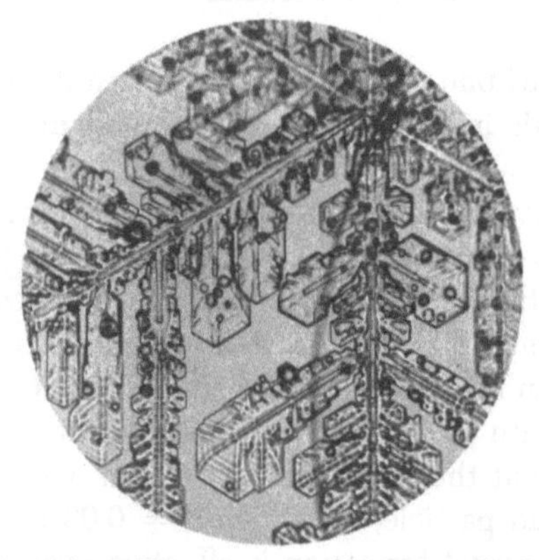

FIG. 435 [1412]. Numerous droplets on dendritic crystal (× 41.5).

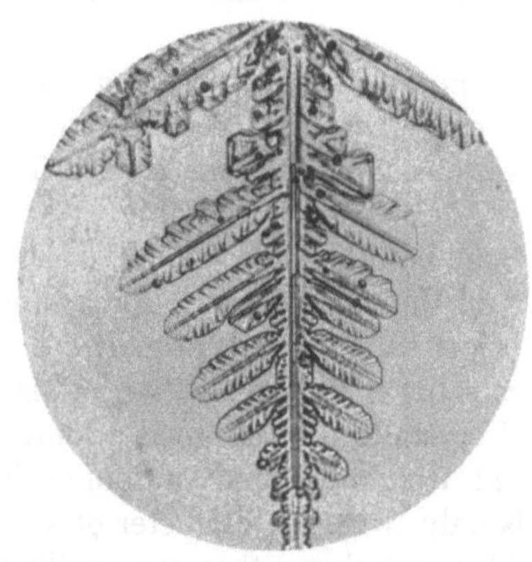

FIG. 436 [1413]. Few droplets on dendritic crystal (× 41.5).

FIG. 437 [1414]. Initial stage of snow crystal; $d = 0.38$ mm (× 128).

FIG. 438 [1415]. Initial stage of snow crystal; $d = 0.34$ mm (× 128).

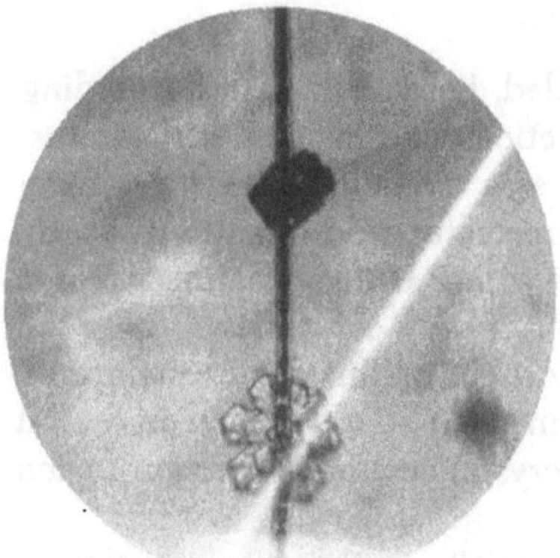

Fig. 439 [1416]. Initial stage of snow crystal; d = 0.07 mm (× 185).

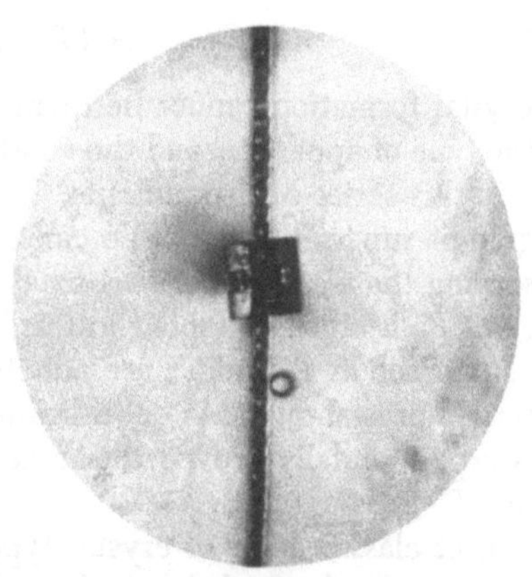

Fig. 440 [1417]. Initial stage of snow crystal; d = 0.06 mm (× 153).

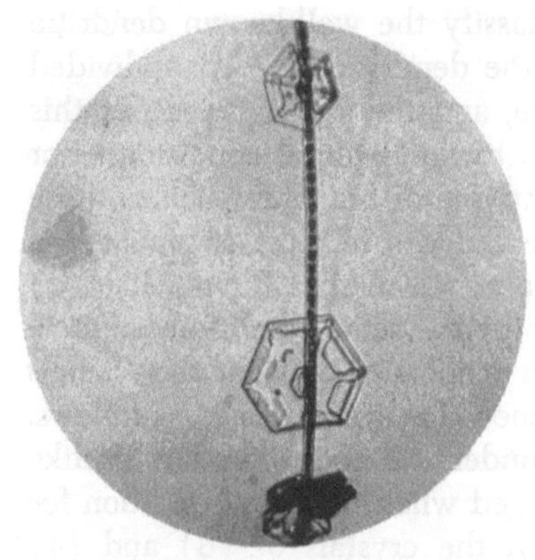

Fig. 441 [1418]. Initial stage of snow crystal; d = 0.10 mm (× 132).

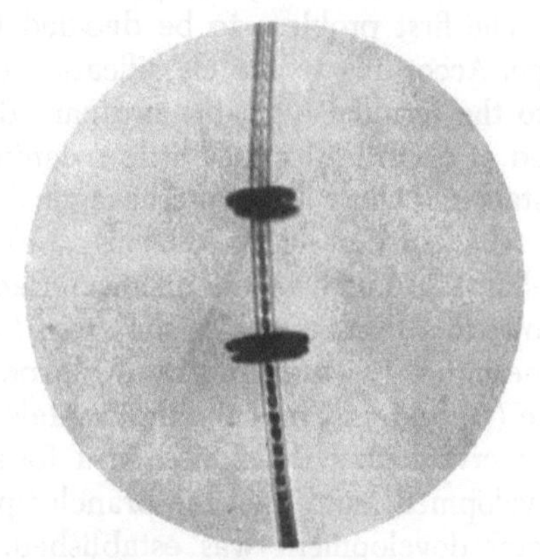

Fig. 442 [1419]. Initial stage of snow crystal; d = 0.06 mm (× 150).

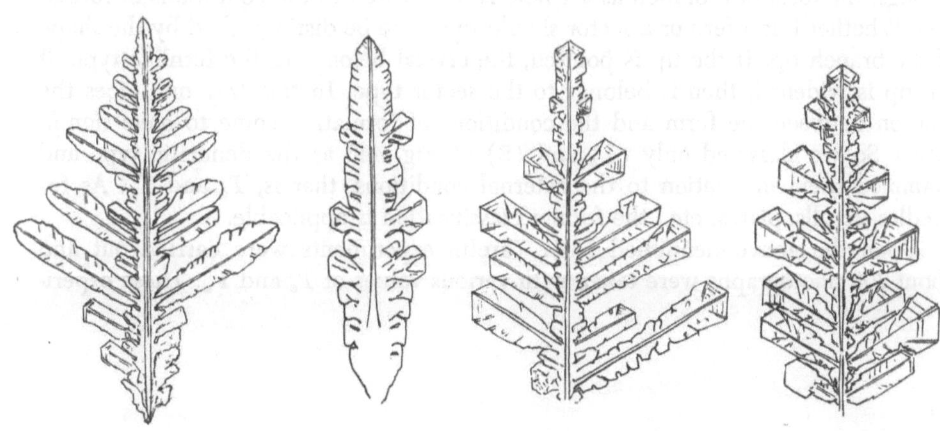

Fig. 443. Distinction between dendritic and sector crystals.

of crystal formation cannot be quantitatively decided, because it varies according to the type of apparatus and the condition of convection of air inside the apparatus. But in this series of experiments, T_w was adopted as a qualitative measure of the degree of supersaturation. We can measure the supersaturation quantitatively by absorbing the water vapor in a fixed amount of air with P_2O_5, as described in Sec. 80, but T_w is chiefly adopted in this book, because it is very convenient.

Inasmuch as T_a and T_w are adopted as the factors controlling the crystal forms, a certain crystal ought to be indicated by a point in the T_a–T_w diagram. In fact, it was experimentally proved that a certain type of crystal occupies a certain region in the T_a–T_w diagram.

Strict classification of crystal types is the prerequisite to the determination of this region. As the final shape of the crystal as a whole is decided by what happens to the crystal during the course of its formation, it is more adequate to discuss the form of one of the branches for determining the region in the T_a–T_w diagram in which a particular crystal type falls.

The first problem to be decided is how to classify the well-known dendritic type. According to the classification used so far, the dendritic type is subdivided into the fernlike type, the ordinary dendritic type, and the stellar type. But this kind of classification has little meaning in relation to the external conditions. For instance, if there are plane hexagonal crystals composed of branches such as those sketched in Fig. 443, apparently (1) ought to be called a fernlike crystal, (2) a stellar, (3) and (4) the ordinary dendritic. But a crystal with a branch like (2) grows to a fernlike one as shown in (1) if kept under the same condition for some time more. It was found by the experiments with artificial snow that the crystals like (3) and (4), namely, the crystals with broadened tips, were formed as follows. The crystal first developed in a form like (1) under the condition for fernlike development, and then the branch tips were widened when the new condition for sector development was established. Accordingly, the crystals of (3) and (4) ought to be regarded as sectors, in view of the final condition for formation. In other words, these crystals would have grown as sectors if the last condition had been set up from the beginning. In deciding the type to which a dendritic crystal belongs, the form of a branch as a whole is not related to the conditions of formation. Whether it is a fern or a sector should therefore be distinguished by the shape of the branch tip. If the tip is pointed, the crystal belongs to the fernlike type; if the tip is widened, then it belongs to the sector type. In this way only does the relation between the form and the conditions of formation come to have significance. So we classified only (1) and (2) of Fig. 443 as the dendritic type and examined them in relation to the external conditions, that is, T_a and T_w. As for needles, scrolls, plates, etc., the former classification is applicable.

Upon the above-mentioned basis, careful experiments were carried out and about 700 photographs were taken with various values of T_a and T_w. These experiments were difficult and at the same time required extraordinary perseverance, but fairly satisfactory results were obtained after two years of hard exertion by M. Hanajima. The value of T_a being a function of the room temperature and T_w, an air thermostat was equipped in the cold chamber, and the experiments were conducted in this thermostat. When a changed value of T_a was required with fixed T_w, the temperature of the thermostat T_t was altered. Instead of changing T_t, the same result was obtained by illuminating the interior of the apparatus with an electric lamp.

The relation between the crystal form and T_a and T_w was examined for 700 crystals, including dendritics, sectors, and others. The results are plotted in the T_a–T_w diagram of Fig. 444. As seen in the figure, various crystal types were divided into seven regions in the T_a–T_w diagram. Crystals were classified into eight types: dendritic, sector, plate, spatial plates, scroll, needle, irregular needle, and column.

(*a*) *Dendritic crystal.* Region I in Fig. 444 is the condition for dendritic growth. It means that the necessary condition for dendritic development is that T_a is between $-14°C$ and $-17°C$ and the degree of supersaturation is above a certain lower limit. Although there is a fairly definite limitation for T_a, the degree of supersaturation covers a pretty wide range. Consequently, contrary to what has hitherto been believed, it was found that the air temperature is an important factor controlling the crystal form. Degree of supersaturation also has some influence. Although T_a stays between $-14°C$ and $-17°C$, if T_w is lowered below a certain limit, the crystal becomes a sector form. If water vapor is too abundant, the crystal becomes the one with frozen droplets attached.

In the lower part of Region I, where the supply of water vapor is comparatively scanty, fernlike crystals are intermixed with sectors. Under such conditions, the crystal becomes a dendritic form when the rate of increase of T_w is more than 1.5 C deg in 30 min, and when it is less, that is, when T_w is almost constant, the crystal becomes a sector. Under such conditions, when the dendritic growth is somewhat difficult, the supply of water vapor must be gradually increased to get a dendritic crystal.

(*b*) *Sector and plate.* As mentioned in Sec. 67, plates are classified into such types as thin plate with parallel lines, transparent ice plate, thick plate with skeleton structure, and plate with projecting ridges. And a variation of the last type is the sector-form crystal. The distinction among the conditions for formation of these various types is not established as yet. However, the points for condition of formation are scattered surrounding Region I, and they are grouped together to make Region II. Further distinction within Region II will have to be made by future investigation.

(*c*) *Spatial plates.* This is a spatial assemblage of small plates. The so-called "flour snow" often observed on high mountains belongs to this type. Crystals of this type are formed when T_a is about $-20°C$ or lower and the supply of water

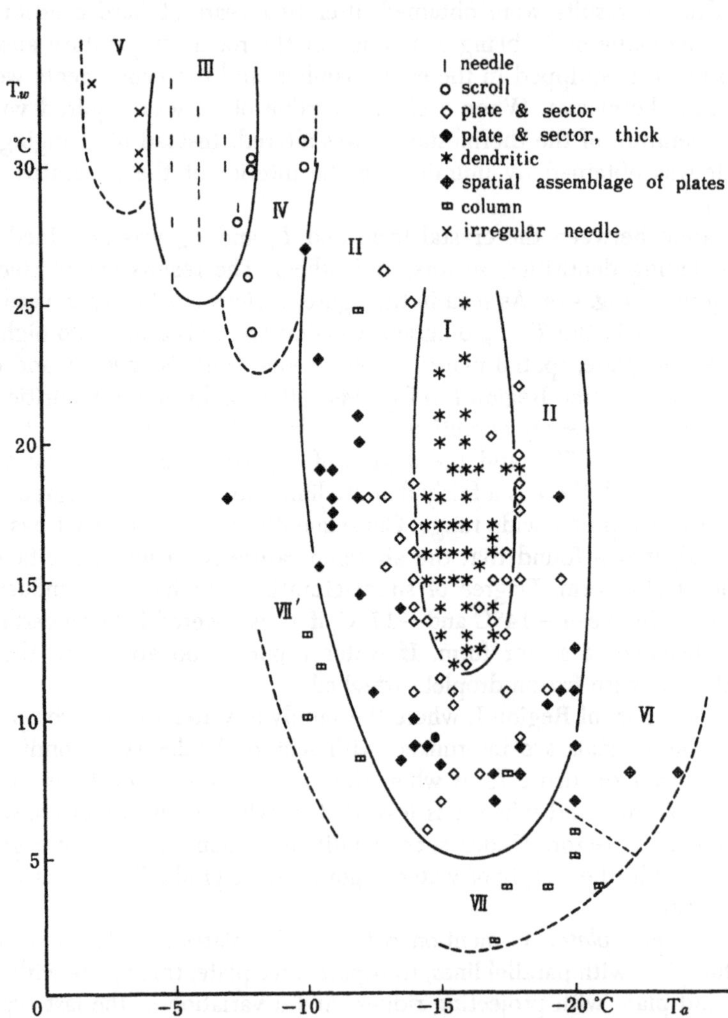

Fig. 444. External conditions controlling the form of snow crystals: T_a–T_w diagram.

vapor is scanty. The scope of the conditions is not sufficiently definite yet, but Region VI covers most of it.

(*d*) *Column.* Columns are formed, as has been generally believed, when the supersaturation of water vapor is small. The range of T_a is fairly wide, from −10°C to −20°C or lower. Although −10°C is about the upper limit, the lower limit will be below −20°C. Region VII is the condition for columns.

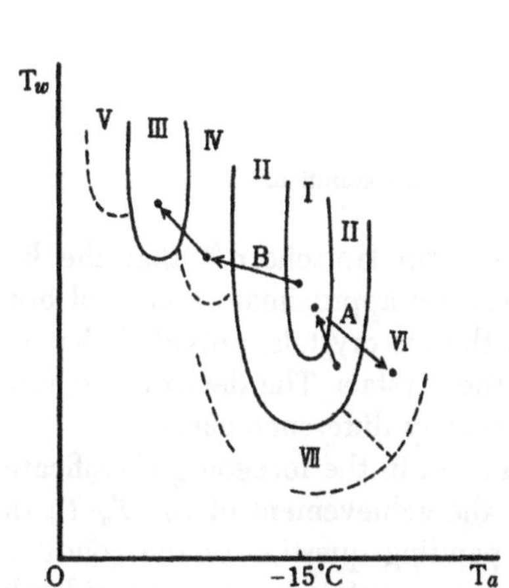

FIG. 445. Use of the T_a–T_w diagram.

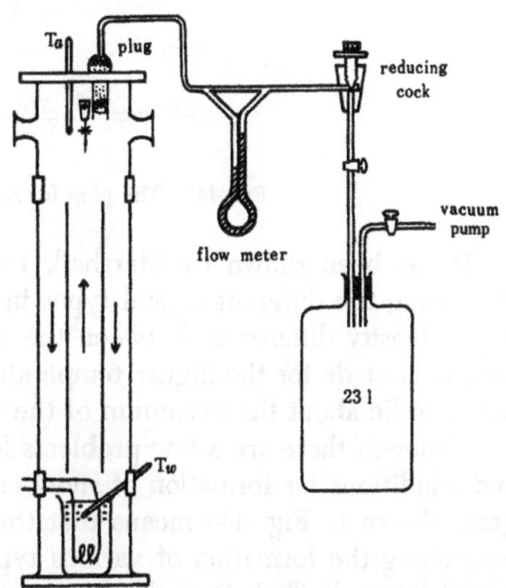

FIG. 446. Method of measuring the degree of supersaturation.

(*e*) *Scrolls.* When T_a becomes higher than the condition for thick plates, the scroll-like crystals begin to grow. That the snow crystal becomes dendritic in Region I, or when T_a is about −15°C, means that at this temperature the crystal develops in the base plane of the hexagonal system of crystallization. When the temperature goes above the limit, the crystal begins to grow also in the direction of the principal axis. Here it becomes a thick plate with skeleton structure. When the temperature becomes still higher and the supply of water vapor is sufficient, development in the direction of the principal axis is accelerated and at about −8°C the scroll-like crystal begins to grow. The scope for scrolls is Region IV in the T_a–T_w diagram.

(*f*) *Needles.* If T_a is raised above the limit for scrolls, the growth in the direction of the principal axis becomes extremely rapid at about −6°C. This yields the needle crystal. The condition for formation of needles is covered by Region III. This region coincides with the meteorological data for natural needles observed on the ground.

(*g*) *Irregular needles.* When the air temperature goes still higher, between about −4°C and −1°C, and the supply of water vapor is abundant, the crystal can no longer grow into a complete needle and irregular combinations of needlelike crystals are obtained. This type, named irregular needles, is covered by Region V. This type has a very irregular structure, as shown in Fig. 345. Details about its structure are still unknown.

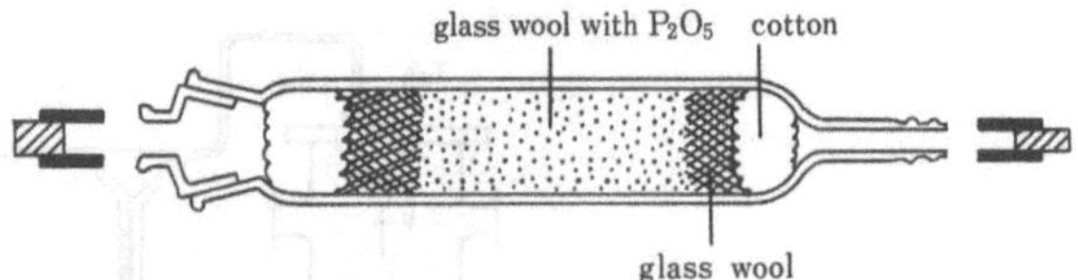

FIG. 447. The plug for measuring the supersaturation.

It has been shown by Marshall, Langleben, and Herschorn [69] that the lines separating the different crystal types in Fig. 444 are approximately lines of equal vapor-density differences between the air and the ice crystals, provided that correction is made for the higher temperature of the crystals. The dendritic region is found to lie about the maximum of the vapor-density difference curve.

Although there are a few problems left unsolved in the foregoing classification and conditions for formation of snow crystals, the achievement of the T_a–T_w diagram shown in Fig. 444 means that the long-pending question of the conditions controlling the formation of various types of snow crystals has to a considerable extent been clarified. It enabled us to indicate the conditions for formation of a certain crystal with a dot on the T_a–T_w diagram, and when the conditions altered during the growth, the course of formation can be shown by an arrow. For instance, in Fig. 445, *A* shows that the crystal first grew into a sector and then dendritic branches extended out from the sector, finally having spatial plates attached. Similarly, *B* indicates that scrolls developed at the branch tips of a dendritic crystal and then needles extended from the scrolls, as in the case of Fig. 351. We found this T_a–T_w diagram very convenient in examining the course of growth, as will be explained in Sec. 83.

80. Method of measuring the degree of supersaturation [70]

The external conditions controlling the form of a snow crystal were found to be the degree of supersaturation and T_a, the temperature of the air where the crystal is made. It was not possible, however, to determine quantitatively the degree of supersaturation, so we took T_w, the temperature of the water in the reservoir, as a measure of the supersaturation, and made the T_a–T_w diagram for representing the relations between the form of the crystal and the external conditions.

As the next step, we proceeded to measure the degree of supersaturation quantitatively. The general procedure will be understood from Fig. 446. A plug containing P_2O_5 is placed in the apparatus near the spot where the snow crystal is formed. The air is drawn through the plug into the tank of about 23 lit capacity, which was previously evacuated. The water content in the 23 lit of air is absorbed by P_2O_5 and measured by weighing on a chemical balance. The water content means the water vapor plus the minute droplets described in Sec. 77.

The details of the plug are seen in Fig. 447. Cotton, glass wool, and glass wool sprinkled with powdered P_2O_5 are put into the plug in the manner shown in the figure. Both ends of the plug are sealed airtight with glass rods and rubber tubes. The plug thus prepared is kept in a desiccator till just before it is to be used.

The details of measuring the degree of supersaturation are as follows. When the snow crystal had developed to the desired size, the plug, which had previously been cooled down to T_a, was quickly introduced into the apparatus. Keeping T_a, T_w, and the rate of air flow as constant as possible, the air was drawn into the flask at the rate of approximately 1 lit/min. The rate of air flow was adjusted to be constant with the aid of the flow meter and the reducing cock. In this series of experiments the increase in weight of the absorbing reagent was between 30 and 60 mg, and the measurement could be done with sufficient accuracy.

For representing the degree of supersaturation, we defined the supersaturation value * s as follows:

$$s = \frac{w + \rho'}{\rho_0},$$

where w is the liquid water content per unit volume, ρ' is the vapor density at saturation over supercooled water, and ρ_0 is the vapor density over ice. The absolute humidity D of the air at temperature T_a is $D = m/V$, where m is the increase in mass of the absorbing reagent and V is the volume of air drawn into the tank. But the temperature of the tank T' is usually different from that of the air where the crystal is made T_a, so a correction is necessary for the capacity of the tank V', that is, 23 lit. Thus,

$$V = V'\frac{273 + T_a}{273 + T'};$$

hence,

$$D = \frac{m}{\dfrac{273 + T_a}{273 + T'}V'}.$$

The value of D_{sat} (gm/m^3), the absolute humidity of saturated vapor at T_a, is calculated from a table; then s (percent) is obtained from the equation

$$\cdot s = 100\frac{D}{D_{sat}}.$$

The method of determining absolute humidity by the use of absorbing reagent is by no means new, glass wool sprinkled with P_2O_5 having been used by Tyndall.[71] But when we use this method in a cold chamber, many small precautions are necessary.

* This s is different from the supersaturation ratio defined on p. 138. Later discussions are based on this new definition of s.

Using this apparatus we measured the relative humidity in a laboratory room at ordinary temperature and compared the result with that measured with an Assmann hygrometer. The two values agreed to within 1 percent. It was verified that when two plugs were used in series, the second plug showed no sensible increase in weight. So we used each time only one plug throughout the experiments.

81. Crystal form and degree of supersaturation [72]

The supersaturation value s of the air where the snow crystal is formed was measured by the foregoing method for various combinations of T_a and T_w. Thus we could determine the form of the crystal as a function of T_a and s. The results were plotted in a T_a–s diagram, as shown in Fig. 448.

In the T_a–T_w diagram shown in Fig. 444, we saw that the air-temperature range for each region was quite narrow and the values of T_w for each region were scattered over a wider range. The phenomenon is more conspicuously seen in the T_a–s diagram. The chief factor controlling the form of the crystal is the air temperature, and a certain type of crystal is obtained for a wide range of supersaturation if the temperature lies within the necessary range.

It has generally been believed that the degree of supersaturation determines the form of the crystal, but this "common-sense" belief was found not to be the truth. That rule is applicable only to a certain narrow range of temperature. The transition from a plate to a dendritic form occurs when the supersaturation exceeds a certain value. But this is true on the condition that the temperature lies between −14°C and −17°C. The critical value of supersaturation for the transition of a plate into a dendritic form was found to be about 110 percent. Above 140 percent the crystal has numerous water droplets attached. Needles and cups or scrolls were known from the former T_a–T_w diagram to be produced with higher values of both T_a and T_w.

The simultaneous increase in T_a and T_w shows that the supersaturation is not much altered. The T_a–s diagram tells us that these crystals are formed if the supersaturation exceeds 105 percent. Plates have at least three varieties and their con-

Table 36. Formation of various types of crystals, in terms of T_a and s.

Crystal type	T_a (°C)	s (percent)
Dendritic	−14 – −17	110–140
Plate and sector	Just beyond dendritic region	110–140
Thick plate	− 8 – −12	100–110
Spatial plates	< −18	
Cup or scroll	− 6 – −18	100–140
Needle	− 5 – − 6	> 105
Irregular needle	> − 5	100–135

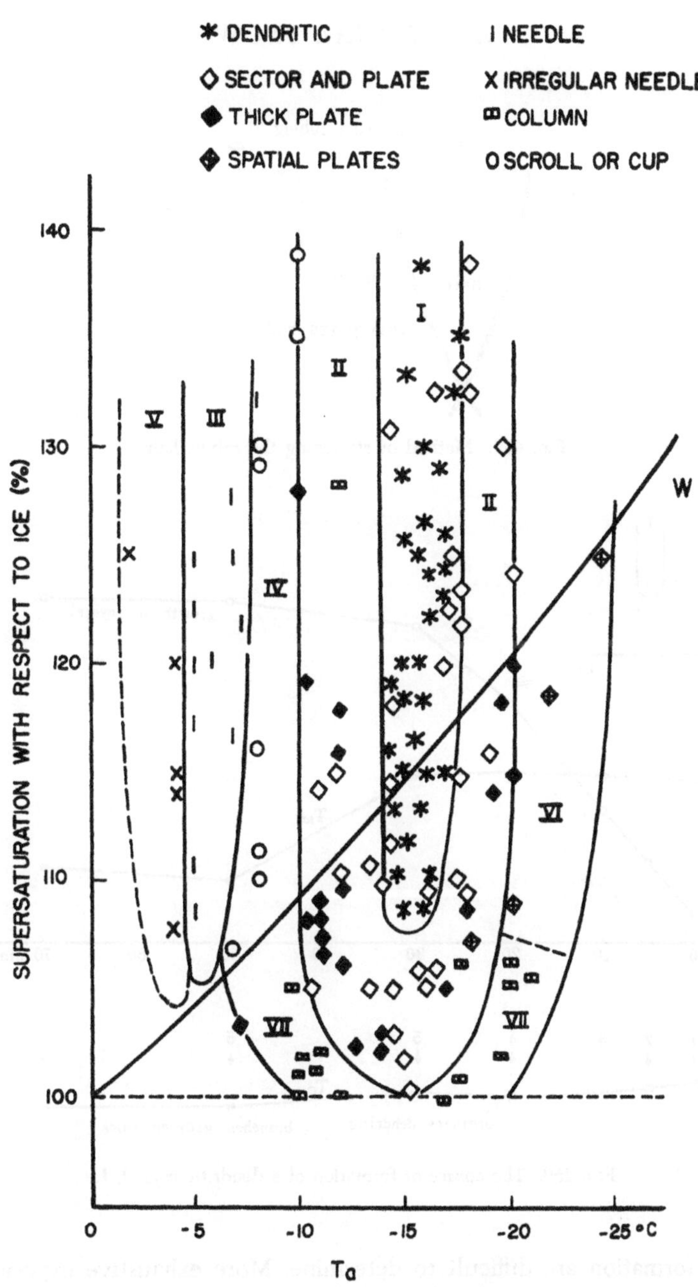

Fig. 448. T_a–s diagram, showing the conditions of formation of various shapes of snow crystals; W is a line giving the saturated vapor pressure with respect to supercooled water.

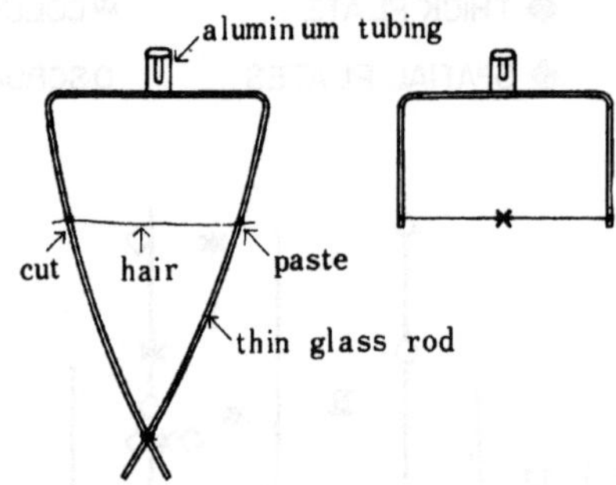

FIG. 449. Method of stretching the rabbit hair.

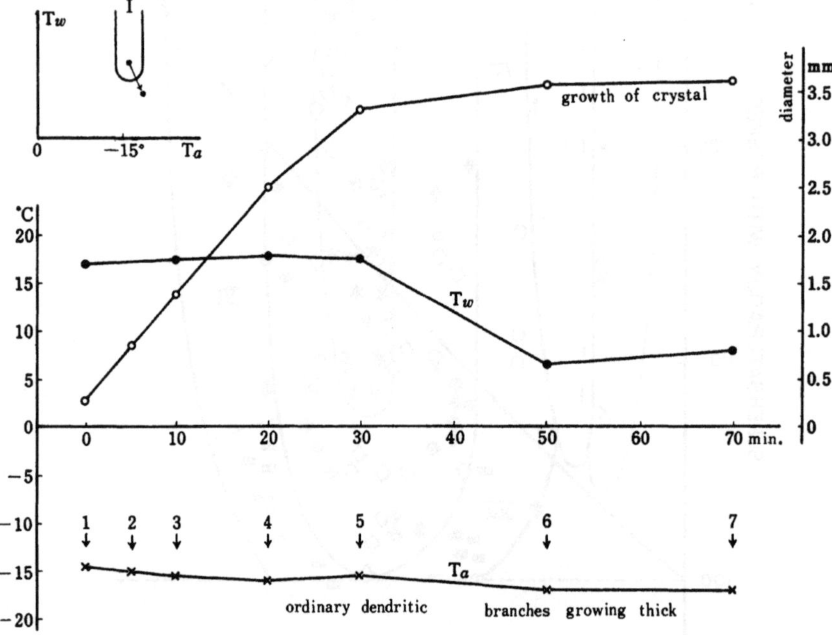

FIG. 450. The course of formation of a dendritic crystal, I.

ditions of formation are difficult to determine. More exhaustive experiments are necessary for settling these points. The spatial assemblage of plates is obtained at T_a lower than $-18°C$. The conditions of formation of various types of snow crystals are roughly summarized in Table 36.

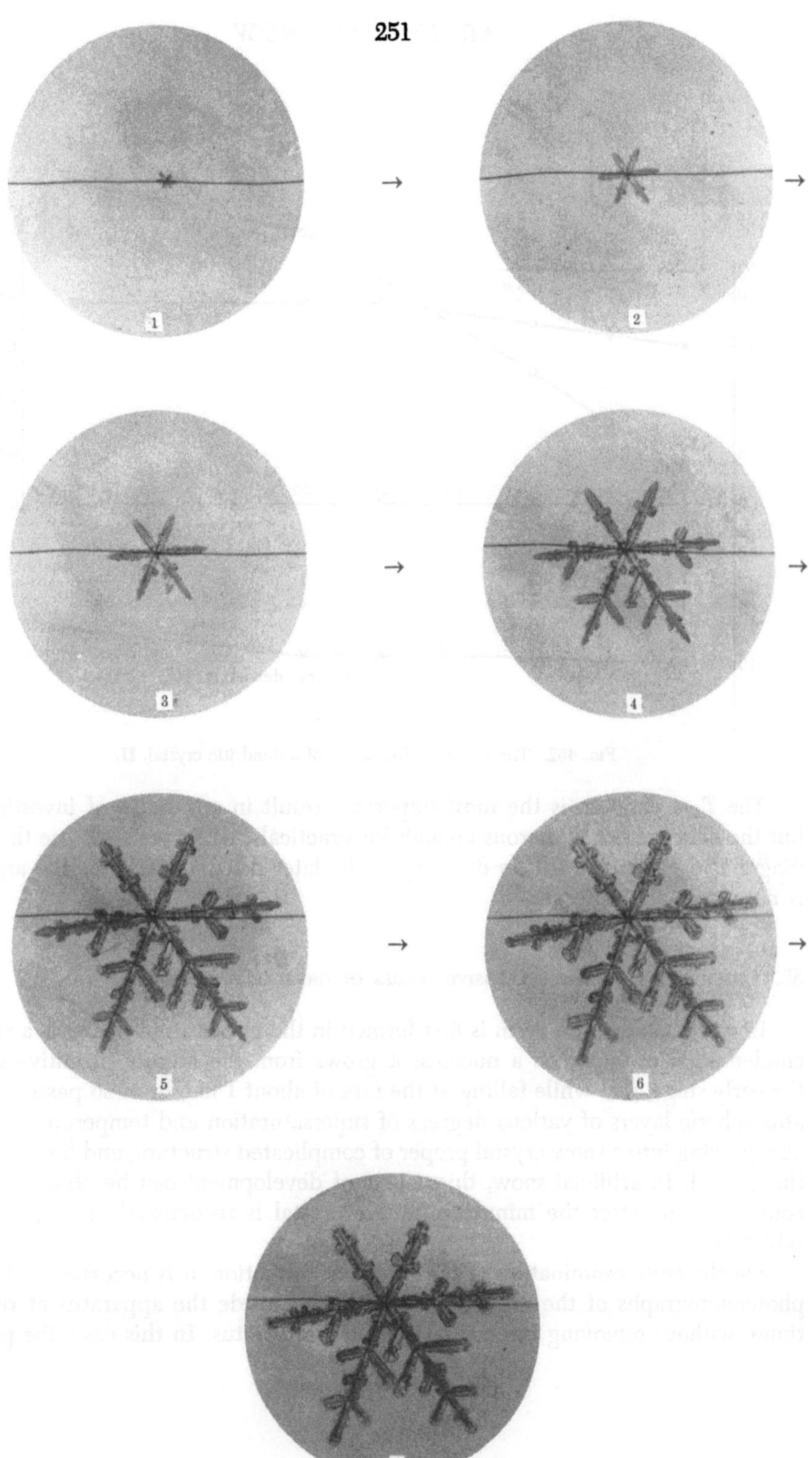

Fig. 451 [1420–26]. Successive stages in the course shown in Fig. 450 (× 12).

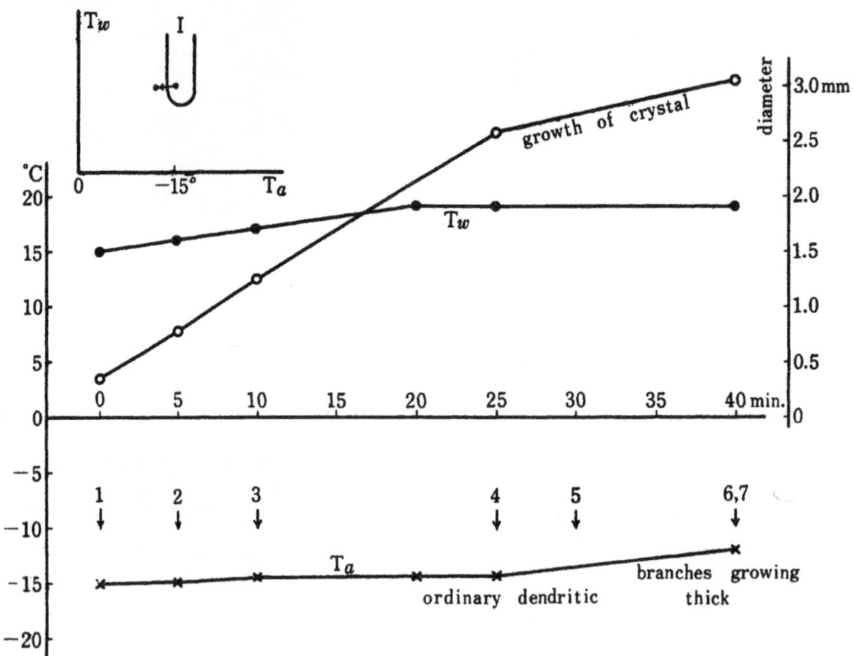

Fig. 452. The course of formation of a dendritic crystal, II.

The T_a–s diagram is the most important result in our series of investigations, but the data are not numerous enough for practical use. So we shall use the T_a–T_w diagram in place of the T_a–s diagram in the later discussions, since the argument is not essentially different.

82. Observation of the successive stages of crystal formation

In natural snow, the germ is first formed in the upper atmosphere as a result of condensation of vapor on a nucleus; it grows from the minute primitive state to the early stage, and while falling at the rate of about 1 km/hr or so passes through atmospheric layers of various degrees of supersaturation and temperatures, gradually growing into a snow crystal proper of complicated structure, and finally reaches the ground. In artificial snow, this course of development can be observed in the course of time after the minute primitive crystal is recognized at a spot on the rabbit hair.

For the close examination of the course of formation, it is necessary to take the photomicrographs of the growing crystal from outside the apparatus at required times, without removing the crystal from the apparatus. In this case, the problem

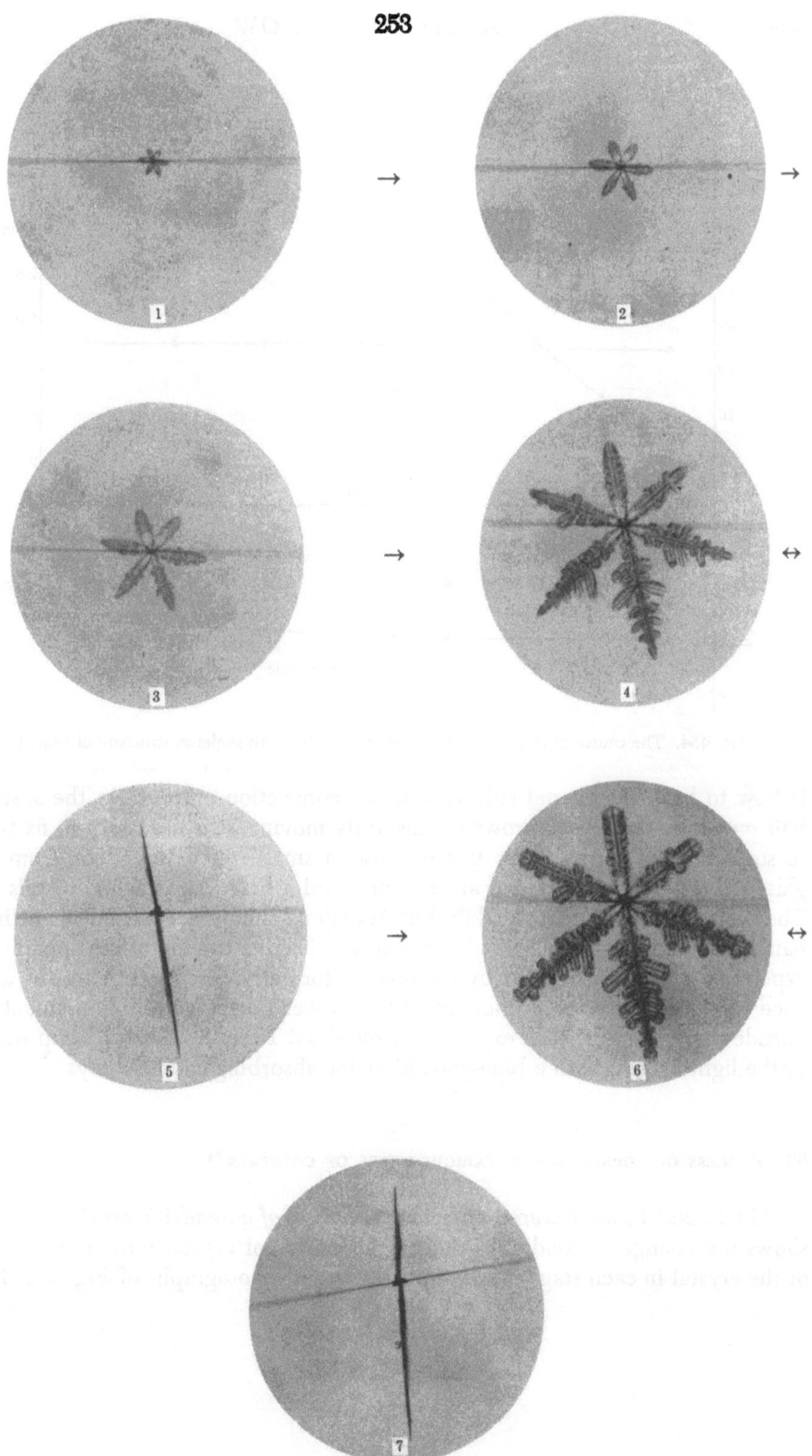

Fig. 453 [1427–33]. Successive stages in the course shown in Fig. 452 (× 14).

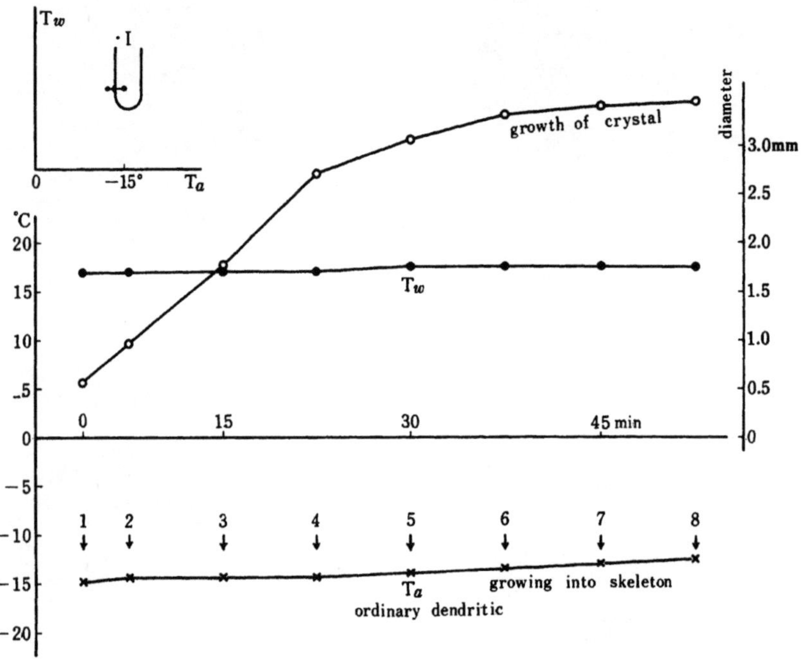

Fig. 454. The course of formation of a dendritic crystal with skeleton structure of branches.

is how to keep the crystal still against the convection current. As the suspended hair on which the crystal grows is constantly moving, it is necessary to fix the hair to some kind of support. For this purpose, a small frame was made from a thin glass rod with the diameter of about 1 mm and a hair was attached to this frame. The work needed a delicate skill, but Hanajima finally succeeded in setting the hair as shown in Fig. 449. The glass frame was fixed to the top of the apparatus and experiments were carried out by the process formerly described. After the appearence of a germ, a series of photomicrographs was taken at suitable moments from outside the apparatus, each exposure being about 1 sec. An electric lamp was used as the light source, with a blue-glass filter for absorbing the heat rays.

83. Process of formation of various types of crystals [73]

(1) *T_a and T_w are lowered after the formation of a dendritic crystal.* Figure 450 shows the change of conditions during the course of crystal formation. The form of the crystal in each stage is seen in the series of photographs of Fig. 451. In Fig.

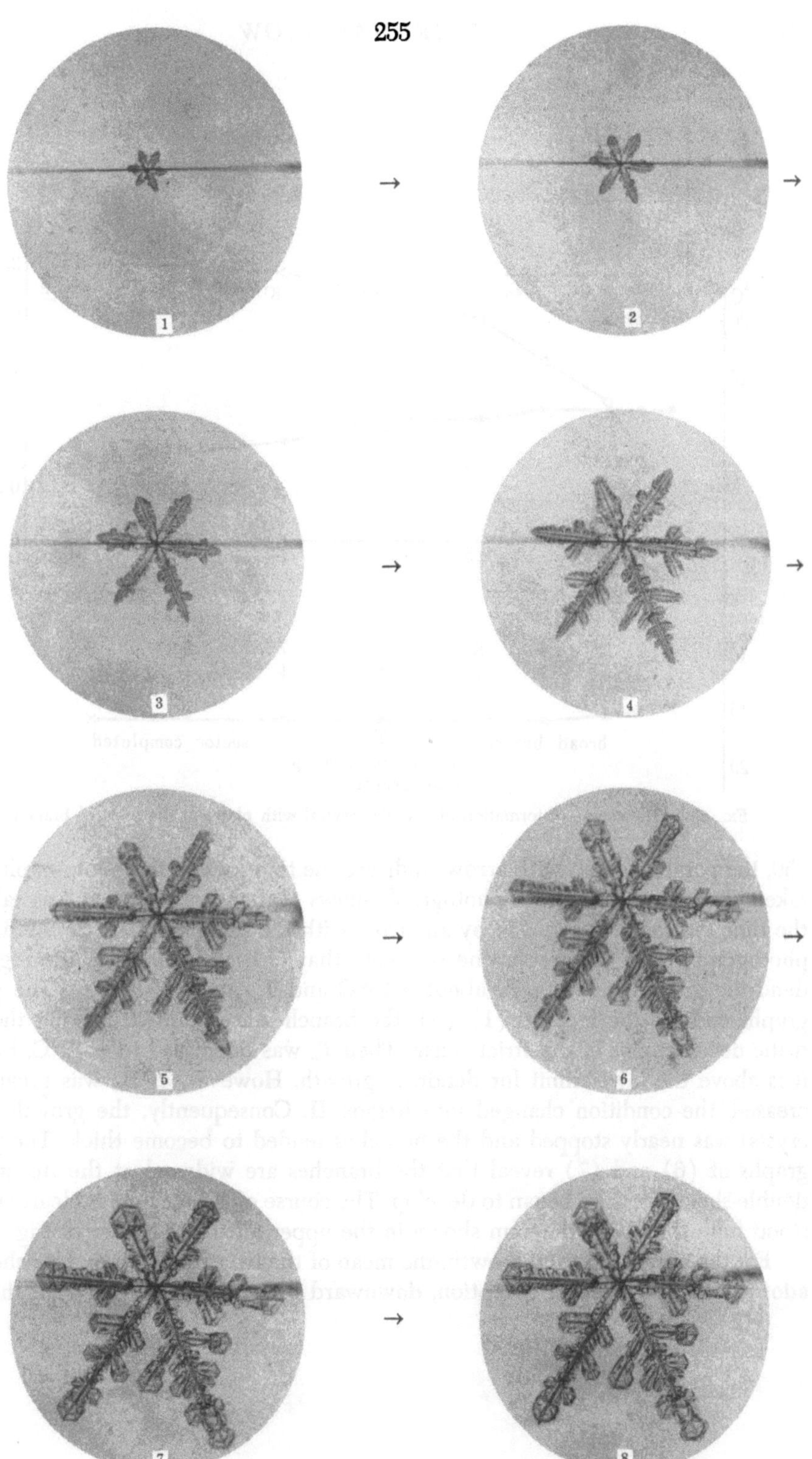

FIG. 455 [1434–41]. Successive stages in the course shown in Fig. 454 ($\times$ 12.5).

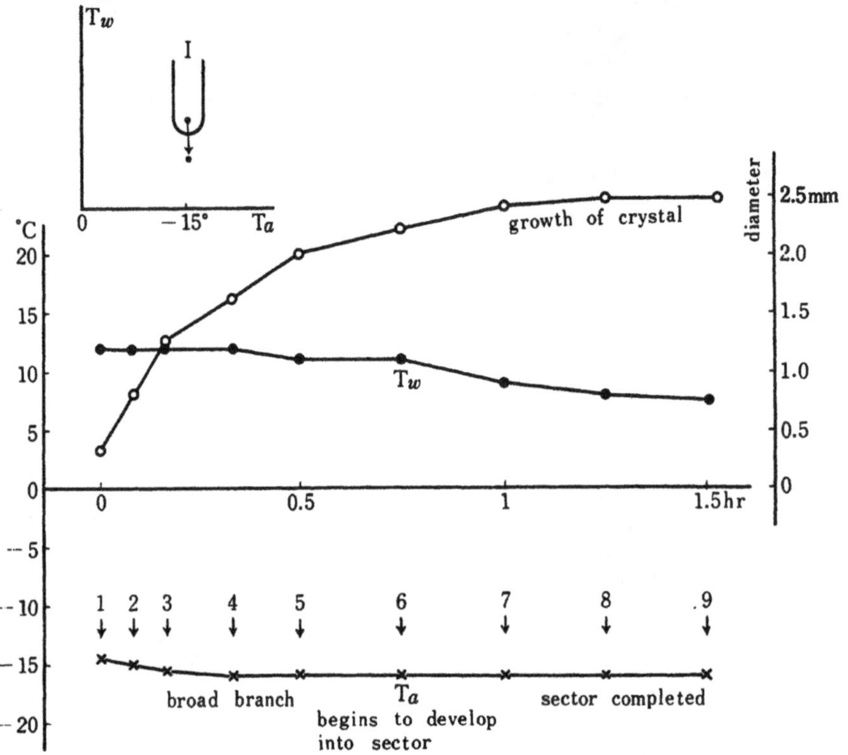

FIG. 456. The course of formation of a stellar crystal with plates at the ends of branches.

450, the points marked with arrows indicate the times when the photographs were taken. The number under a photograph shows that the photograph was taken at the time indicated in Fig. 450 by an arrow with the same number. Comparing the photographs with the graph, one will note that (1)–(5) belong to the region of dendritic growth, namely, T_a about $-15°C$ and T_w about $+17.5°C$. The photographs endorse the fact. In (1)–(5), the branches are pointed, making the dendritic development in the strict sense. Then T_a was decreased to $-17°C$, but still it is above the lower limit for dendritic growth. However, as T_w was greatly decreased, the condition changed into Region II. Consequently, the growth of the crystal was nearly stopped and the branches tended to become thick. The photographs of (6) and (7) reveal that the branches are widened at the tip and the double-sheet structure began to develop. The course of this change is clearly understood from the T_a-T_w diagram shown in the upper left-hand corner of Fig. 450.

For the curve of crystal growth, the mean of the lengths of the six branches was adopted. Almost without exception, downward branches are longer than the rest.

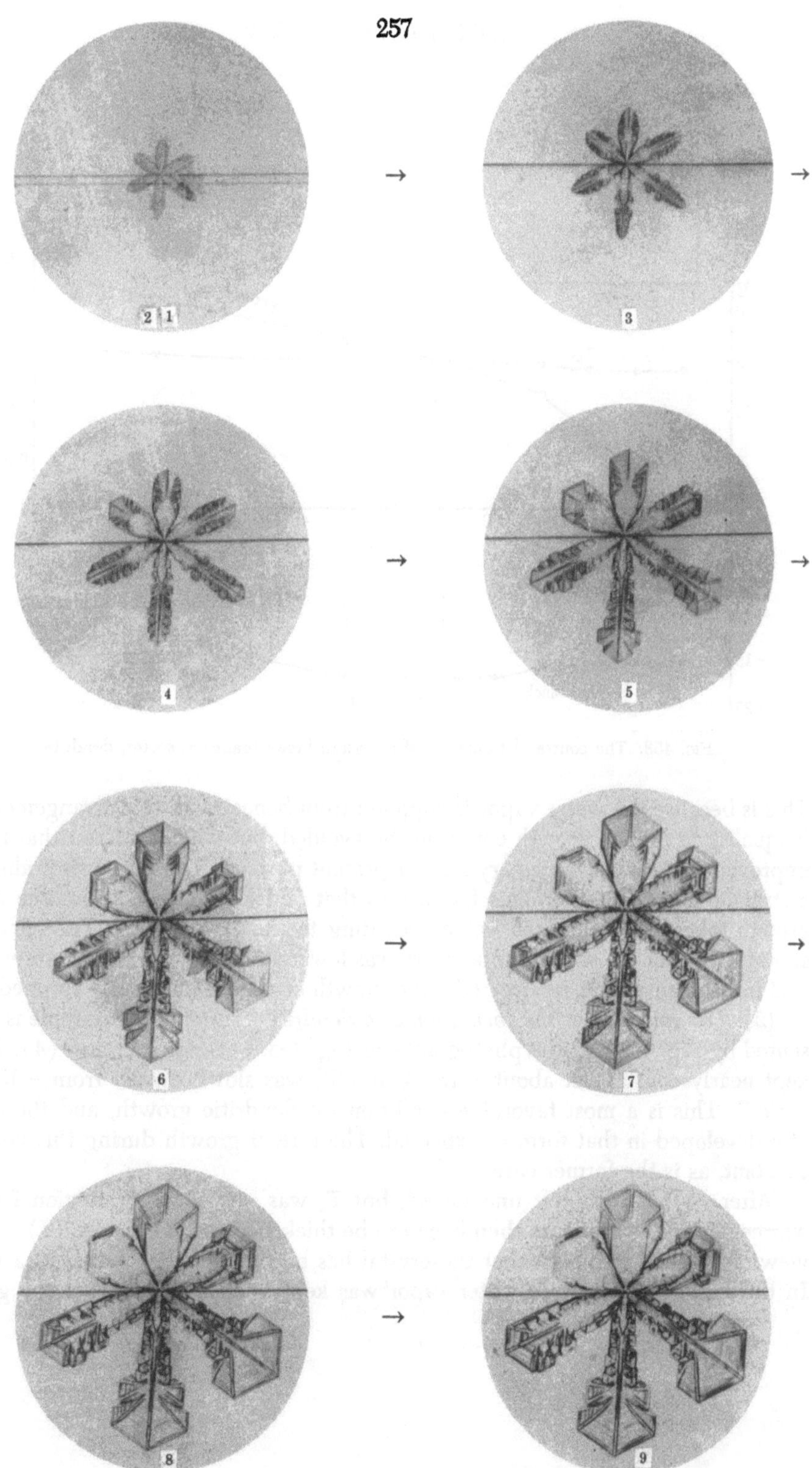

Fig. 457 [1442–49]. Successive stages in the course shown in Fig. 456 (× 16).

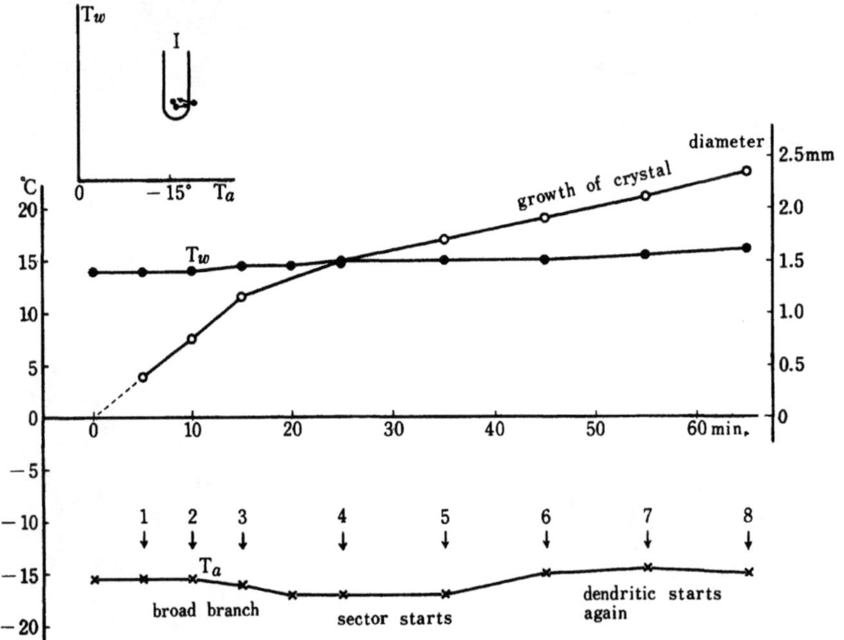

FIG. 458. The course of formation of a crystal; broad branches, sector, dendritic.

This is because the water vapor is supplied from beneath. In this arrangement this inequality in branch growth could not be avoided, but we considered that further improvement was not necessary for our present purpose. The curve rises almost in a straight line up to (5). This fact shows that under given conditions the rate of growth takes a certain fixed value according to the crystal type. The curve also shows that after the water temperature was lowered and the supply of water vapor was made scanty, that is, after (5), the growth of the crystal nearly stopped.

(2) *T_a is raised after the formation of a dendritic crystal.* One example is represented by Fig. 452 and the photographs of Fig. 453. Between (1) and (4), T_a was kept nearly constant at about −15°C, and T_w was slowly raised from +15°C to +19°C. This is a most favorable condition for dendritic growth, and the crystal also developed in that form as expected. The rate of growth during this period is constant, as is the former case.

After (4), T_w was left unchanged, but T_a was raised out of Region I to the warmer side. The branches then began to be thickened, as (6) shows; (7) is a side view of (6) and it is seen that the crystal has become a little thicker as a whole. In this case, the supply of water vapor was kept unchanged, so that the growth

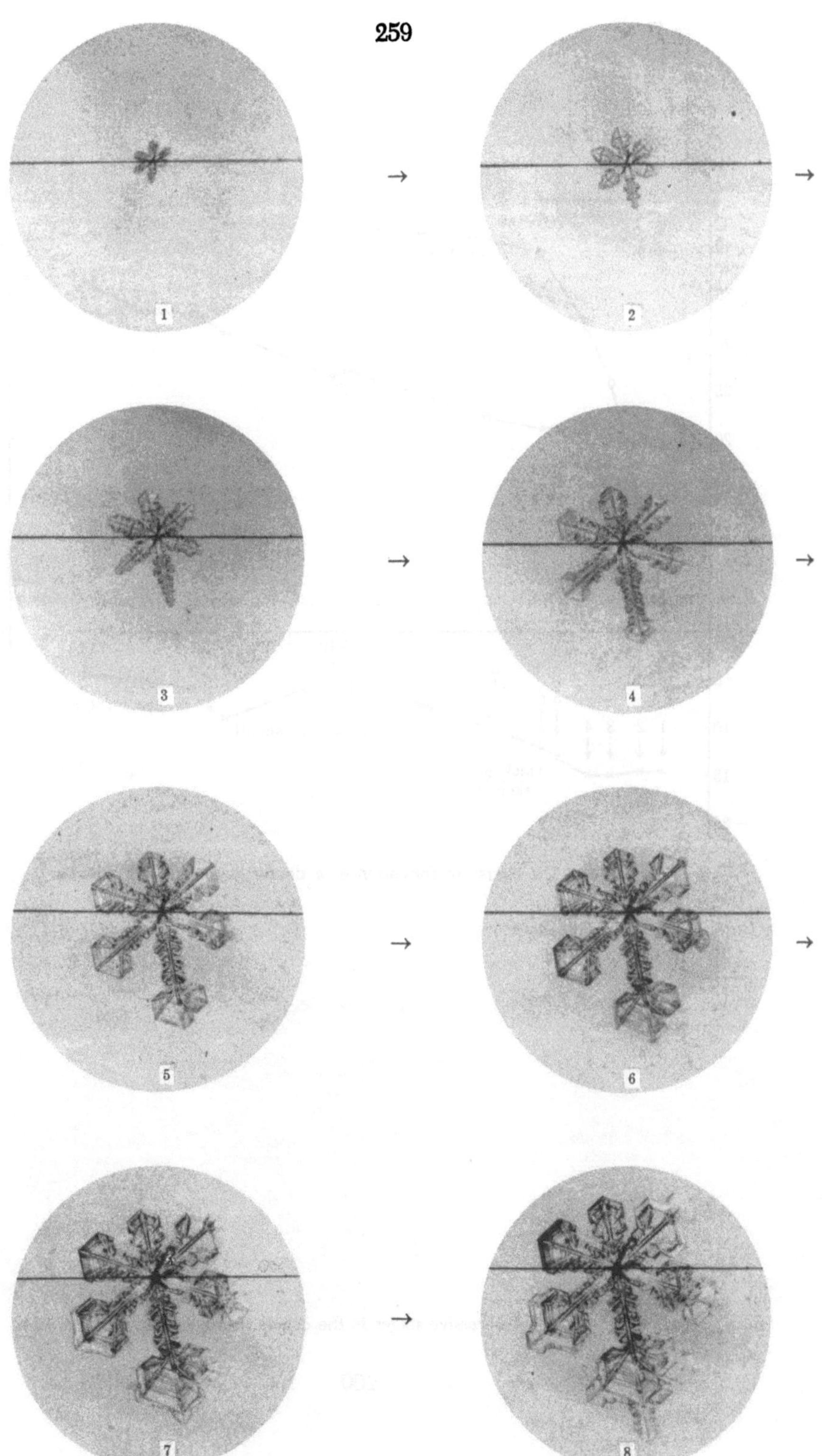

Fig. 459 [1450–57]. Successive stages in the course shown in Fig. 458 (× 17).

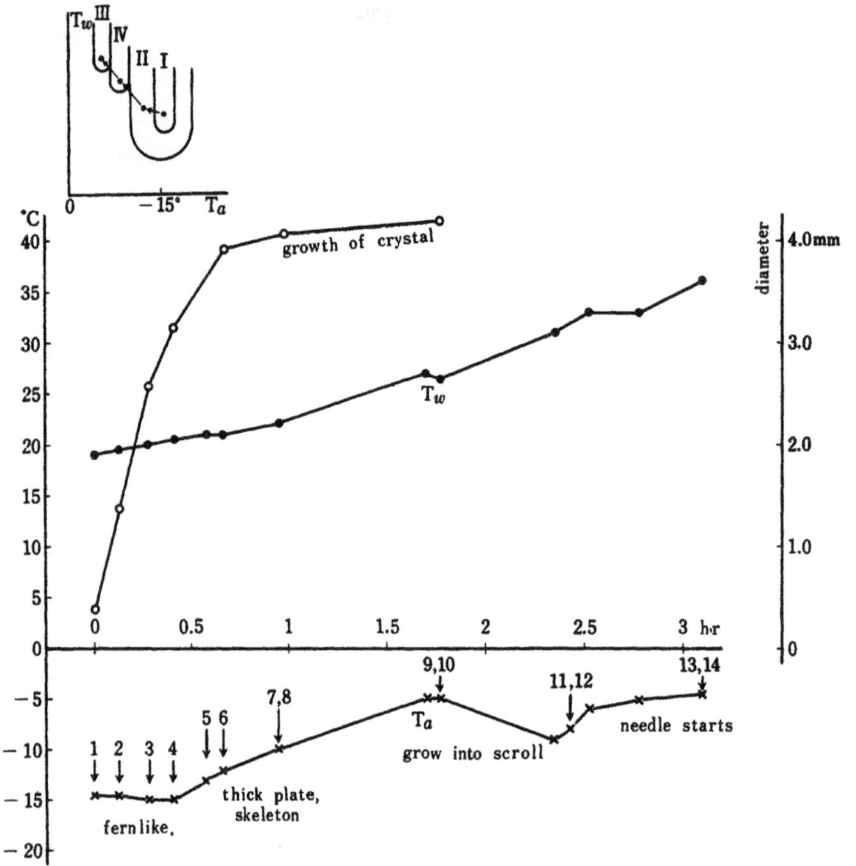

FIG. 460. The course of formation of a dendritic crystal with needles.

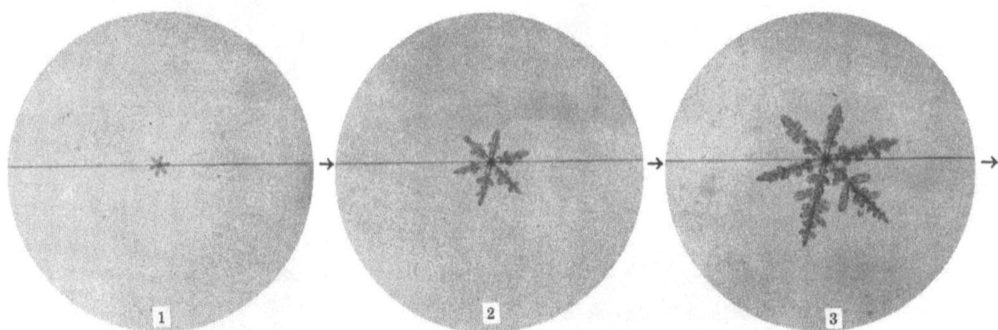

FIG. 461 [1458–71]. Successive stages in the course shown in Fig. 460 (× 9.3).

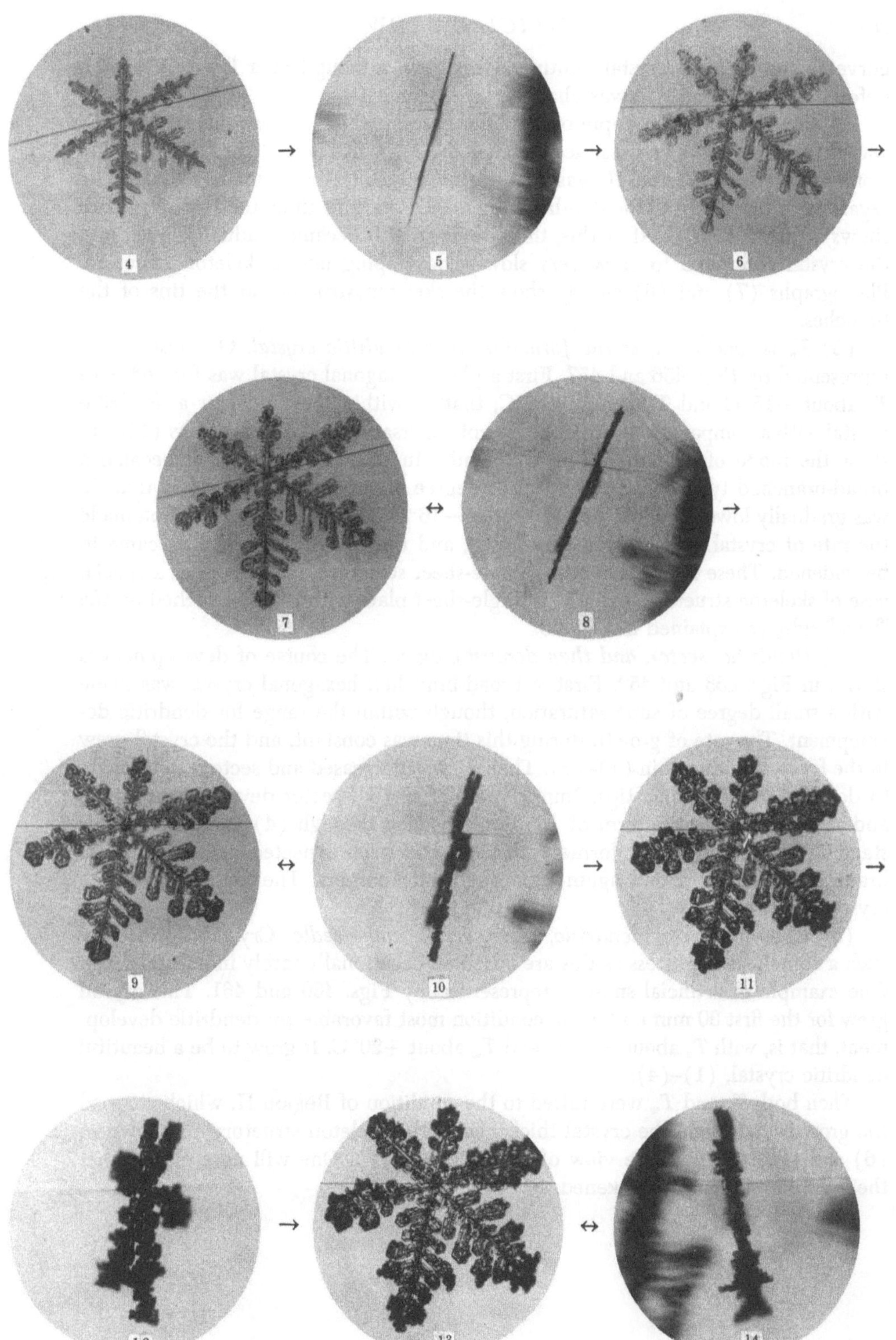

Fig. 461 (*continued*).

curve shows that the crystal continued to grow, although at a lower rate than before, after the condition was changed into Region II.

Another and better example of a similar case in which T_a was raised while T_w was kept as constant as possible is shown by Figs. 454 and 455. The transition from Region I to Region II was made gradually but stages up to (5) can be regarded as belonging to the dendritic region. During this time, the rate of growth shows a constant value. After this, the rate of growth became gradually lower and the crystal continued to grow very slowly, developing into a skeleton structure. Photographs (7) and (8) clearly show the skeleton structure at the tips of the branches.

(3) *T_w is lowered after the formation of a dendritic crystal.* One example is represented by Figs. 456 and 457. First a plane hexagonal crystal was formed with T_a about $-15°C$ and T_w about $+12°C$, that is, within the range for a dendritic crystal with a comparatively small degree of supersaturation. Photographs (1)–(4) show the mode of growth during this while. In this case the crystal became a broad-branched type owing to the small degree of supersaturation. After that T_w was gradually lowered while T_a was kept at $-16°C$. This change in condition made the rate of crystal growth gradually lower, and the tips of the branches came to be widened. These sectors have no double-sheet structure, but represent a special case of skeleton structure, namely, a single-sheet plate with ridges attached on the "back" side, as explained in Sec. 68.

(4) *Dendritic, sector, and then dendritic again.* The course of development is shown in Figs. 458 and 459. First, a broad-branched hexagonal crystal was made with a small degree of supersaturation, though within the range for dendritic development. The rate of growth during this time was constant, and the crystal grew in the form reproduced in (1)–(3). Then T_a was decreased and sectors were made to develop at the branch tips. Immediately after (3), sector development began and the crystal took the form of (6), after passing through (4) and (5). At this stage T_a was raised to the former value. As had been expected, broad dendritic branches began to extend again from the sector corners. The final form of the crystal is shown in (8).

(5) *Combination of dendritic, plate, scroll, and needle.* Crystals formed by such a complicated process as this are observed occasionally rarely in natural snow. One example of artificial snow is represented by Figs. 460 and 461. This crystal grew for the first 30 min under the condition most favorable for dendritic development, that is, with T_a about $-15°C$ and T_w about $+20°C$. It grew to be a beautiful dendritic crystal, (1)–(4).

Then both T_a and T_w were raised to the condition of Region II, which stopped the growth and made the crystal thicker with the skeleton structure, as shown in (6) and (7); (8) is a side view of the crystal in (7). One will note clearly that the crystal is somewhat thickened.

After this stage T_a and T_w were still more raised, and the condition of Region IV favorable for scroll development was established. The crystal came to appear dark by transmitted light, which shows that scrolls extended at the edge of skeletons. This stage is shown in (11). In (12) the crystal of (11) is viewed aslant, and the scroll development is clearly observed.

Finally, T_a and T_w were further raised and the condition of Region III was established, which is for needles. As was expected, needles began to grow from the scroll edges. Photographs (13) and (14) show the final form of the crystal. In the side view (14), the extended needles are clearly shown.

We observe in natural snow many crystals that are thick and appear white by reflected light and dark by transmitted light. Some of them are those with cloud particles attached, but others are those thickened by scroll or needle development, as in the present example. It is easily understood that when the temperature is low in the upper air, approaching 0°C near the ground, there are more opportunities for crystals of this type to fall.

As is mentioned in the above five cases, the relation between the crystal type and the condition of formation is fairly clear today. We can, therefore, produce artificially any crystal belonging to a certain type or combination of several types. The method is simple. As clarified by this study on the course of formation, the only requirement is to establish the necessary conditions one after another.

CHAPTER 7

Comparison of Natural and Artificial Snow Crystals

84. Process of formation of natural snow crystals

As already described in this book, there are several theories of the mode of development of natural snow from the form of crystal. Among them the theory of Wegener, assuming the transformation of a dendritic form into a plate under conditions of low supersaturation, and other theories following his, seem to have lost their foundation. The experimental criticism of Wegener's theory of crystal growth described in Sec. 76 shows that the process supposed by him cannot be observed in our experiments on artificial snow. The crystal in the form of a plate with dendritic extensions, however, was verified in our experiments to have been produced by the formation of a plate in the upper atmosphere and the growth of dendritic branches from the plate while the crystal was falling through the lower layers of the atmosphere where conditions were suitable for dendritic development.

From the experiments on the formation of various types of crystals, Sec. 83, we found that a given snow crystal was made by adding the characteristic features that correspond to the variations in the external conditions. Examining the natural snow crystals from this point of view, we can find many examples showing stepped development in the ordinary dendritic crystals. Examples are shown in Figs. 462–465.

One will see in the central part of the large dendritic crystal of Fig. 463 a form just like the one shown in Fig. 462. The crystal of Fig. 462 was born in a layer of a certain condition which lay near the ground, and it fell to the ground in this form. For the crystal of Fig. 463, a similar condition must have existed in the upper atmosphere; the resulting crystal, shown in the central part of the figure, then fell through a lower layer that was suitable for the dendritic development.

A similar phenomenon is observed in the case of Figs. 464 and 465. A form very similar to that of the crystal in Fig. 464 is observed in the central part of the crystal

FIG. 462 [1478]. Natural snow; the first stage (× 33).

FIG. 463 [1479]. Natural snow; further development from Fig. 462.

FIG. 464 [1480]. Natural snow; the first stage (× 30).

FIG. 465 [1481]. Natural snow; further development from Fig. 464 (× 14.5).

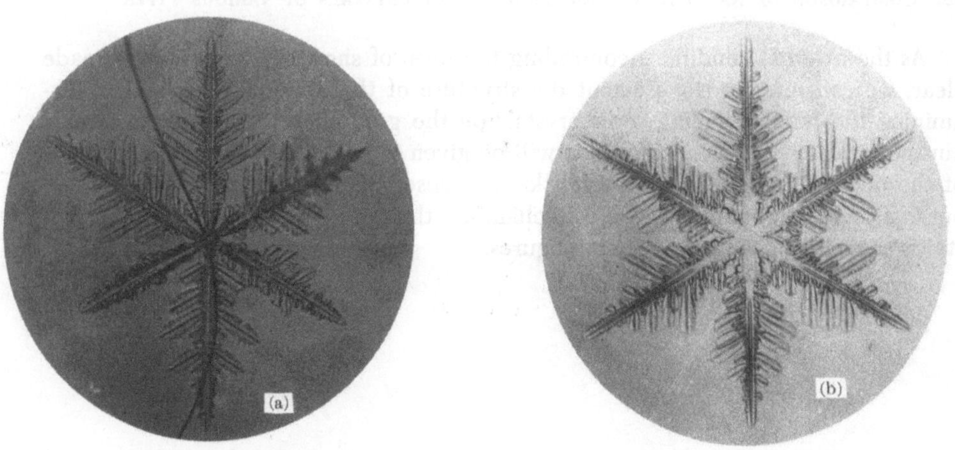

FIG. 466. Comparison of (*a*) [1482] artificial and (*b*) [1483] natural snow, I (× 34).

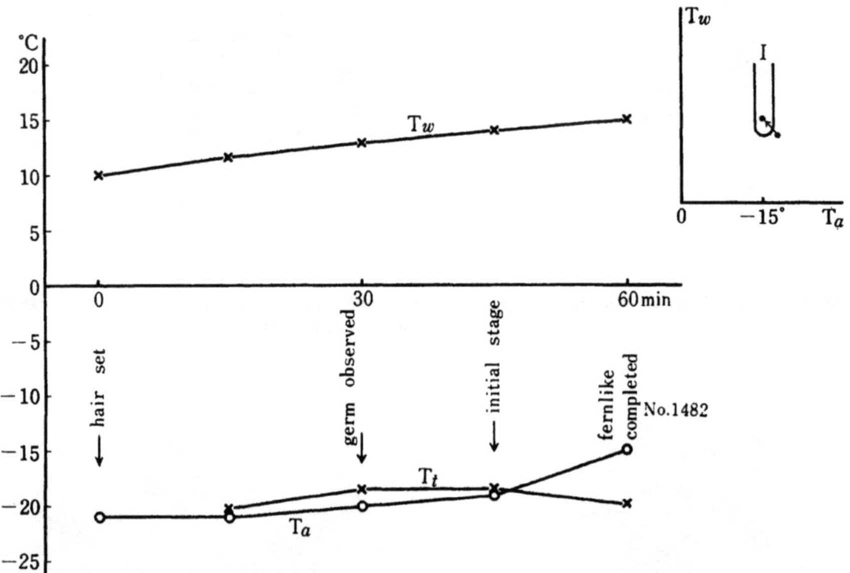

FIG. 467. The course of formation of Fig. 466(*a*).

in Fig. 465. The secondary development in Fig. 465, that is, the dendritic portion of the branch, is much longer than in Fig. 463. We may infer that the lower layer of the atmosphere suitable for the dendritic development must have been thicker in this case than in the case of Fig. 463.

Many other examples can be chosen among the photographs collected in this book, which show a similar stepped development of branches.

85. COMPARISON OF NATURAL AND ARTIFICIAL SNOW CRYSTALS OF VARIOUS TYPES

As the external conditions controlling the form of snow crystals are now made clear, we can infer to some extent the structure of the upper atmosphere by examining the form of falling snow crystals on the ground. With respect to twelve kinds of crystals, some consideration will be given to the meteorological conditions of the upper atmosphere when each kind is observed. Experiments were carried out in a thermostat placed in the cold chamber; the temperature inside the thermostat is shown by T_t in the following figures.

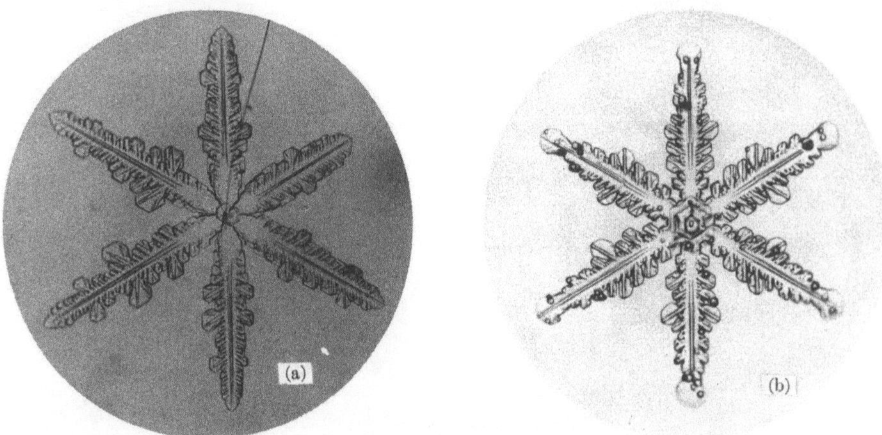

FIG. 468. Comparison of (*a*) [1484] artificial and (*b*) [1485] natural snow, II (× 34).

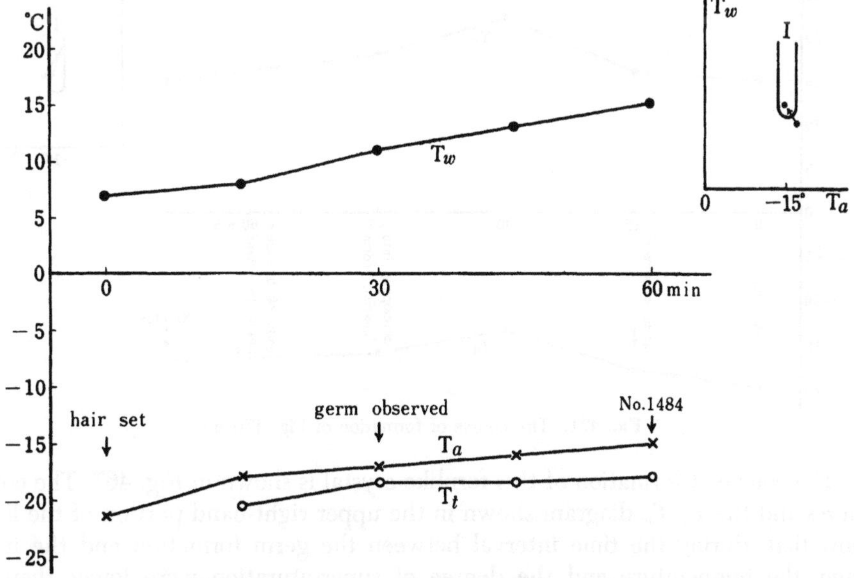

FIG. 469. The course of formation of Fig. 468(*a*).

(1) ***Fernlike hexagonal crystal.*** Natural and artificial snow crystals of fernlike type are reproduced in Figs. 466(*a*) and (*b*). Comparing these two photographs, one will see that the form and structure of the branches are very similar, although the central part is a little different.

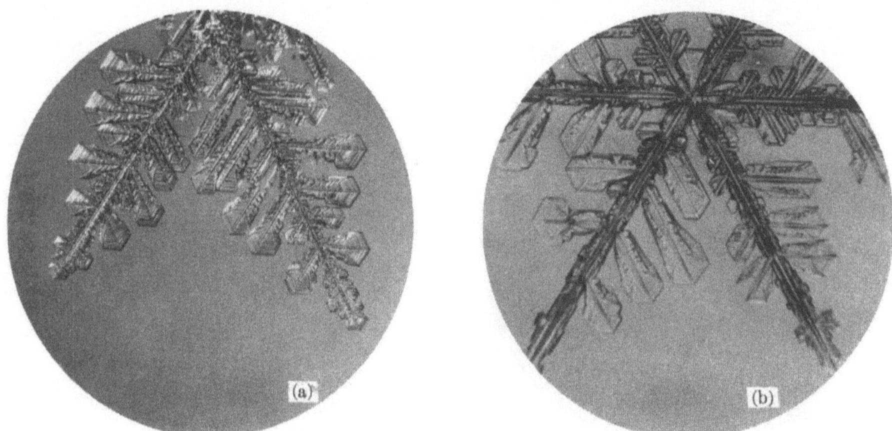

FIG. 470. Comparison of (*a*) [1486] artificial (× 38) and (*b*) [1487] natural snow, III.

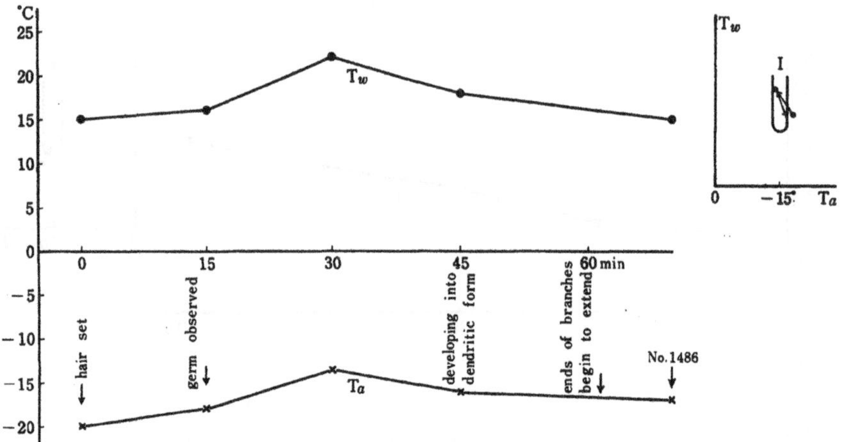

FIG. 471. The course of formation of Fig. 470(*a*).

The course of formation of this fernlike crystal is shown in Fig. 467. The course curves and the T_a–T_w diagram shown in the upper right-hand portion of the figure show that, during the time interval between the germ formation and the initial stage, the temperature and the degree of supersaturation were lower than the critical values for dendritic development; that is, T_a and T_w were outside Region I. The initial stage of this crystal was thus the irregular assemblage of small sectors. Then both T_a and T_w were increased so that the conditions were most favorable for dendritic development. The crystal rapidly grew in a fernlike form, and it developed in the form shown in Fig. 446(*a*) in about 15 min. The rate of growth was very high in this case. From these data for the formation of an artificial snow crystal, one can infer that there must exist an atmospheric layer with ample moisture and a temperature about $-15°C$ near the ground when a crystal like that reproduced in Fig. 466(*b*) is observed on the ground.

(2) *Stellar crystal.* Figures 468(*a*) and (*b*) and Fig. 469 show an example of the stellar crystal. This type of crystal has hitherto been distinguished from the dendritic one, but the pointed form of the ends of branches shows that this must be treated as a dendritic crystal for the discussion of its condition of formation, as already described in detail in Sec. 79. The stellar crystal shown in Fig. 468(*a*) will develop shortly into an ordinary dendritic form, if the conditions are left as they are. The T_a–T_w diagram in Fig. 469 is similar to that in Fig. 467.

The natural snow crystal as shown in Fig. 468(*b*) is thus considered to be an earlier form of the ordinary dendritic crystal. It is quite natural that this sort of crystal is often observed falling mixed with dendritic ones. When there is an atmospheric layer suitable for dendritic development, a crystal born near the ground reaches it in this stellar form.

(3) *Broadening of the ends of dendritic branches.* The natural snow crystal shown in Fig. 470(*b*) is often observed among the large dendritic crystals, and appears to show no marked characteristics. Close examination, however, of the ends of the branches indicates that they are somewhat broadened. This sort of crystal is to be considered as the result of a slight sublimation of the crystal similar to that reproduced in Fig. 470(*a*).

The artificial crystal of Fig. 470(*a*) is formed under the conditions shown in Fig. 471. Putting aside the conditions for the initial stage, this crystal developed to its full extent into a dendritic form; then both T_a and T_w were decreased so that the point in the T_a–T_w diagram was displaced to the border of Region I. The broadening of the ends of the branches took place under this condition. The crystal shown in Fig. 470(*b*) is thus inferred to have fallen when a slight inversion of temperature existed in the atmosphere near the ground.

(4) *Sector-form crystal.* Comparison of the natural and artificial crystals reproduced in Figs. 472(*a*) and (*b*) shows such a resemblance that there is scarcely any doubt as to the coincidence of the conditions of formation for these two crystals.

The course curves and T_a–T_w diagram in Fig. 473 show that the temperature is suitable for dendritic development but the supersaturation is not high enough, resulting in a sector-form development. The rate of growth of sector-form crystals is fairly low. It took about 1.5 hr for the development of the crystal to the stage of Fig. 472(*a*). The natural snow of Fig. 472(*b*) seems to have taken about 1 hr for its growth. We can expect a less supersaturated layer of considerable thickness existing near the ground in this case.

(5) *Stellar crystal with plates at the ends of branches.* Crystals like that shown in Fig. 474(*b*) are frequently observed in nature, and are chosen as one type in the

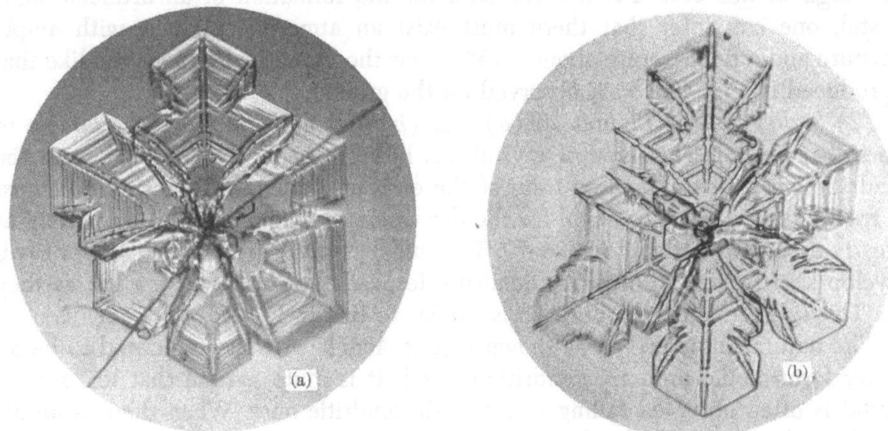

Fig. 472. Comparison of (*a*) [1488] artificial (× 28) and (*b*) [1489] natural snow, IV (× 52).

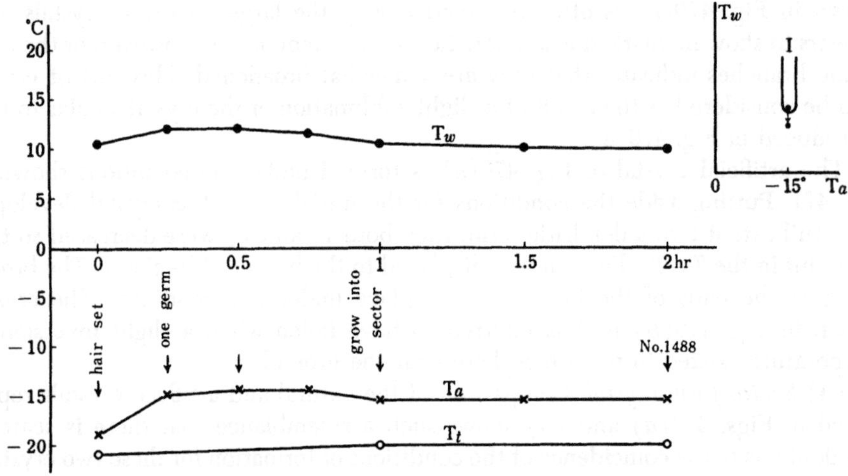

Fig. 473. The course of formation of Fig. 472(*a*).

classification of hexagonal plane crystals. The crystal of Fig. 474(*a*) produced artificially belongs certainly to this type. The fine stripes in the plate portion are less marked in the case of natural snow. This can be explained by assuming that the natural crystal has passed, in approaching the ground, through a thin layer of atmosphere in which sublimation has taken place.

The course of formation is shown in Fig. 475. In the earlier stage the conditions were favorable for stellar or dendritic development; then both T_a and T_w were

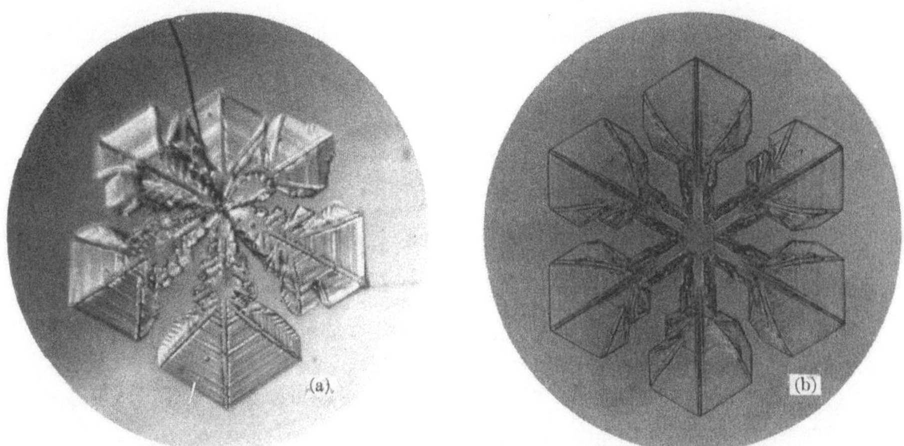

FIG. 474. Comparison of (*a*) [1490] artificial (× 31) and (*b*) [1491] natural snow, V (× 43).

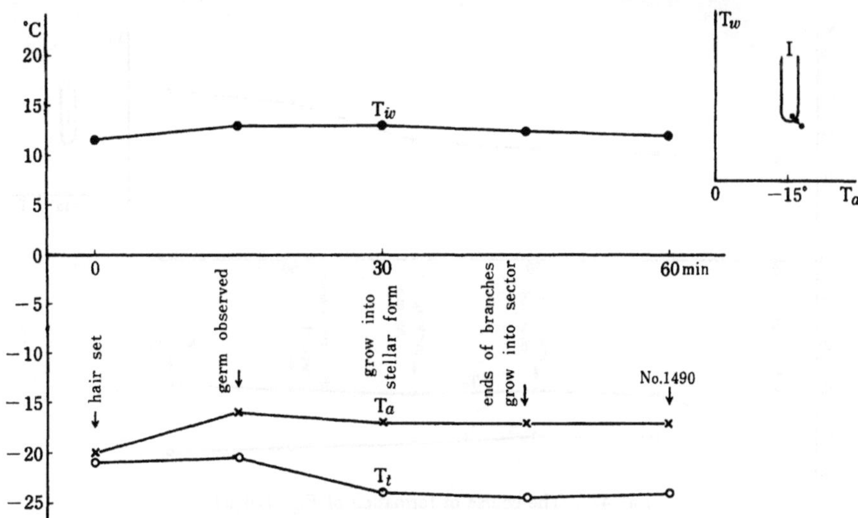

FIG. 475. The course of formation of Fig. 474(*a*).

decreased to the border of Region I in the T_a–T_w diagram and the crystal was kept for a considerable time in this condition. In general when a crystal is kept under a nearly stationary condition at the point of lower temperature and less supersaturation in the border of Region I, a small hexagonal crystal of broad branches is obtained in the earlier stage and then the broadening of the ends of branches into the sector form is observed in the later stage. This tendency can be explained as the result of decreasing the vapor supply per unit area of the crystal surface. The

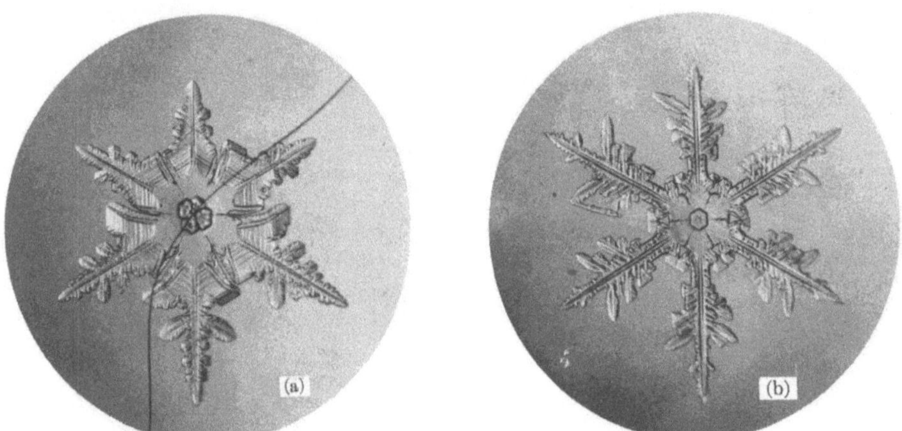

FIG. 476. Comparison of (*a*) [1492] artificial (× 22) and (*b*) [1493] natural snow, VI.

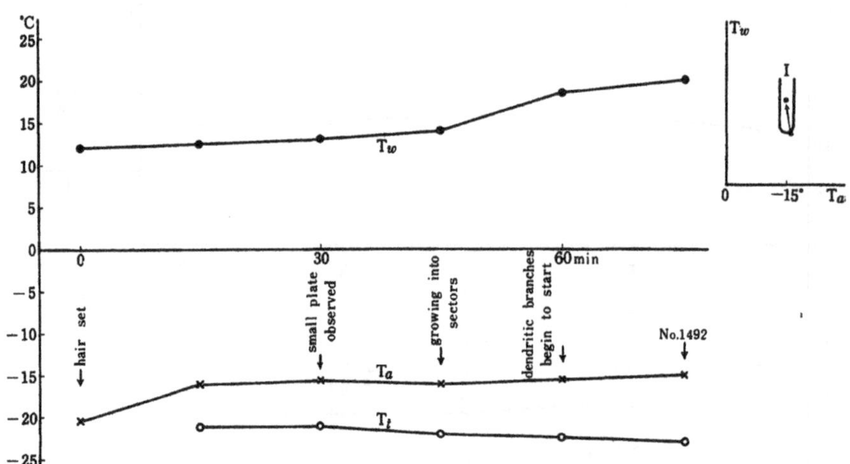

FIG. 477. The course of formation of Fig. 476(*a*).

crystal of the form illustrated in Fig. 474(*b*) seems to show that a layer of considerable thickness of uniform slight supersaturation exists near the ground and above it a thin layer suitable for dendritic development.

(6) *Sectors with dendritic extensions.* One example is shown in Figs. 476(*a*) and (*b*) and Fig. 477. The artificial crystal of Fig. 476(*a*) was made as follows. In the earlier stage the temperature was suitable for dendritic development, but the supersaturation was not high enough, so that the crystal first developed into a plate and then into the sector form. The sector part was made in nearly 30 min.

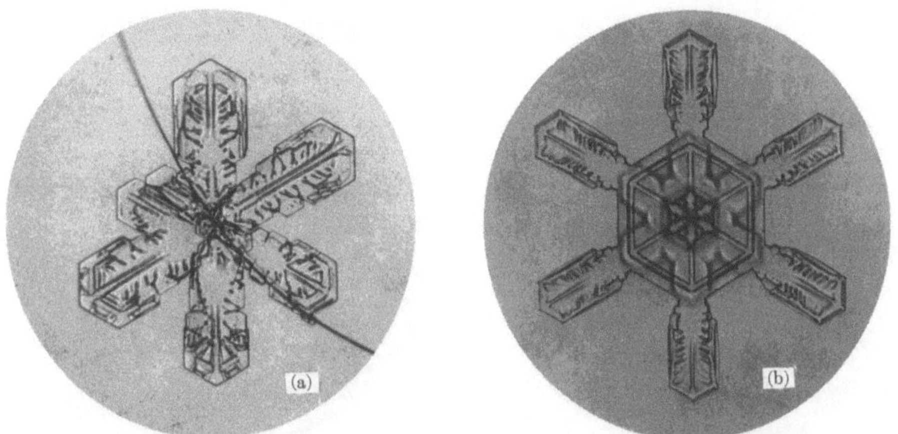

Fig. 478. Comparison of (*a*) [1494] artificial (× 27) and (*b*) [1495] natural snow, VII (× 35.5).

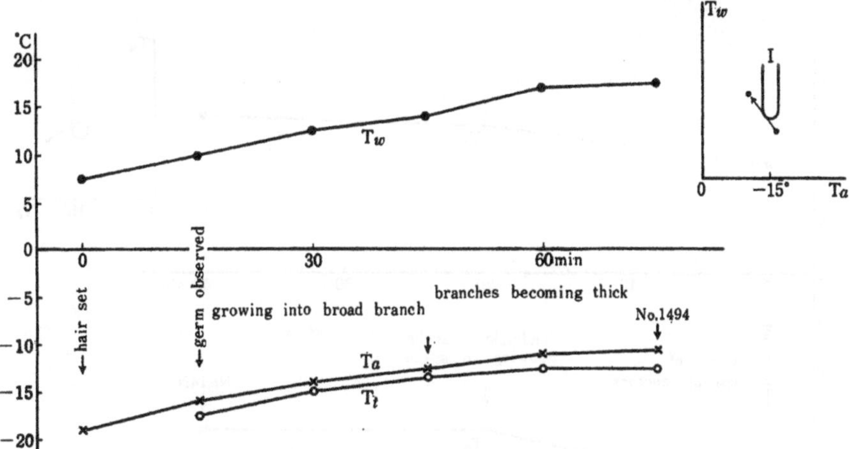

Fig. 479. The course of formation of Fig. 478(*a*).

Then T_a and T_w were both increased, and the dendritic branches extended from the corners of the sectors. The rate of growth of the dendritic portions was fairly high, the crystal having developed to the form shown in Fig. 476(*a*) in about 15 min after the completion of the sectors. The structure of the atmosphere when the crystal of Fig. 476(*b*) was observed on the ground will be inferred from these data.

(7) *Branches with double-sheet structure.* The crystal of Fig. 478(*b*) is a tsuzumi type, one plane crystal of which is a hexagonal plate with extensions of broad branches and the other a smaller stellar crystal with broad branches. As the focus is adjusted on the former, the latter is seen as a diffuse image. The broad

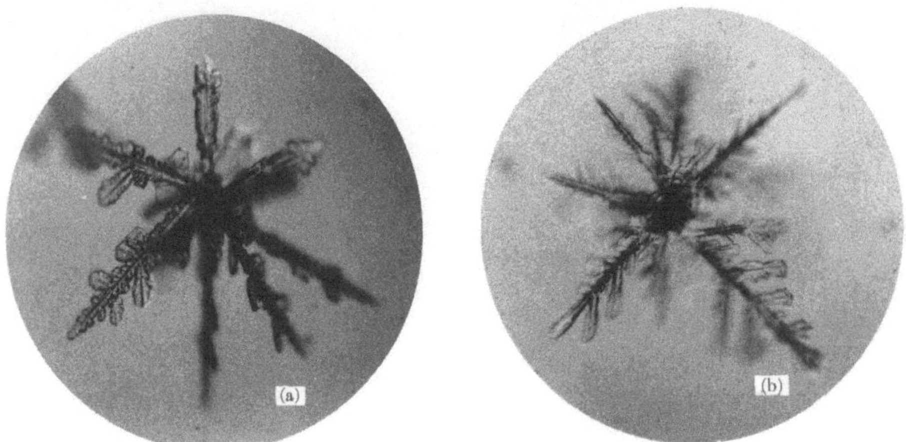

Fig. 480. Comparison of (*a*) [1496] artificial (× 28) and (*b*) [1497] natural snow, VIII (× 20).

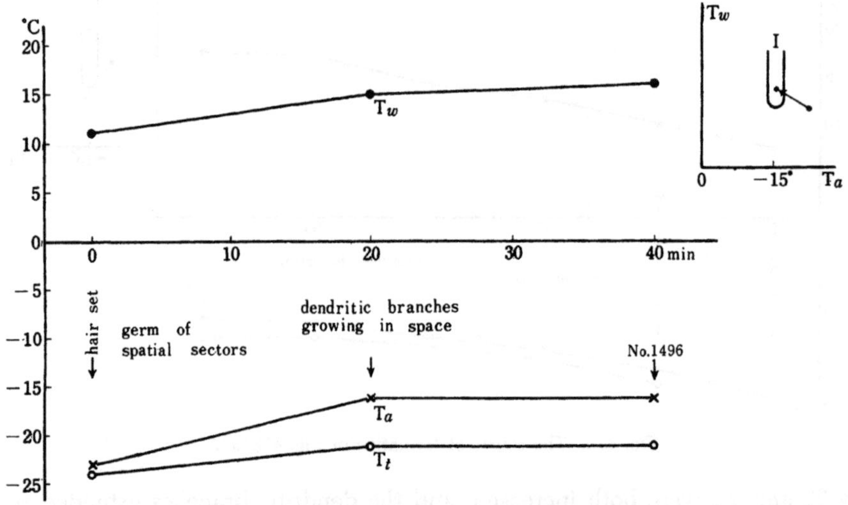

Fig. 481. The course of formation of Fig. 480(*a*).

branches extended out from the hexagonal plate show a characteristic design due to double-sheet structure as described in Sec. 6.

Paying attention only to the structure of these branches, the crystals of Figs. 478(*a*) and (*b*) are quite similar. The course of formation of this type of branches is seen in Fig. 479. After the formation of a stellar crystal with broad branches, both T_a and T_w were increased. The development of the branches into a double-

Fig. 482. Comparison of (*a*) [1498] artificial (× 14) and (*b*) [1499] natural snow, IX.

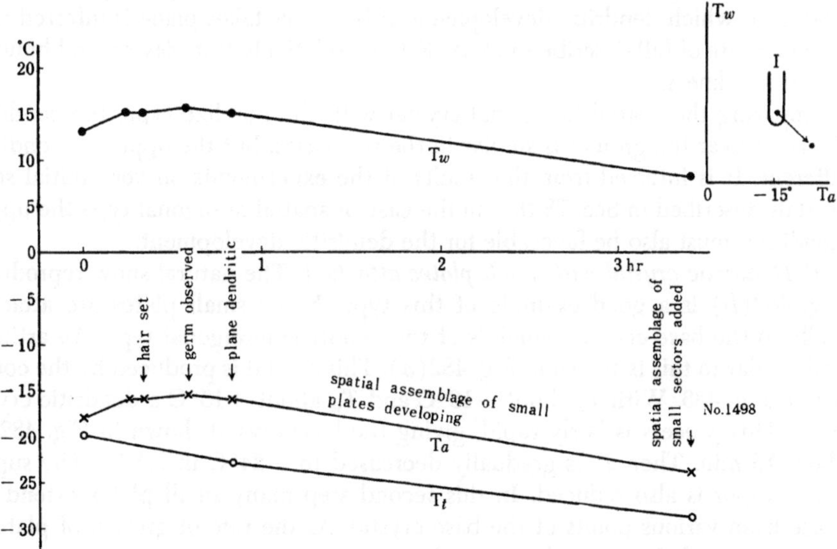

Fig. 483. The course of formation of Fig. 482(*a*).

sheet structure, a sort of skeleton form, began at $T_a = -13°C$. As T_a was increased gradually to $-11°C$, the characteristic design developed to its full extent. In this case the ample supply of water vapor to the crystal is another condition.

From these data it is inferred that there will exist near the ground a layer of temperature between $-11°C$ and $-13°C$ and of fairly high supersaturation, when a crystal like that in Fig. 478(*b*) is observed. Above this layer there will be conditions suitable for the formation of the tsuzumi type of crystal. The condition for tsuzumi formation is described in (12) of this section.

(8) *Burrlike crystal.* The condition for the formation of a burrlike crystal, or a radiating type of spatial assemblage of dendritic branches, is also made clear. This sort of crystal has at the center an assemblage of small sectors or short columns, and the dendritic branches radiate from them in space. The early stage in the form of an assemblage of sectors or columns is obtained at low values of T_a and T_w. For the crystal of Fig. 480(*a*), T_a was $-20°C$ or thereabout and T_w was about $+12°C$. After the formation of this early stage, both T_a and T_w were increased to the conditions favorable for the dendritic development, that is, T_a was $-16°C$ and T_w was $+15°C$. Dendritic branches rapidly grew in space, giving in 20 min the crystal shown in Fig. 480(*a*).

We may interpret the form of crystal shown in Fig. 480(*b*) as follows: in the upper atmosphere there exists a layer at a temperature of $-20°C$ or less and of low supersaturation; the crystal is born in this layer, and the minute crystal thus formed passes while falling through a layer at about $-15°C$ and of ample supersaturation; this layer, in which dendritic development of branches takes place is inferred from the data on rate of fall described in Sec. 36 to be relatively thin, say several hundred meters in thickness.

Comparing the spatial hexagonal crystal with the burrlike type, the condition of the layer near the ground is similar in the two cases, but the upper-air condition is different. It is inferred from the results of the experiments on very initial stage of crystals described in Sec. 78 that in the case of spatial hexagonal type the upper-air condition must also be favorable for the dendritic development.

(9) *Dendritic crystal with small plates attached.* The natural snow reproduced in Fig. 482(*b*) is a good example of this type. Many small plates are attached spatially to the base crystal, which is of the dendritic hexagonal type. An artificial crystal similar to this is shown in Fig. 482(*a*). This crystal is produced by the course shown in Fig. 483. With T_a about $-16°C$ and T_w about $+15°C$ a dendritic crystal is made. This process is fairly rapid, giving the base crystal shown in Fig. 482(*a*) in about 15 min. Then T_a is gradually decreased to $-24°C$ in 2.5 hr. The supply of water vapor is also reduced. In this second step many small plates extend out in space from various points of the base crystal. As the rate of growth of plates is very small, it took 2.5 hr in this example.

When this type of natural snow is observed, we may expect a thick layer of temperature inversion. The layer near the ground must be at a temperature of nearly $-20°C$ and be less supersaturated. The thickness of this layer is estimated from the data on rate of fall to be about 2 km, and above this cold layer there exists

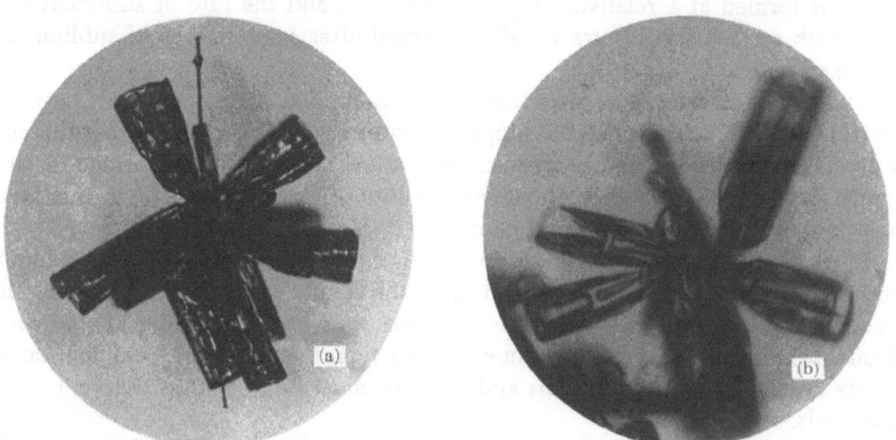

FIG. 484. Comparison of (*a*) [1500] artificial (× 27) and (*b*) [1501] natural snow, X (× 34).

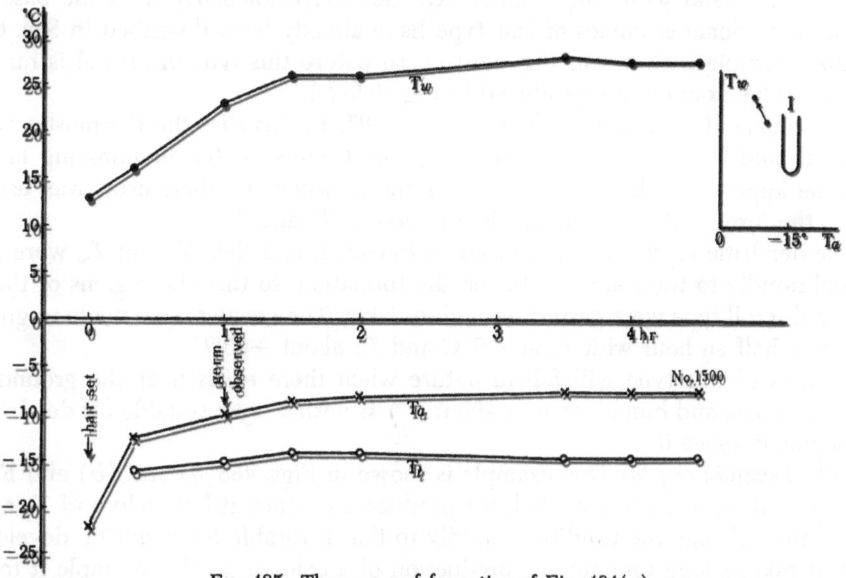

FIG. 485. The course of formation of Fig. 484(*a*).

a warmer layer at about −15°C and well supersaturated. This warmer layer may not be so thick, say several hundred meters.

(10) *Assemblage of bullets.* The artificial and natural crystals of this type are shown in Figs. 484(*a*) and (*b*). The outline of the image of the natural crystal is somewhat rounded. This is considered to be the result of sublimation. This sort of

crystal is formed at a relatively high temperature, and the rate of sublimation is fairly high so that crystals are usually observed after some degree of sublimation has taken place.

The course curves show that this crystal is formed at temperatures between −9°C and −7°C and with ample supply of water vapor. It is inferred that this sort of crystal is observed in nature when there is a layer of well-supersaturated air at a temperature of −8°C or thereabout. As the rate of growth is not high, this layer must be quite thick in the natural case.

Stüve says in his report of flight in snowfall that the combinations of columns were observed only when the temperature in the clouds was −5°C.[74] It is generally accepted that columns fall when the temperature is low. According to the result of our experiment the type mentioned in his report will be the combination of bullets instead of columns. Bullets and columns must be treated as different types of crystals.

(11) *Dendritic crystal with long needles extended perpendicularly.* As our artificial-snow technique improved, many peculiar crystals were produced. The dendritic plane crystal with long needles extended perpendicularly from the base is one of them. Some examples of this type have already been described in Sec. 63. Another example is shown in Fig. 486(*a*). In nature this type of crystal is rarely observed. One example is reproduced in Fig. 486(*b*).

The course of formation is shown in Fig. 487. In this case the thermostat was not used, and T_a was varied by changing the distance of the illuminating lamp from the apparatus. The result was the same as when the thermostat was used; that is, the form of the crystal was determined by T_a and T_w.

The dendritic crystal was produced in Region I, and then T_a and T_w were increased rapidly to the condition for needle formation, so that the regions of thick plate and scroll type were passed through quickly. The needles were made to grow for nearly half an hour with T_a at −6°C and T_w about +35°C.

Crystals of this type will fall in nature when there exists near the ground a relatively warm and humid layer at about −5°C with a layer suitable for dendritic development above it.

(12) *Tsuzumi crystal.* One example is shown in Figs. 488(*a*) and (*b*) and Fig. 489. For making a tsuzumi crystal, we produce a column at low values of T_a and T_w and then change the condition rapidly to that favorable for dendritic development. It takes a long time for the production of a column. In this example it took about 5 hr. Usually a slender column is obtained in the earlier stage and it gradually grows thick into an ordinary column. As the intermediate stage from columnar to dendritic, the transition stage is observed as already described in Sec. 71. After the start of dendritic branches, the rate of growth is quite high. Natural snow of the tsuzumi type is considered to be produced when a relatively thin layer suitable for dendritic development exists near the ground and the upper atmosphere con-

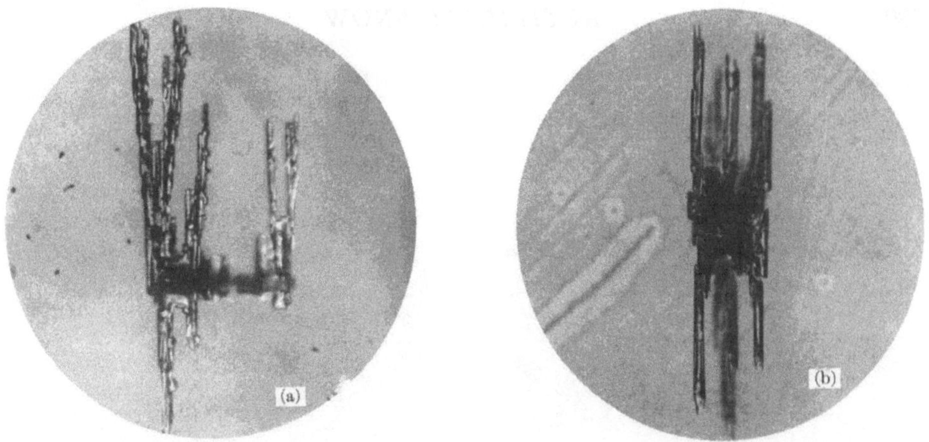

Fig. 486. Comparison of (*a*) [1502] artificial (× 22.5) and (*b*) [1503] natural snow, XI (× 18.5).

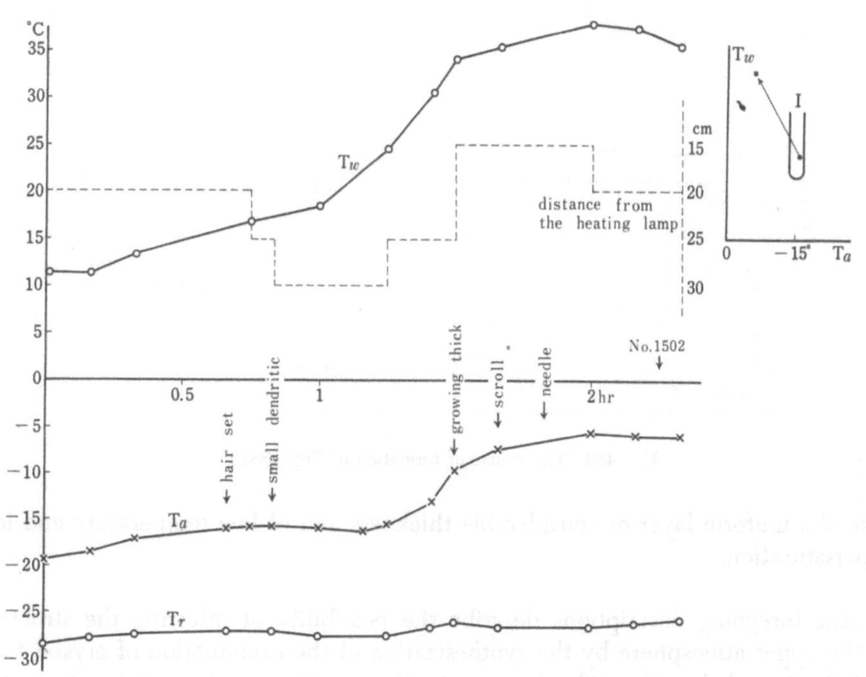

Fig. 487. The course of formation of Fig. 486(*a*).

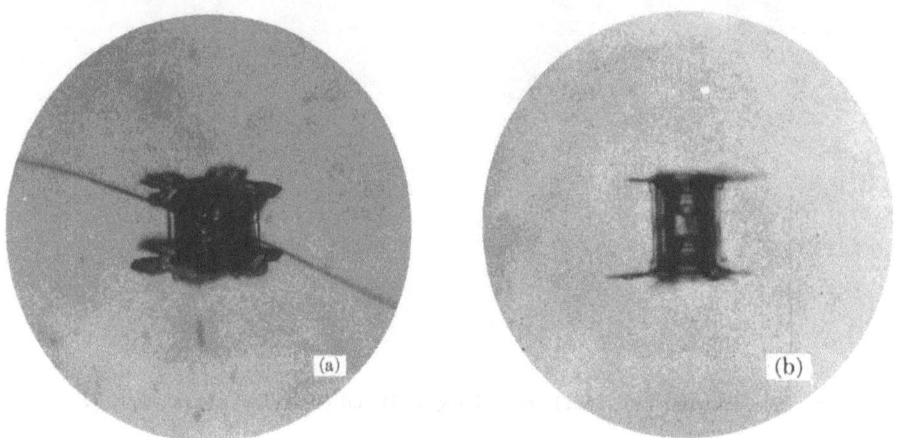

FIG. 488. Comparison of (*a*) [1504] artificial (× 32.5) and (*b*) [1505] natural snow, XII (× 27).

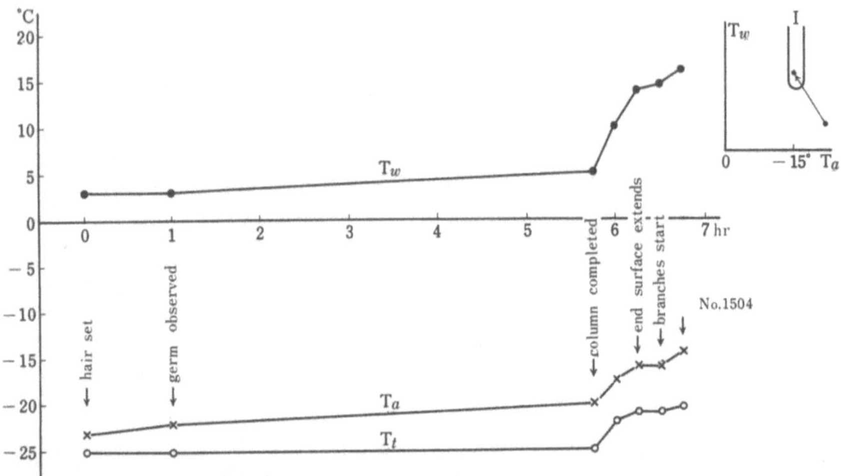

FIG. 489. The course of formation of Fig. 488(*a*).

sists of a uniform layer of considerable thickness and of low temperature and low supersaturation.

The foregoing descriptions describe the possibility of inferring the structure of the upper atmosphere by the synthesization of the examination of crystal form and the knowledge of artificial snow. In this section we did not touch on the question of the wind. The horizontal component of wind velocity has no sensible influence upon our argument, but the upward current and the turbulence will have a strong effect upon the above statement.

86. Process of formation of natural and artificial crystals of tsuzumi type

The process of formation of natural snow crystals can best be studied by examining the stages of crystal development directly by flying through a snowfall, or by catching the crystals on a suitable oil film as described in Sec. 78 by the use of a balloon or a kite. In our case no such facilities were available, so we tried to await in a mountain district a fortuitous chance when the sequence of crystal growth is observed. When the atmosphere is displaced as a whole without sensible change of structure, we can expect at a high spot on a mountain such a chance when the proper conditions of upper and lower layers of the atmosphere are observed successively in time.

This observation was made at Mount Tokachi in February 1940. At the altitude of 1030 m, observations of crystals were carried out for every snowfall. During one month's stay, we met with a chance when the process of formation of bullets with dendritic crystals was observed. Photographs showing this process are reproduced in Figs. 490–495.

At first bullet-type crystals were observed, one example being shown in Fig. 490. Shortly crystals of bullet type with a small plate attached to the basal end began to fall mixed with the simple bullets. One example of this type is shown in Fig. 491. Having carried on the observation further, we noticed that some of the plates were replaced with broad branches. General aspects of this transition will be understood from the photographs of Figs. 492–494. After this stage the type of crystal changed into that of bullet with dendritic branches as shown in Fig. 495. This final form continued for some time and the snowfall consisted of these peculiar crystals to the end. The process observed is as expected from our knowledge obtained in the experiments on artificial snow. Bullets with dendritic branches are similar to tsuzumi crystals from the point of view of their structures, that is, a combination of columnar and plane crystals.

The process of formation of a tsuzumi crystal which is similar to that of natural snow above described is shown in Figs. 496 and 497.

In the earlier stage, a column grows very slowly at T_a nearly $-18°C$ and T_w nearly $+3°C$, that is, in the condition of low temperature and low supersaturation. The column is slender in form at the initial stage, and then grows thick without marked longitudinal elongation. In this experiment it took nearly 5 hr to get the size that is observable in the central part of the natural tsuzumi crystal. This stage is shown in Fig. 497 (3).

Then the temperature of the water in the reservoir was raised rather quickly

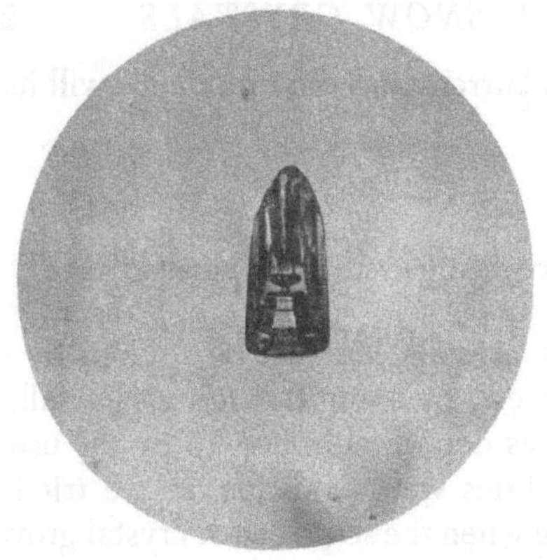

Fig. 490 [1472]. Development of bullets with dendritic crystals in natural snow, I (× 40.5).

Fig. 491 [1473]. Development of bullets with dendritic crystals in natural snow, II (× 40).

Fig. 492 [1474]. Development of bullets with dendritic crystals in natural snow, III (× 34.5).

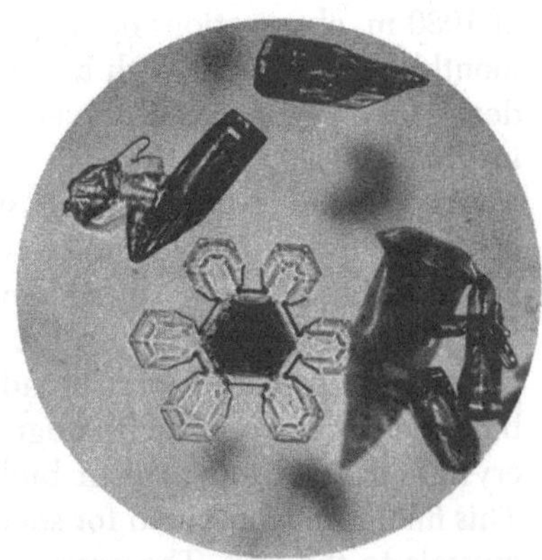

Fig. 493 [1475]. Development of bullets with dendritic crystals in natural snow, IV (× 34.5).

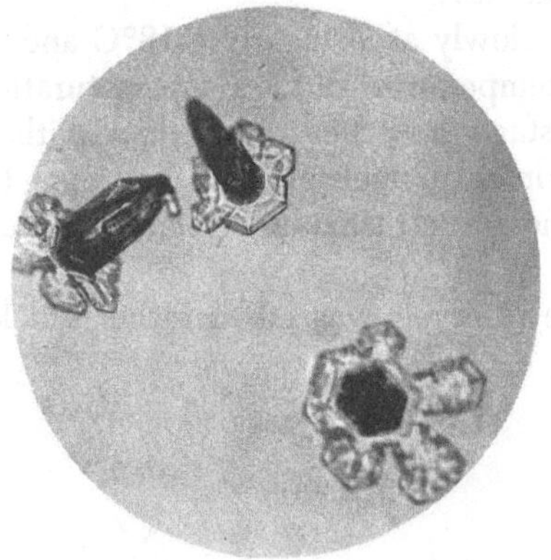

Fig. 494 [1476]. Development of bullets with dendritic crystals in natural snow, V (× 33).

Fig. 495 [1477]. Development of bullets with dendritic crystals in natural snow, VI (× 33).

in order to increase the temperature and the rate of supply of water vapor, the temperature of the air in the apparatus also having been brought up as a result; T_a became $-15°C$ with T_w at $+10°C$. This stage is in Region II in Fig. 444, that is, the region for plate formation. In this condition a plate develops at each end of the column, as shown in Fig. 497 (4). Thin transparent ice plates extend out from the longitudinal ridges of the hexagonal column. We named this the transition stage, as already described in detail in Sec. 71. After this T_w was increased to the value favorable for dendritic development. The crystal finally developed into a completed tsuzumi form as shown in Fig. 497 (7) and (8), through stages (5) and (6).

87. Natural snow crystals predicted from the investigation of artificial snow

Thus far the experiments were conducted in order to produce all sorts of natural snow crystals in the laboratory. In the progress of the research, we became aware of the fact that some rare types of crystals that have not yet been observed in nature could be made artificially in the cold chamber. In February 1940 we made an expedition to Mount Tokachi and stayed there for a month, in order to see whether those unknown crystals are observed also in nature on rare occasions. The result was affirmative. Examples are shown in Figs. 498–504, in which the photographs on the right are the artificial ones and those on the left are the corresponding natural snow crystals.

(1) *Cup crystal.* Figure 498(*b*) is an artificial cup crystal. This type of crystal is not so difficult to produce artificially, and has frequently been observed among natural frost crystals. It has, however, not previously been proved to exist in natural snow. Figure 498(*a*) is the first photograph of a cup crystal of natural snow, which was taken in this expedition to Mount Tokachi. This crystal was observed when the minute ice crystals were seen at the spot of observation. Minute ice crystals were observed for some time, and then they grew into the early stage. This cup crystal was found at that time mixed with various types of small crystals. We had noticed once a cup crystal in the field of the microscope, but could not take a photograph on that occasion.

(2) *Sheathlike crystal.* Figure 499(*a*) is an example of a sheathlike crystal of natural snow and shows a similar structure to that of the artificial one, Fig. 499(*b*). This crystal was observed in the same snowfall in which the cup crystal above mentioned was found.

(3) *Peculiar crystal with needles attached.* Figure 500(*b*) is a peculiar crystal consisting of an irregular assemblage of columns with many short needles attached. The condition of formation of this extraordinary crystal is not yet clarified in our experiment on artificial snow, and it cannot be made in a reproducible manner.

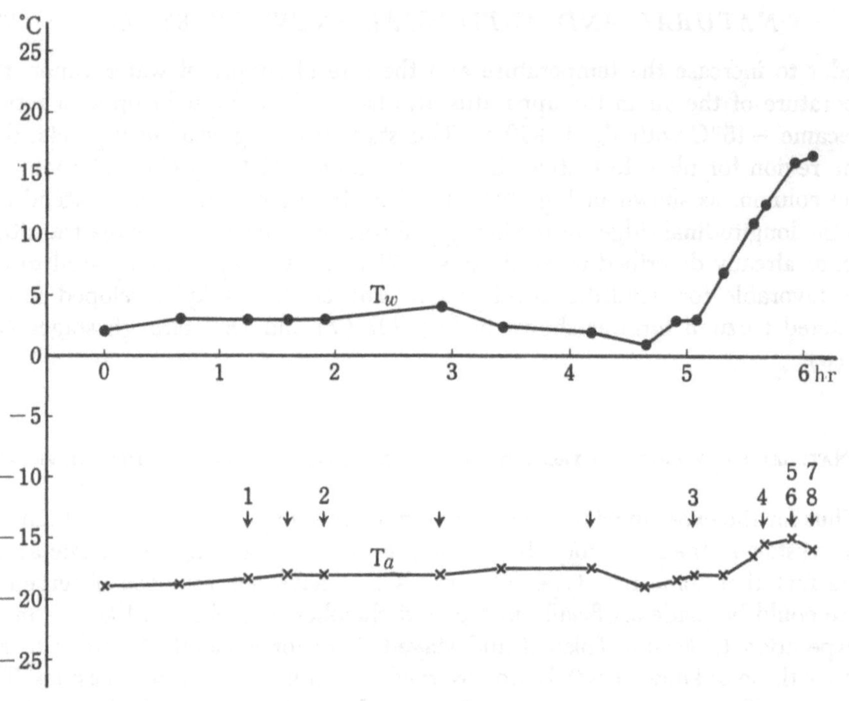

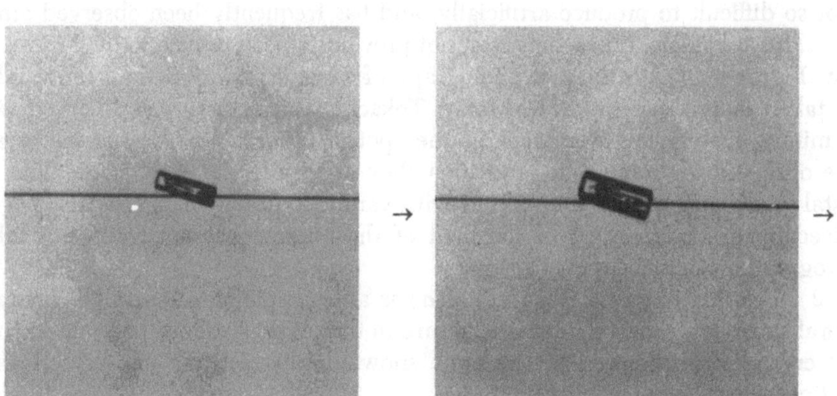

Fig. 497 [1506–13]. Successive stages in the course shown in Fig. 496 (× 33).

↓

↓

↓

FIG. 497 (*continued*).

↓

↓

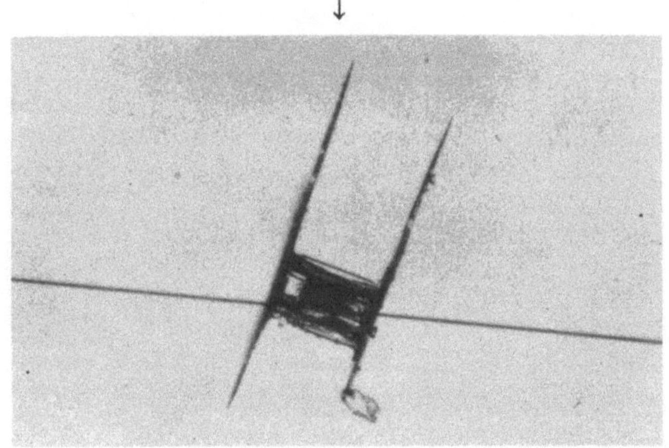

↕

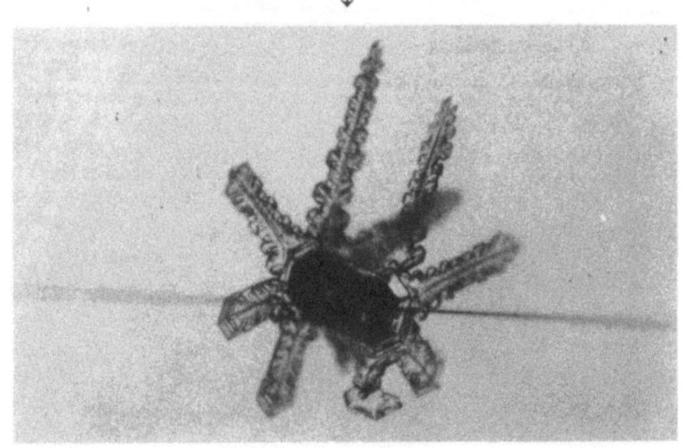

Fig. 497 (*continued*).

TABLE 37. Formation of various rare crystals.

Fig. No.	Photo. No.	T_a (°C) Initial	Middle	Final	T_w (°C) Initial	Middle	Final	Time for growth (hr)	(min)
499(*b*)	1517	−26	−26	−25	+ 4.7	+ 7.9	+ 9.2	4	10
500(*b*)	1519	−15	− 8	− 6	+35	+35	+35	ca. 1	30
501(*b*)	1521	−25	—	−30	0	—	< 0	70	
503(*b*)	1525	−18	−19	−17	+ 9	+13	+14	2	30

Among the natural snow crystals, we have found one very similar to this crystal, which is shown in Fig. 500(*a*).

(4) *Hexagonal prism of nearly solid structure.* The hexagonal prism shown in Fig. 501(*b*) was made by a very slow process. The condition of formation of this crystal has already been described, in Sec. 74. This is a prism of nearly solid structure, although a trace of skeleton development is observable. A natural snow crystal very similar to this one in both shape and structure was observed at Mount Tokachi. It is reproduced in Fig. 501(*a*).

(5) *Hexagonal plate with skeleton structure.* Thick hexagonal plates with characteristic skeleton structure are frequently obtained in the experiment on artificial snow. Many examples have already been shown, and Fig. 502(*b*) is one of them. Several crystals of this type were observed at Mount Tokachi. A representative one is shown in Fig. 502(*a*).

(6) *Hexagonal crystal with a small plate at the center.* In Sec. 7 we have already cited many examples of this type of crystal. They were picked up from the collection of photographs of natural snow, after we became aware of the existence of this kind from the results of the artificial-snow studies. The similarity can be distinctly seen from their side views, as shown in Figs. 503(*a*) and (*b*).

(7) *Transparent ice plate.* Transparent ice plates with a trace of design are sometimes obtained in artificial snow. We see at least one plate of this nature in the crystal of Fig. 504(*b*). The corresponding natural snow is shown in Fig. 504(*a*).

The conditions of formation for some of above-mentioned crystals are tabulated in Table 37.

Today we have no doubt about the existence of such extraordinary crystals as are described in this article in natural snow. These crystals were not reported in the literature and we had not found any one of them before we were led to predict their existence from the result of our artificial-snow studies. It is quite interesting that these rare crystals were easily found after we predicted their existence.

88. Snow crystals made at low temperatures

In our artificial-snow experiments, when the temperature goes below −20°C, hexagonal dendritic crystals are no longer obtained and the crystals take the form

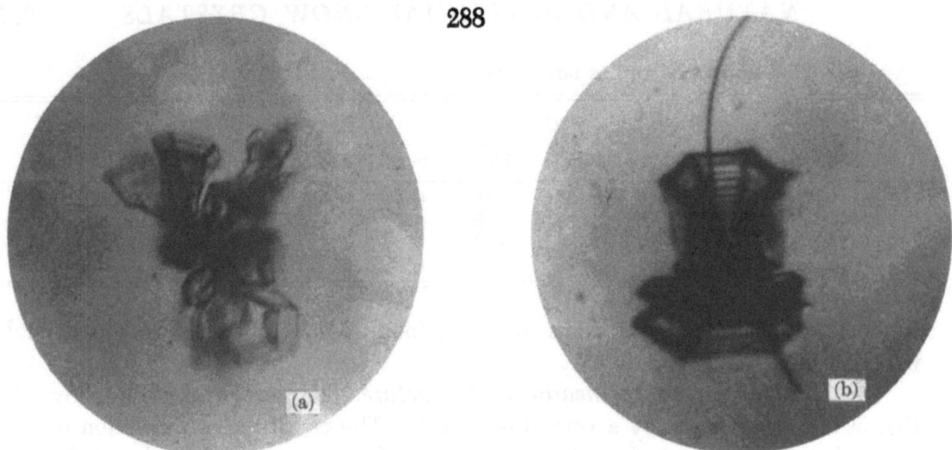

FIG. 498. Cup crystal: (*a*) [1514] natural (× 37); (*b*) [1515] artificial.

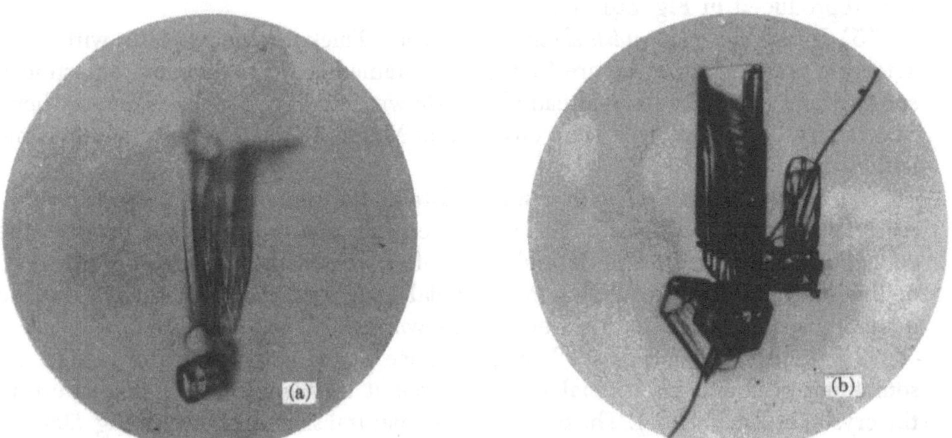

FIG. 499. Sheathlike crystal: (*a*) [1516] natural (× 37); (*b*) [1517] artificial (× 39).

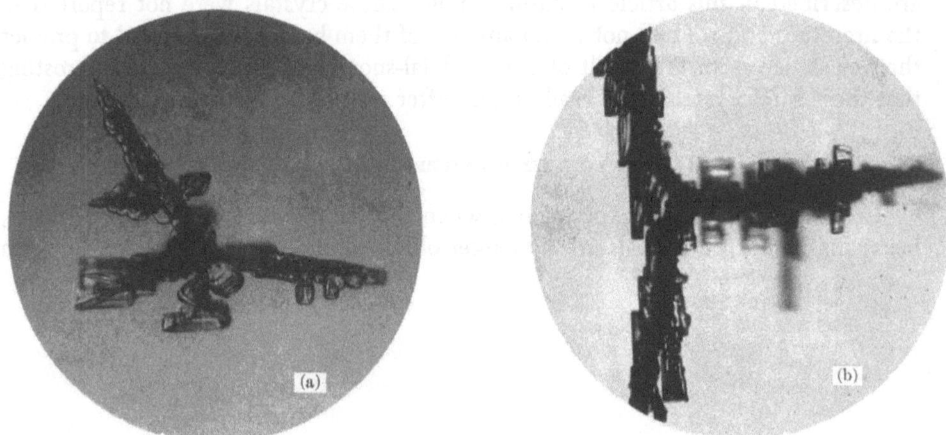

FIG. 500. Peculiar crystal with needles attached: (*a*) [1518] natural (× 33); (*b*) [1519] artificial (× 22).

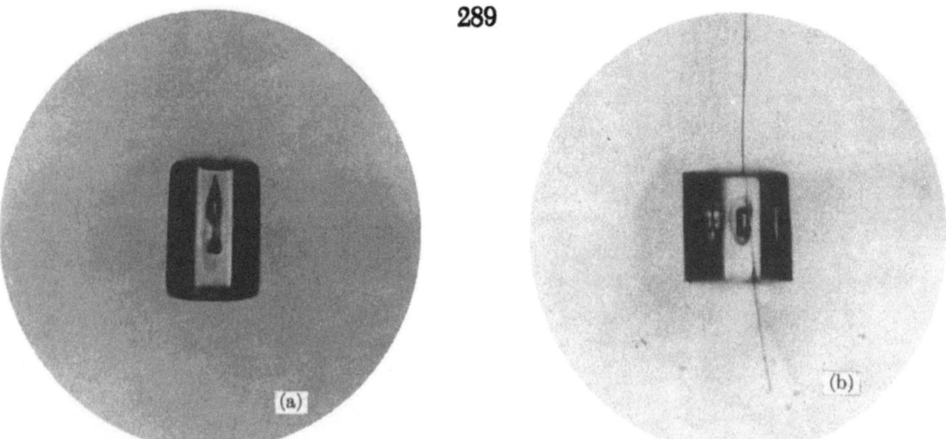

Fig. 501. Solid hexagonal prism with a trace of skeleton structure: (*a*) [1520] natural (× 75); (*b*) [1521] artificial (× 40).

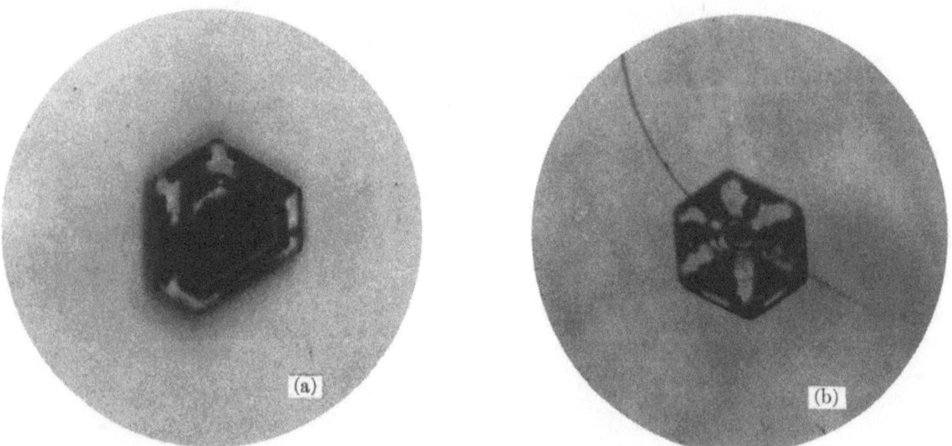

Fig. 502. Hexagonal short prism with skeleton structure: (*a*) [1522] natural (× 90); (*b*) [1523] artificial (× 30).

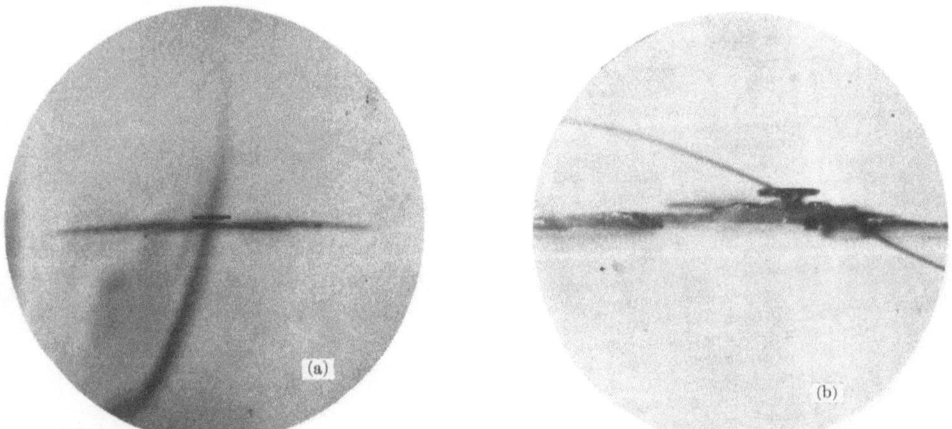

Fig. 503. Hexagonal crystal with a small plate at the center: (*a*) [1524] natural (×24); (*b*) [1525] artificial (×35).

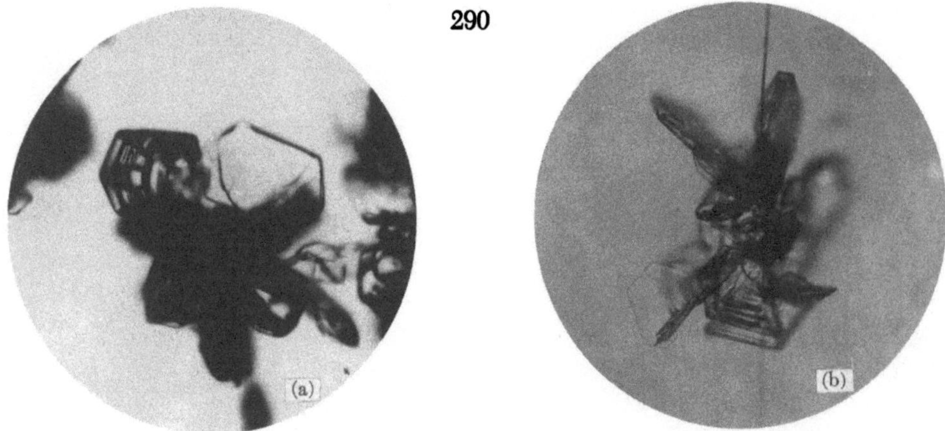

FIG. 504. Transparent ice plate: (*a*) [1526] natural; (*b*) [1527] artificial (× 36).

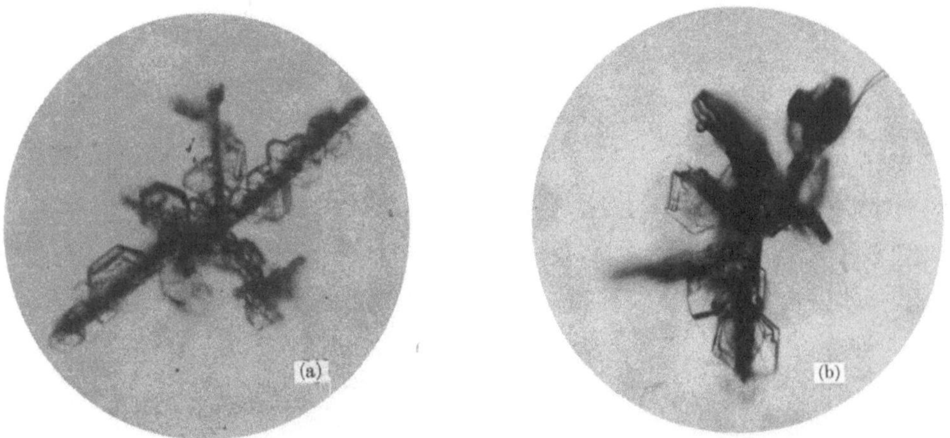

FIG. 505. Crystal made at low temperature: (*a*) [1528] natural (× 35); (*b*) [1529] artificial.

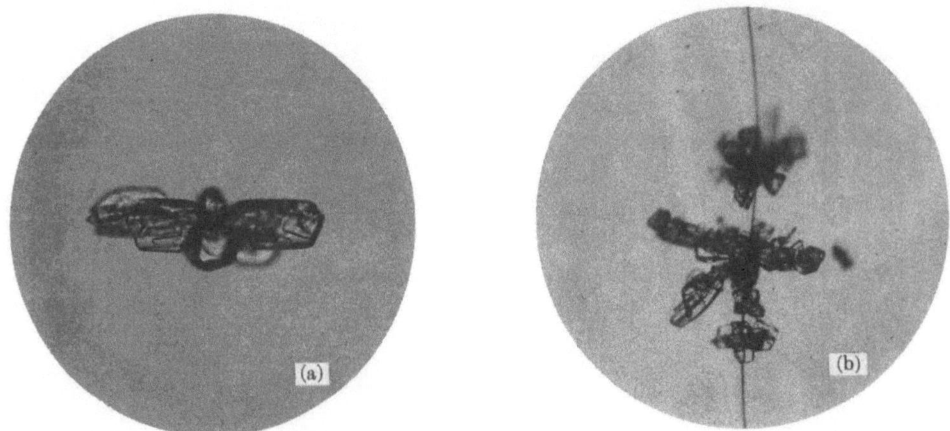

FIG. 506. Crystal made at low temperature: (*a*) [1530] natural (× 50); (*b*) [1531] artificial (× 19).

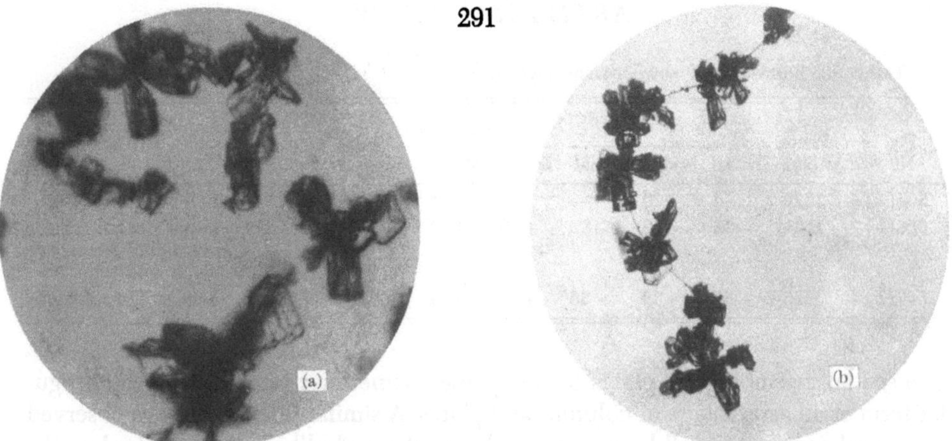

FIG. 507. Flour snow: (*a*) [1532] natural (× 34); (*b*) [1533] artificial (× 15).

FIG. 508. Broomlike snow: (*a*) [1534] natural (× 17.5); (*b*) [1535] artificial (× 15).

TABLE 38. Formation of powder-snow crystals.

Fig. No.	Photo. No.	T_a (°C) Initial	T_a (°C) Middle	T_a (°C) Final	T_w (°C) Initial	T_w (°C) Middle	T_w (°C) Final	Time for growth (hr)	Remarks
505(*b*)	1529	—	−20	—	—	+ 4	—	ca. 15	Left overnight
506(*b*)	1531	−32	−30	−32	+10	+12	+11	3	Initial stage made overnight; 6-mm stop diaphragm
507(*b*)	1533	−36	−38	−36	+ 7	+ 4	+ 7	3.5	1-cm stop diaphragm

of a spatial assemblage of plates or side planes, sometimes representing the irregular form of an assemblage of columns and plates. A similar phenomenon is observed in natural snow; it is well known that only small powderlike crystals fall when the cold is severe.

This so-called "powder snow" has many varieties, some of them being reproduced, together with the corresponding artificial crystals, in Figs. 505–508. The conditions of formation of the artificial crystals are shown in Table 38. All experiments were made by introducing a stop diaphragm in the apparatus, in order to reduce the upward convection of water vapor. At these low temperatures it was usually observed that many germs appeared on the filament and it was very difficult to get a few isolated germs. When the upward convection was reduced by the use of a stop diaphragm, it was possible rather easily to obtain several isolated germs on the filament.

(1) *Assemblage of columns with extended side planes.* Examples are shown in Figs. 505 and 506. The artificial crystal of Fig. 506(*b*) was made by leaving the filament overnight in the apparatus for getting the germ and then exposing to water vapor, the temperature of the air being kept at nearly −30°C. The form of the crystal shows an early stage of an assemblage of columns with side planes. Comparing this with the crystal observed at Mount Tokachi, Fig. 506(*a*), one will see no essential difference between their forms and structures. Figures 505(*a*) and (*b*) are similar crystals to Figs. 506(*a*) and (*b*), showing more developed states. Similar crystals are found in Plates 130 and 131.

(2) *"Flour snow."* The columnar crystals associated with small planes, which are observed in the Alpine zone and are commonly called "flour snow" by skiers, are shown in Figs. 507(*a*) and (*b*). The artificial one, Fig. 507(*b*), was made by reducing the supply of water vapor and keeping the temperature of the air very low, as shown in Table 38. These conditions will be consistent with the supposed mechanism of the formation of the "flour snow" observable in nature.

(3) *Broomlike snow.* Figure 508(*b*) shows a peculiar form of irregular snow. In this case the filament was set at the center of the circular hole of the stop diaphragm, so that the filament was just in the position where the water vapor brought

upward by natural convection converges. The temperature of the air was kept very low.

A typical example of this sort of natural snow is reproduced in Fig. 508(*a*). This photograph was taken by K. Nakata in South Sakhalin, where the climate is colder than at the place of our observations. A close examination of the photograph shows that this crystal is not an irregular particle, but is made up of many dendritic branches growing in thick clusters and covered with numerous droplets. We may call this type "broomlike snow."

CHAPTER 8

Recent Researches on the Formation of Snow Crystals

89. Snow crystals and ice crystals in the atmosphere

Recent progress in research on the solid precipitation of water in the atmosphere led us to consider that it may be conveniently classified into two categories, the snow crystal proper and the ice crystal. The snow crystal proper is the one observed on the earth's surface, and what we call snow crystals in daily use belong to this kind. This snow crystal was born at a high altitude, and grew into the ordinary form observed on the earth's surface while falling through various strata of the atmosphere.

The initial stage of this crystal can be seen in the cirrus cloud or ice fog. This is also the crystal of ice formed in the air by condensation of water vapor by sublimation, and it is not different in nature from the snow crystal proper. However, in size it is quite different, so it is convenient to consider this germ of snow as different from the snow crystal proper. Let us call it an "ice crystal." Many examples are seen in the photographs reproduced in this book, which are of fernlike type with a minute hexagonal plate at the center. The latter is the ice crystal, and it is observed by itself as the initial stage of the snow crystal, as described in Sec. 21. In the general discussions on the form and structure of the snow crystal, we ought to note the dendritic nature of its branches before we go into the discussion about the small hexagonal plate at the center, especially when the relation between the crystal form and the meteorologic conditions is discussed.

The crystals formed from supercooled water droplets by seeding are also to be called ice crystals. Although we cannot draw a distinct line between snow and ice crystals, it will be convenient to group them as shown in Table 39, in which the average sizes of snow and ice crystals together with those of cloud particles and nuclei are given.

In the course of formation of a snow crystal in nature, first the ice crystal is

TABLE 39. Dimensions of snow and ice crystals and related particles.

Classification	Average size (mm)	Average size (μ)
Snow crystals		
Fernlike crystal	4.0	4,000
Ordinary dendritic crystal	2.5	2,500
Plate with extensions	1.5	1,500
Simple plate	0.8	800
Column	.5	500
Ice crystals observed in nature (initial stage of snow crystals)		
Ice fog or cirrus particle: large	.3	300
small	.1	100
Ice crystals obtained by seeding		
Ice crystal transformed from water droplet by seeding: large	.05	50
small	.01	10
Cloud particles		
Fog particle	.03	30
Ordinary cloud particle	.01	10
Minute droplet	.001	1
Nuclei		
Center nucleus found in snow crystal	.001	1
Condensation nucleus or aerosol	.0001	0.1
AgI particle used for seeding	.0001	.1

formed. This process may be the spontaneous transformation of a supercooled water droplet at cirrus temperature, or the result of natural or artificial seeding at higher temperatures, or the direct condensation of water vapor on a solid nucleus. The minute fragment of ice crystal may be included in the seeding materials as the most efficient one. All these processes will give the same result in the sense that small ice crystals or germs of snow crystals make their appearance in the atmosphere. The form and structure of the snow crystal that is grown from this germ are determined by the meteorologic conditions met *after* this germ is made. The condition controlling the form of the ice crystal may be different from that of the snow crystal, especially when the seeding process is employed. This point is one of the problems to be solved by future investigations. The classification of the crystals of ice formed in the air into snow and ice crystals will be useful in eliminating the possible confusion between the two.

90. THE RELATION BETWEEN THE FORM OF A SNOW CRYSTAL AND THE EXTERNAL CONDITIONS

In Sec. 81 it was concluded that the elements, at least the chief elements, that control the form of a snow crystal, are the temperature T_a and the degree of supersaturation s. The results of the experiments are shown in Fig. 448. It has been generally believed that the crystal form is determined by the degree of supersaturation, but not by the temperature. According to aufm Kampe, Weickmann, and Kelly,[75] however, the temperature may be substituted for the supersaturation in the case of ice, when the crystallization is considered to take place at water saturation, which is a function of temperature. This idea that the crystallization of ice in the air is controlled by the vapor-density difference of water and ice saturation, which is a function of temperature, was taken up by Houghton [76] in his theory of the rate of growth of ice crystals, and showed some successful results.

The supersaturation in our case does not necessarily mean the excess of water vapor in the air. As described in Sec. 77, the upward current of air in the snow-making apparatus was found to contain a large number of minute droplets of diameter about 1 μ or less. One example is shown in Fig. 432. A known volume of this air was taken from the neighborhood of the spot of crystal formation, and the total water content was measured by a gravimetric method using P_2O_5, as described in Sec. 80. The supersaturation value s is taken as

$$s = \frac{w + \rho'}{\rho_0}$$

where w is the liquid water content per unit volume, ρ' is the vapor density at saturation over supercooled water, and ρ_0 is the vapor density over ice. A T_a–s diagram, which shows the shape of snow crystals as a function of temperature and supersaturation, is shown in Fig. 509.

The region above the water saturation curve W in Fig. 448 must be considered to be in the condition of cloud formation. The numerous minute droplets of about 1-μ diameter, which are observed in the upward convection current in our apparatus, are produced in this region. These minute droplets show a peculiar behavior; they do not freeze to the surface of a snow crystal in droplets but spread over the ice surface. Details have already been given in Sec. 77. As a result these minute droplets behave like vapor molecules in the process of condensation. Further discussions on these droplets will be given in Sec. 93.

The liquid-water content is calculated from the equation

$$w = s\,\rho_0 - \rho'$$

for the region above the water saturation curve W, and the curves of equal water content are shown in Fig. 509. At the point A in the figure, that is, in the most favorable condition for dendritic development of the crystal, the water content is 0.06 g/m^3. In our experiment the probable diameter of minute droplets in this case is about 1 μ. The number of droplets per cubic centimeter n is calculated from the equation

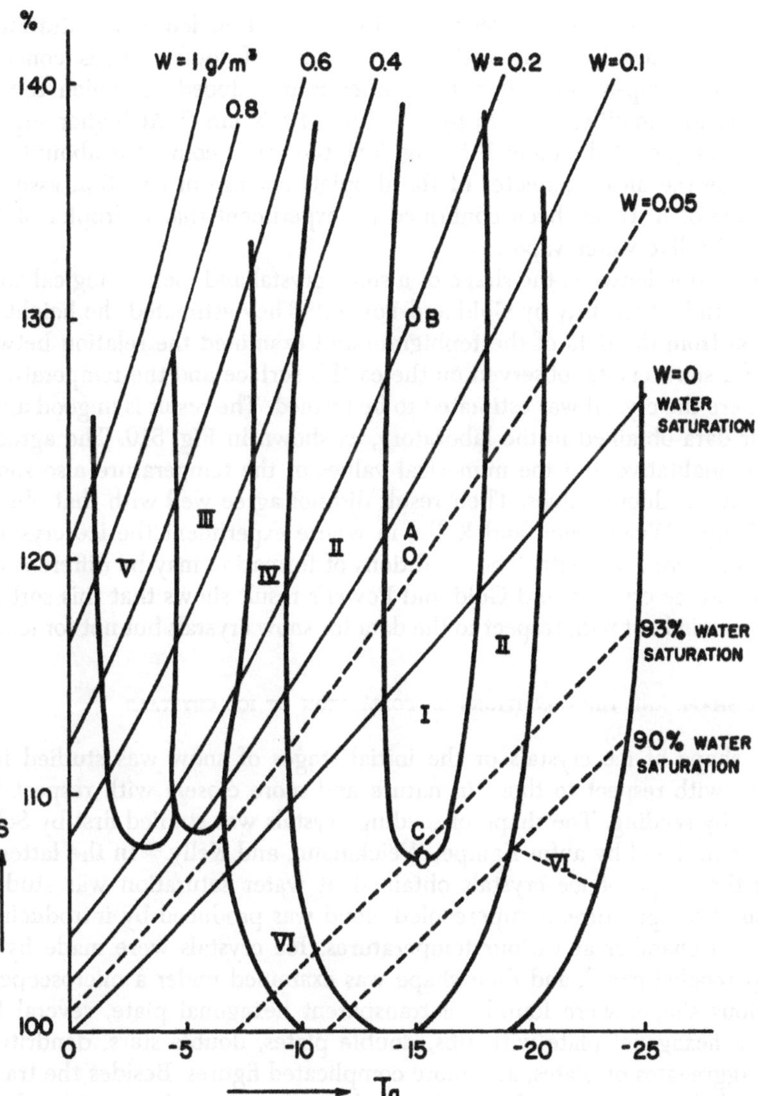

Fig. 509. T_a–s diagram, showing the conditions of snow formation as a function of temperature and supersaturation.

$$w = \frac{4\pi}{3} n \rho r^3,$$

and we get

$$n = 1.2 \times 10^5 \times \text{cm}^{-3}.$$

This value is in fairly good agreement with the total nucleus concentration in city air. In Mason and Ludlam's article;[77] the value of total nucleus concentration measured by Simpson in different localities is introduced, in which the average value measured in cities is given by $n = 1.5 \times 10^5 \times \text{cm}^{-3}$. At higher supersaturation, for example, at the point *B* in Fig. 509, the water content is about 0.2 g/cm^3. In this case the mean diameter of the droplets comes out as 1.5 μ, assuming the same value of *n*. It has been confirmed by experiment that a droplet of this size behaves also like water vapor.

The relation between the shape of a snow crystal and meteorological conditions has been studied recently by Gold and Power.[78] They estimated the height of snow formation from the data of the tephigram and examined the relation between the shape of a snow crystal observed on the earth's surface and the temperature of the layer where the crystal was estimated to be formed. The result is in good agreement with our data obtained in the laboratory, as shown in Fig. 510. The agreement is not only qualitative, but the numerical values of the temperature also show good coincidence with our values. Their result did not agree well with that obtained by aufm Kampe, Weickmann, and Kelly, in whose experiment the ice crystals made by seeding were dealt with. The conditions of formation may be different for snow crystals and ice crystals, and Gold and Power's result shows that this sort of study must be carried out with respect to the data for snow crystals but not for ice crystals.

91. The shape and the conditions of formation of ice crystals

The shape of ice crystals or the initial stages of snow was studied in a few examples with respect to those in nature and more closely with respect to those obtained by seeding. The shape of seeding crystals was studied first by Schaefer [79] and more in detail by aufm Kampe, Weickmann, and Kelly.[80] In the latter investigations the shape of ice crystals obtained at water saturation was studied as a function of temperature. A supercooled cloud was produced by introducing steam into a cold chamber at various temperatures. Ice crystals were made by seeding this supercooled cloud, and their shape was examined under a microscope.

Various shapes were found – a transparent hexagonal plate, several kinds of prisms, a hexagonal plate with ribs, double plates, double stars, dendritic forms, spatial aggregates of plates, and more complicated figures. Besides the transparent plates and the aggregates of crystals, all these shapes can be considered as variations of a hexagonal prism of skeleton structure, which we consider to be a fundamental form of ice crystal. A good example of this kind of crystal is shown in Fig.

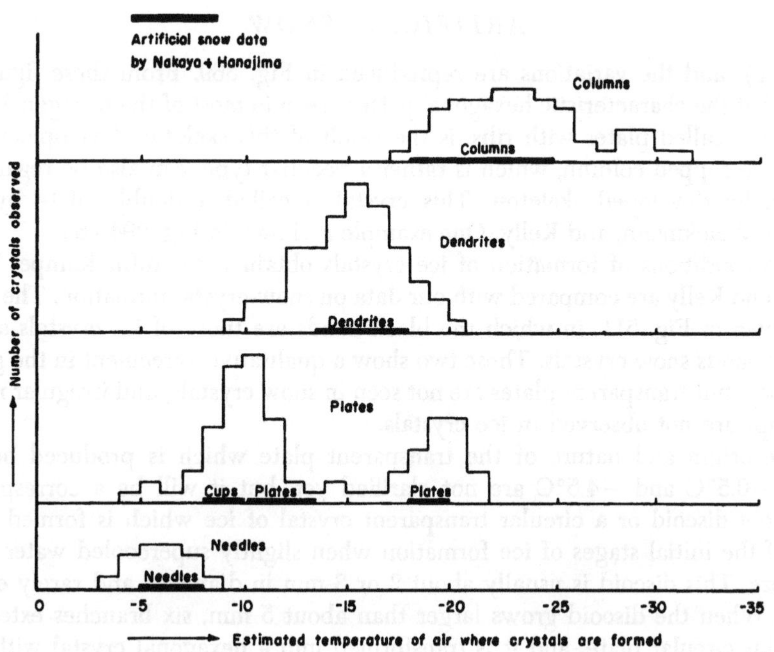

FIG. 510. The relation between the shape of snow crystals and the meteorologic conditions.

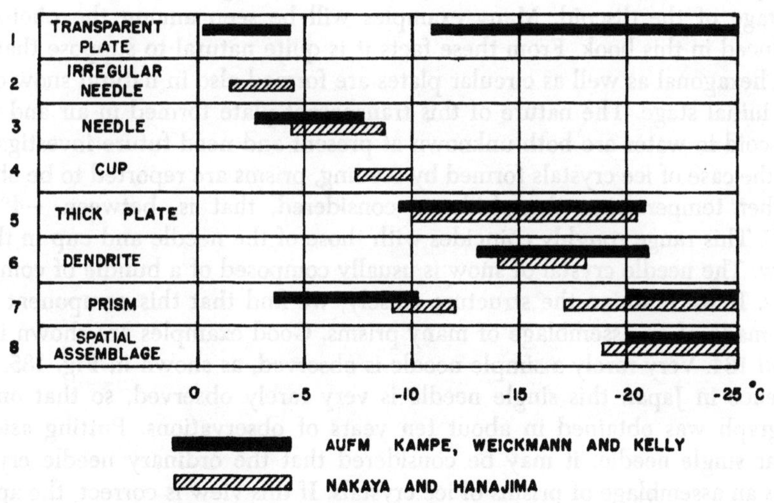

FIG. 511. Comparison of the conditions for snow-crystal and ice-crystal formation.

294 (11), and the variations are reproduced in Fig. 369. From these figures we know that the characteristic hexagonal pattern seen in most of the hexagonal plates, which are called plates with ribs, is the result of this skeleton development. The miniature capped column, which is rather a peculiar type, can also be regarded as a specially developed skeleton. This crystal is called a double plate by aufm Kampe, Weickmann, and Kelly. One example is shown in Fig. 294 (6).

The conditions of formation of ice crystals obtained by aufm Kampe, Weickmann, and Kelly are compared with our data on snow-crystal formation. The results are shown in Fig. 511, in which the black bands are those of ice crystals and the shaded bands snow crystals. These two show a qualitative agreement in the general tendency, but transparent plates are not seen in snow crystals, and irregular needles and cups are not observed in ice crystals.

The origin and nature of the transparent plate which is produced between about −0.5°C and −4.5°C are not clarified yet, but it will be a corresponding form to a discoid or a circular transparent crystal of ice which is formed as one type of the initial stages of ice formation when slightly supercooled water begins to freeze. This discoid is usually about 2 or 3 mm in diameter and rarely exceeds 5 mm. When the discoid grows larger than about 5 mm, six branches extend out from this circular plate, and it is transformed into a hexagonal crystal with a circular area at the center. The fact that the transparent plate is observed only among ice crystals and not among snow crystals is concordant with the case of ice formation by freezing of water. Examining the photomicrographs of natural snow crystals, we can find many crystals that have transparent portions at the center. This transparent portion may be circular or it may have a hexagonal pattern, which is the next stage of the discoid. Many examples will be seen among the photographs reproduced in this book. From these facts it is quite natural to suppose that transparent hexagonal as well as circular plates are formed also in natural snow crystals in the initial stage. The nature of this transparent plate formed in air and that of the discoid in water are both unknown at present and need future investigation.

In the case of ice crystals formed by seeding, prisms are reported to be obtained at higher temperatures than hitherto considered, that is, between −4°C and −10°C. This range roughly coincides with those of the needle and cup in the case of snow. The needle crystal of snow is usually composed of a bundle of component needles. If we examine the structure closely, we find that this component needle is also made of an assemblage of many prisms. Good examples are shown in Figs. 133 and 134. Very rarely a simple needle is observed, as shown in Fig. 135. In our experience in Japan this single needle is very rarely observed, so that only one photograph was obtained in about ten years of observations. Putting aside this peculiar single needle, it may be considered that the ordinary needle crystal of snow is an assemblage of prisms of ice crystals. If this view is correct, the apparent discrepancy between the conditions of formation of snow and ice crystals will be explained satisfactorily. Needle crystals are also reported in the case of ice crystals formed by seeding, but we cannot discuss this point further because no photograph is shown and no explanation is given about its structure.

92. The mechanism of snow-crystal formation

It is briefly mentioned in Sec. 90 that the minute droplets of about 1-μ diameter, which are almost always observable in the air under the condition of snow-crystal formation in our snow-making apparatus, show a peculiar behavior. They do not freeze to the crystal in the form of droplets, but they suddenly disappear at the moment when they are brought into contact with the crystal. It appears as if the water molecules of the droplet spread on the surface of the ice.

This phenomenon is more easily observed in the case of window hoar, that is, the ice crystals formed on a window pane by "sublimation." In this case also, the minute droplets play an important role in the development of hoar crystals. In most cases the glass surface is previously covered with numerous supercooled droplets and the hoar crystal develops through the field of scattered droplets. When a tip of the crystal approaches to the neighborhood of droplets, most of the latter vanish or diminish their size markedly by evaporation, which is caused by the vapor-pressure difference of ice and water. It is, however, often observed that some remaining minute droplets are caught by the streamer of hoar crystals. The behavior of a droplet in this case is very similar to that in the case of snow. The droplet does not freeze to the crystal in the form of a droplet, but it shows an appearance as if it spreads on the crystal surface. This phenomenon was studied by Yosida,[81] and microphotographs showing the process are reproduced in his paper. In the case of hoar crystals, this phenomenon occurs for larger droplets than in the case of snow. In one example the diameter of the droplet which showed this phenomenon was 14 μ.

This phenomenon may be explained by the sudden evaporation of a droplet, as pointed out by aufm Kampe, Weickmann, and Kedesdy.[82] They suggest that a droplet of diameter 1 μ can evaporate at $-10°C$ within 0.7 sec at water saturation. This is one way of explaining this phenomenon, but there is another interpretation. That is the supposed surface diffusion of H_2O molecules on the surface of ice crystals. A similar phenomenon was already known for growing mercury crystals by the experiment of Volmer and Estermann.[83] This surface diffusion is considered by many investigators to take place on the surface of ice crystals. For example, de Quervain [84] tried to confirm the existence of this phenomenon by experiment. He kept a crystal of snow in paraffin oil at $-5°C$ and could observe the deformation in its shape. His idea is excellent, although this experiment can hardly establish the fact, because the water solubility of paraffin oil was not considered seriously. The

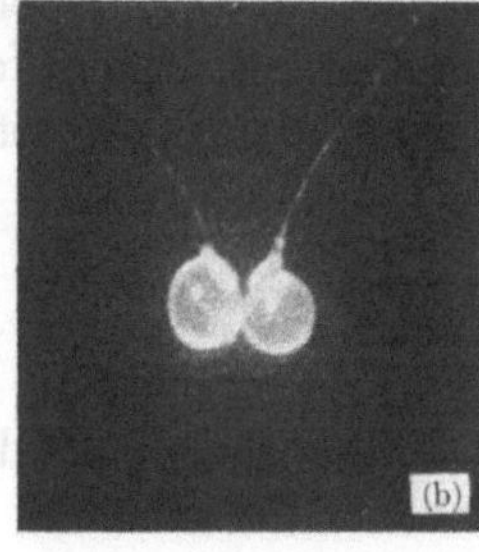
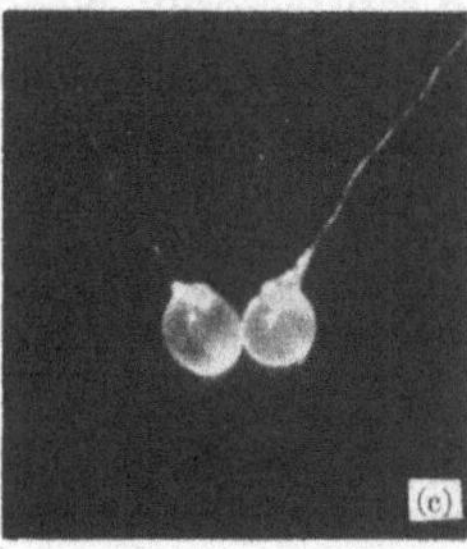

FIG. 512. Rotation of ice balls: (*a*) start of experiment; (*b*) just before rotation; (*c*) after rotation. Mean diameter of ball, 1.45 mm; temperature, −5.5°C.

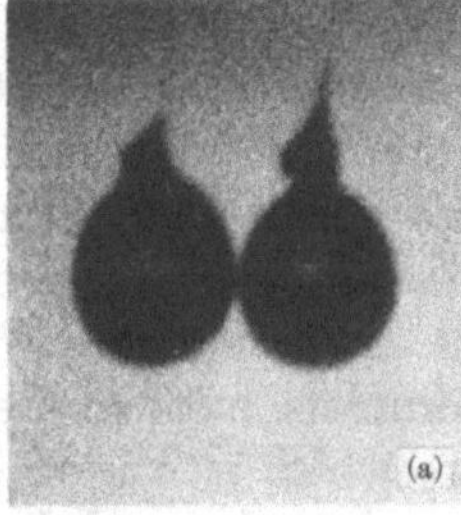
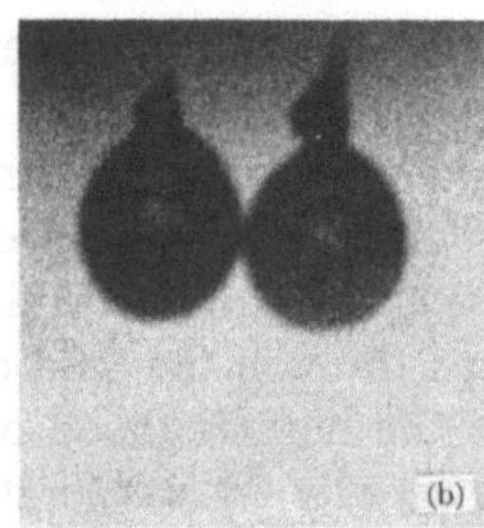
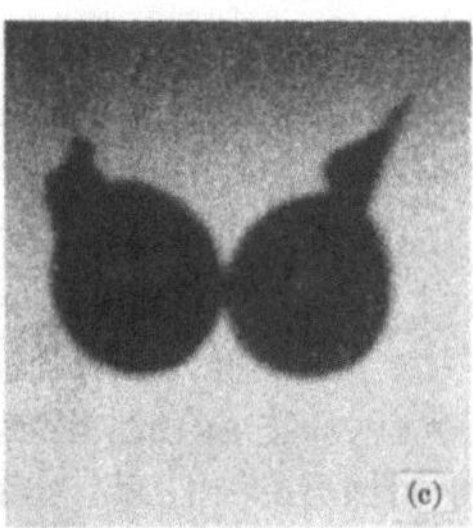

FIG. 513. Rotation of ice balls: (*a*) start of experiment; (*b*) just before rotation; (*c*) after rotation. Mean diameter of ball, 1.95 mm; temperature, −3.0°C.

surface nature of ice crystals has been an old problem since the days of Faraday and Tyndall. This problem was recently taken up by Weyl,[85] and the existence of a "liquid" water film on the ice surface was suggested. The metamorphism of crystals in snow cover on the ground, the mechanism of hardening of snow, and allied phenomena are also considered to be the result of the behavior of molecules on the surface layer of ice. The author is now inclined to attribute the phenomenon of a sudden disappearance of a minute droplet in contact with an ice surface to the surface diffusion of H_2O molecules, although the latter fact itself is not experimentally well established yet.

Recently a simple experiment showing the existence of a liquid-like film on the surface of ice was carried out in our laboratory. Two small ice balls were suspended by very thin filaments. One of the filaments could be moved with a screw motion so that the top of the filament was displaced horizontally. The normal adhesive force is measured by this method from the inclination of the filaments when the spheres are separated. The experiments were carried out in a metal box with a suitable window for taking photomicrographs. The temperature in the box was varied between −0.5°C and −16°C. Usually the two spheres were separated at a certain angle of inclination of the filament. Sometimes, however, it was observed

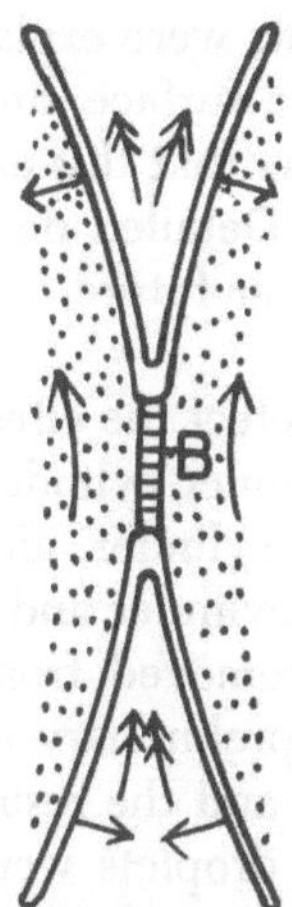

Fig. 514. Model of the point of contact between ice spheres.

that the ice spheres showed a rotation before the separation. Two examples are shown in Figs. 512 and 513. In both cases, (*a*) shows the initial state, (*b*) the state just before the rotation, and (*c*) the one after the rotation.

This rotation was not always observed, but it was not rare. For example, in one series of experiments this rotation phenomenon was observed in 9 cases among 38. The rotation took place more frequently in the range of temperature near the freezing point, but it was once observed at −7.0°C, which was the lowest temperature in the case of ice spheres made with distilled water. It sometimes happens, though seldom, that the spheres rotate two or three times in succession and then separate. This successive rotation takes place more often when a 0.1-percent NaCl solution is used.

In order to explain this peculiar phenomenon of the rotation of the sphere, a model shown in Fig. 514 is considered, in which *B* indicates the supposed bridge of ice at the point of contact. This solid bridge or bond must be present, because the rotation does not take place until a critical moment of rotation is applied. The thick lines in the figures show the surface films of a liquid-like nature; without assuming the existence of this film the explanation of this rotation phenomenon is extremely difficult. The boundary layer of a liquid-like nature is believed to have the ability to freeze when bounded by ice on both sides. So the infinitesimal portion of the film at the point of contact will be frozen into ice. This will give the ice bridge or bond *B*. If the adhesion of ice is chiefly due to this bridge, which is a solid material, the spheres will not show any rotation but will separate at a certain critical inclination of the filament. The rotation of the sphere must be due to an action of the boundary layer which has a liquid-like nature. The adhesion must be caused by the surface tension of the film, after the bridge is broken. All the results obtained in this series of experiments were explained nicely by assuming the existence of a liquid-like film on the ice surface, and no other explanation seemed to be adequate. So the author considers that this experiment shows the existence of a liquid-like film on the ice surface. Detailed descriptions will be published in the *Journal of Colloid Science* in the near future.

An apparatus was designed to detect the effect of removing the minute droplets from the upward air current. The inner cylinder of glass in our old snow-making apparatus was replaced by a brass cylinder, and a thin wire was stretched along the axis of the cylinder. The brass cylinder and the wire are connected to a high-tension source. The droplets are removed from the upward air current by the well-known Cotler's method. The preliminary experiment was carried out in the laboratory of Hokkaido University, and the result showed that the growth of dendritic crystals stopped when these droplets were removed, the value of s having come down about 20 percent; for example, it dropped from 125 percent to 105 percent when these minute droplets were removed. The experiments are now being continued, and the detailed report will be published in some journal.

The delicate structure and design of a snow crystal are due to its sharp edges and minute ridges. It is easily shown that these edges and ridges are transformed into a more rounded form at temperatures below freezing, even if the crystal is kept in air slightly supersaturated with respect to ice. Warm water vapor is sent into a bottle at low temperature, and the frost crystals are made to grow all over the inside of this bottle. If the bottle is closed, the air inside must be slightly supersaturated with respect to the plane surface of ice, because the whole space is surrounded by crystalline frost with pointed edges. A snow crystal of fine dendritic shape is hung in this closed bottle by a thin filament from its cap, and left at low temperature, say $-10°C$. Under these circumstances the crystal also undergoes a sensible sublimation metamorphosis in a few hours, and we cannot keep its original shape and structure. Even for keeping the original form we need more supersaturation. So it is not surprising that such high values of s are necessary, as shown in Fig. 509, for the growth of this sort of crystals.

The measurement of total water content in the snow-forming air showed that most of the snow crystals are formed in the range above water saturation. The excess water was found to exist in the form of minute droplets about 1 μ in diameter, which behave like water vapor in the process of condensation. The term supersaturation used in this book means the total water content, and the condensation of these minute droplets is included in the term sublimation, which essentially means the condensation of vapor only.

The foregoing descriptions are entirely founded on our knowledge of artificial snow crystals, but the author hopes to extend this view to show that a similar mechanism will take place in the case of formation of snow crystals in nature. This idea is supported by Kumai's investigations on snow crystal nuclei.[86] By the use of an electron microscope, he found almost always a relatively large solid nucleus existing at the center of a snow crystal. Besides this center nucleus, which is the order of 1 μ in diameter and is regarded as the nucleus of the germ of the snow crystal or the ice crystal, he found numerous condensation nuclei of much smaller size, the mean diameter having been measured as about 0.1 μ. An electron photomicrograph of any portion of various snow crystals was found to be full of these condensation nuclei. The mean concentration of these condensation nuclei per unit area of crystal is roughly 100 per square micron for thicker crystals, and 10 per square micron for thinner ones. This value coincides roughly with the number of droplets of which the crystal is assumed to be made up. According to this view, these minute droplets are considered to play an important role in the phenomenon of snow-crystal growth. The difficulty is that we do not know whether such minute droplets do exist abundantly in the natural atmosphere or not. Unfortunately, up to the present, such minute droplets have not been found in natural clouds, as pointed out by aufm Kampe, Weickmann, and Kedesdy in their comment to Kumai's paper. However, it does not mean that these minute droplets do not exist, because the present techniques cannot catch droplets of this size. Dessens observed minute droplets 1 μ or less in diameter by the use of spider's threads as the collector, but it was during fine weather.[87] The precise determination of droplet distribution within clouds will answer this question.

One important problem left untouched is the formation of some snow crystals in the range between water and ice saturation, as seen in Fig. 509. Especially the fact that the region for dendritic growth in Fig. 509 extends down the curve of water saturation seems to be contradictory to the theory above described. We consider, however, that cloud formation will take place even in this region. It is not conceivable in an equilibrium state, but in the growing cloud or fog many results of measuring the humidity often show the values less than 100 percent. It is not rare that the humidity in a fog comes down to 90 percent or less. The lower limit of dendritic formation is 93 percent, as seen in Fig. 509. If the humidity is 90 percent in this case, the difference of 3 percent will mean that 1×10^5 droplets 1 μ in diameter must exist in 1 cm^3 of the air.

In conclusion, the author presents this new theory as a working hypothesis for the formation of snow crystals, on which subject we have had so little knowledge up to this time.

93. Concluding remarks

This book is a summary of our twenty years investigations on the form of snow crystals and the conditions under which they are formed. Snow crystals observed in nature show a very great variety in form and structure. It seems that snow is a unique example of a single chemical compound exhibiting so much variation in crystal habit.

The earlier classifications were inclined to attach importance to the regular crystals or the simpler forms. The feature of our investigation was that we regarded all sorts of crystals as of equal importance, including the malformed crystals, those consisting of a spatial assemblage of branches or plates, and the irregular particles, which were, as a matter of fact, no less frequently observed among the natural snow crystals.

With this consideration in view, we took about 2500 photomicrographs and made the general classification. As the principle of classification, importance was laid on the structure of the crystal, not on its apparent form.

For undertaking the general classication of snow crystals, the land of Hokkaido was very favorable, because the variety of crystals observed in this district was very great. For example, the pyramid shape of snow crystal is considered as the rarest type, but we were able to take photomicrographs of two examples of this type at Mount Tokachi. Further we succeeded in finding in natural snow such peculiar crystals as the cup, the sheathlike crystal, and the skeleton prism, whose existence was predicted by our artificial-snow experiments. Our general classification is considered to have been carried out under the most favorable conditions of climate.

The latter half of our investigations was devoted to the artificial production of snow crystals. The only report of artificial snow crystals prior to our work was that of J. M. Adams, who made a minute hexagonal column of ice in 1930 by mixing cold air with warm wet air. His crystal corresponds to our "early stage," and could better be called an artificial ice crystal. Our studies are concerned with the process of developing into a snow crystal proper, starting from the early stage, or the ice crystal. The problem of formation of this early stage from a nucleus is not treated in this book. The recent researches on artificial snowfall by I. Langmuir and V. J. Schaefer of the General Electric Laboratory are directed to the problem of making the early stage of snow crystals. Our investigations do not overlap with their experiments, because their researches solved the question of formation of an ice cloud by seeding and our experiments are confined to the problem of development of the snow crystal proper from the ice-cloud particle.

The most important result of this series of experiments was that all types of natural snow crystals were produced artificially in the laboratory and their conditions of formation were clarified. Some peculiar types of crystals which had not been known in natural snow were also produced in our artificial-snow experiments. Later they were found in natural snow. The external conditions controlling the forms of snow crystals were found to be the air temperature and the degree of supersaturation, the former having been known to be more important. This result does not agree with the previous conceptions of crystal formation, but it seems to be the truth. Recently this phenomenon has been explained by H. G. Houghton by considering that the mode of the crystallization of ice in air is controlled by the vapor-density difference of water and ice saturation, which is a function of temperature.

As the relation between the crystal form and the external conditions became known by the artificial-snow experiments, we were led to estimate the upper-air conditions by examining the shapes of snow crystal on the ground. These estimates must be checked by flights in snowfall, but we were not favored with such facilities. The relation between the forms of snow crystals and meteorological conditions was studied quite recently by L. W. Gold and B. A. Power. They estimated the height of snow formation from the data of tephigrams and examined the relation with the temperature of the layer where the crystal was estimated to be formed. The result is in good agreement with our data obtained in the laboratory.

Besides the meteorological applications, we promoted research in the line of the study of crystal habits. In the massive pile of the earlier literatures in the field of crystallography we see exclusively studies on crystal systems, and none on crystal habit. It need scarcely be said that the theory of the crystal lattice cannot solve the problem of the macroscopic form of a crystal. No research on crystal habits has been reported, except the fragmentary studies of Lehmann, Vogel, and a few other scientists. The reason is doubtless the difficulty that exists in not being able to produce abundant variations in crystal form, as can be done for ice. Fortunately, we succeeded in the artificial production of crystals of ice in air, which gives the most complicated varieties in form, and were able to determine the conditions of formation for each of them. So we were allowed to get into the untrodden field of the physics of form to some extent.

We leave here one important problem unsolved, namely, the problem of supersaturation. As M. G. Bennett said in "Some problems of modern meteorology, No. 13," published by the Royal Meteorological Society, London, in 1934, "the evidence is merely negative as to whether supersaturation does or does not exist, and positive evidence is urgently required." This question is still left unanswered. It is very unlikely that more water vapor than the critical value of saturation exists in purely gaseous state in the natural atmosphere. Strictly speaking, the existence of supersaturation of water vapor in air cannot be expected, except in some special cases such as the instant of adiabatic expansion in a Wilson cloud chamber. In this special case also, the duration of the supersaturation is extremely short, say 10^{-3} or 10^{-4} sec. Supersaturation observable in the natural atmosphere is considered to mean the existence of minute droplets in saturated air. In our experiments on the artificial production of snow crystals, minute water droplets 1 to 2 μ in diameter are always observable abundantly in the region where snow crystals are formed. It was also found that they act in a similar manner to vapor for the contribution to ice-crystal formation; that is, they do not freeze to the ice crystal in the form of droplets but spread over the ice surface at the moment when they are brought in contact with ice. As a result these minute droplets show, like water vapor, a sublimation phenomenon in the process of condensation. This phenomenon is considered to be caused at least partly by the surface diffusion of water molecules on the surface of ice. According to this view, these minute droplets play an important role in the phenomenon of snow-crystal growth. We presume that a similar phenomenon will exist in the case of snow formation in nature. The difficulty remains, however, that we do not know whether such minute droplets do exist abundantly in the natural atmosphere or not. The precise determination of droplet distribution within clouds in which snow crystals are growing will answer this question.

Appendix

Bibliography

References

APPENDIX

CLASSIFICATION OF SNOW CRYSTALS FOR PRACTICAL PURPOSES

For this purpose the tentative snow classification proposed by the International Commission of Snow and Ice is considered to be most convenient. In this classification all the solid precipitations are included.

Practical Classification of Solid Precipitation.

	Code	Graphic symbol	Term	Remarks
Type of particle	F1	⬡	Plates	P1a, P1b, P1c, P4 of General classification of snow crystals.
	F2	✱	Stellar crystals	P1d, P1e, P1f, P1g, P1h, P1i, P2a, P2b, P2c, P3a, P3b, P4
	F3	▭	Columns	C1a, C1b, C1c, C2a, C2b
	F4	↔	Needles	N1a, N1b, N2
	F5	⊗	Spacial dendrites	P5a, P5b
	F6	⊢⊣	Capped columns	CP1a, CP1b, CP1c, CP2a, CP2b
	F7	⌒⌒	Irregular crystals	CP3, S, I1, I2, I3
	F8	✱ over △	Graupel (Snow pellet)	R4a, R4b, R4c
	F9	△ with dot	Sleet (Ice pellet)	U. S. definition; frozen raindrops, fairly small and transparent
	F0	△ a) ▲ b)	Hail a) small hail (opaque) b) hail (transparent)	Solid precipitation formed by the successive freezing of water layers
Additional characteristics	m	✱ (broken)	Broken	Broken crystals of type 1, 2, etc.
	r	✱ (rimed)	Rimed	R1, R2, R3
	f	(✱)	Flake	Clusters of crystals of type 1, 2, etc.
	w	✱ with dots	Wet	Wet or partially melted crystals of type 1, 2, etc.
	—	✱ with dot	English sleet	Snow and rain falling together.
Size of particle	D			The size of particle means the greatest extension of a particle (or average when many are considered) measured in millimeters. For a cluster of crystals it refers to the average size of the crystals composing the flake.

BIBLIOGRAPHY

Most of the contents of this book have been published in ten numbers of the *Journal of the Faculty of Science, Hokkaido University*, each time a subject was finished, and in three papers in Japanese, as follows.

No. 1. U. Nakaya and T. Iijima, "Snow crystals observed in 1933 at Sapporo and some relations with meteorological conditions" (1933).

No. 2. U. Nakaya and K. Hashikura, "Classification and explanation of snow crystals observed in winter of 1933–34 at Mt. Tokachi and at Sapporo" (1934).

No. 3. U. Nakaya and T. Terada, Jr., "On the electrical nature of snow particles" (1934).

No. 4. U. Nakaya and T. Terada, Jr., "Simultaneous observations of the mass, falling velocity and form of individual snow crystals" (1935).

No. 5. U. Nakaya, "On the correspondence of snow and rime crystals" (1935).

No. 6. U. Nakaya and I. Satō, "On the artificial production of frost crystals, with reference to the mechanism of formation of snow crystals" (1935).

No. 7. U. Nakaya, Y. Sekido, and M. Tada, "Notes on irregular snow crystals and snow pellets" (1936).

No. 8. U. Nakaya and Y. Sekido, "General classification of snow crystals and their frequency of occurrence" (1936).

No. 9. U. Nakaya, I. Satō, and Y. Sekido, "Preliminary experiments on the artificial production of snow crystals" (1938).

No. 10. U. Nakaya, Y. Toda, and S. Maruyama, "Further experiments on the artificial production of snow crystals" (1938).

No. 11. M. Hanajima, *Kishō-shūshi* (*Meteorological Magazine*) **20**, 238 (1942) (in Japanese).

No. 12. M. Hanajima, *Kishō-shūshi* **22**, 121 (1944) (in Japanese).

No. 13. M. Hanajima, *Teion-kagaku* (*Low Temperature Science*) **2** (1945) (in Japanese).

REFERENCES

1. G. Hellmann, *Schneekrystalle* (Berlin, 1893).
2. See *Oeuvres de Descartes* (Paris, 1902), vol. 6, p. 298.
3. William Scoresby, *An account of the arctic regions, with a history and description of the northern whale-fishery* (Edinburgh, 1820).
4. Suketoshi Yajima, *Kagaku* (Science), Vol. 11, No. 3 (in Japanese).
5. Masanaga Kamada, *Kekka Zuihitsu.*
6. Reference 1.
7. G. Nordenskiöld, *Nature 48*, 592 (1893).
8. W. A. Bentley and W. J. Humphreys, *Snow crystals* (McGraw-Hill, New York, 1931).
9. G. Stüve, *Beitr. Phys. Atmosphäre 15*, 170 (1929).

10. G. Hellmann, *Met. Zeit. 11*, 281 (1894).
11. Takematsu Okada, photographs in *Meteorology* (ed. 2, Iwanami Shoten Co.).
12. Reference 8, p. 5.
13. K. Wegener, *Thermodynamik der Atmosphäre* (ed. 3, Leipzig, 1928), p. 248.
14. J. C. Shedd, *Month. Weath. Rev. 47*, 691 (1919).
15. Reference 14.
16. A. B. Dobrowolski, *Historja naturalna lodu* (Warsaw, 1922), p. 248.
17. W. J. Humphreys, *Physics of the air* (ed. 2, New York, 1929), p. 516.
18. Reference 16, p. 182.
19. G. Stüve, *Gerlands Beitr. Geophys. 32*, 326 (1931).
20. Reference 13.
21. C. Kassner, *Met. Zeit. 17*, 225 (1900).
22. Reference 16, p. 144.
23. Reference 8, p. 210.
24. P. Czermak, *Sitzber. Akad. Wiss. Wien, Math. naturw. Klasse, Abt. IIa, 109, 185* (1900).
25. E. Barkow, *Met. Zeit. 25*, 456 (1908).
26. K. Wegener, *Met. Zeit. 26*, 272 (1909).
27. G. Greim, *Met. Zeit. 23*, 320 (1906).
28. Reference 1.
29. G. Nordenskiöld, *Met. Zeit. 11*, 346 (1894).
30. Reference 8, p. 5.
31. G. Kalb, *Zentr. Mineral. Geol.*, A, (1921) 129.
32. Reference 14.
33. Reference 1.
34. Reference 32, p. 692.
35. Reference 26, p. 284.
36. O. Lehmann, *Molekularphysik* (Leipzig, 1888), vol. 1, p. 333.
37. H. Vogel, *Zeit. anorg. Chem. 116*, 21 (1921).
38. U. Nakaya and Y. Toda, *Kagaku* (Science) 5, 459 (1935) (in Japanese).
39. T. Terada, *Rept. Aeronaut. Research Inst., Tokyo Imp. Univ.* 3, 1 (1928); T. Terada and M. Tamano, *ibid.* 4, 447 (1929).
40. W. Schmidt, *Sitzber. Akad. Wiss. Wien, Math.-naturw. Klasse, Abt IIa, 118*, 71 (1909).
41. A. Wagner, *Met. Zeit. 26*, 371 (1909).
42. V. Hagemann, *Gerlands Beitr. Geophys. 46*, 261 (1936).
43. G. C. Simpson, *Proc. Roy. Soc. (London)* A *83*, 394 (1910).
44. J. A. MacClelland and J. J. Nolan, *Proc. Roy. Irish Acad.* A 29, 81 (1912); *30*, 61 (1912).
45. K. Kähler, *Met. Zeit. 40*, 203 (1922).
46. P. Gschwend, *Jahrb. Radioaktivität u. Elektrotechnik 17*, 62 (1920).
47. K. Kähler and C. Dorno, *Ann. Physik 77*, 71 (1925).
48. G. Seligman, *Snow structure and ski fields* (London, 1930), pp. 46–77.
49. Reference 19.
50. Reference 14.
51. Reference 13, p. 285.
52. Reference 8, p. 7.
53. J. M. Adams, *Phys. Rev. 35*, 113 (1930); *Proc. Roy. Soc. (London) A 128*, 588 (1930).
54. Reference 8, p. 8.
55. Reference 48, pp. 62–65.
56. Reference 48, p. 75.

57. Reference 48, p. 76.
58. Nakaya and Sugaya, "Frozen soil in tundra," *Riken-ihō* (*Bull. Inst. Phys. Chem. Research, Tokyo*) *21*, 819 (1942) (in Japanese).
59. W. H. Barnes, *Proc. Roy. Soc.* (*London*) *A 125*, 670 (1929).
60. E. Brandenburger, *Zeit. Krist. 73*, 429 (1930).
61. O. Mügge, *Zentr. Mineral. Geol. 137* (1918).
62. N. J. Seljakov, *Compt. rend. acad. sci. U.R.S.S.* (1936) No. 7; (1937) No. 4.
63. E. Cohen and C. J. G. van der Horst, *Zeit. physik. Chem. 40*, 231 (1938).
64. Reference 14.
65. Reference 13.
66. W. Findeisen, *Met. Zeit. 56*, 429 (1939).
67. M. Hanajima, *Kishō-shūshi* (*Meteorological Magazine*) *22*, 121 (1944) (in Japanese), p. 126.
68. M. Hanajima, *Kishō-shūshi* (*Meteorological Magazine*) *20*, 238 (1942) (in Japanese).
69. J. S. Marshall, M. P. Langleben, and M. Herschorn, "Structure of snow crystals as a function of vapor density excess," McGill University Science Report MW-9 (1952).
70. Reference 67, p. 121.
71. A. M. Tyndall, *Proc. Phys. Soc.* (*London*) *A 34*, 72 (1922).
72. Reference 67, p. 123.
73. M. Hanajima, *Teion-kagaku* (*Low-Temperature Science*) 2 (1945) (in Japanese).
74. Reference 19.
75. H. J. aufm Kampe, H. K. Weickmann, and J. J. Kelly, *J. Meteorology 8*, 168–174 (1951).
76. H. G. Houghton, *J. Meteorology 7*, 363–369 (1950).
77. B. J. Mason and F. H. Ludlum, "Microphysics of clouds," *Reports on progress in physics 14*, 153 (1951).
78. L. W. Gold and B. A. Power, *J. Meteorology 9*, 447 (1952).
79. V. J. Schaefer, *Chem. Revs. 44*, 291–320 (1949).
80. Reference 75.
81. Z. Yosida, *J. Fac. Sci.*, Hokkaido University, Sapporo, Series II, vol. 3, No. 2, p. 43–55.
82. H. J. aufm Kampe, H. K. Weickmann, and H. H. Kedesdy, *J. Meteorology 9*, 374–375 (1952).
83. M. Volmer and I. Estermann, *Z. Physik 7*, 13 (1921).
84. M. R. de Quervain, *Experientia 1*, 207–211 (1945).
85. W. A. Weyl, *J. Colloid Science 6*, 389–405 (1951).
86. M. Kumai, *J. Meteorology 8*, 151–156 (1951).
87. H. Dessens, *Quart. J. Roy. Met. Soc. 75*, 23–27 (1949).

Plates

19 (×30.5)

20 (×15.5)

PLATE 2

24 (×34)

25 (×33)

26 (×38)

27 (×37)

28 (×31)

29

PLATE 6

33 (×35.5)

34 (×34)

PLATE 8

35 (×33)

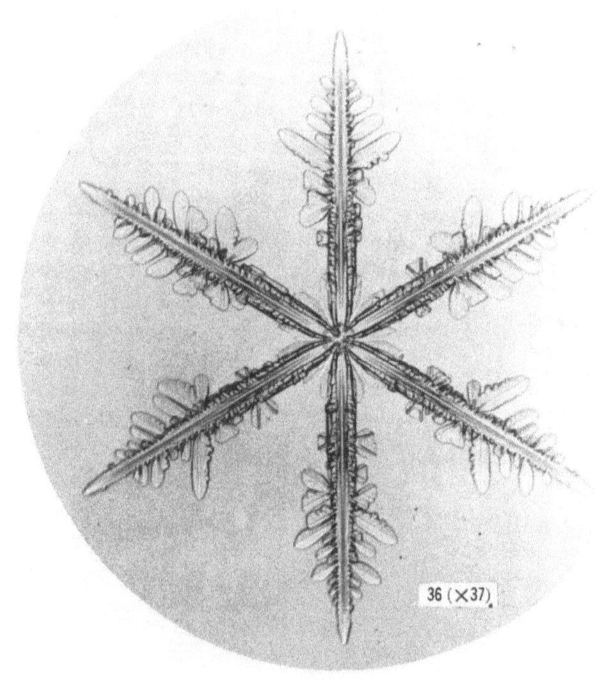
36 (×37)

37 (×19)
38 (×14)
39 (×15)
40 (×18.5)
41 (×16)
42 (×10)

PLATE 10

49 (×20)
50 (×16)
51 (×19.5)
52 (×20)
53 (×13.5)
54 (×20)

PLATE 12

62 (×34)

63

PLATE 14

66 (×20)

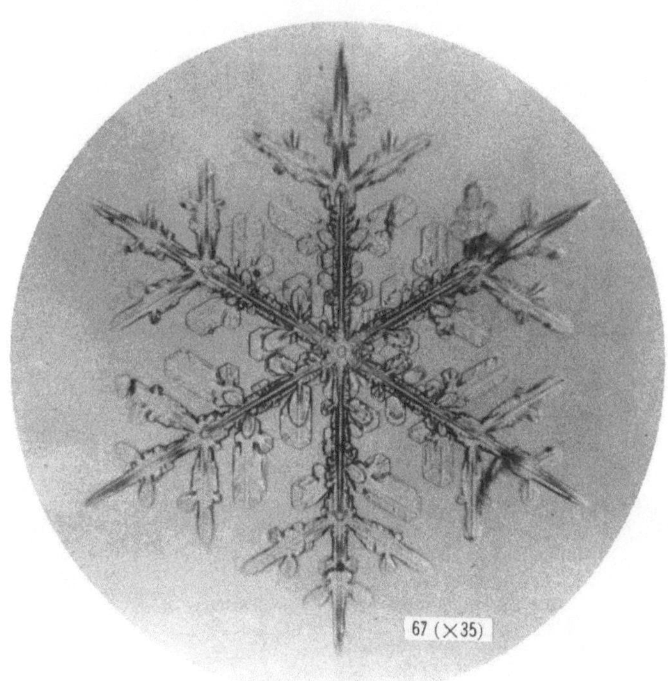
67 (×35)

PLATE 16

71 (×33.5)

72 (×40)

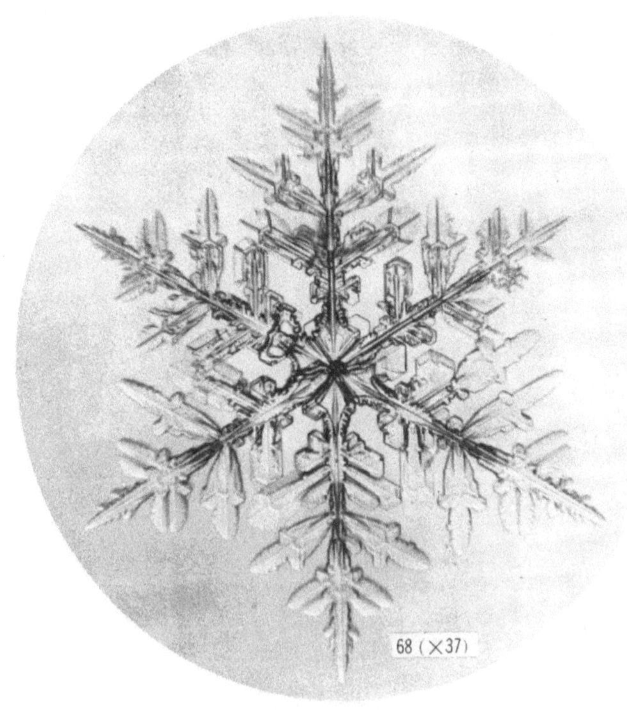

68 (×37)

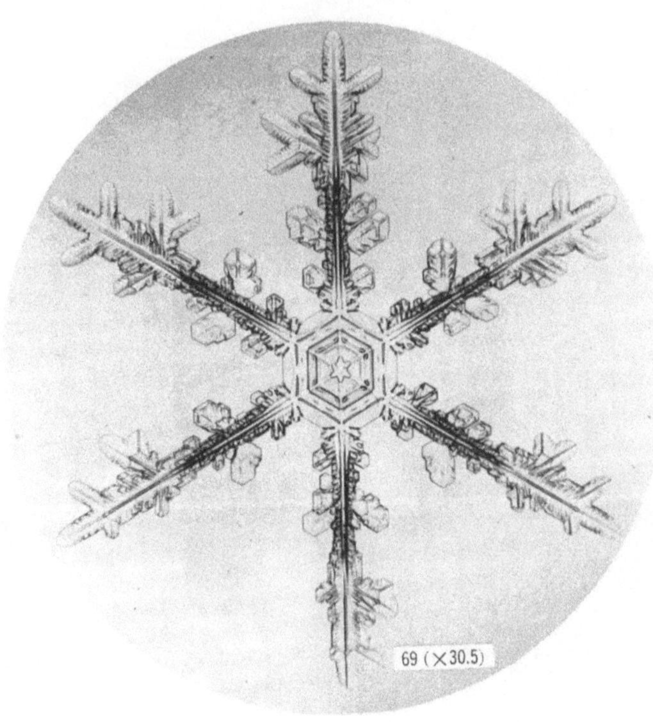

69 (×30.5)

PLATE 18

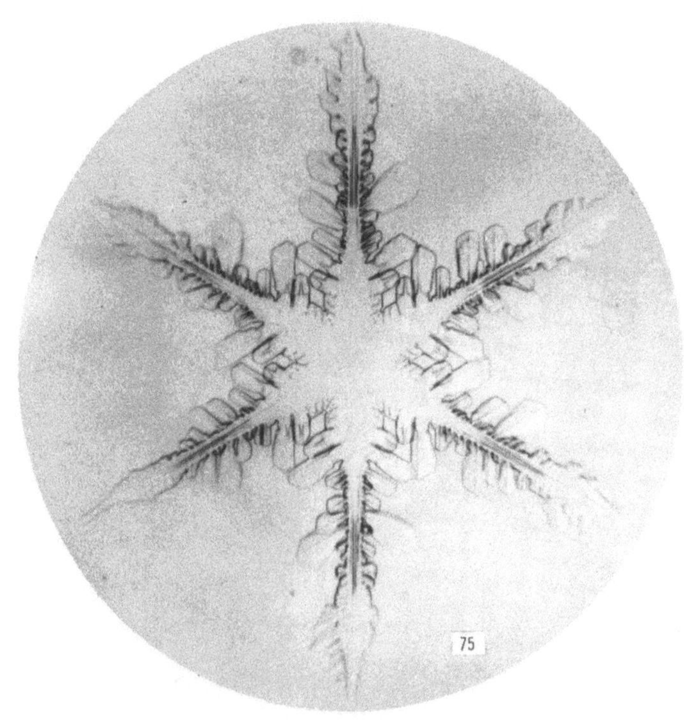

77 (×31.5)
78 (×20)
79 (×24)
80 (×19)
81 (×35)
82 (×11.5)

PLATE 20

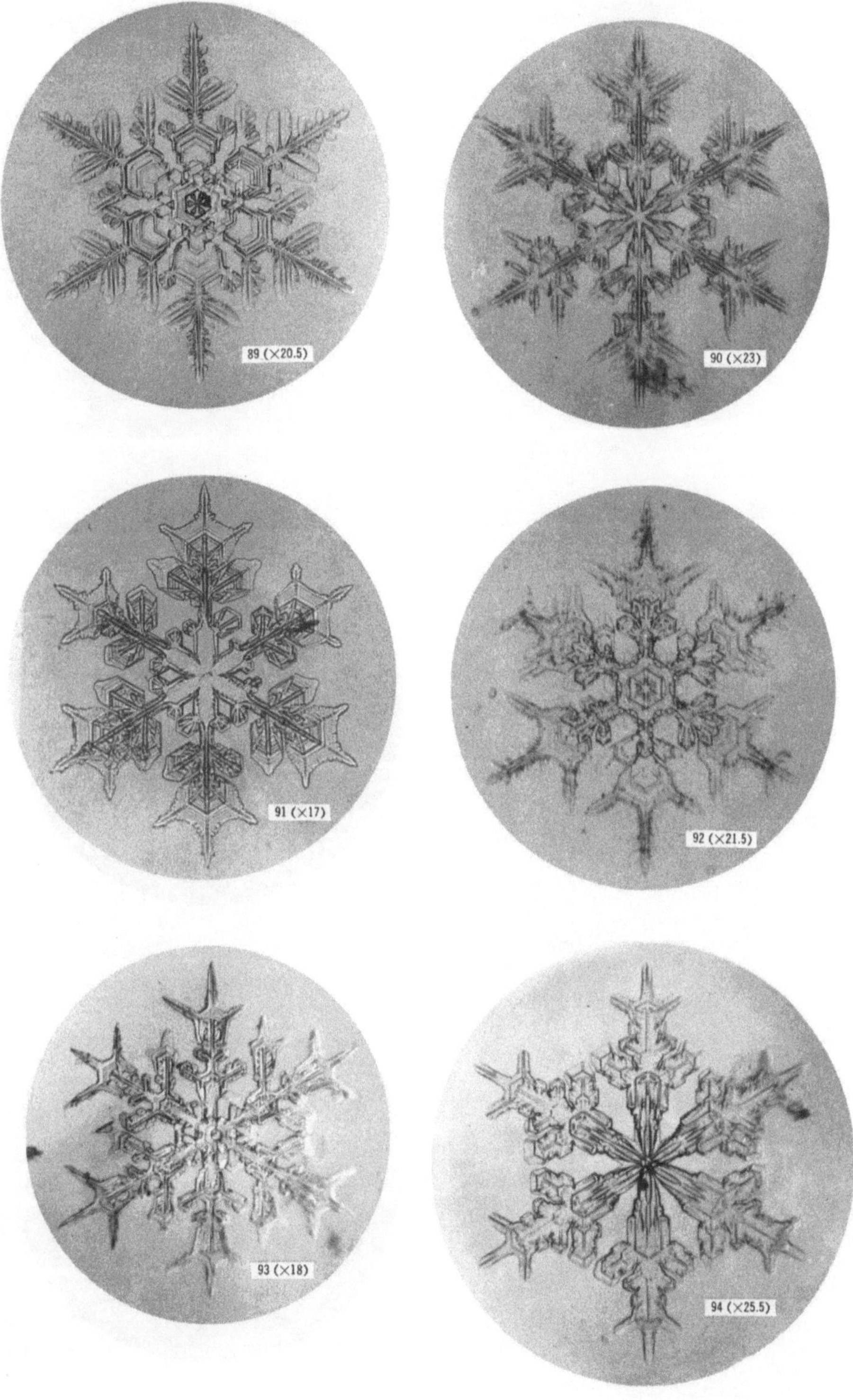
89 (×20.5)
90 (×23)
91 (×17)
92 (×21.5)
93 (×18)
94 (×25.5)

PLATE 22

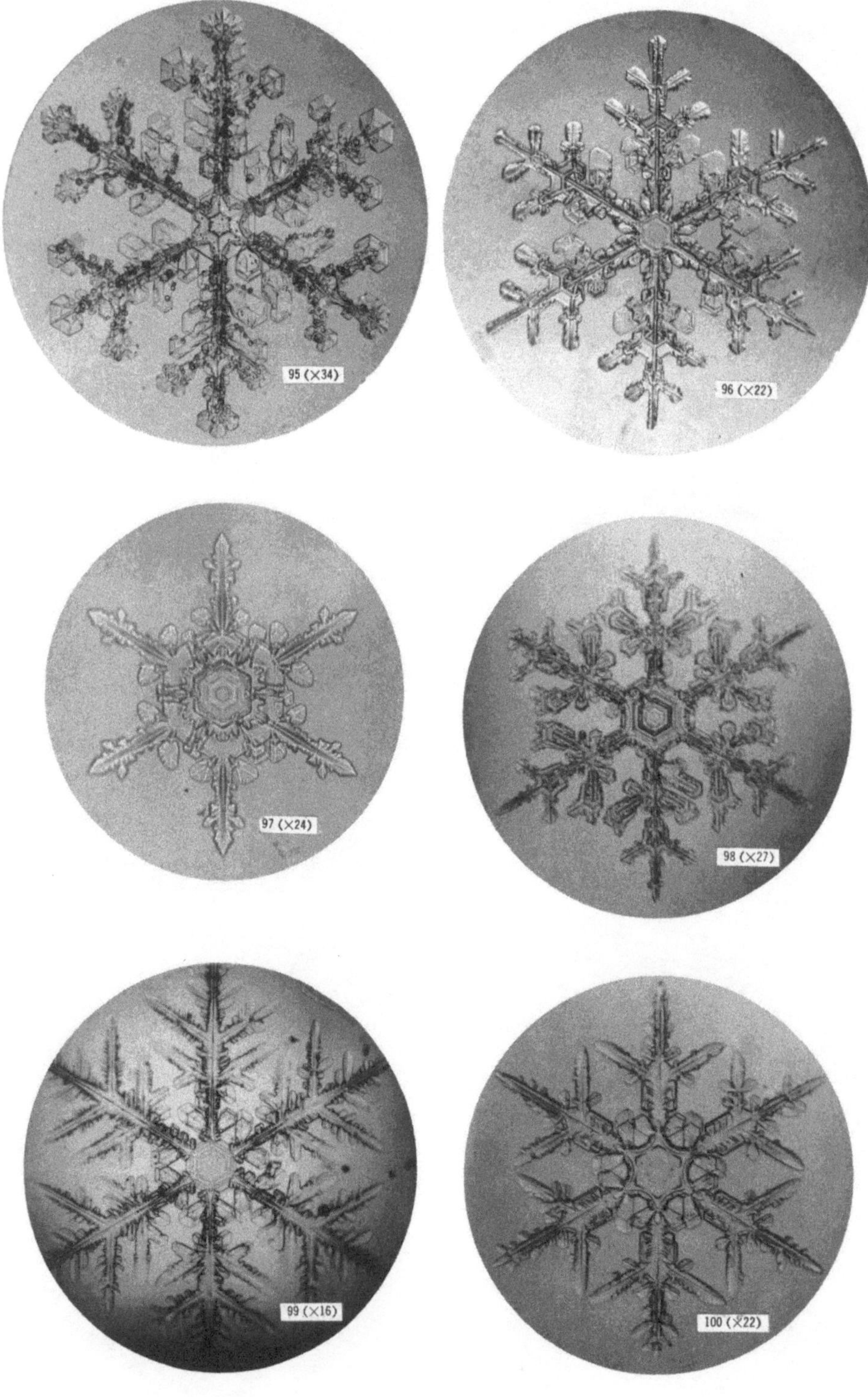

101 (×25)
102 (×14.5)
103 (×20)
104 (×25)
105 (×24)
106 (×20)

PLATE 24

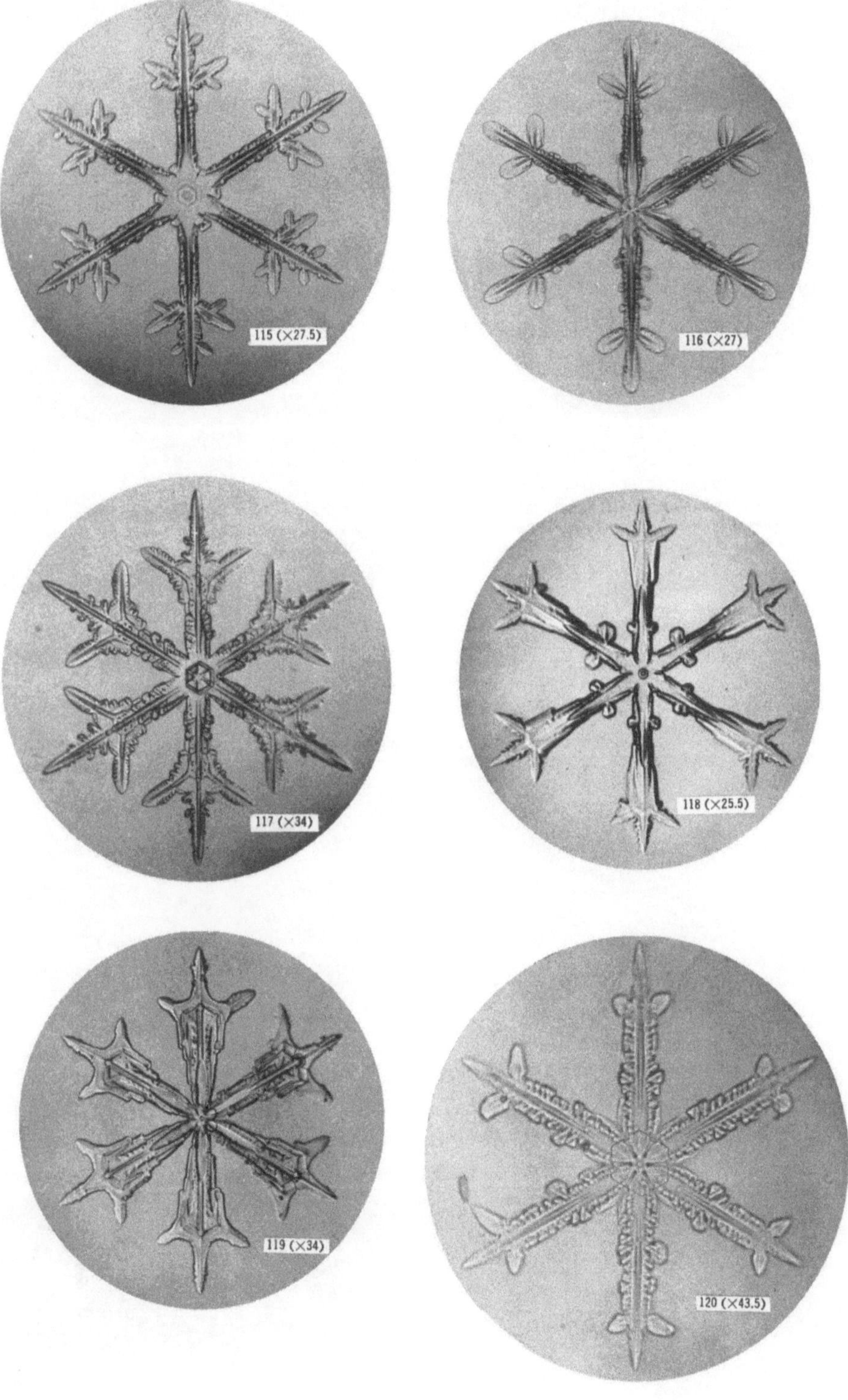
115 (×27.5)
116 (×27)
117 (×34)
118 (×25.5)
119 (×34)
120 (×43.5)

PLATE 26

127 (×55.5)
128 (×24)
129 (×30)
130 (×64)
131 (×47)
132 (×54)

PLATE 28

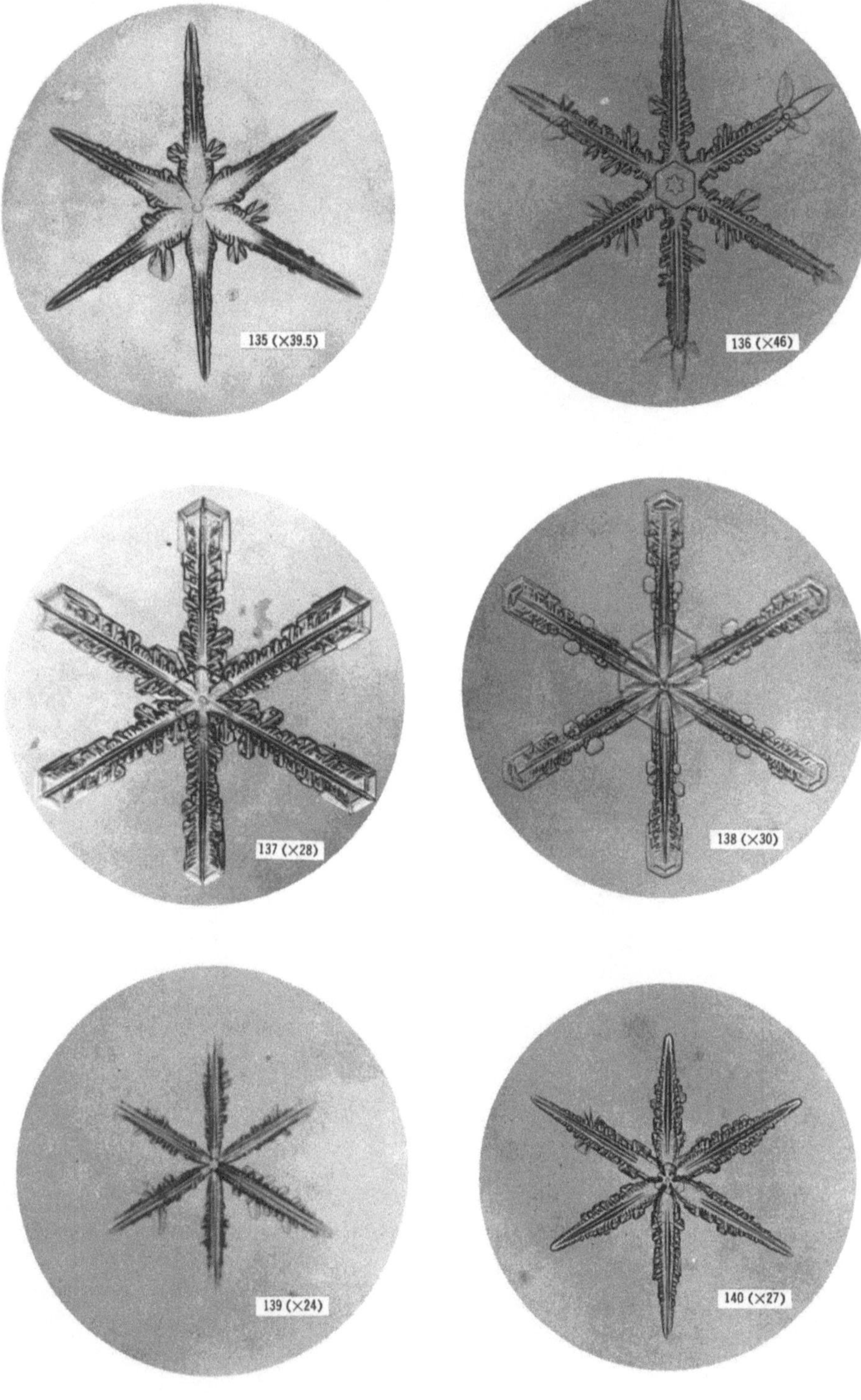

PLATE 29

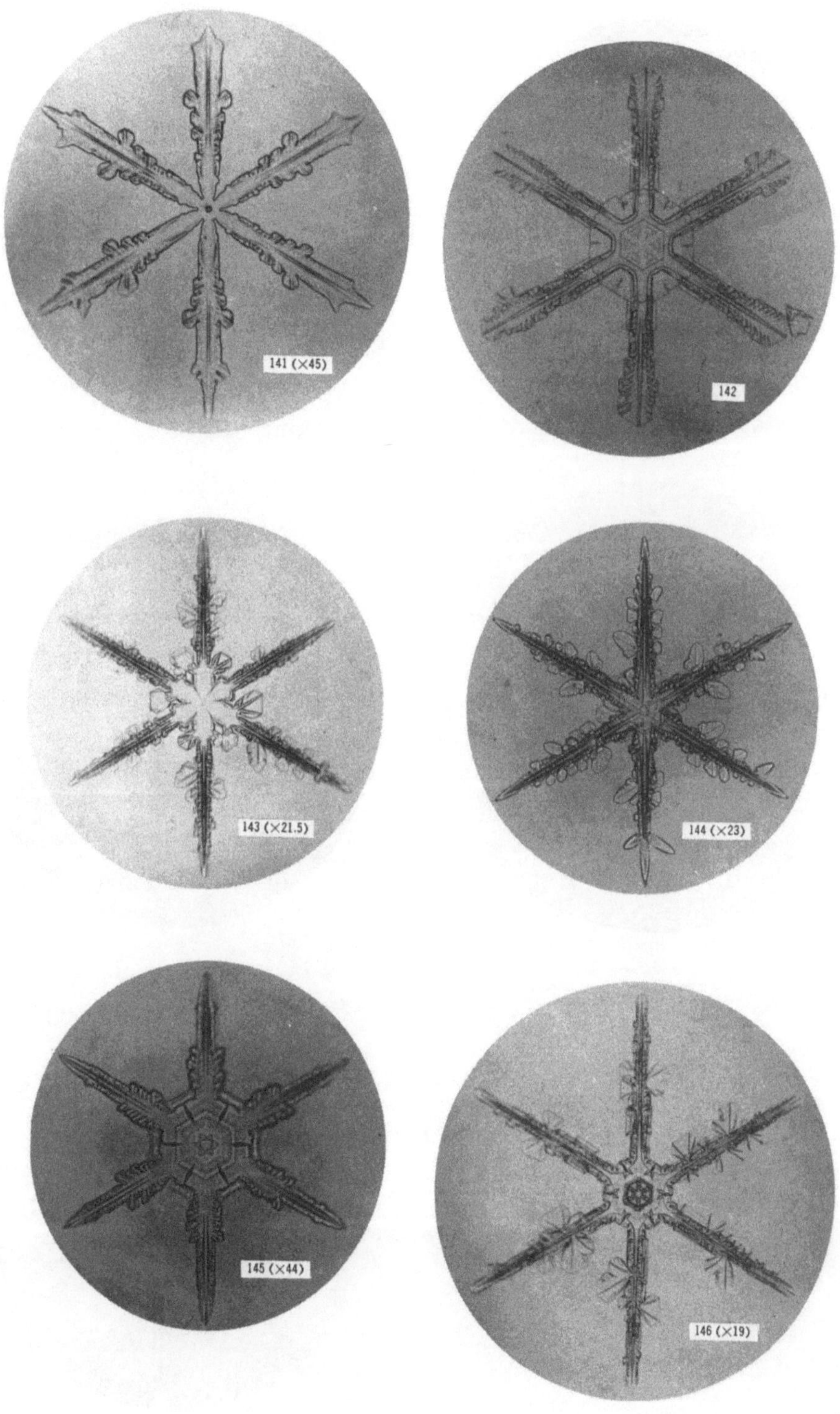

141 (×45)
142
143 (×21.5)
144 (×23)
145 (×44)
146 (×19)

PLATE 30

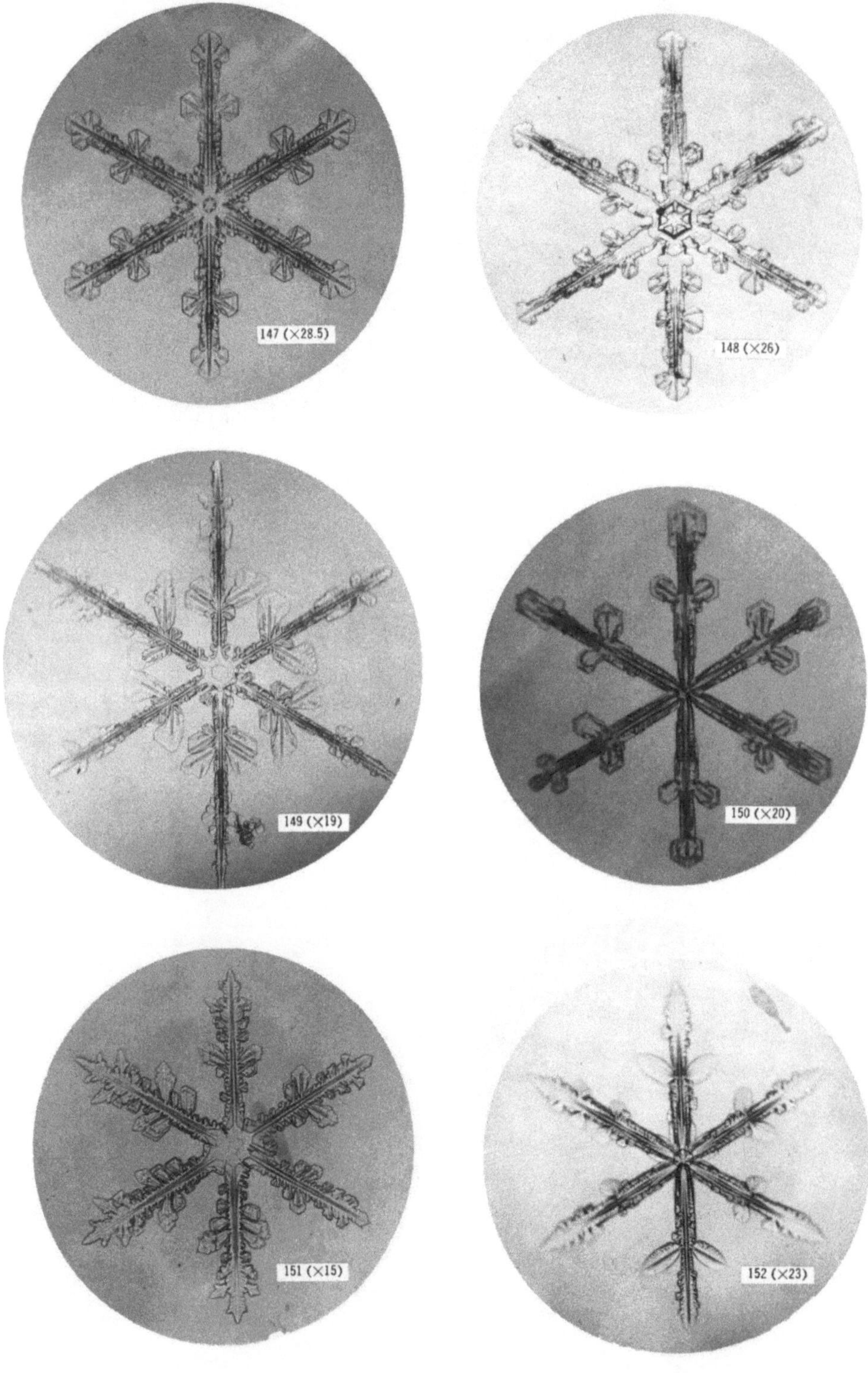

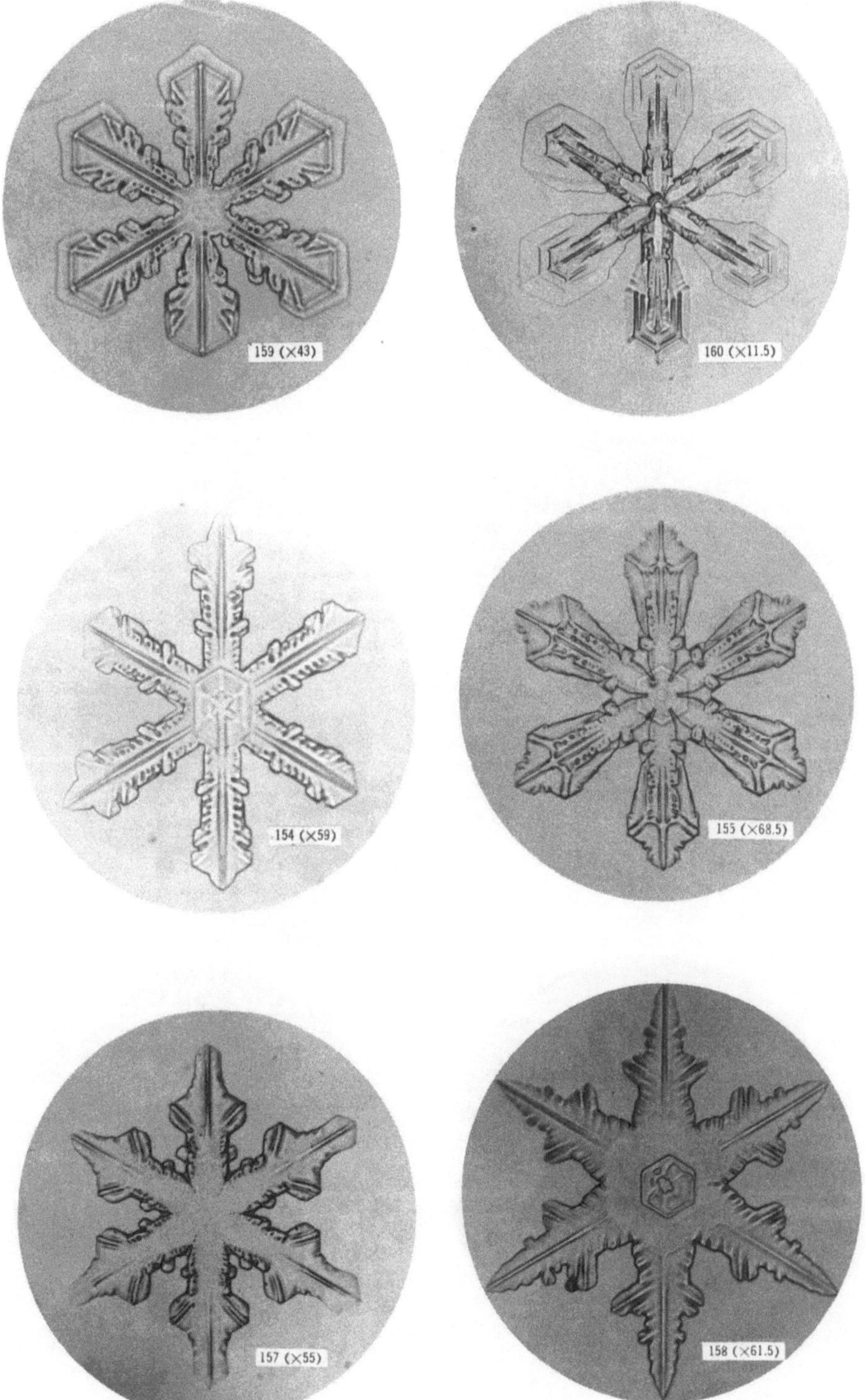
159 (×43)
160 (×11.5)
154 (×59)
155 (×68.5)
157 (×55)
158 (×61.5)

PLATE 32

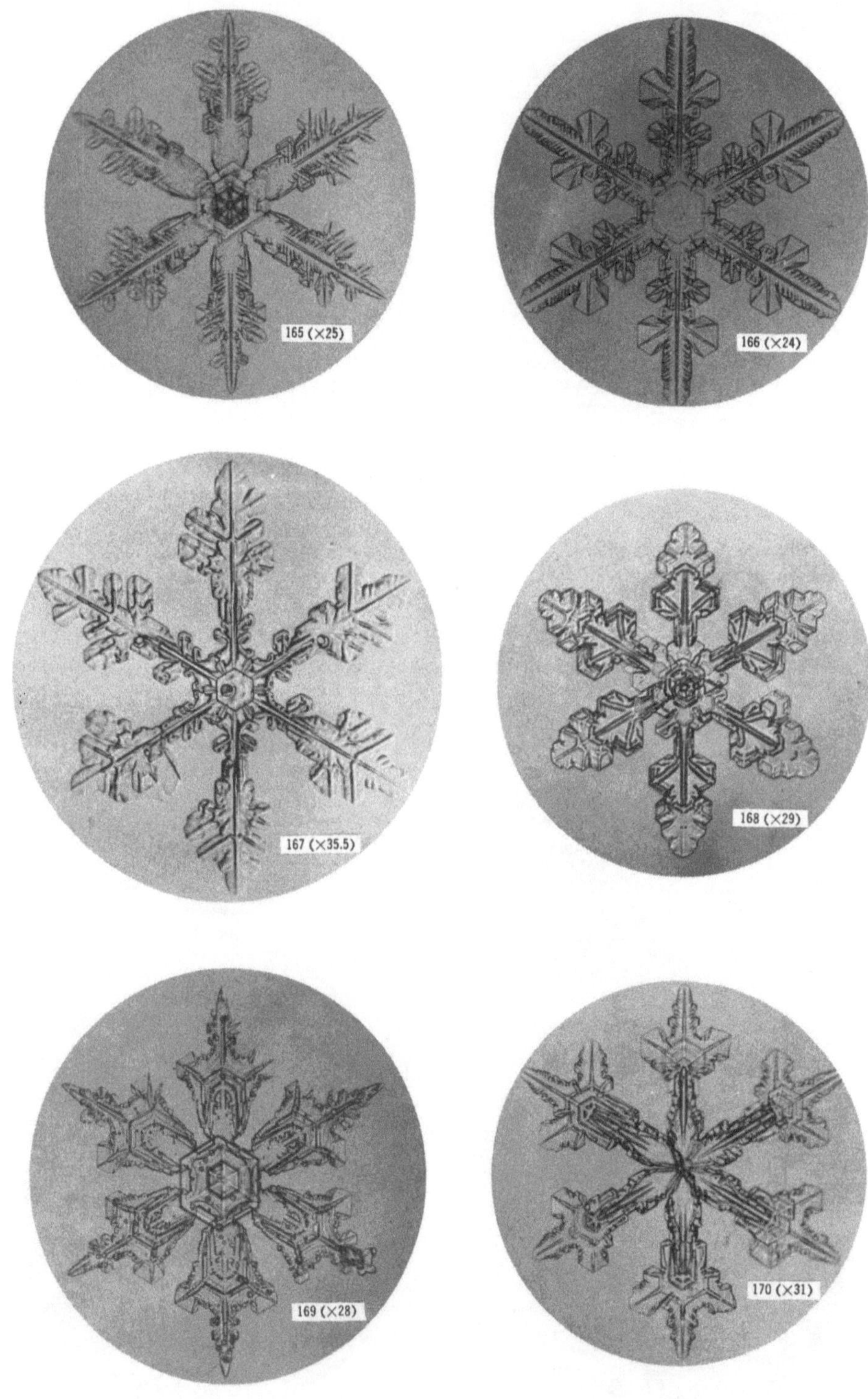

171 (×25)
172 (×25)
173 (×30.5)
174 (×37)
161 (×32)
176 (×40)

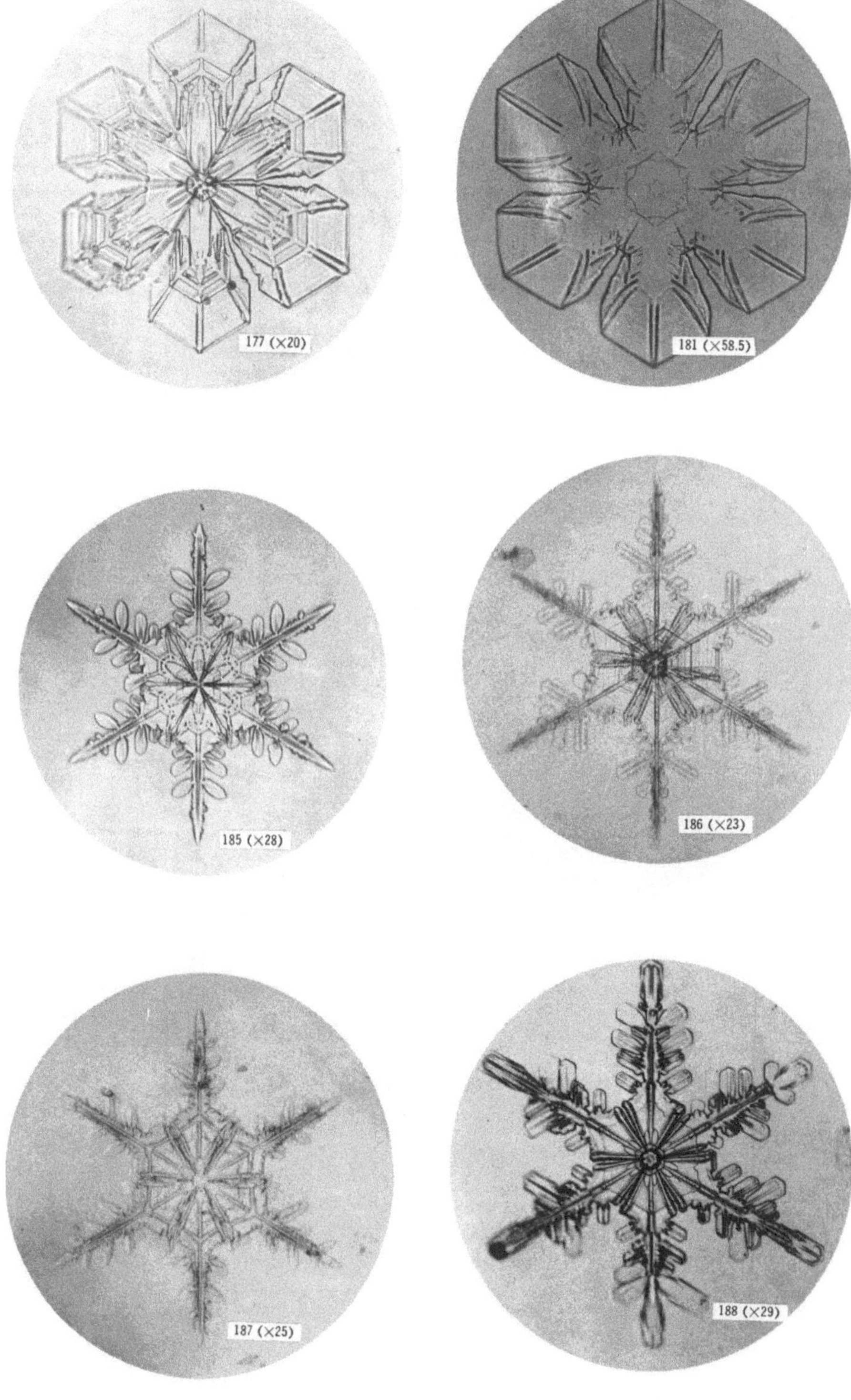
177 (×20)
181 (×58.5)
185 (×28)
186 (×23)
187 (×25)
188 (×29)

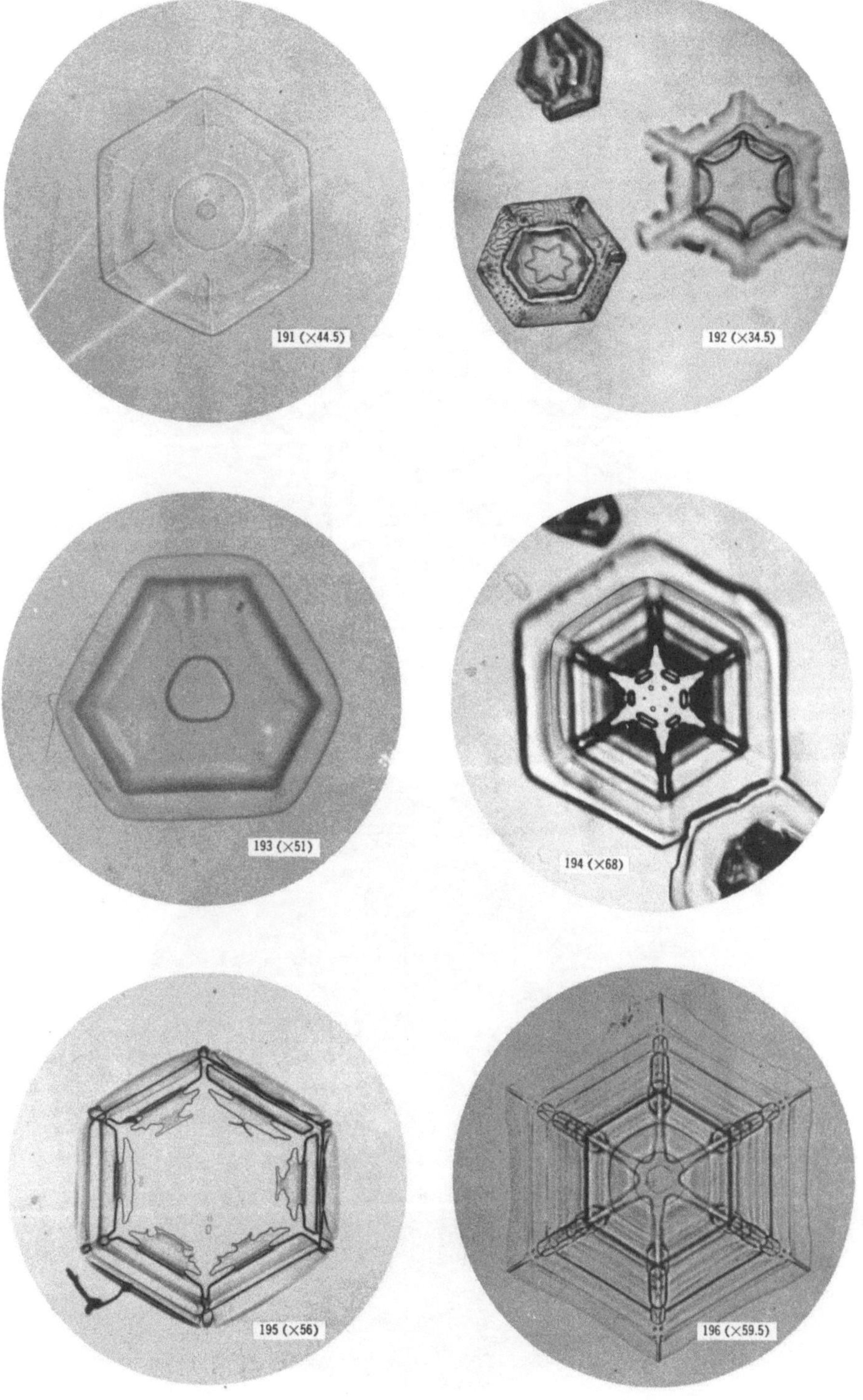
191 (×44.5)
192 (×34.5)
193 (×51)
194 (×68)
195 (×56)
196 (×59.5)

PLATE 36

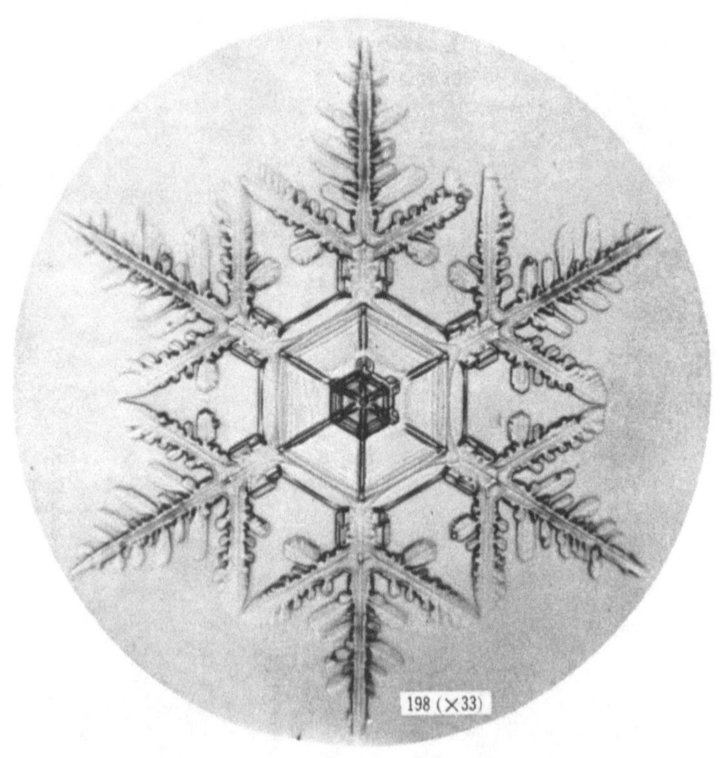
198 (×33)

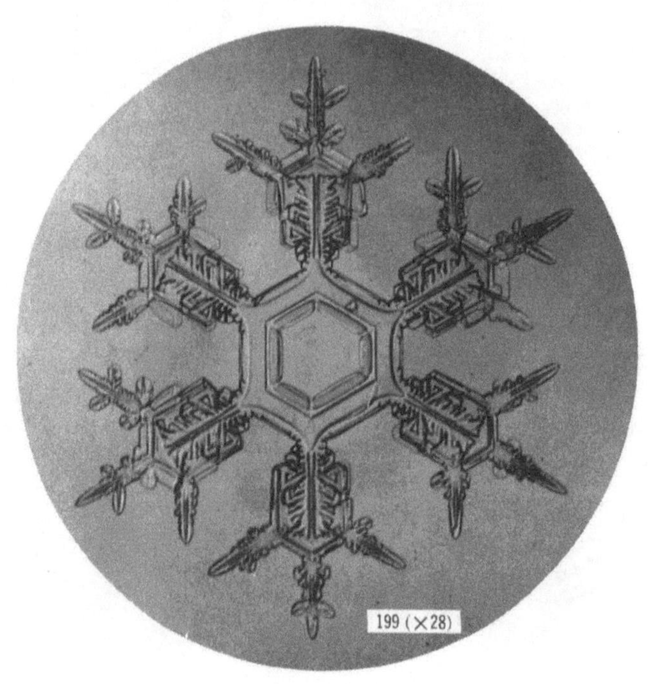
199 (×28)

200 (×37)

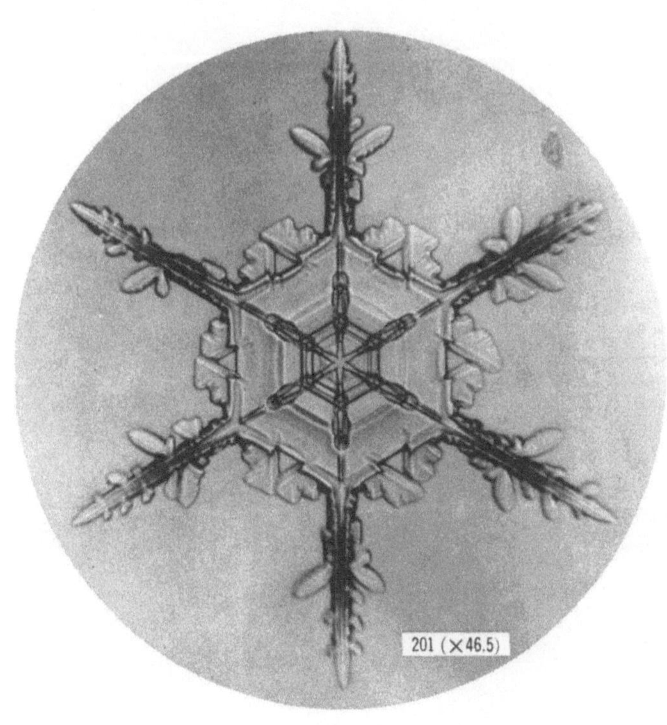
201 (×46.5)

PLATE 38

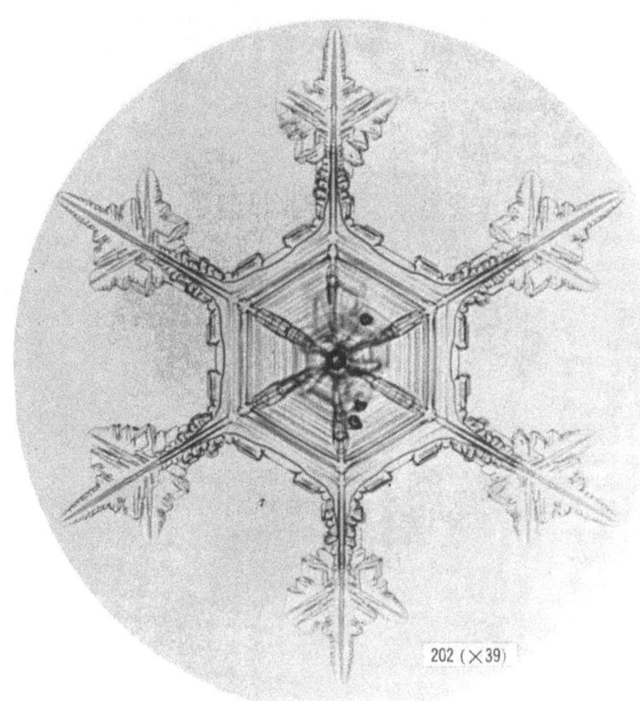

202 (×39)

204 (×41)

205 (×45)

206 (×44)

PLATE 40

207 (×50)

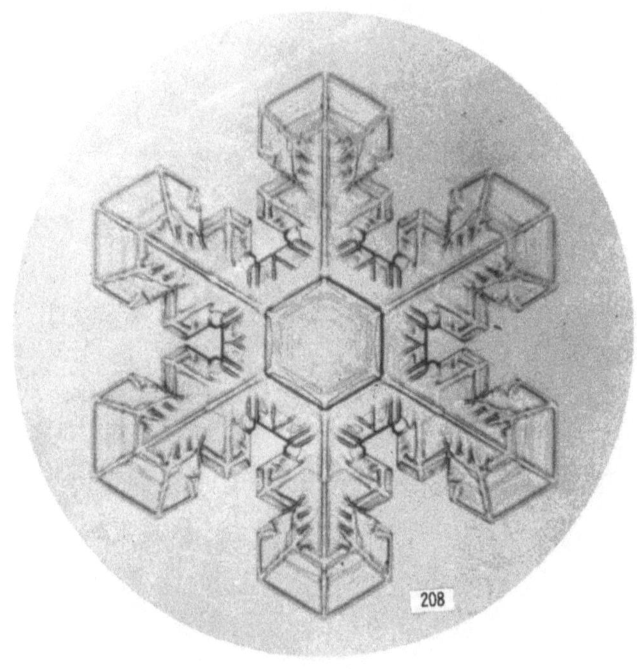
208

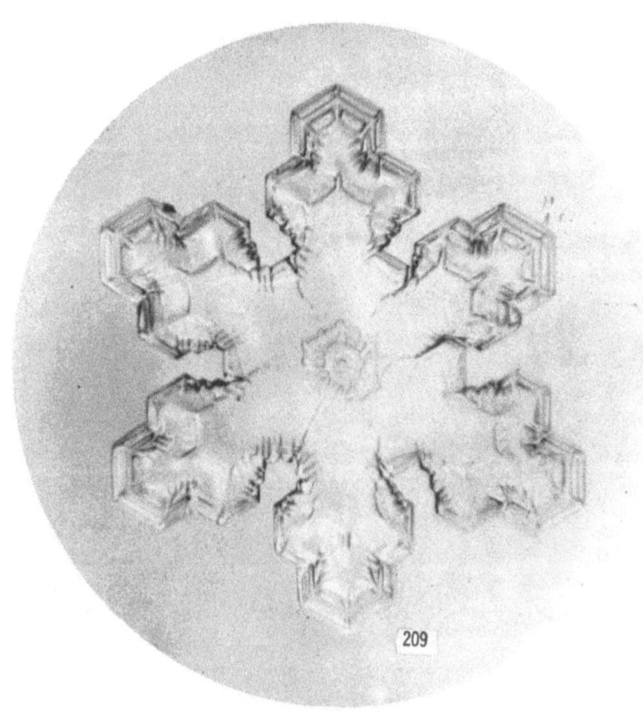
209

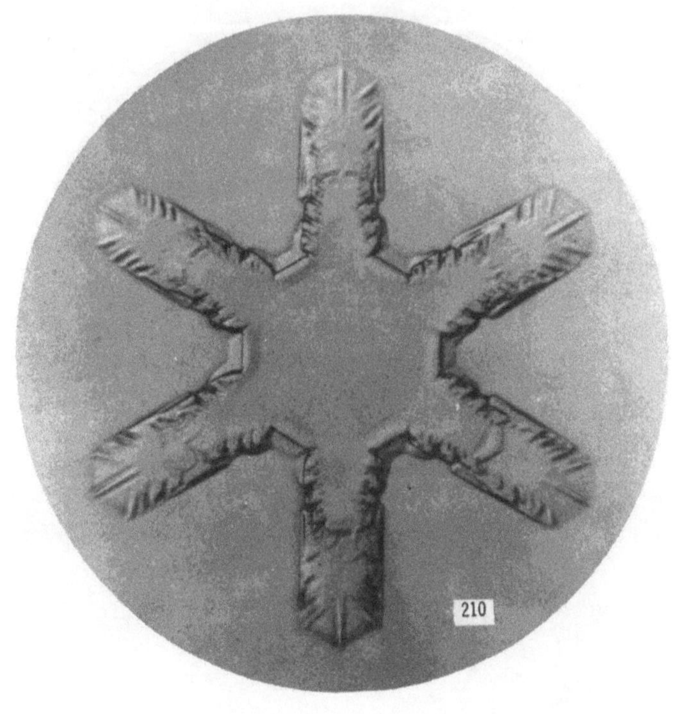
210

PLATE 42

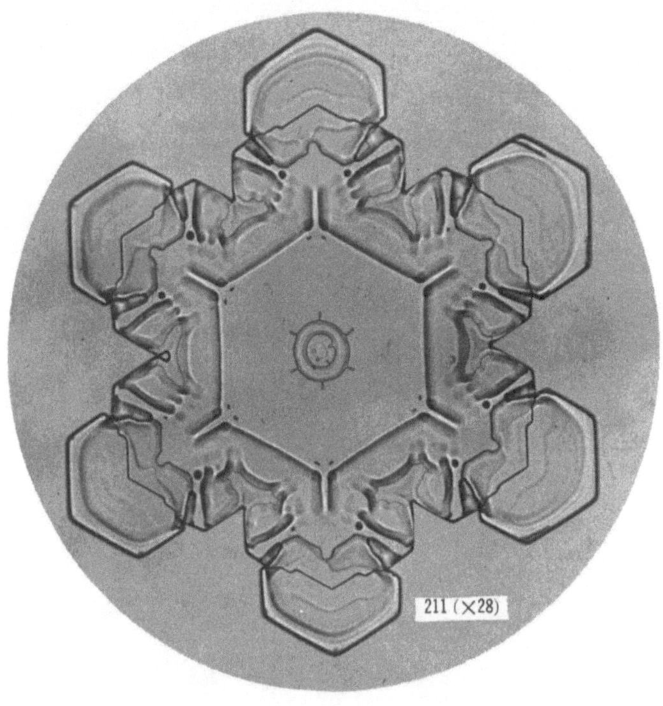

211 (×28)

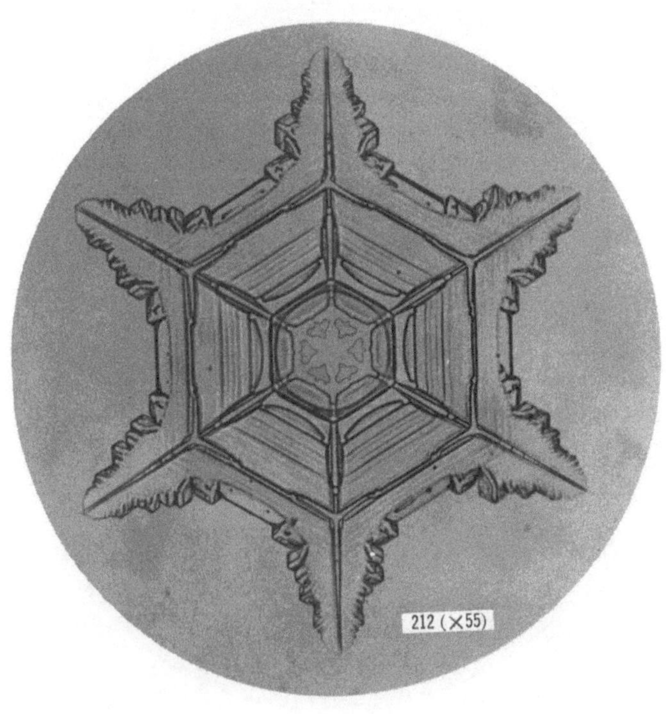

212 (×55)

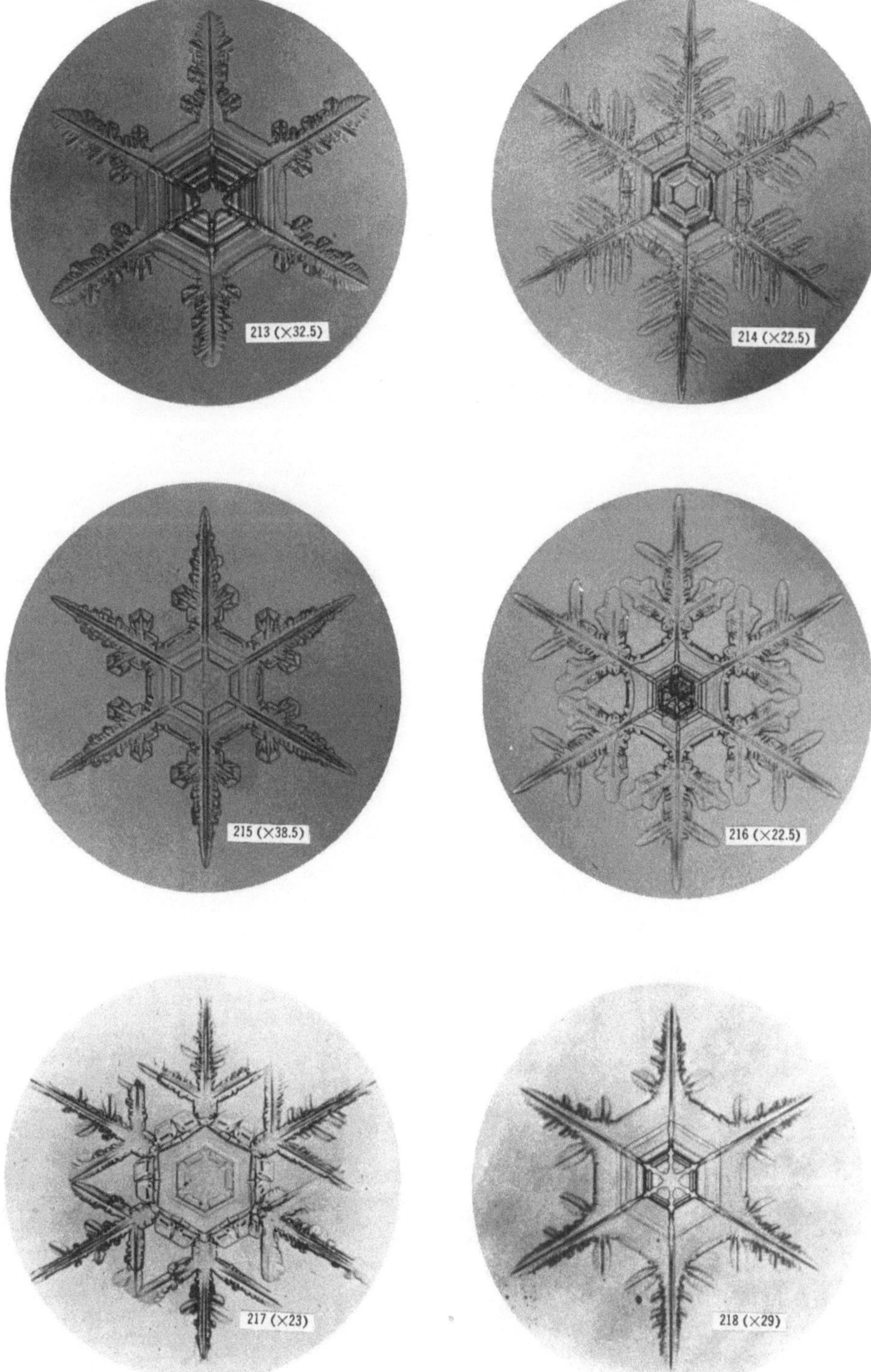
213 (×32.5)
214 (×22.5)
215 (×38.5)
216 (×22.5)
217 (×23)
218 (×29)

219 (×19)
220 (×23.5)
221 (×22.5)
222 (×28)
223 (×19)
224 (×28)

225 (×32)
226 (×26.5)
227 (×25.5)
228 (×34)
229 (×23.5)
230 (×21)

231 (×31.5)
232 (×24)
233 (×35)
234 (×49)
235 (×49)
236 (×44)

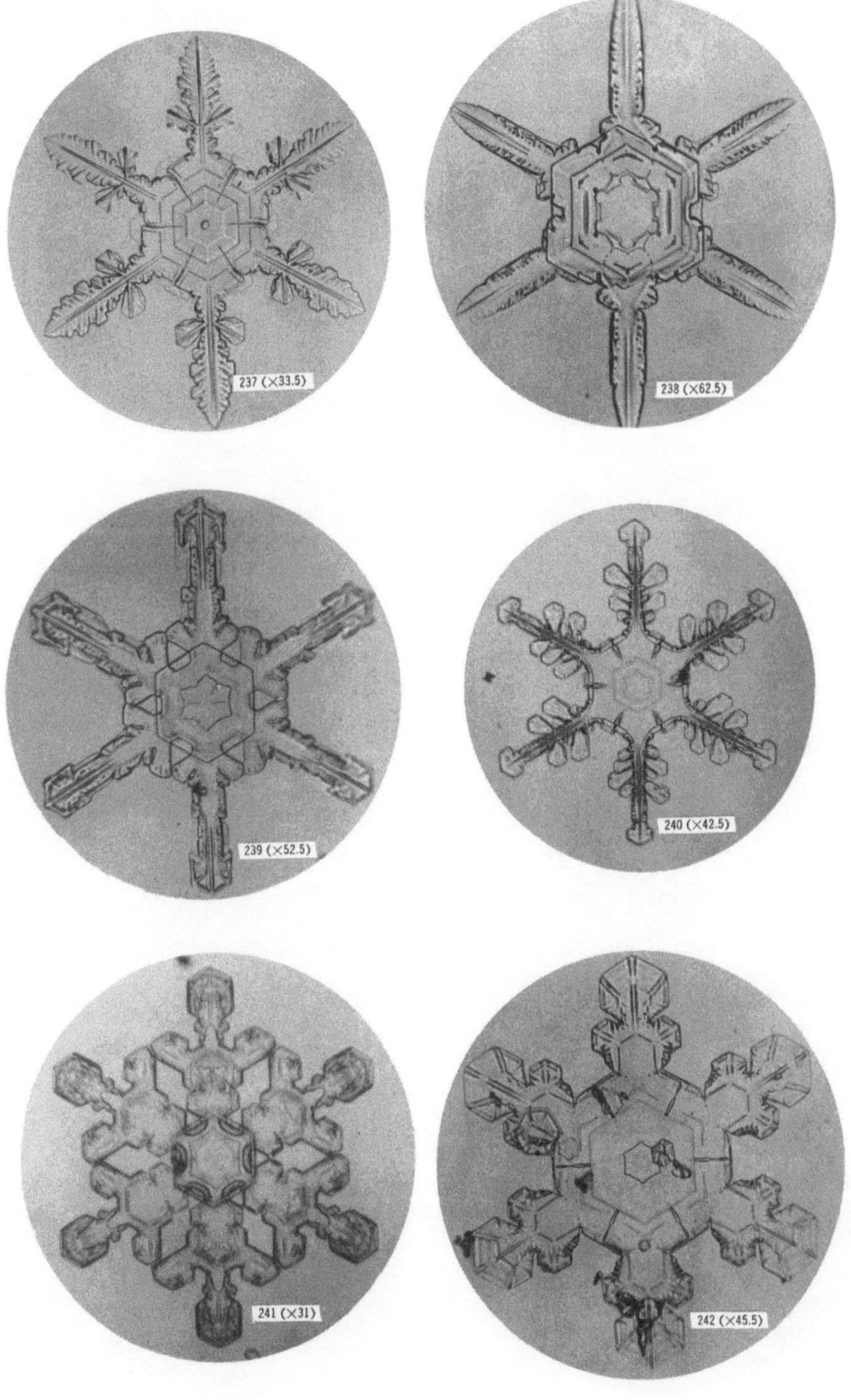
237 (×33.5)
238 (×62.5)
239 (×52.5)
240 (×42.5)
241 (×31)
242 (×45.5)

243 (×27.5)
244 (×27.5)
245 (×35)
246 (×37)
247 (×22)
248 (×23.5)

249 (×35)
250 (×23)
251 (×37)
252 (×78)
253 (×34)
254 (×38)

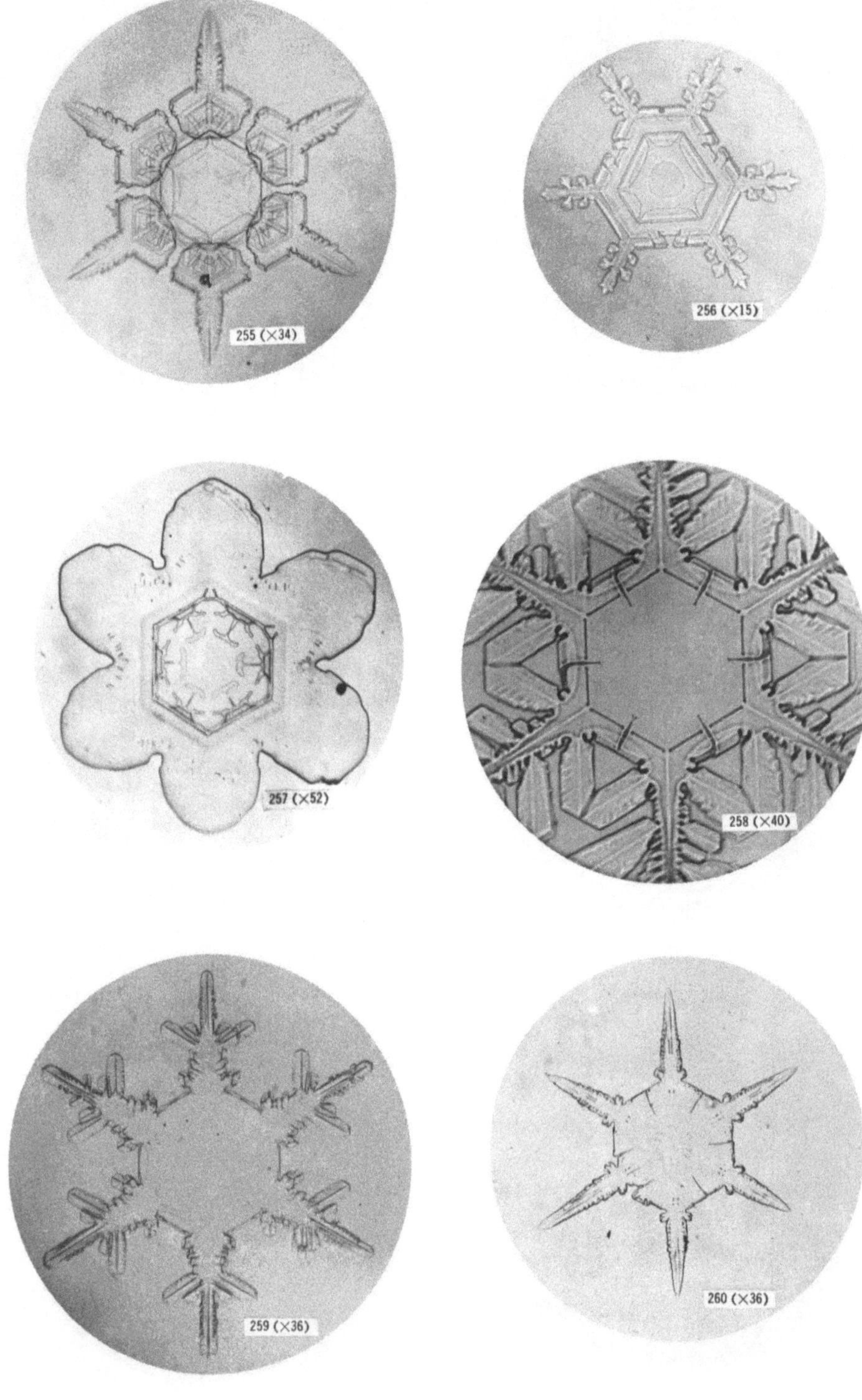
255 (×34)
256 (×15)
257 (×52)
258 (×40)
259 (×36)
260 (×36)

261 (×26)
262 (×47)
263 (×29.5)
264
265 (×42.5)
266 (×11)

PLATE 52

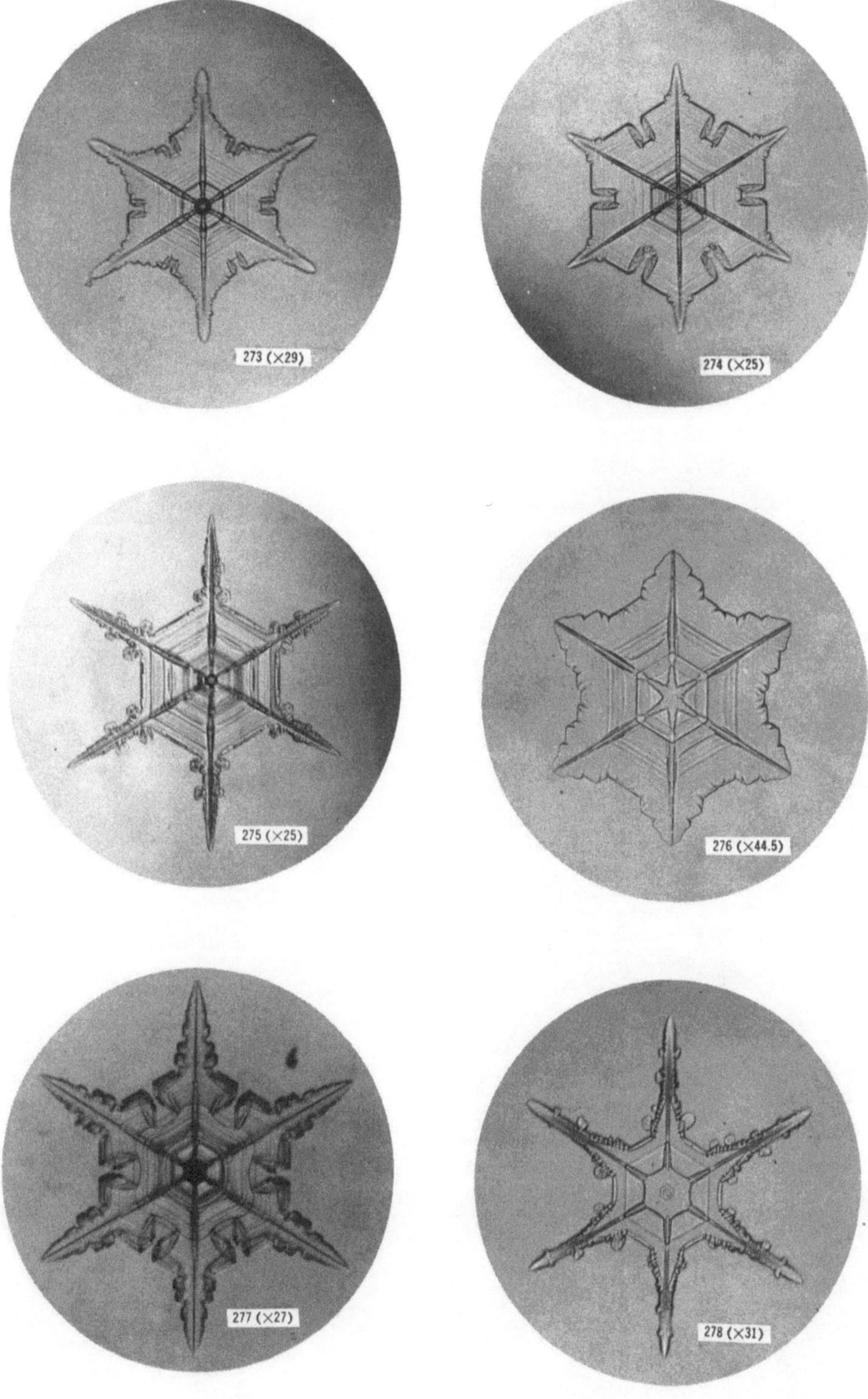
273 (×29)
274 (×25)
275 (×25)
276 (×44.5)
277 (×27)
278 (×31)

PLATE 54

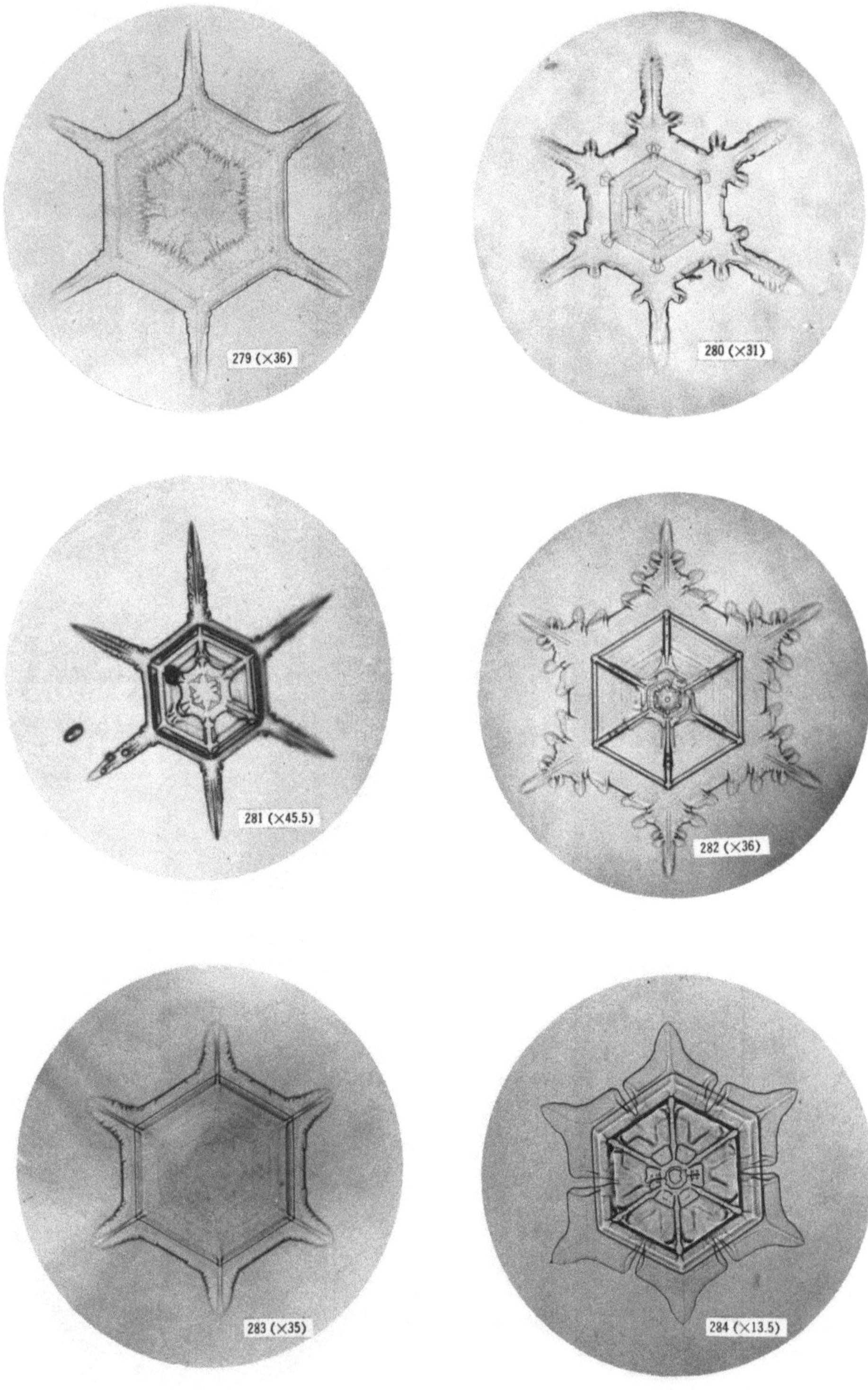

287 (×57)
288 (×28.5)
289 (×42.5)
290 (×16.5)
291 (×51)
292 (×42.5)

PLATE 56

293 (×33)
294 (×39.5)
295 (×75)
296 (×40.5)
297 (×30)
298 (×51)

299 (×14.5)

300 (×29)

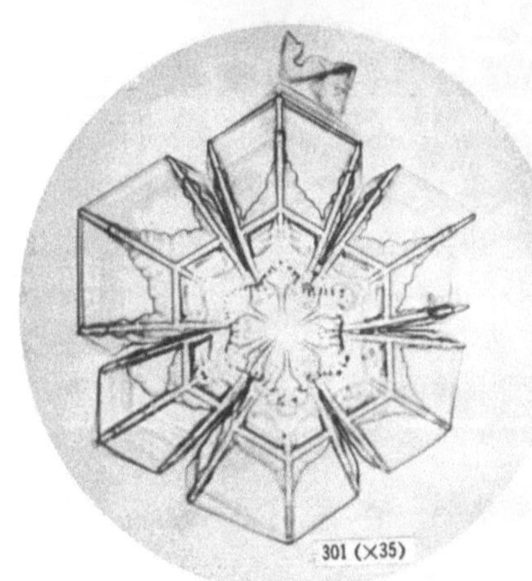
301 (×35)

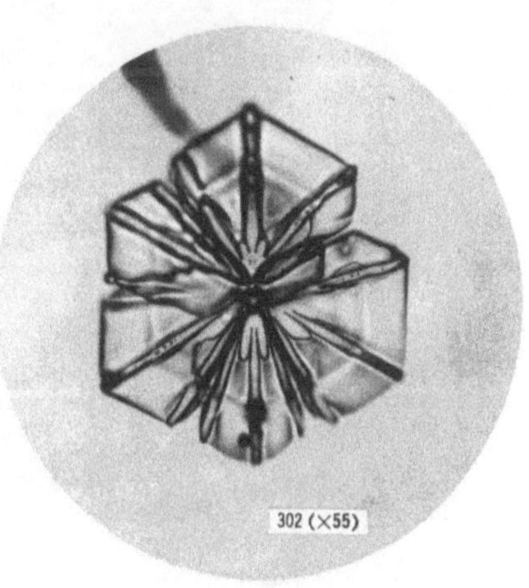
302 (×55)

303 (×36)

304 (×36)

305 (×22.5)

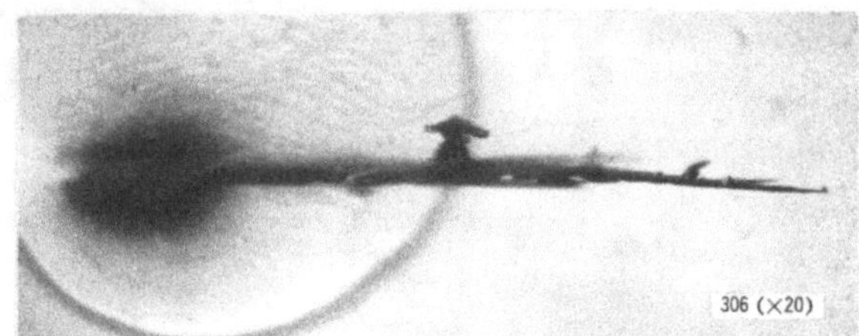
306 (×20)

313 (×29.5)

314 (×43)

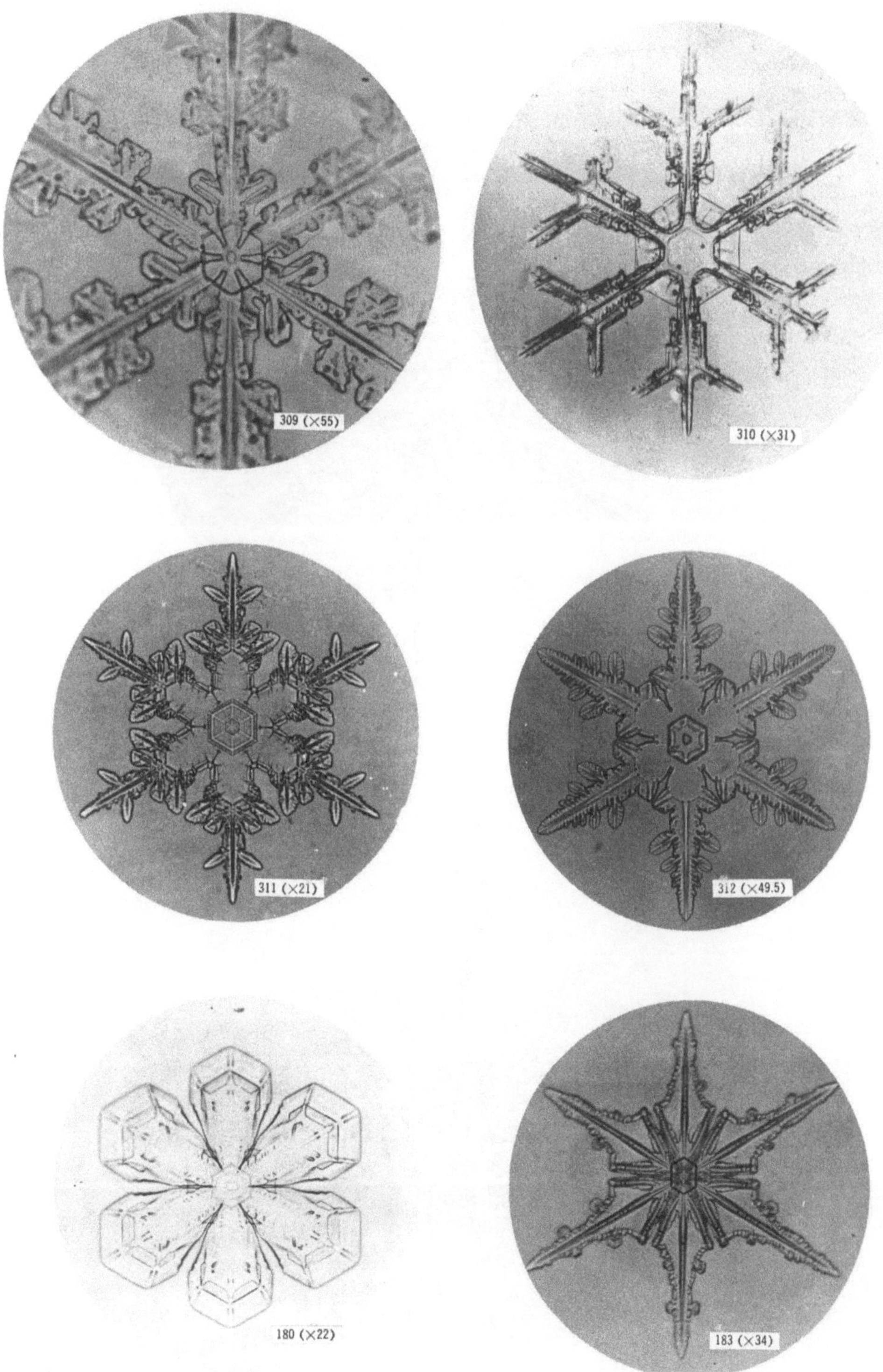
309 (×55)
310 (×31)
311 (×21)
312 (×49.5)
180 (×22)
183 (×34)

PLATE 60

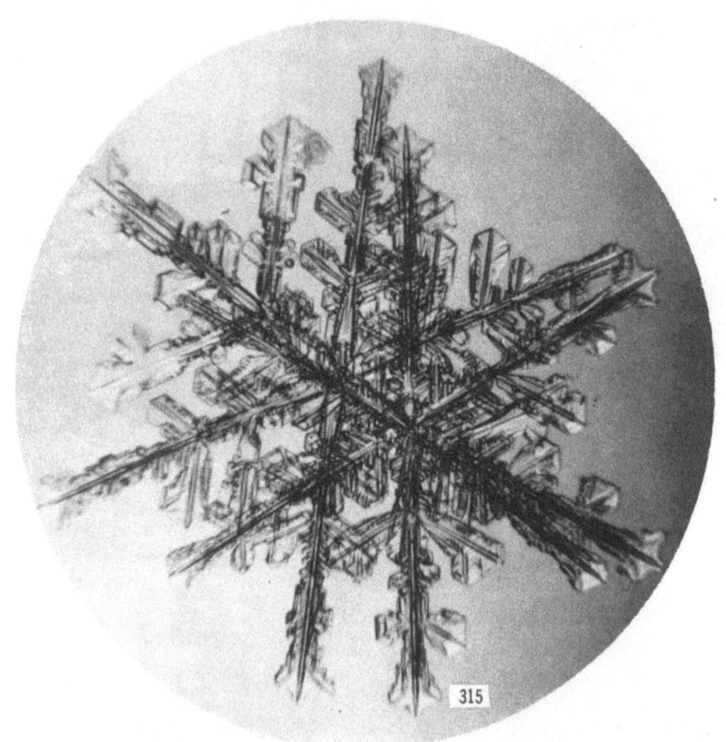
315

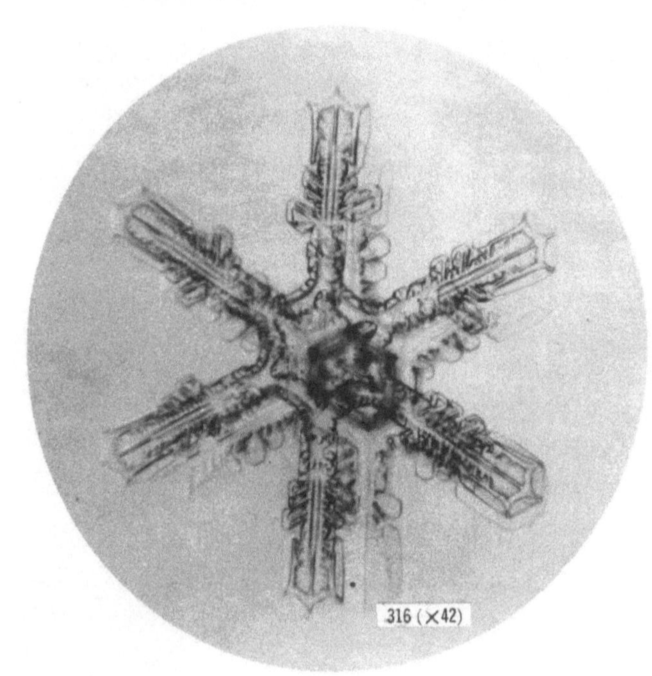
316 (×42)

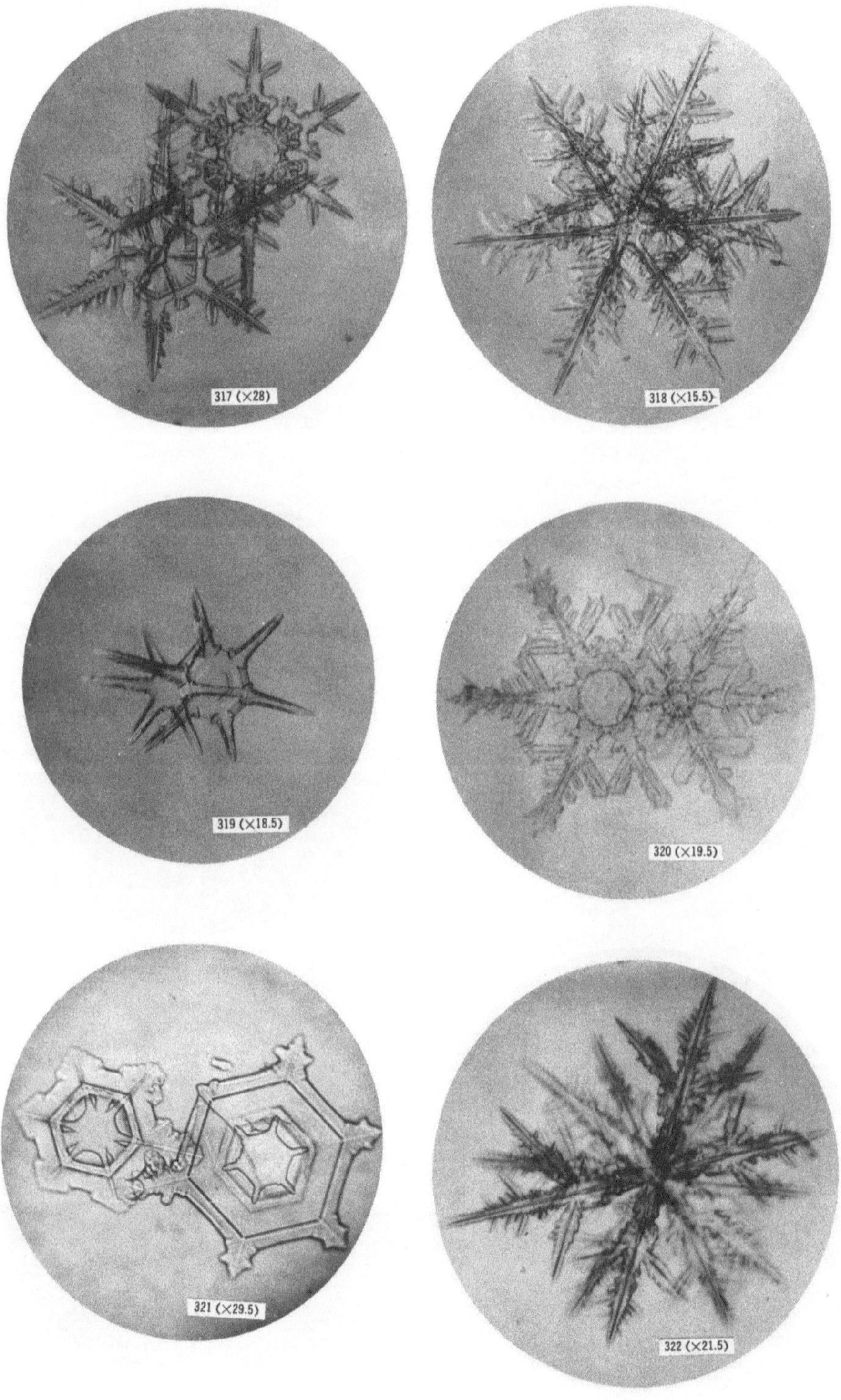
317 (×28)
318 (×15.5)
319 (×18.5)
320 (×19.5)
321 (×29.5)
322 (×21.5)

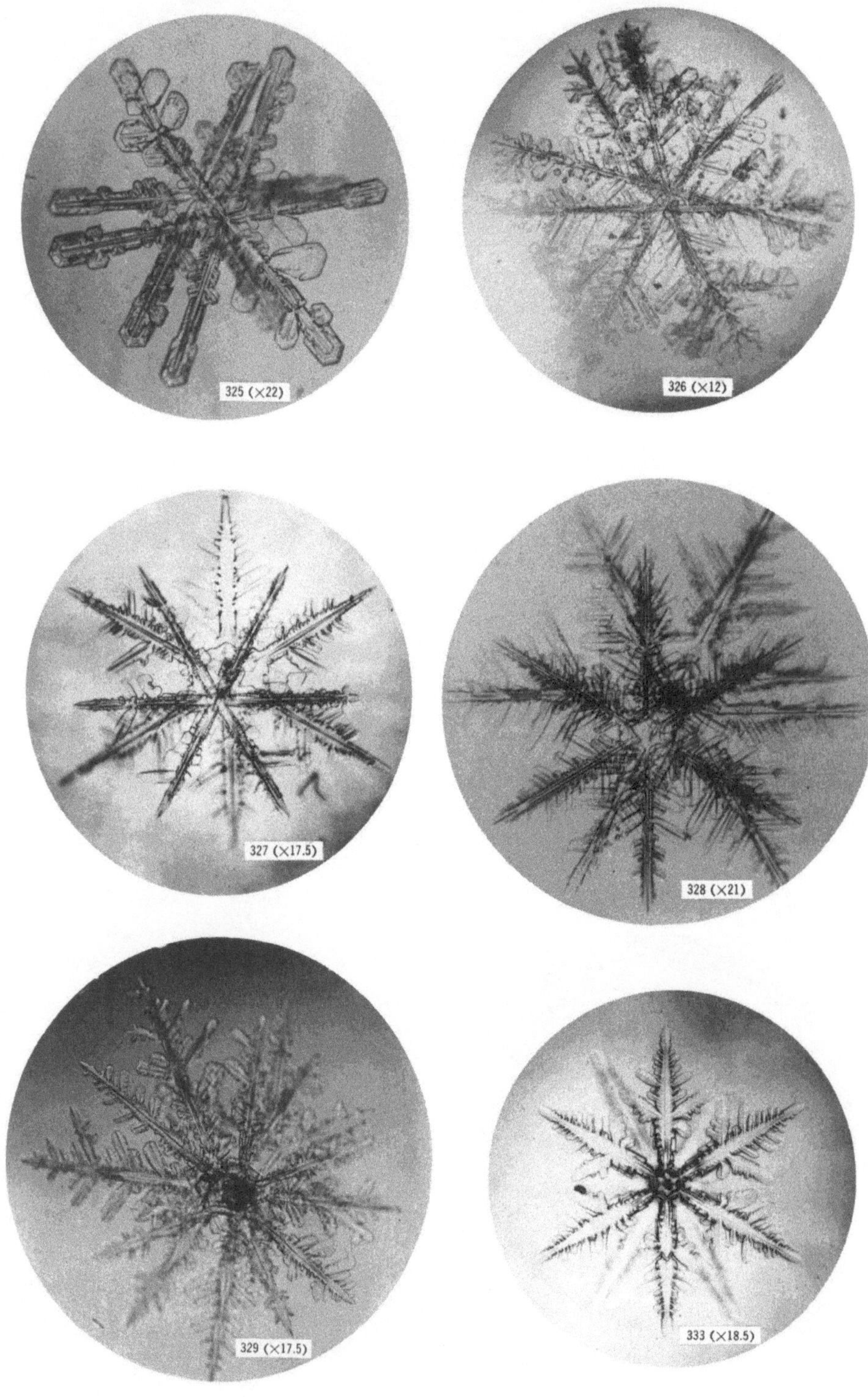
325 (×22)
326 (×12)
327 (×17.5)
328 (×21)
329 (×17.5)
333 (×18.5)

336 (×35)

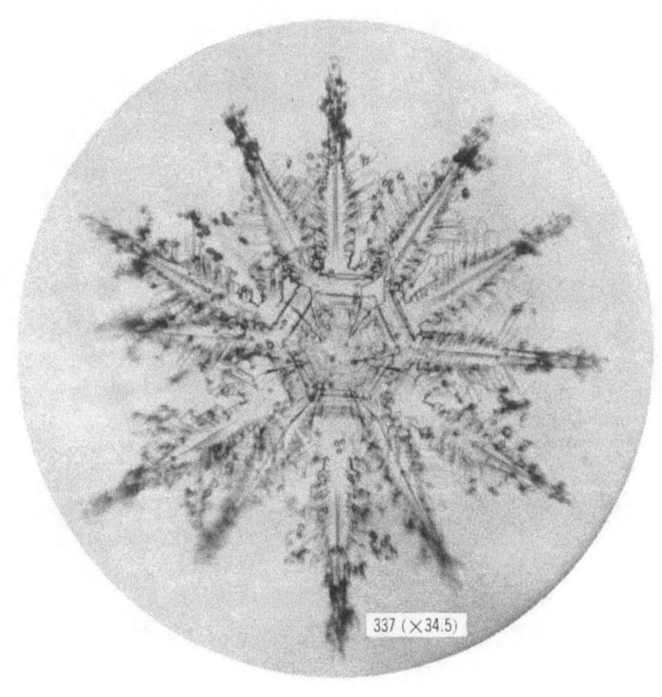

337 (×34.5)

339 (×37)

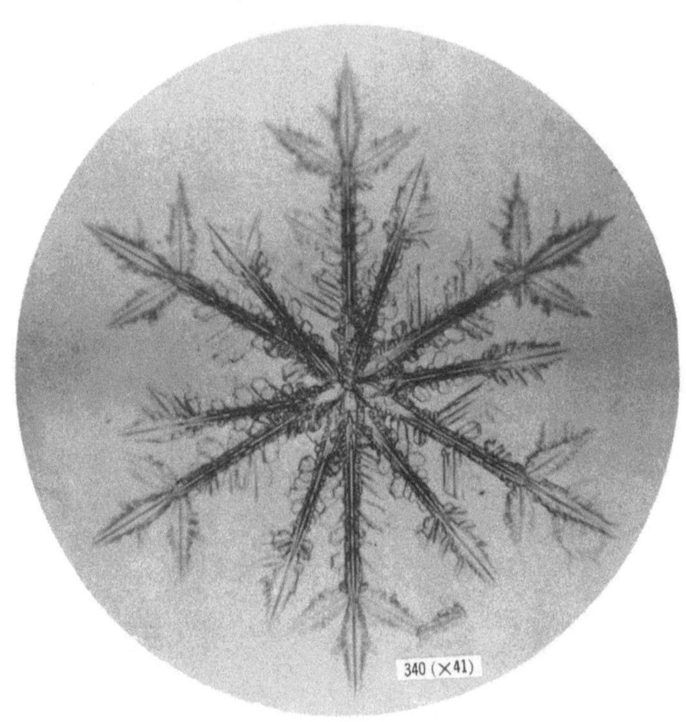
340 (×41)

341 (×21)
342 (×19)
343 (×20)
334 (×15)
345
346 (×21)

PLATE 66

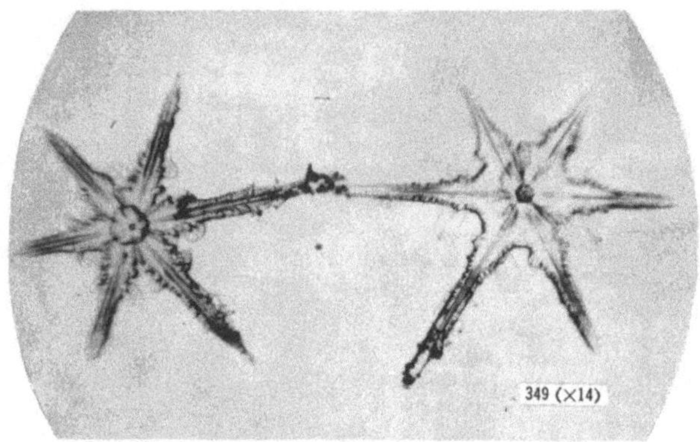

349 (×14)

352 (×27)

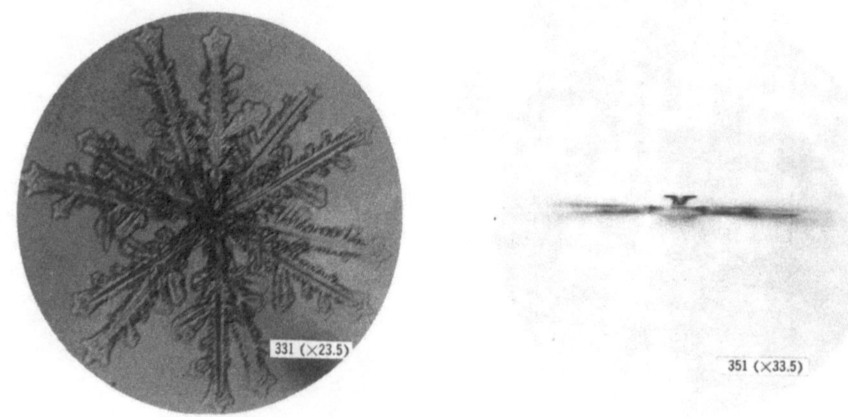

331 (×23.5)

351 (×33.5)

355 (×21.5)

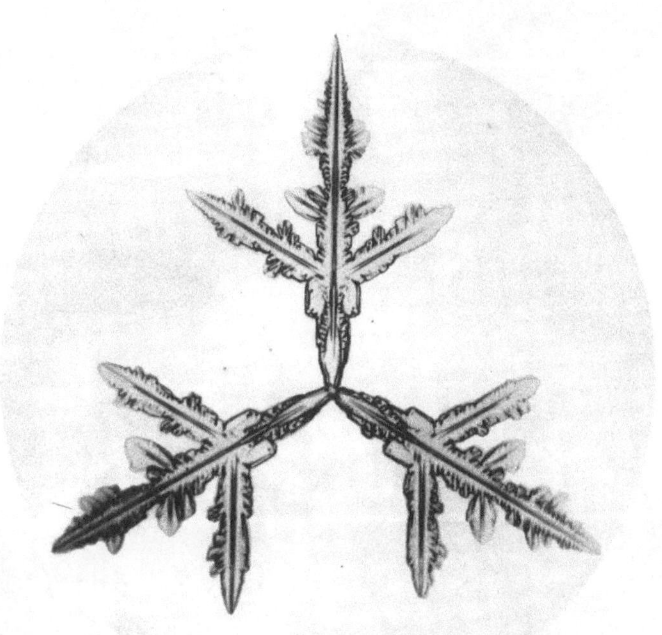

358 (×28.5)

PLATE 68

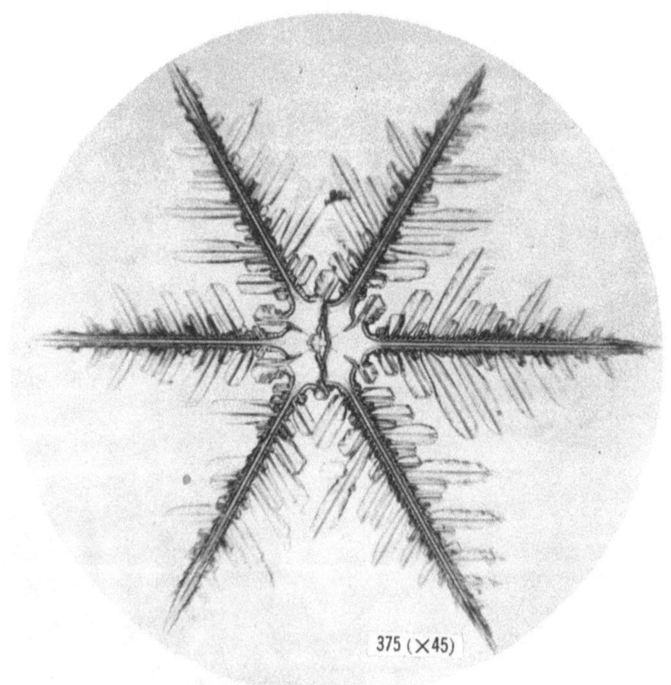

375 (×45)

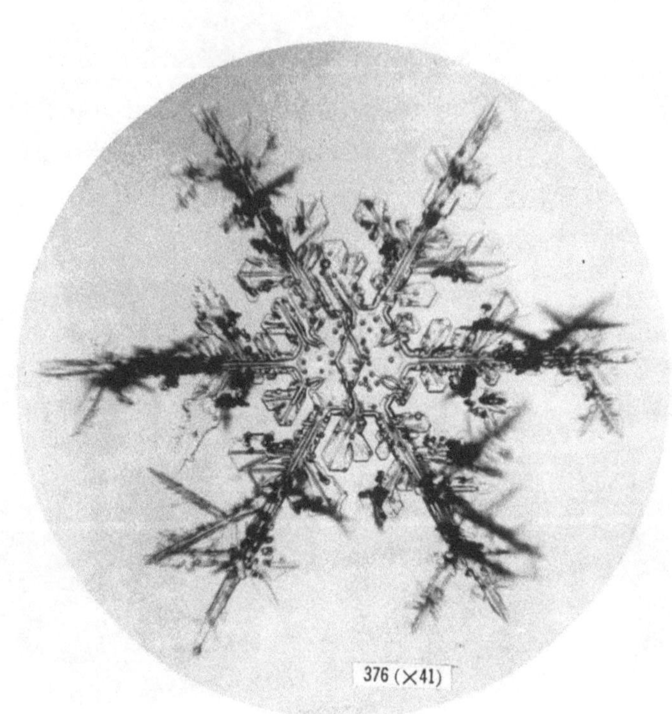

376 (×41)

PLATE 70

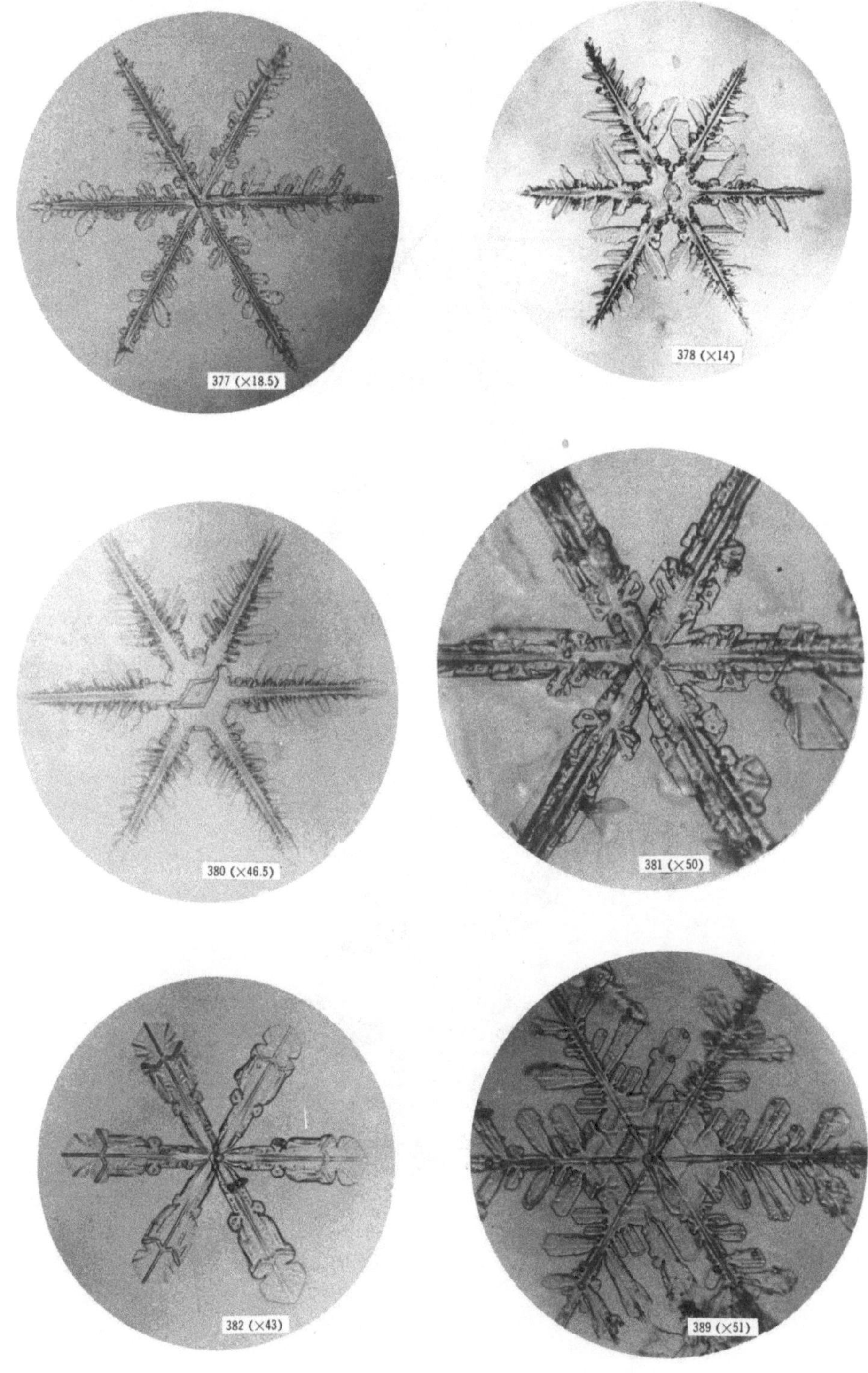

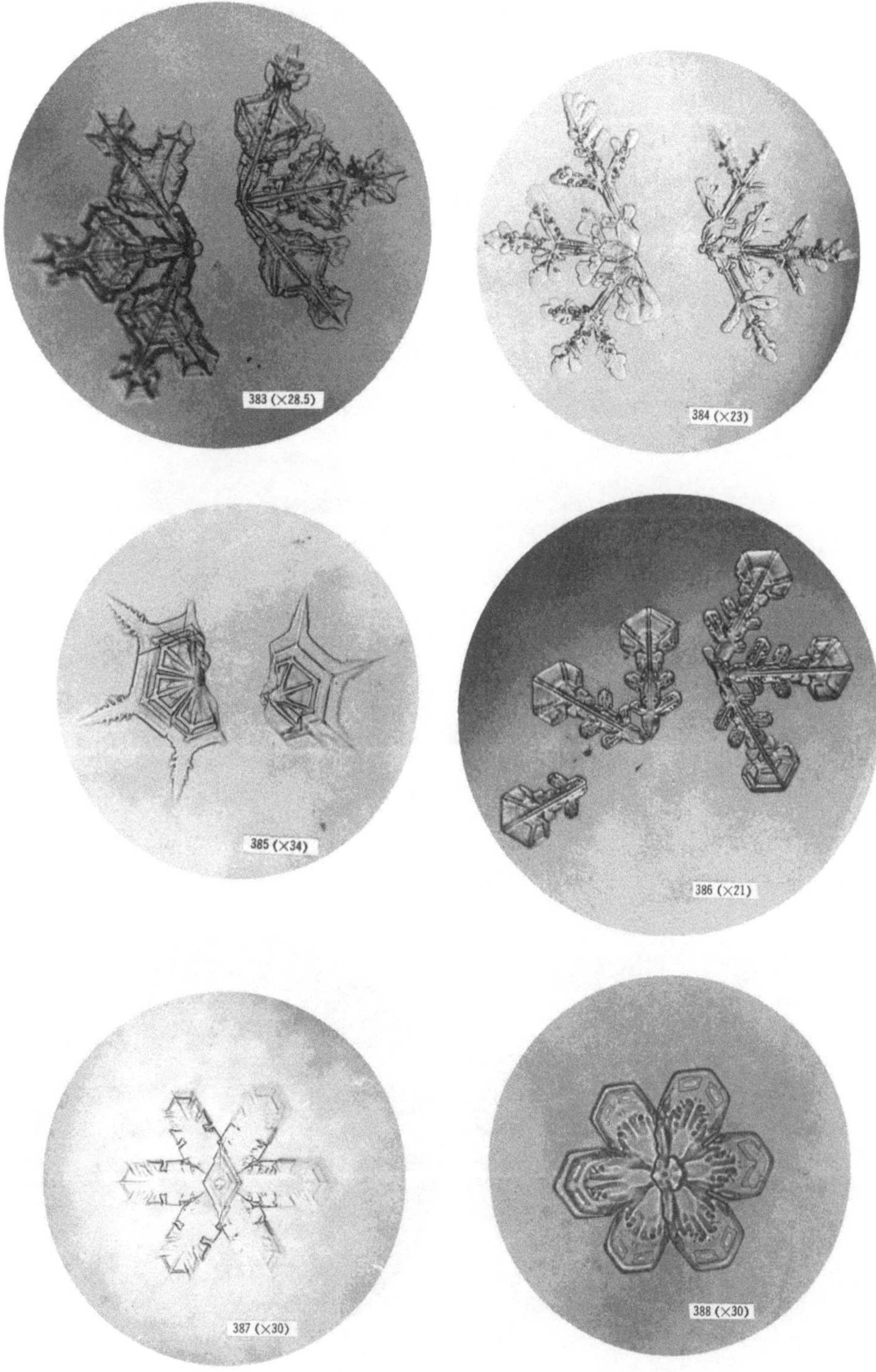
383 (×28.5)
384 (×23)
385 (×34)
386 (×21)
387 (×30)
388 (×30)

PLATE 72

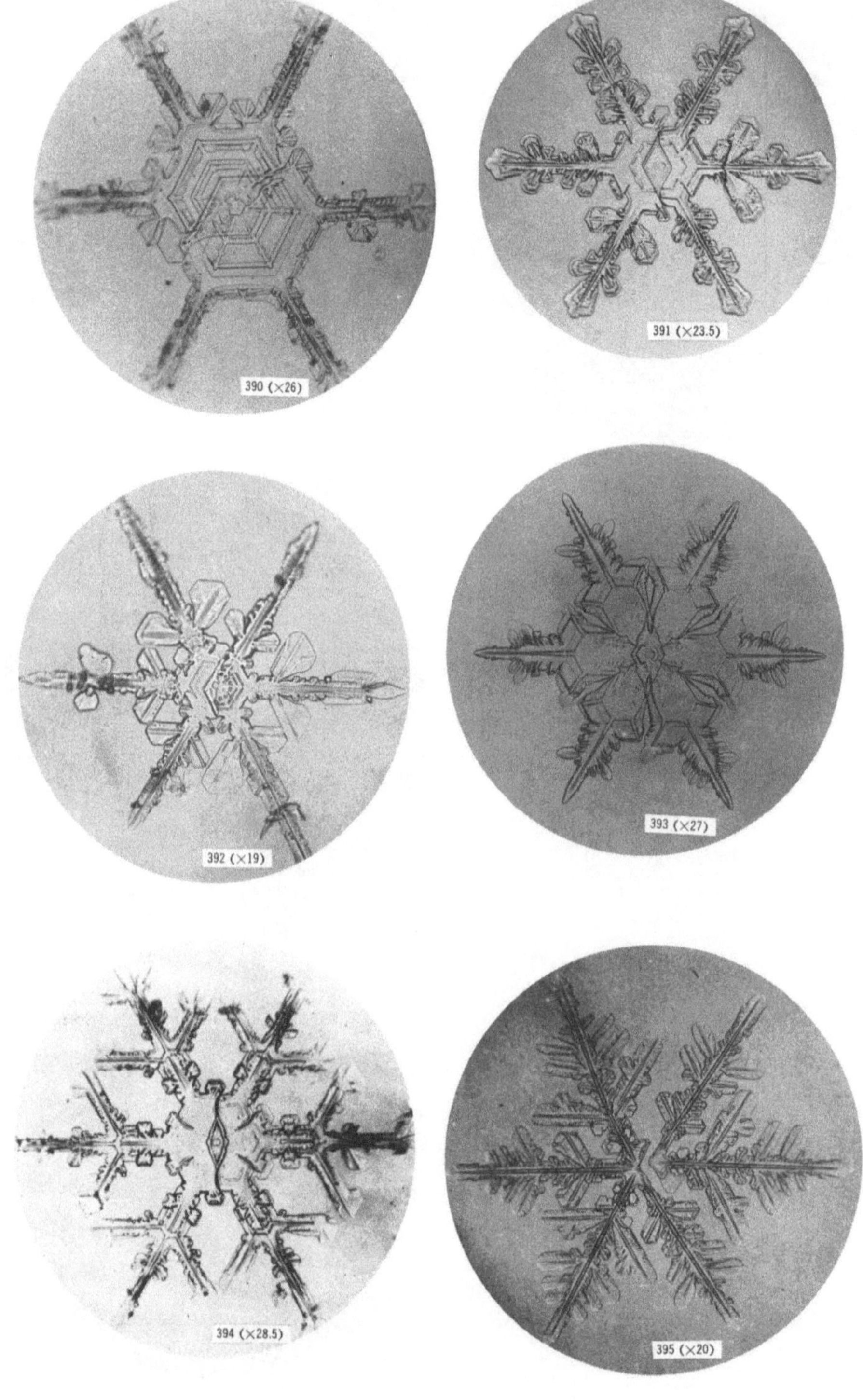

397 (×20)
398 (×17.5)
399 (×25)
400 (×31)
396 (×22)
408 (×41)

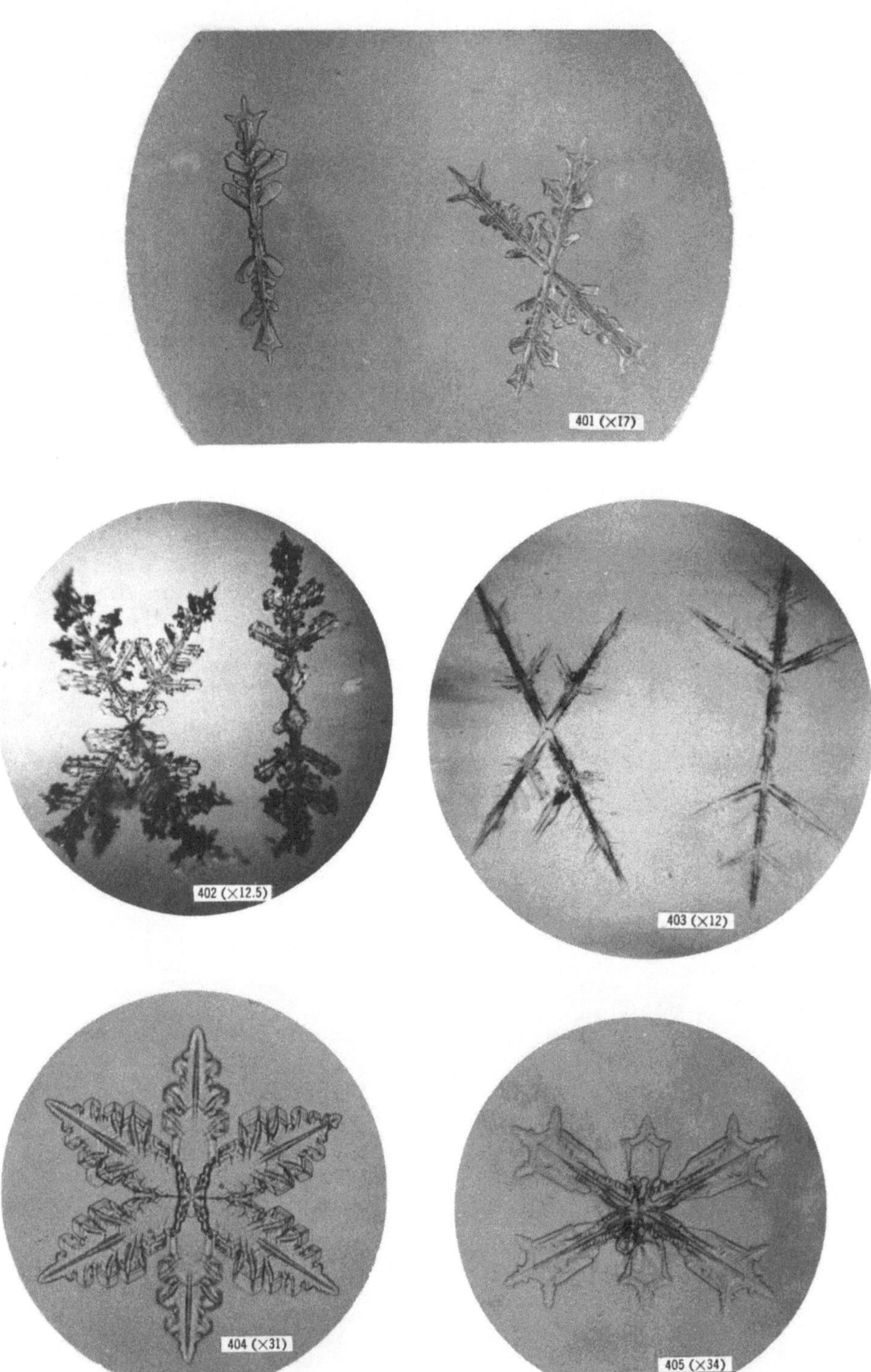
401 (×17)
402 (×12.5)
403 (×12)
404 (×31)
405 (×34)

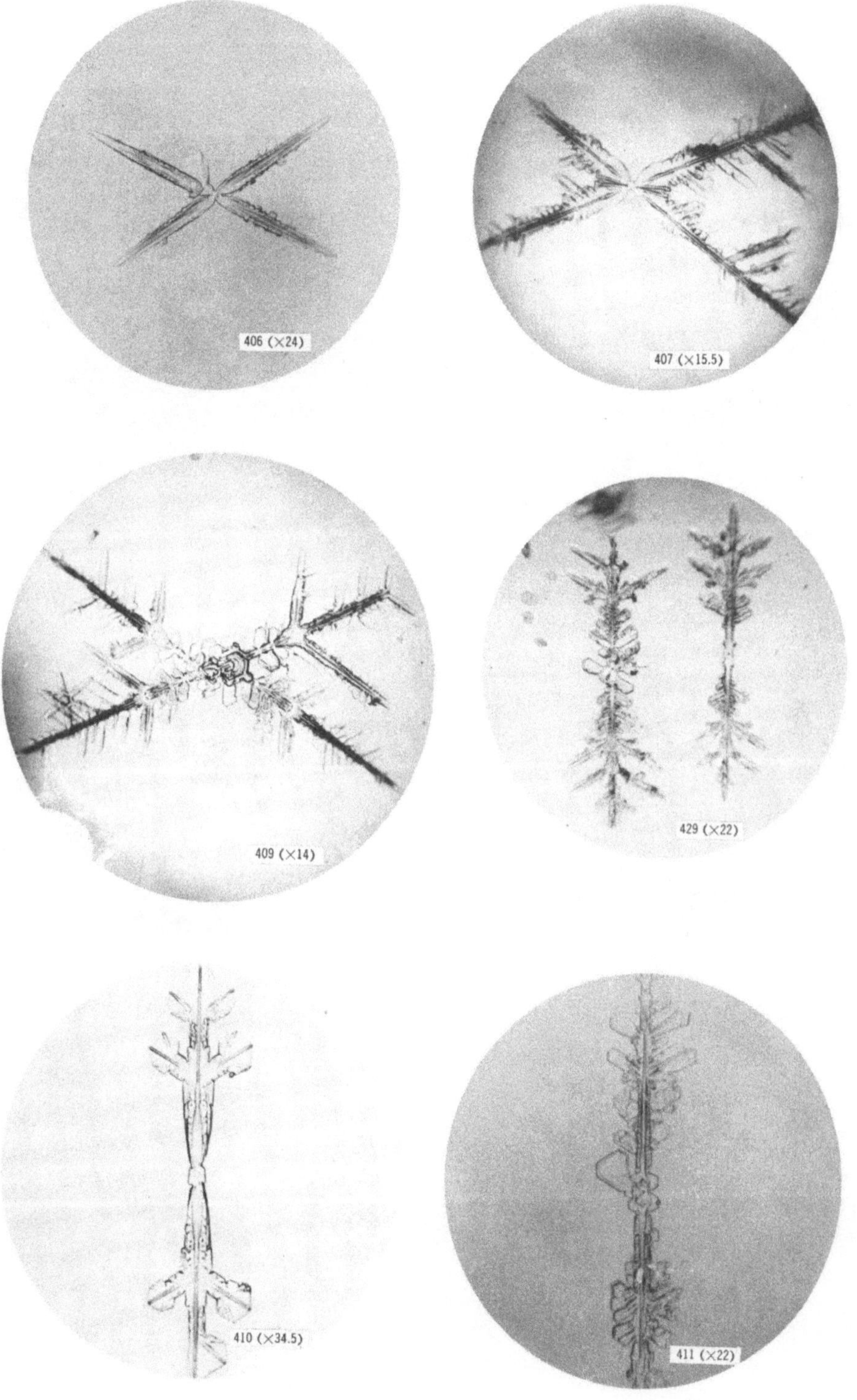
406 (×24)
407 (×15.5)
409 (×14)
429 (×22)
410 (×34.5)
411 (×22)

PLATE 76

418 (×18.5)
419 (×16)
420 (×26)
422 (×21.5)
424 (×58.5)
428 (×36)

430 (×44)

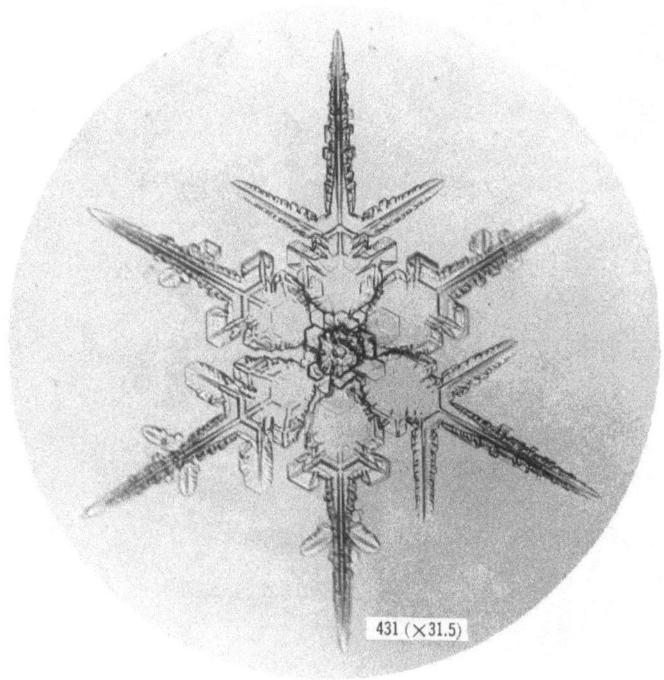
431 (×31.5)

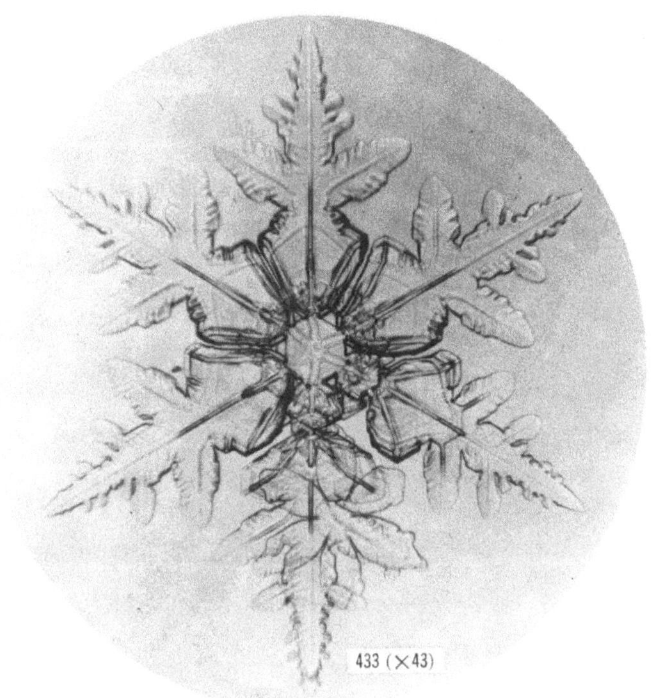

433 (×43)

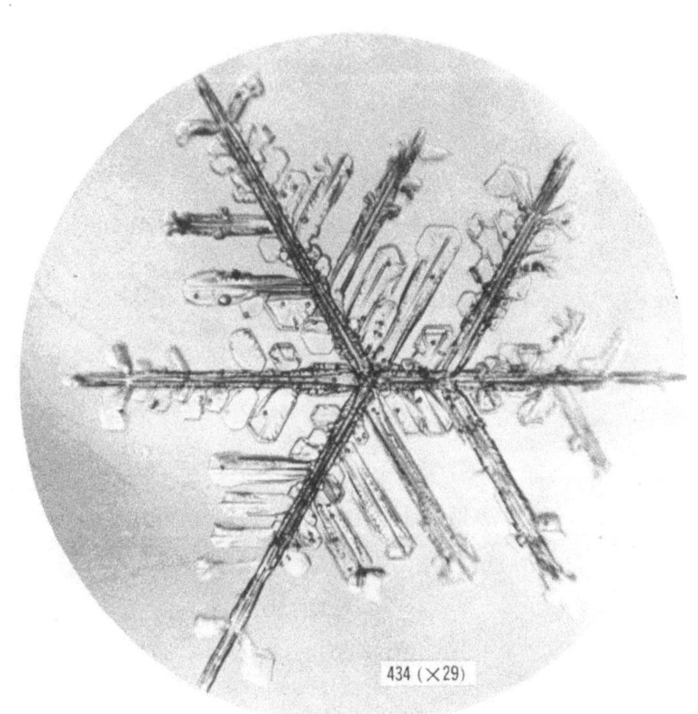

434 (×29)

PLATE 80

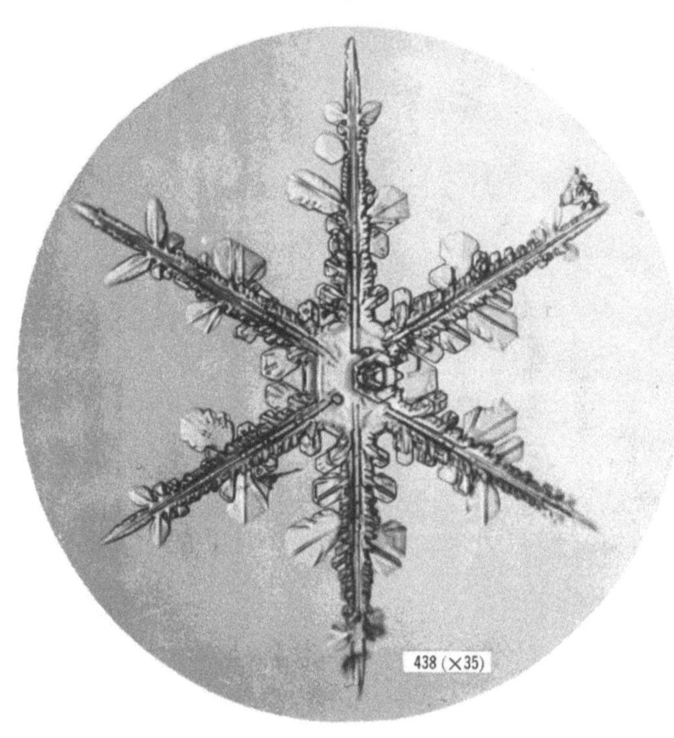

438 (×35)

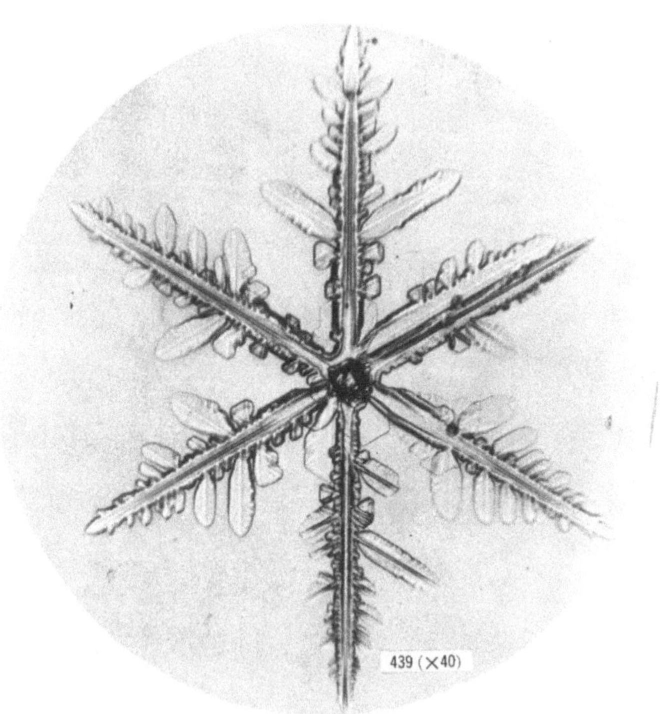

439 (×40)

440 (×25)
441 (×22)
442 (×26.5)
443 (×23)
444 (×32.5)
445 (×18.5)

PLATE 82

452 (×30)
453 (×34)
454 (×36)
455 (×19)
456 (×38)
457 (×59)

PLATE 84

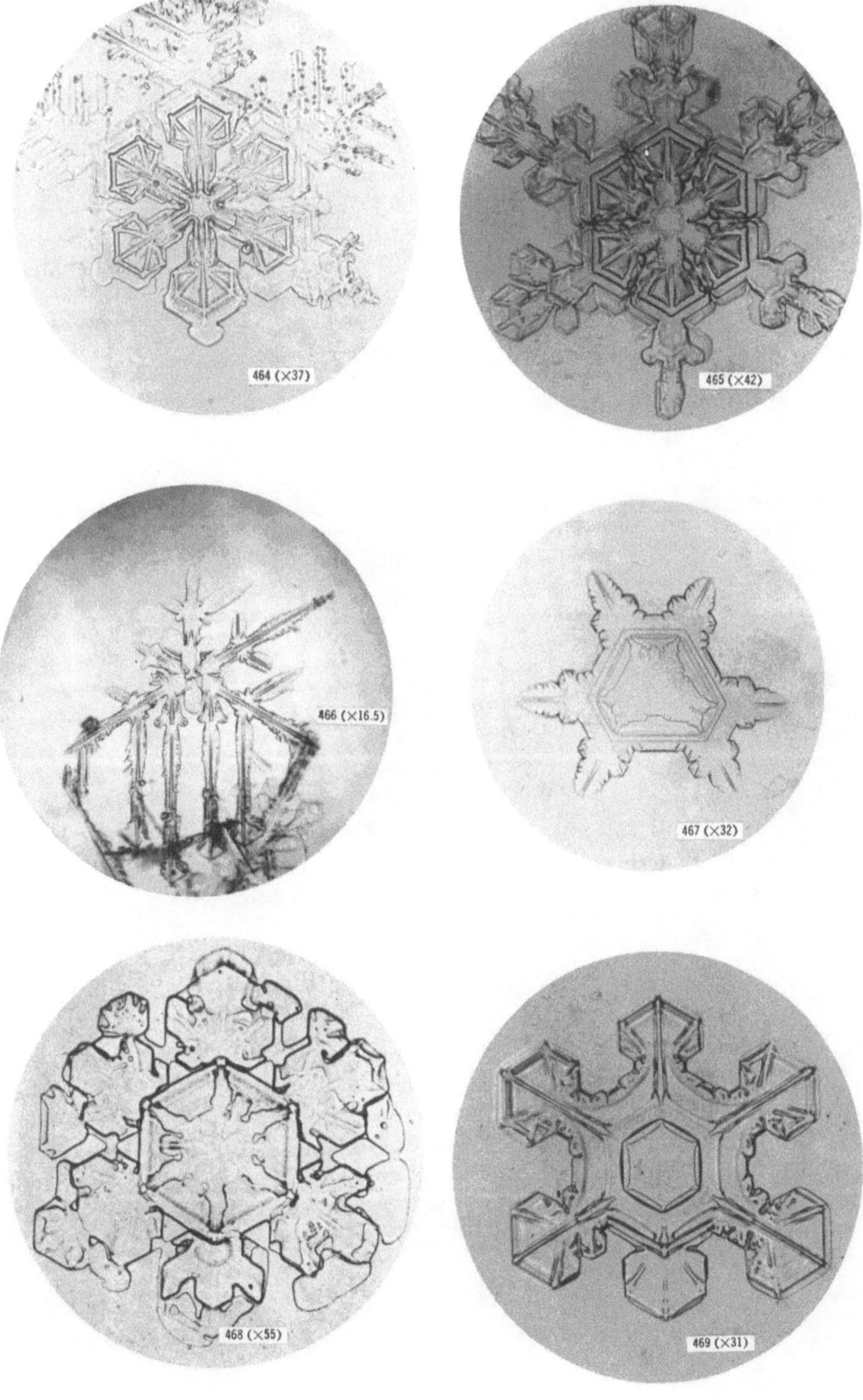
464 (×37)
465 (×42)
466 (×16.5)
467 (×32)
468 (×55)
469 (×31)

PLATE 86

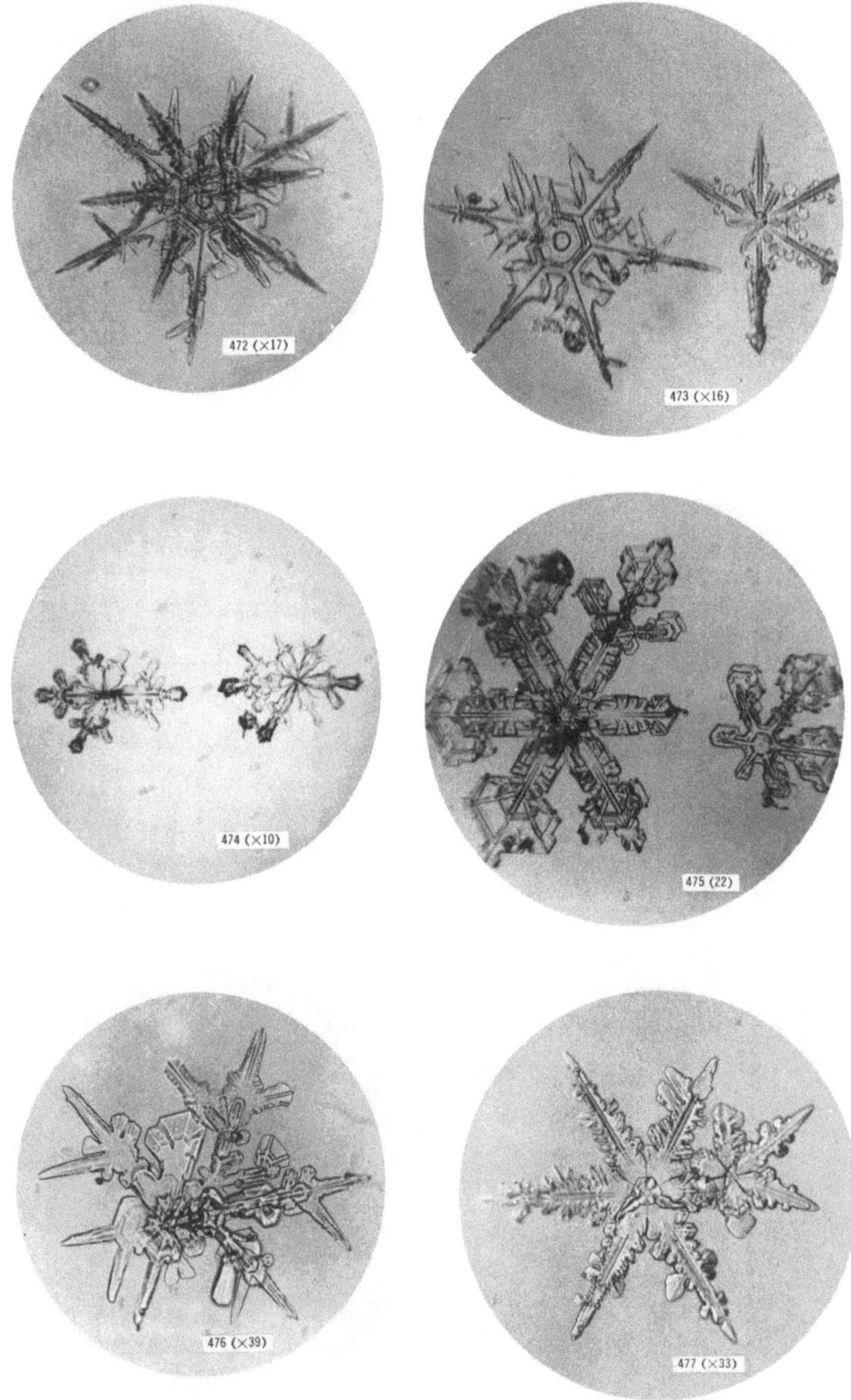

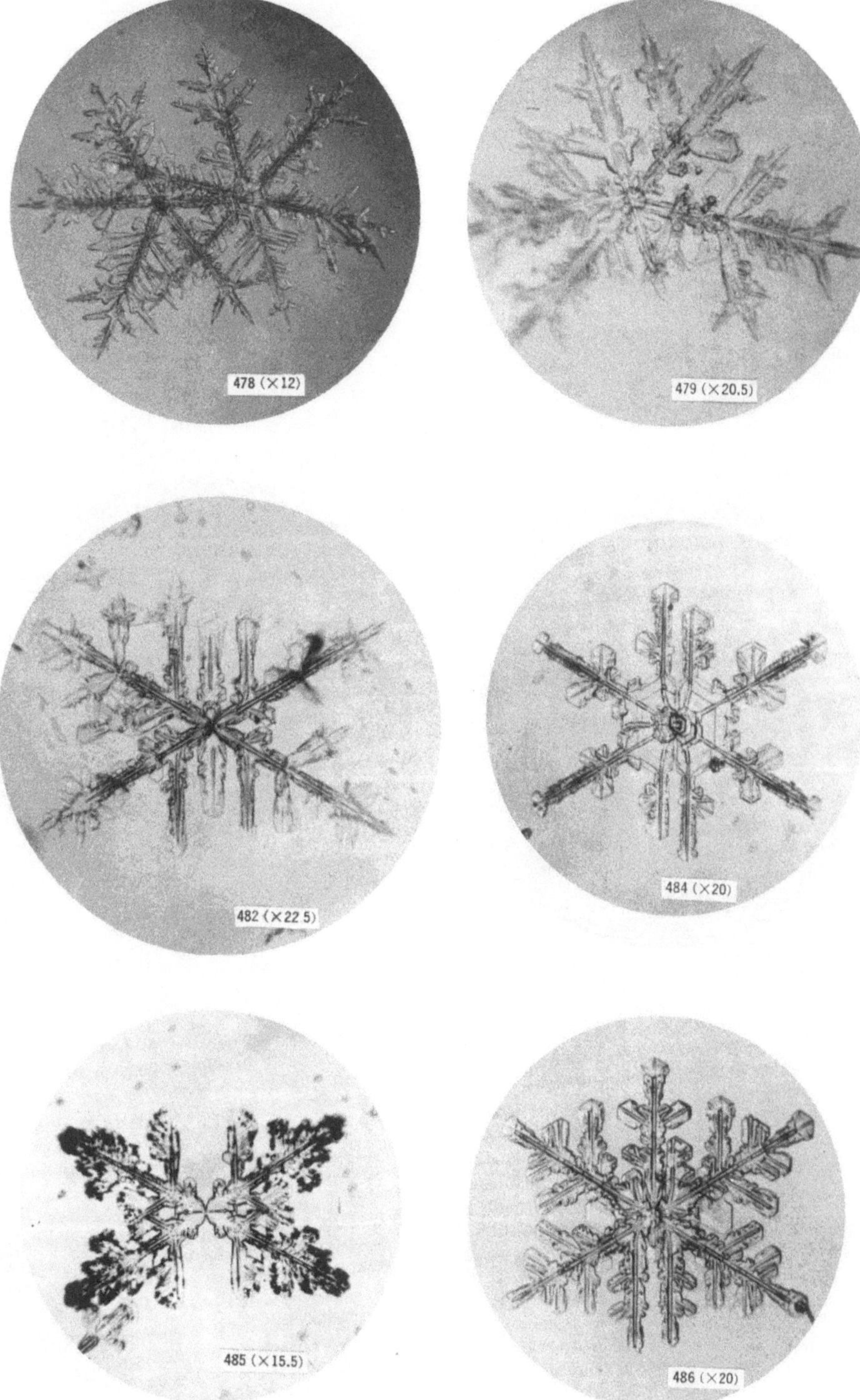
478 (×12)
479 (×20.5)
482 (×22.5)
484 (×20)
485 (×15.5)
486 (×20)

PLATE 88

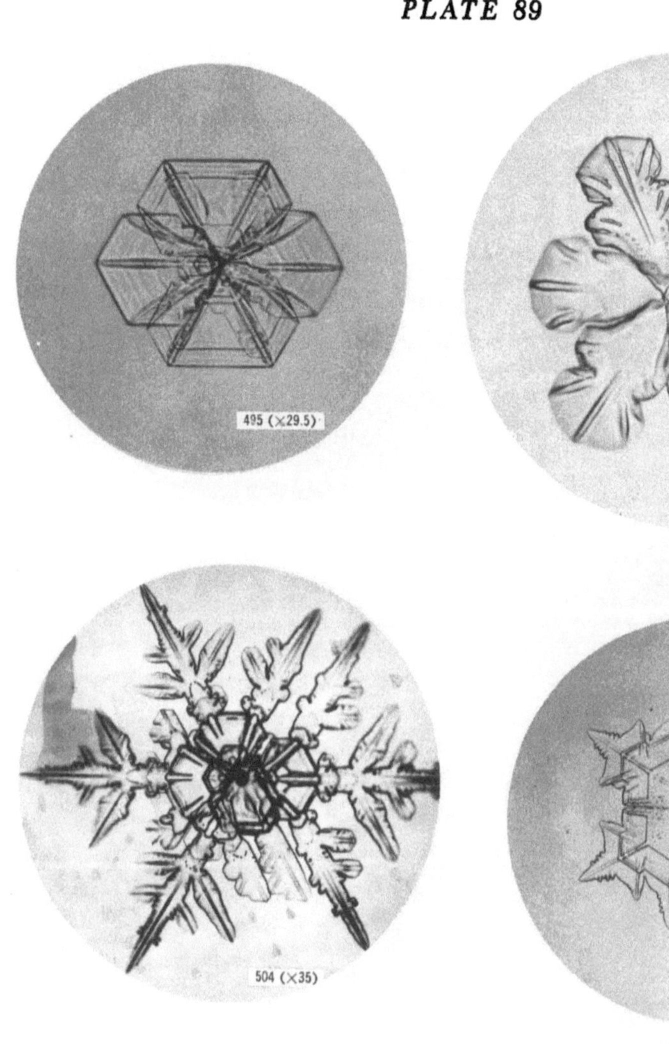
495 (×29.5)
496 (×56)
504 (×35)
497 (×26)

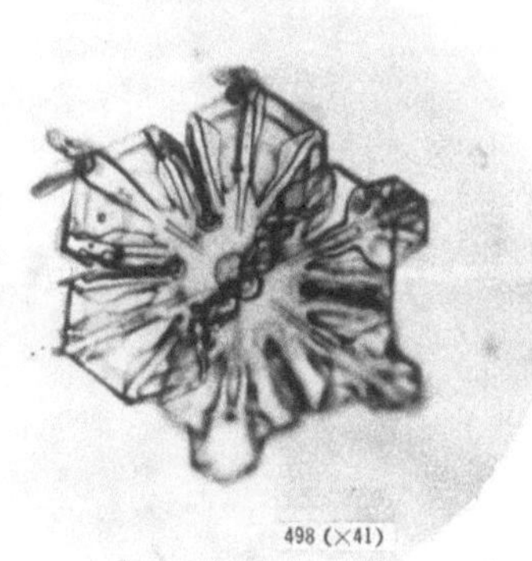
498 (×41)

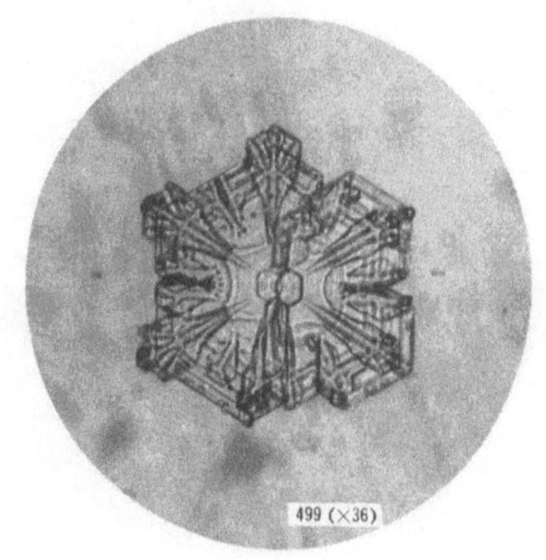
499 (×36)

PLATE 90

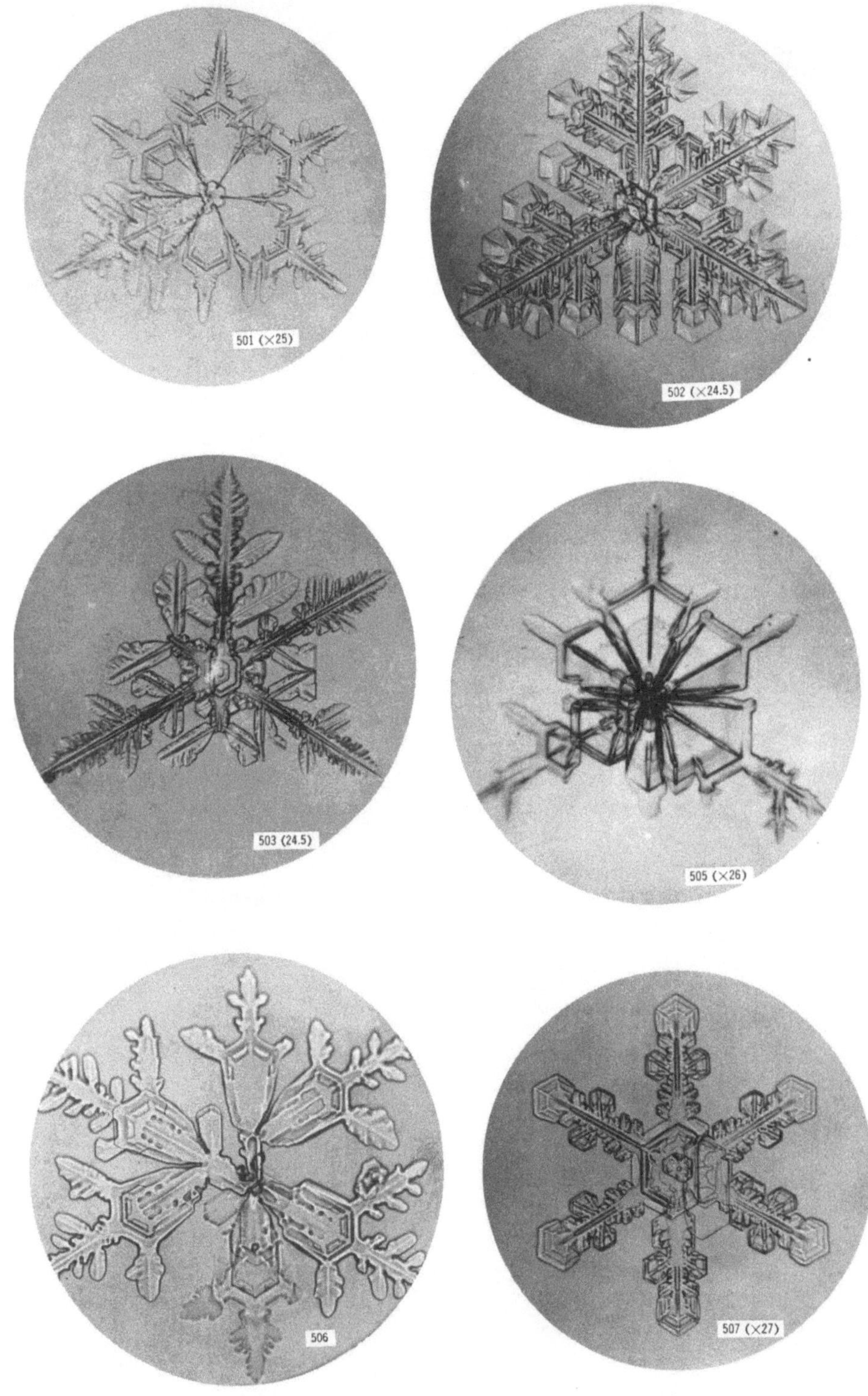

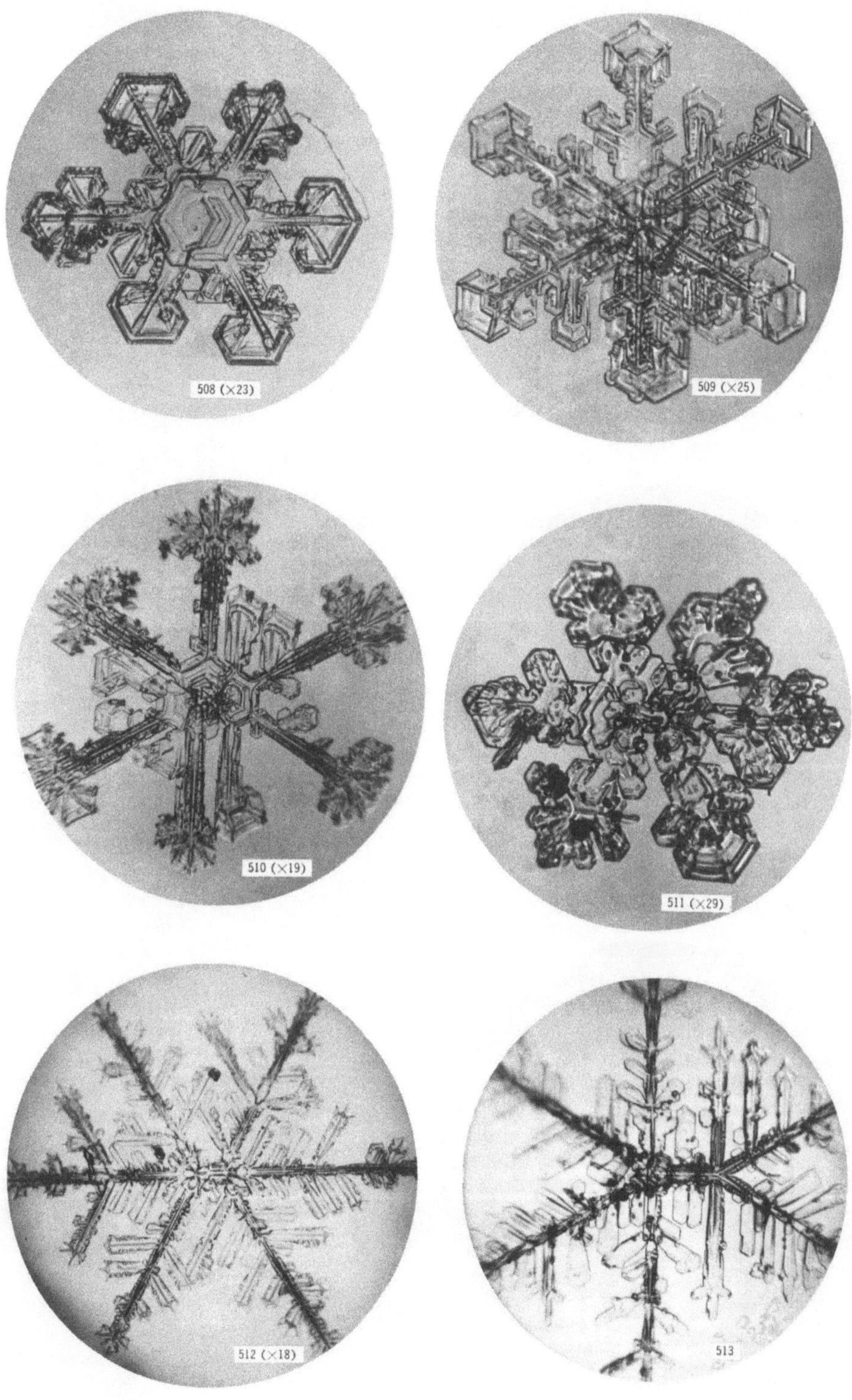
508 (×23)
509 (×25)
510 (×19)
511 (×29)
512 (×18)
513

PLATE 92

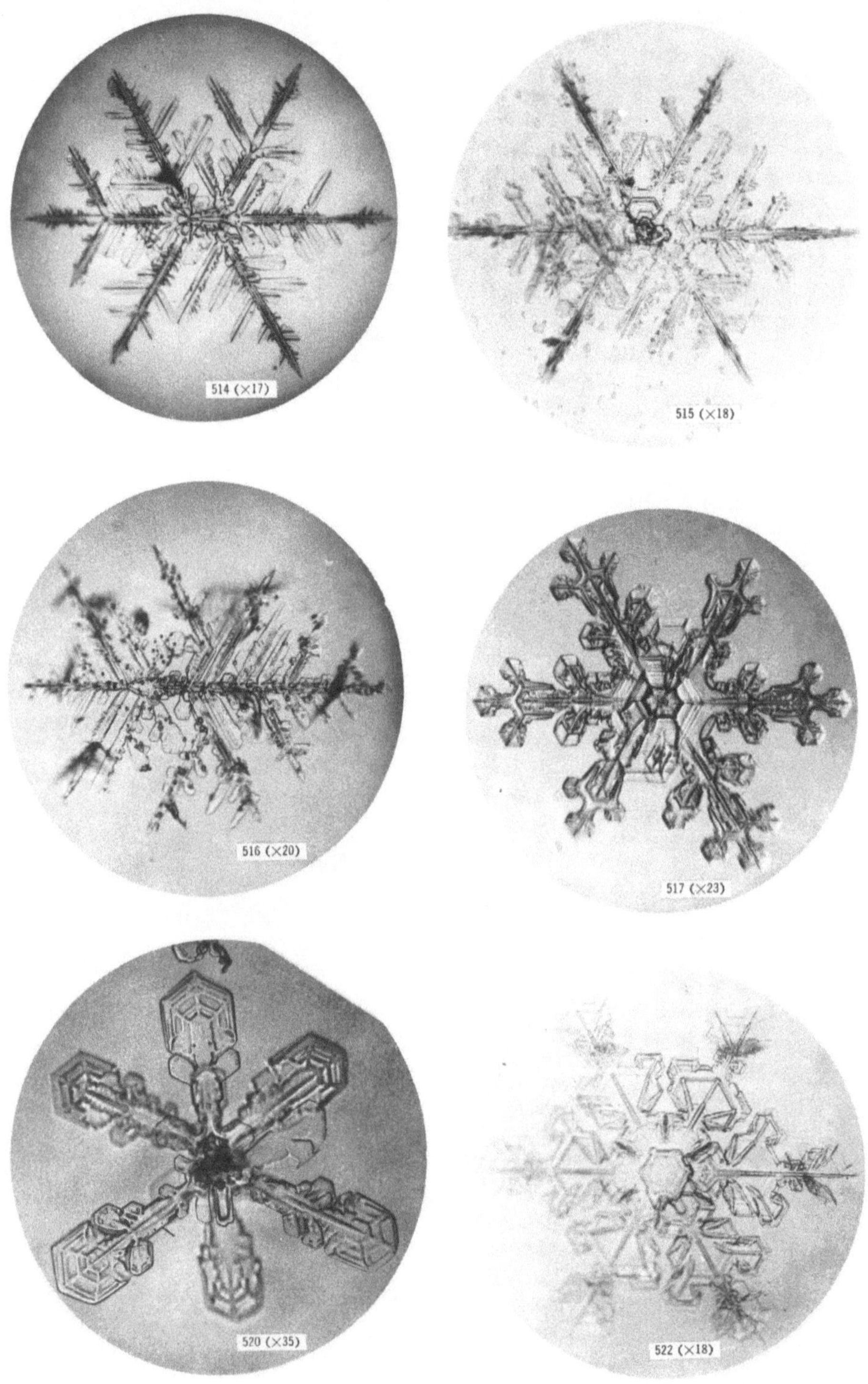

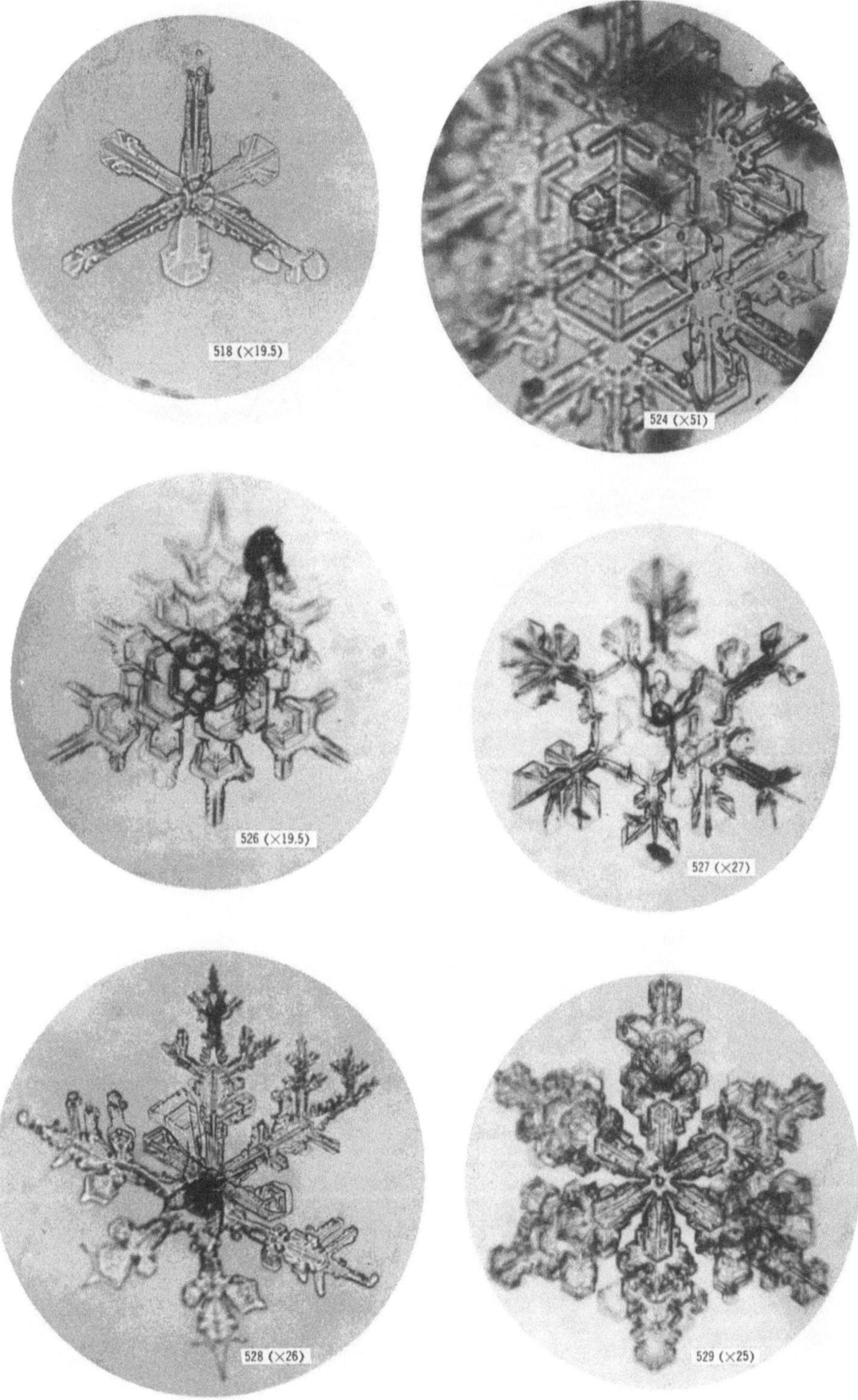
518 (×19.5)
524 (×51)
526 (×19.5)
527 (×27)
528 (×26)
529 (×25)

PLATE 94

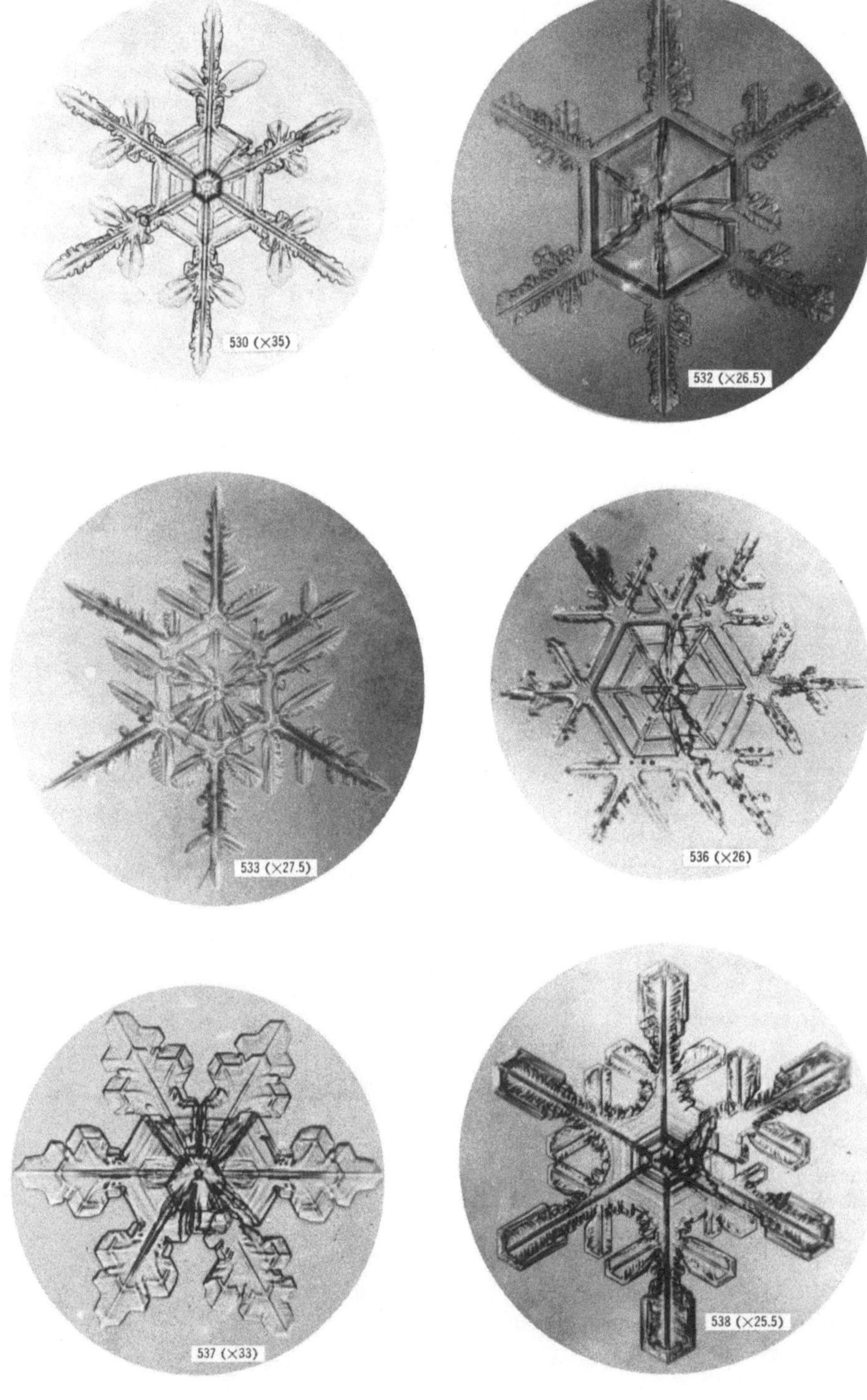

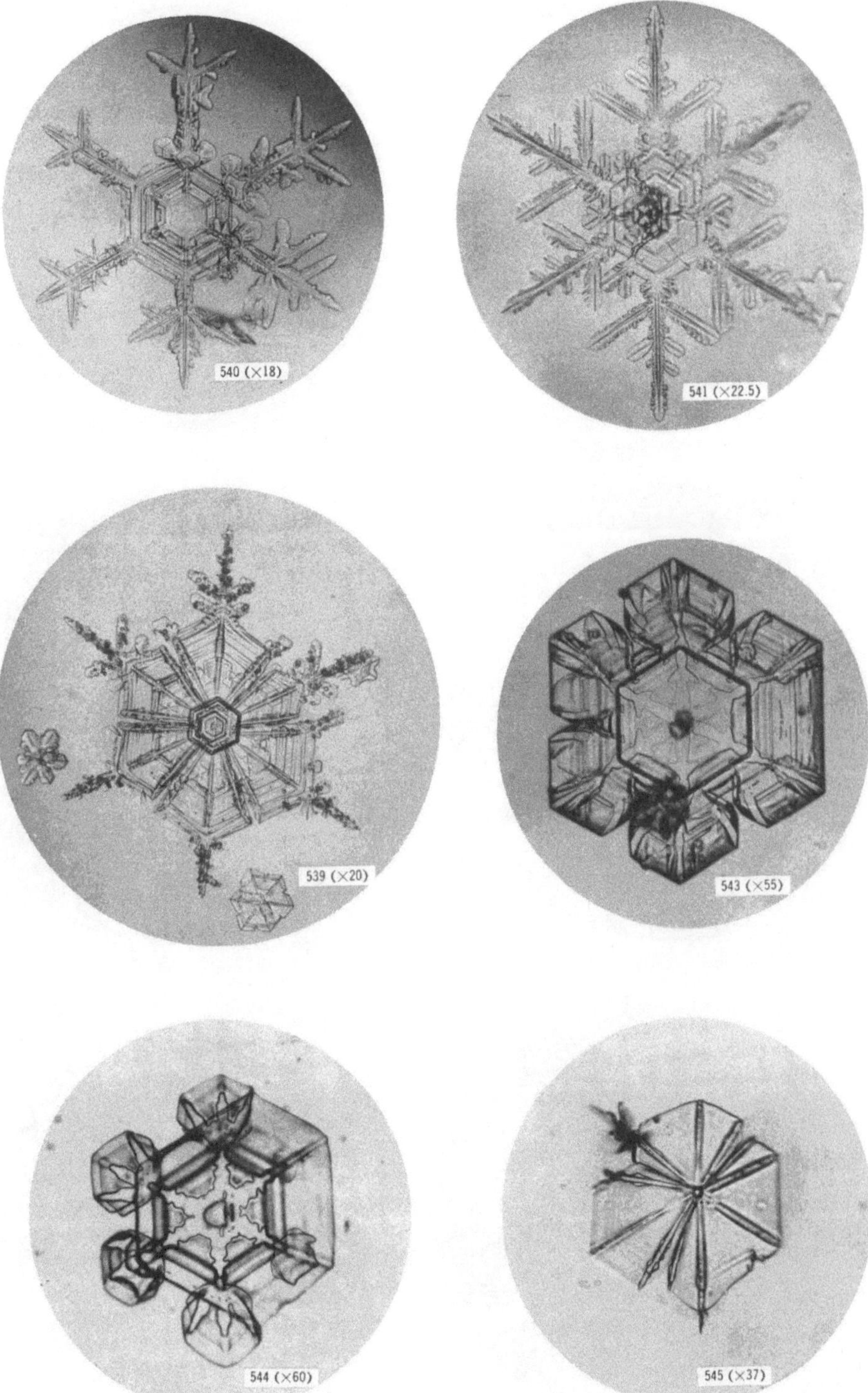
540 (×18)
541 (×22.5)
539 (×20)
543 (×55)
544 (×60)
545 (×37)

PLATE 96

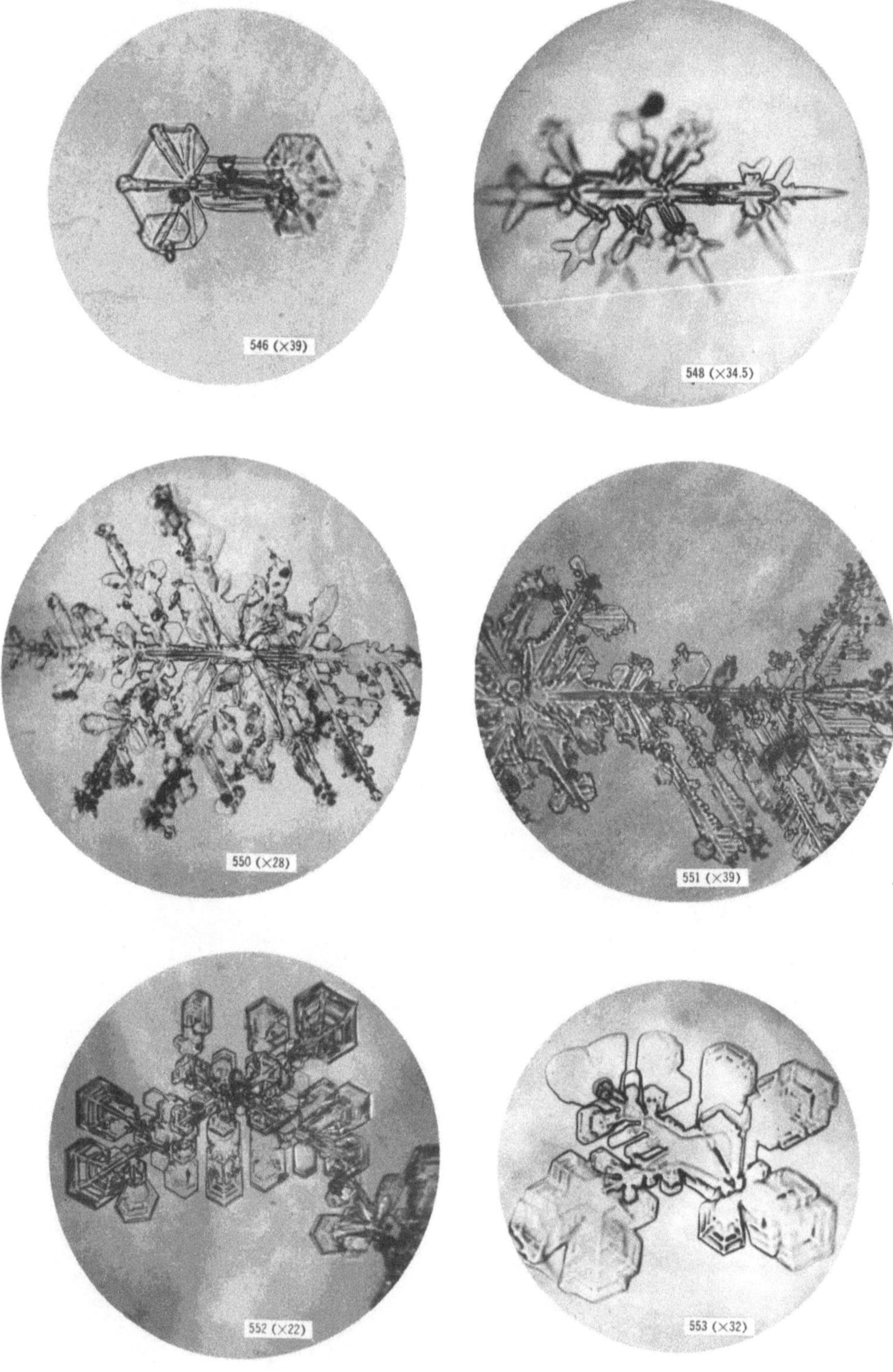

546 (×39)

548 (×34.5)

550 (×28)

551 (×39)

552 (×22)

553 (×32)

554 (×59.5)

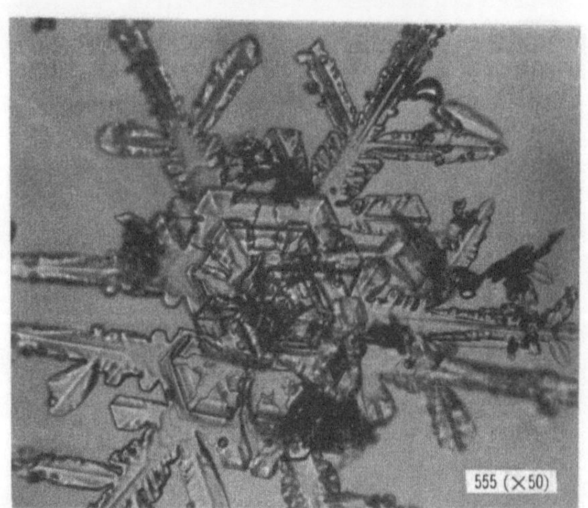
555 (×50)

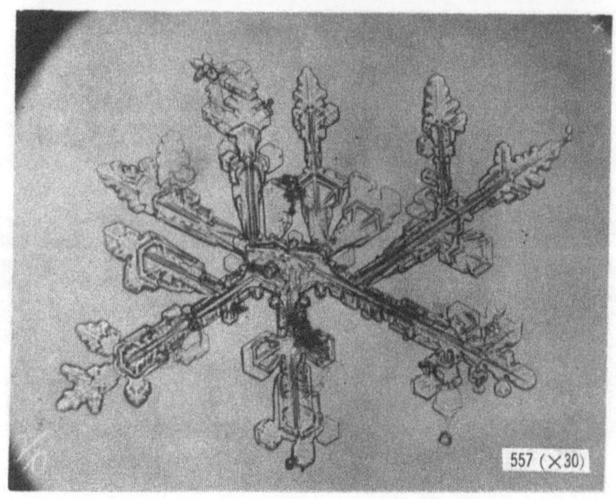
557 (×30)

PLATE 98

573 (×18)
574 (×18)
575 (×18)
576 (×18)
577 (×17.5)
577 (×17.5)

PLATE 100

585 (×36.5)
589 (×32)
590 (×17)
591 (×63.5)
592 (×33.5)
593 (×10)

PLATE 102

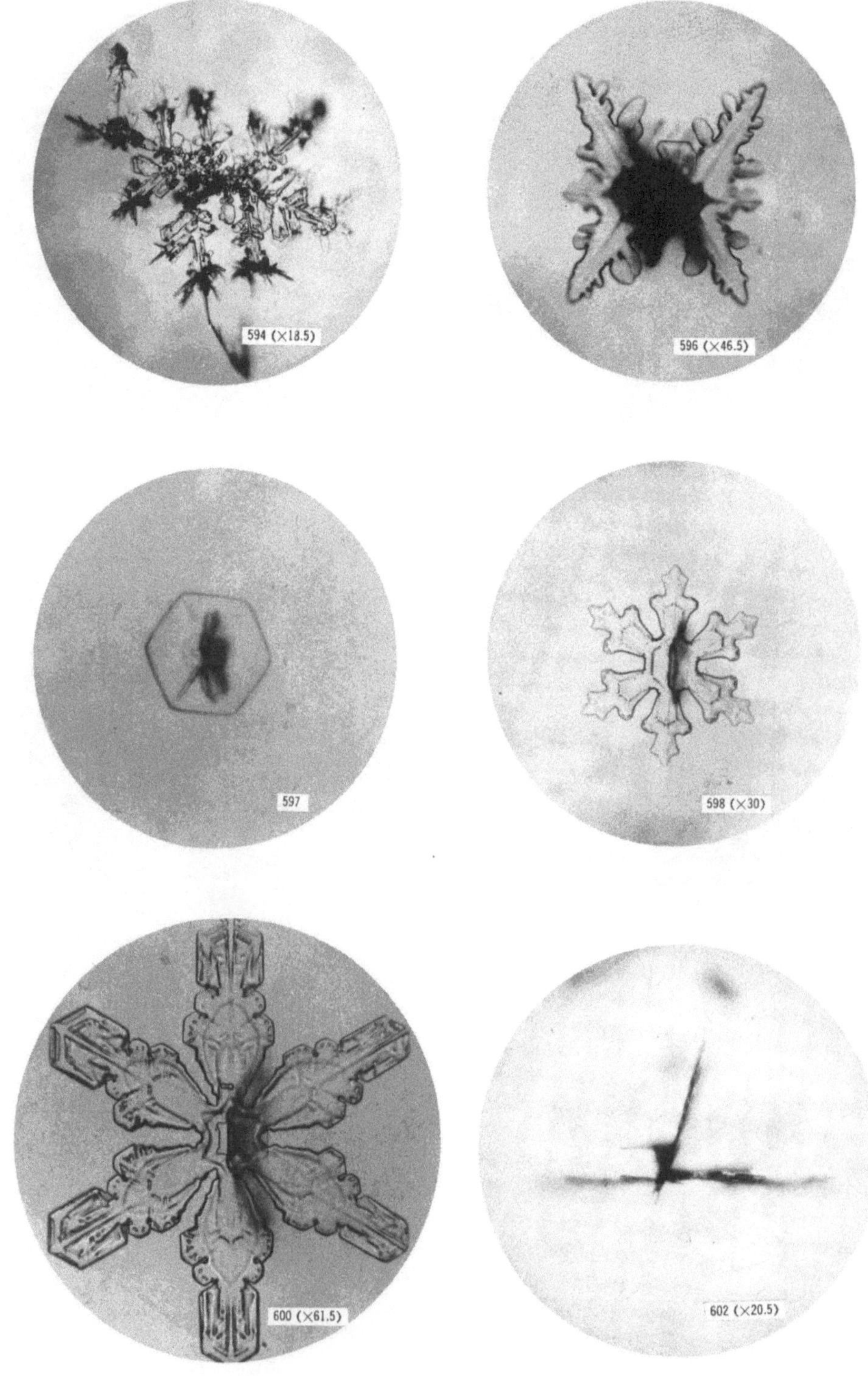

603 (×16)
604 (×18)
605 (×18.5)
606 (×18.5)
611 (×18)
612 (×29)

PLATE 104

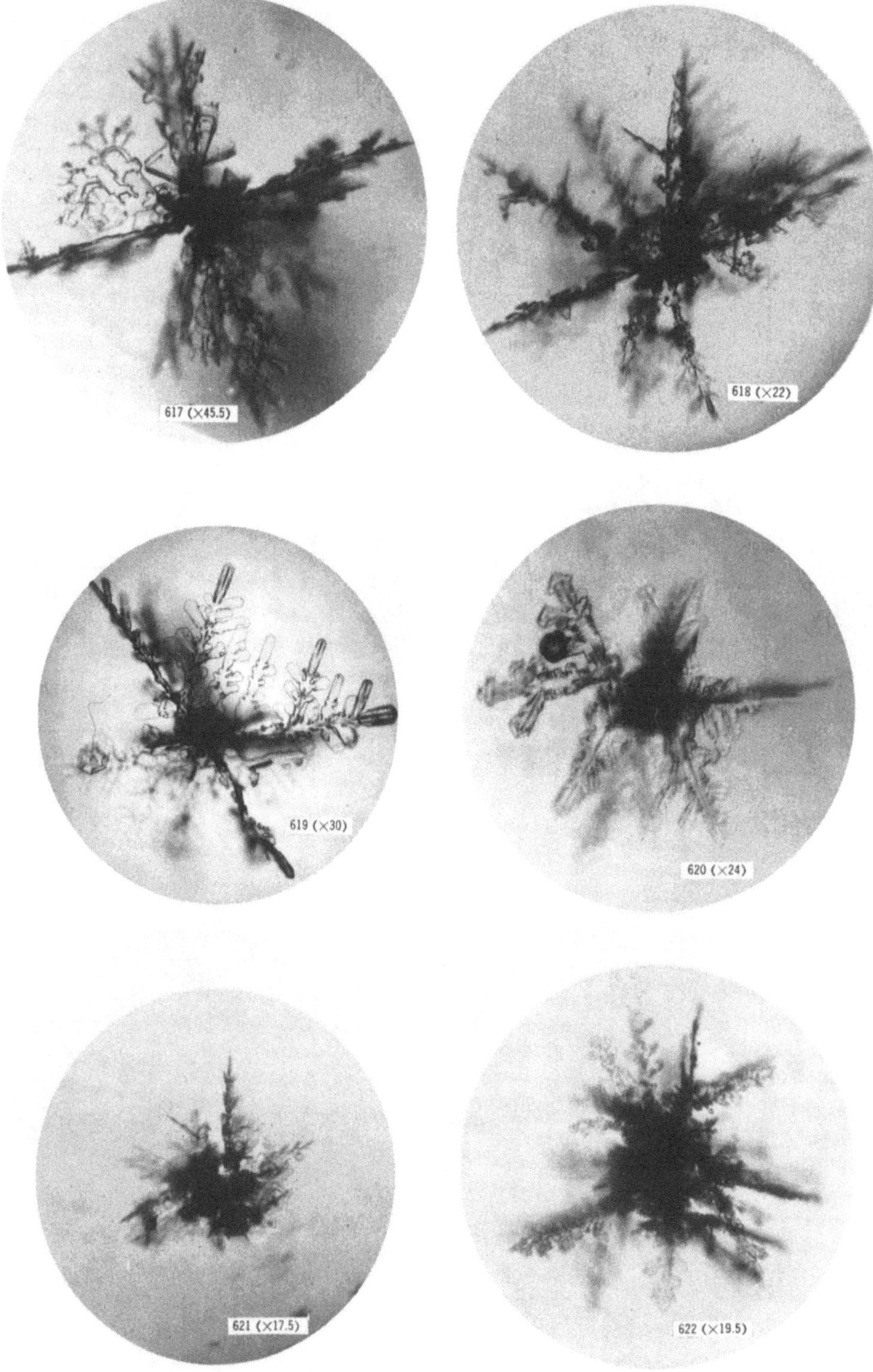

623 (×17.5)
624 (×42)
625 (×11)
626 (×29.5)
627 (×10.5)
628 (×19.5)

PLATE 106

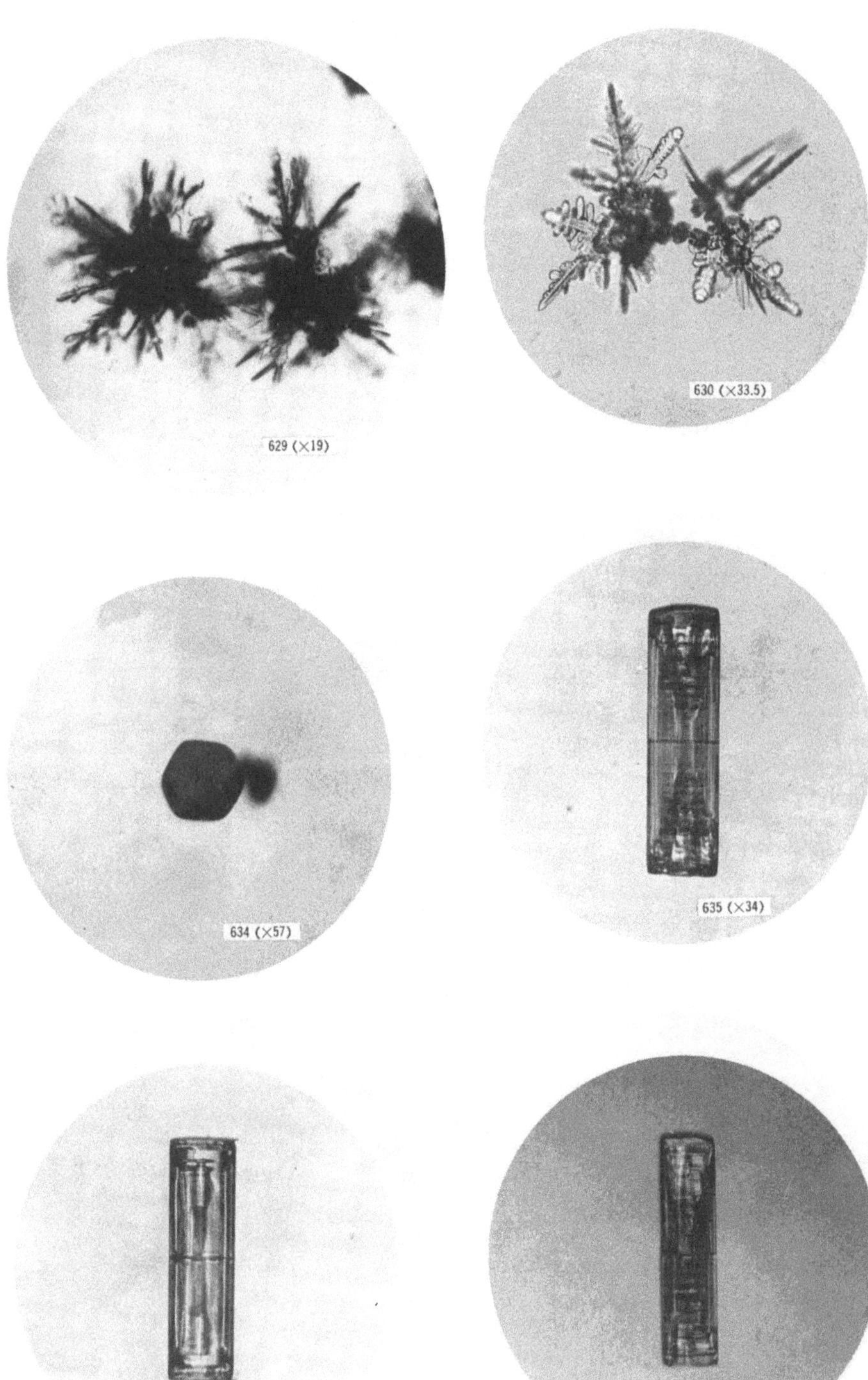

629 (×19)

630 (×33.5)

634 (×57)

635 (×34)

636 (×39)

650 (×33)

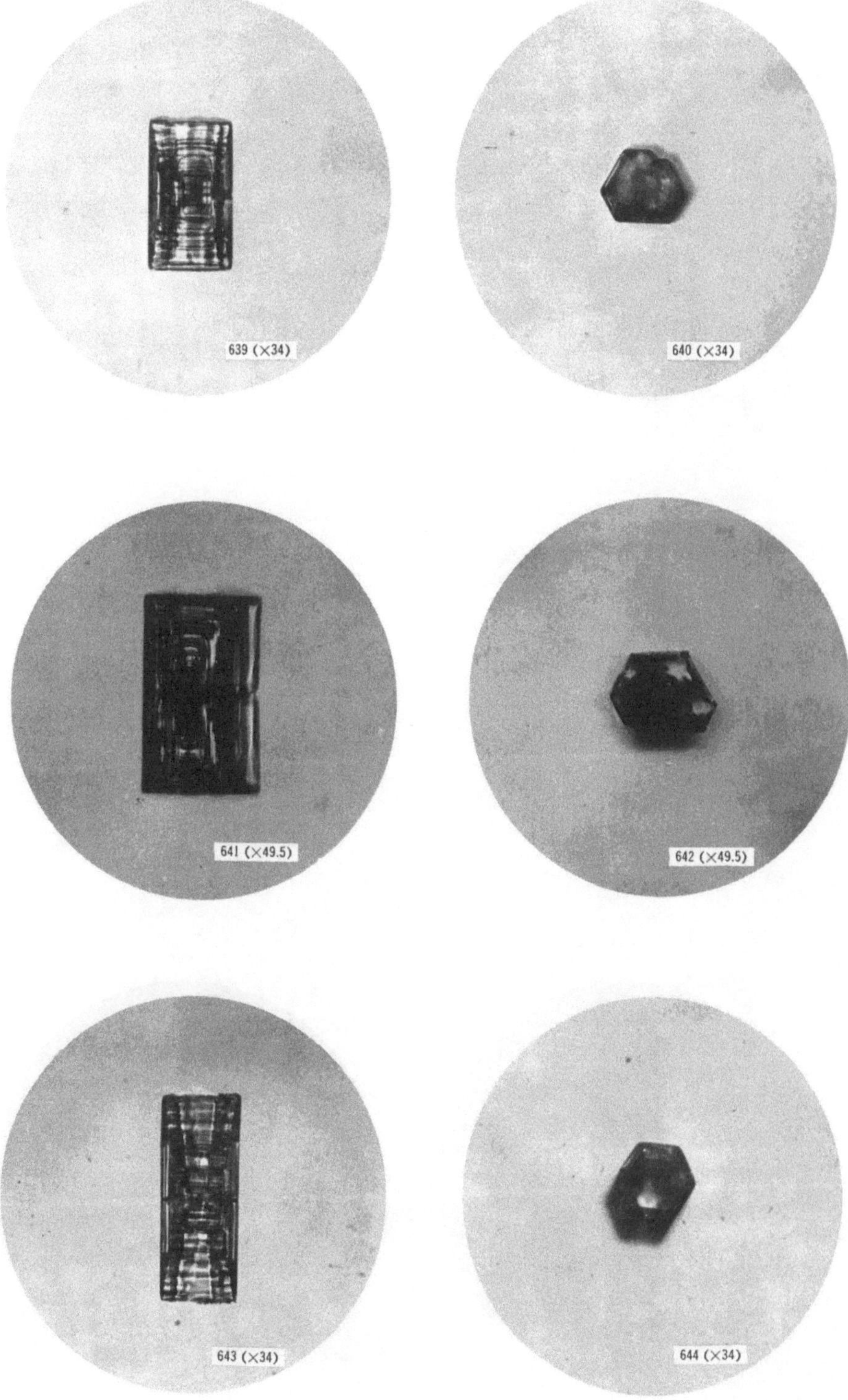
639 (×34)
640 (×34)
641 (×49.5)
642 (×49.5)
643 (×34)
644 (×34)

PLATE 108

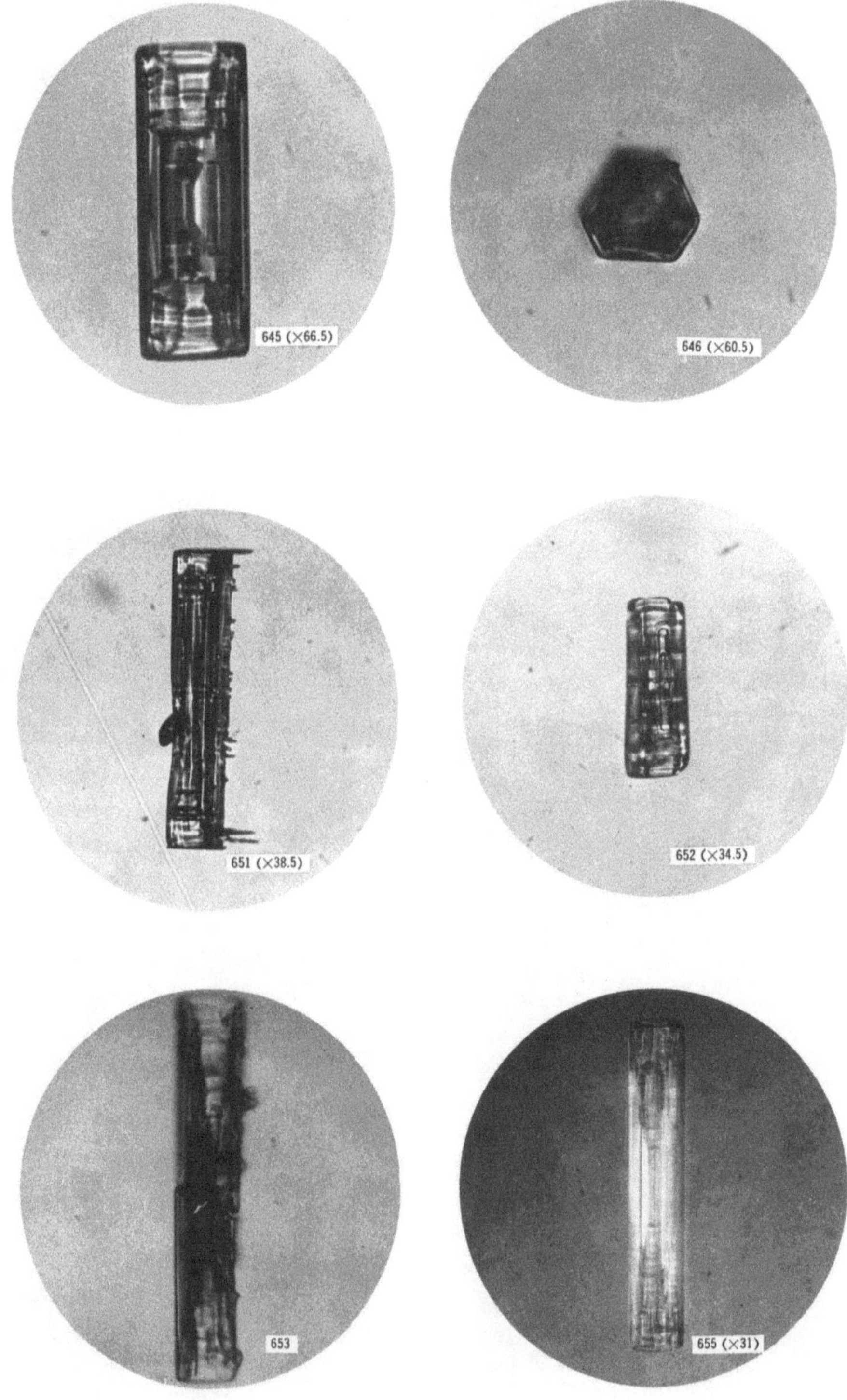

656 (×92)

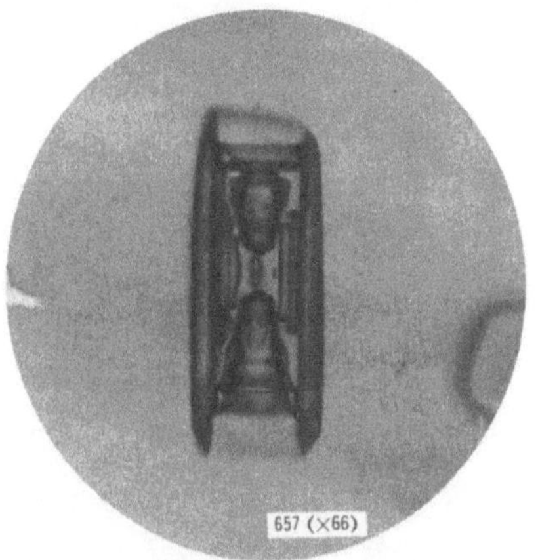
657 (×66)

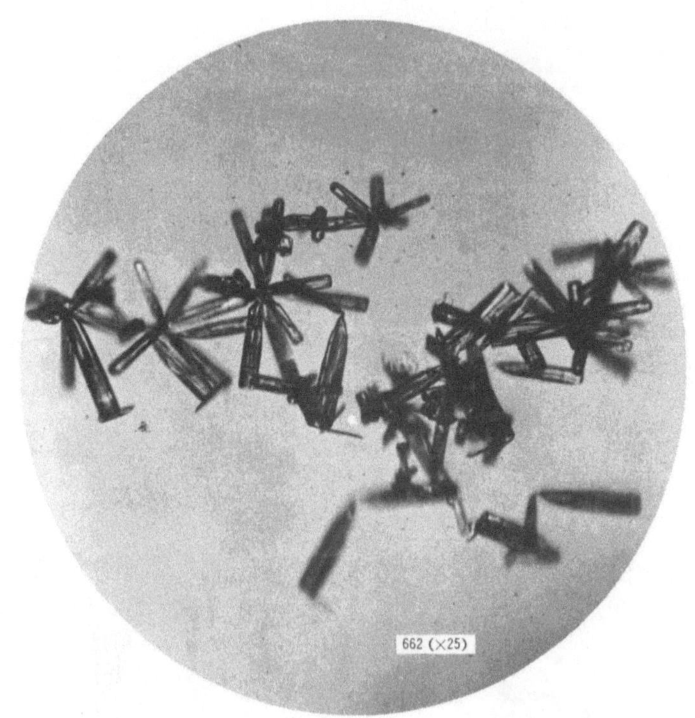
662 (×25)

PLATE 110

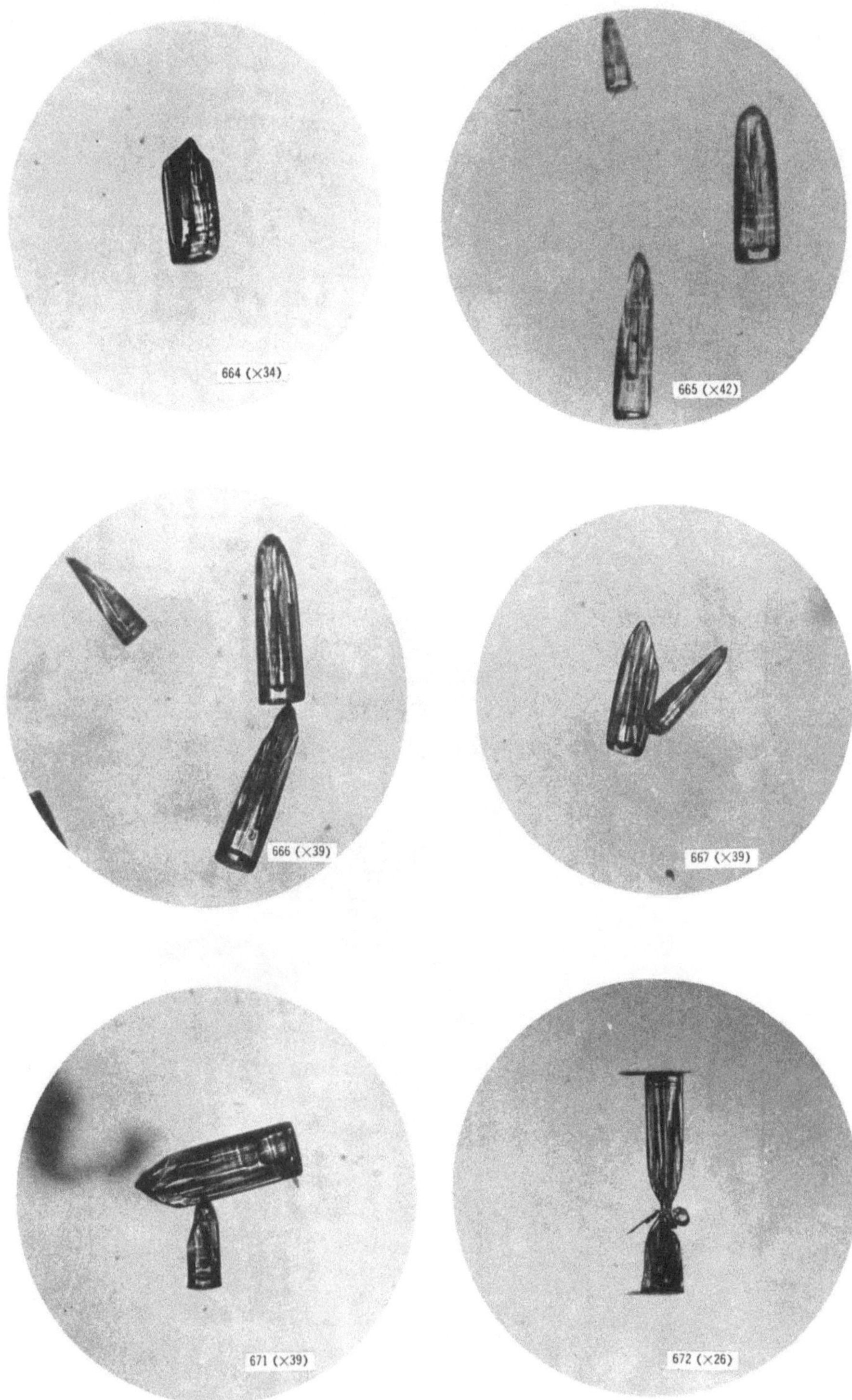

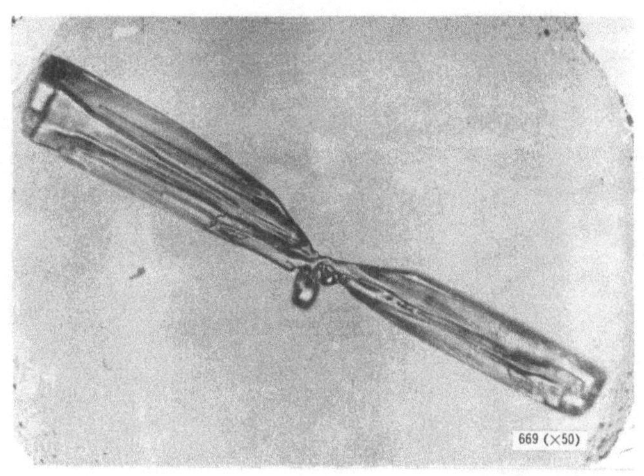
669 (×50)

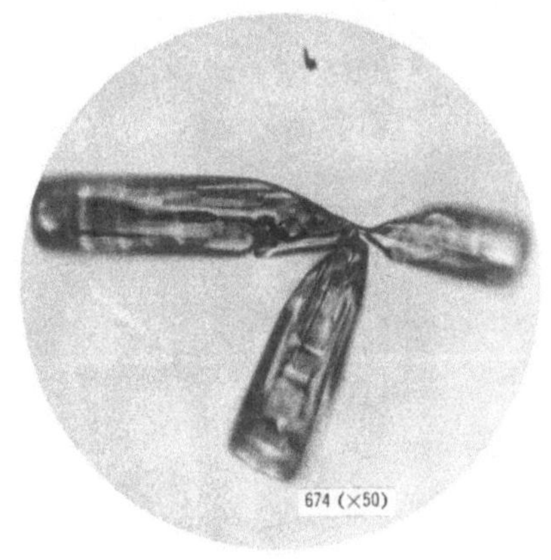
674 (×50)

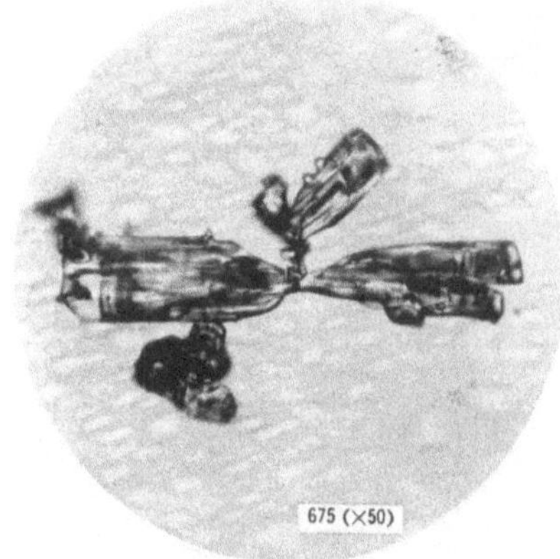
675 (×50)

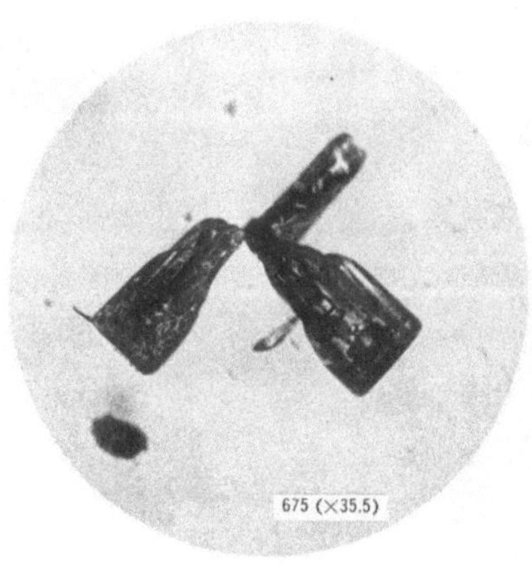
675 (×35.5)

677 (×51)

PLATE 112

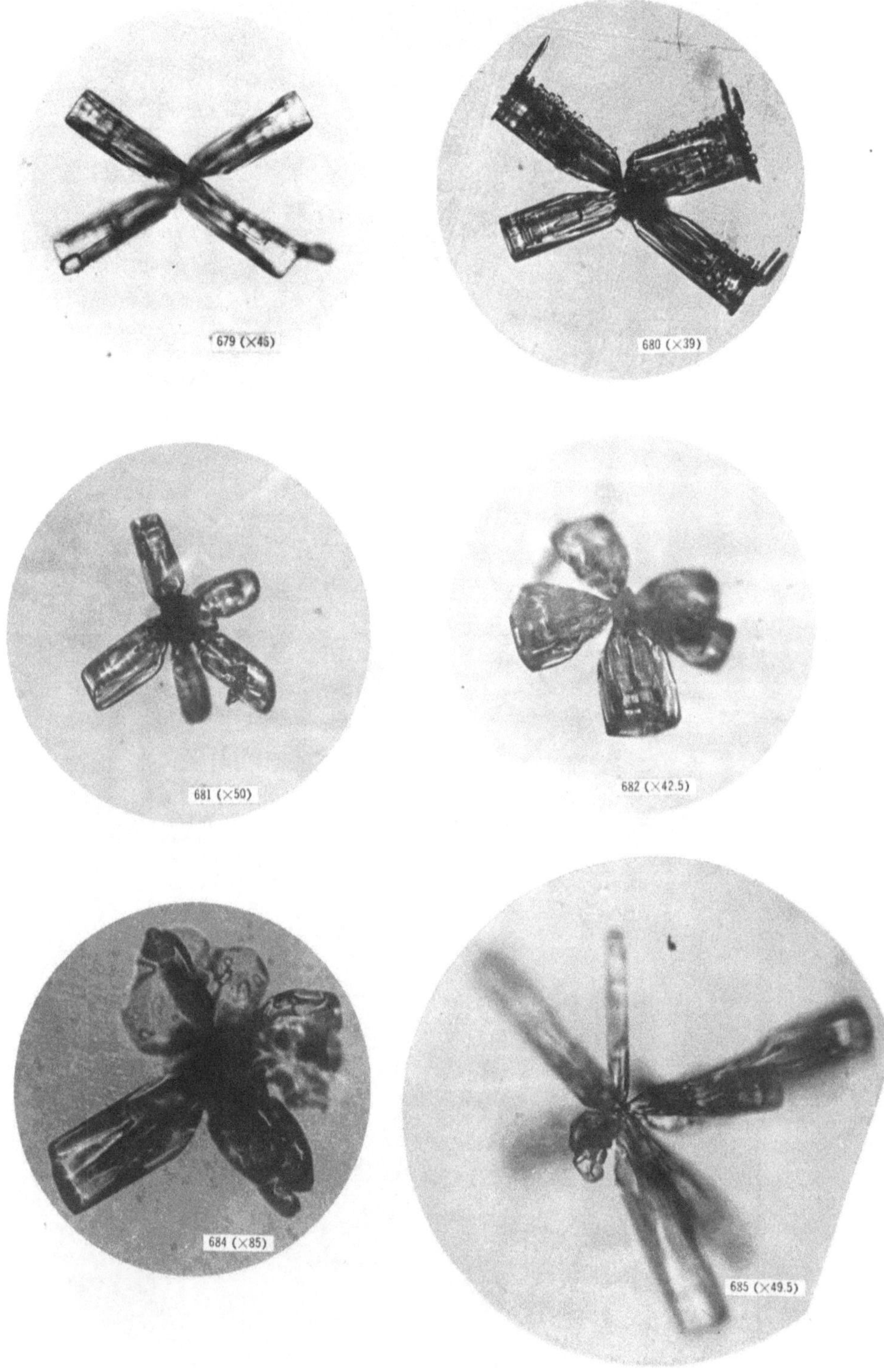

679 (×45)

680 (×39)

681 (×50)

682 (×42.5)

684 (×85)

685 (×49.5)

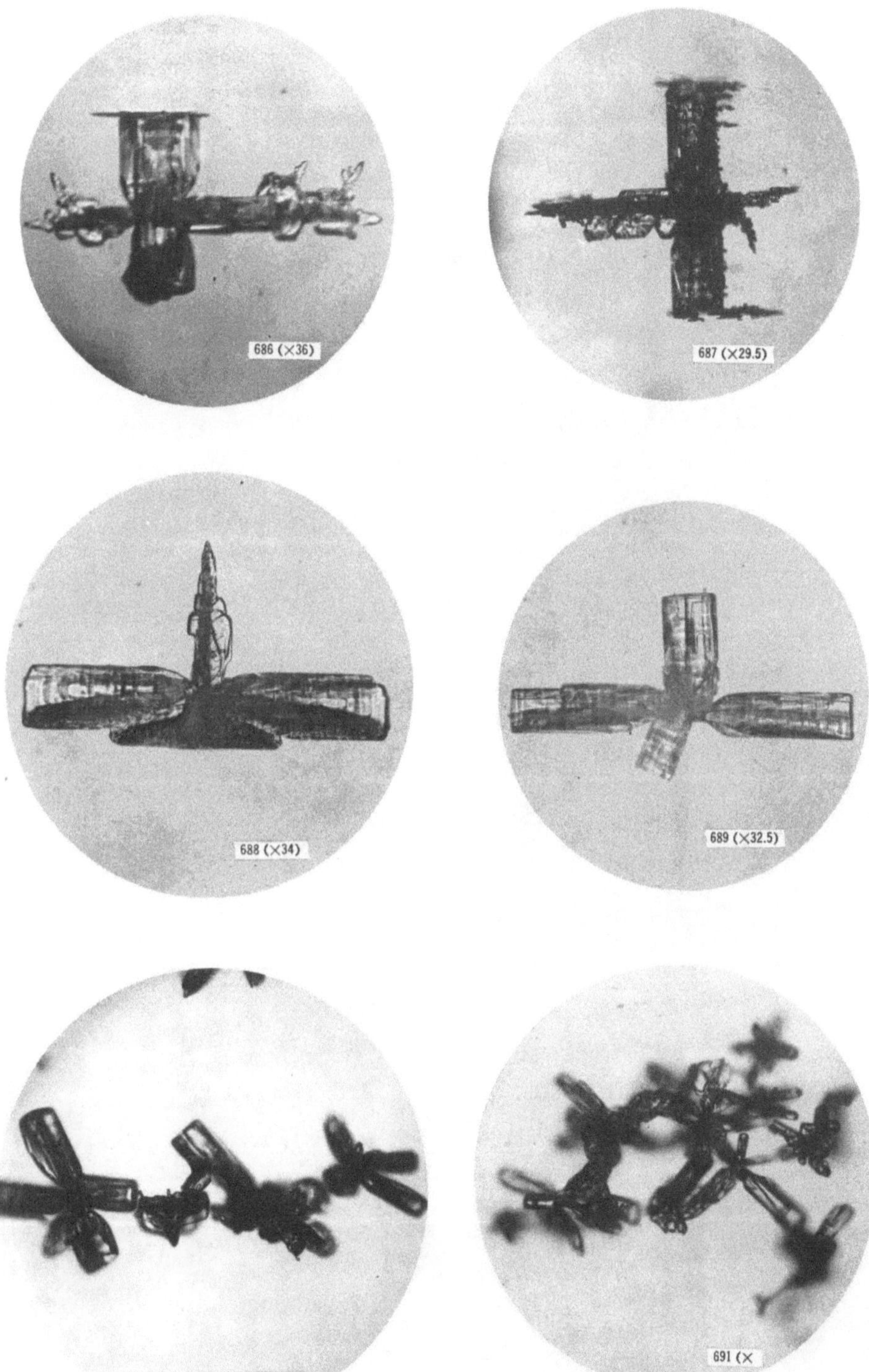
686 (×36)
687 (×29.5)
688 (×34)
689 (×32.5)
690 (×33)
691 (×

PLATE 114

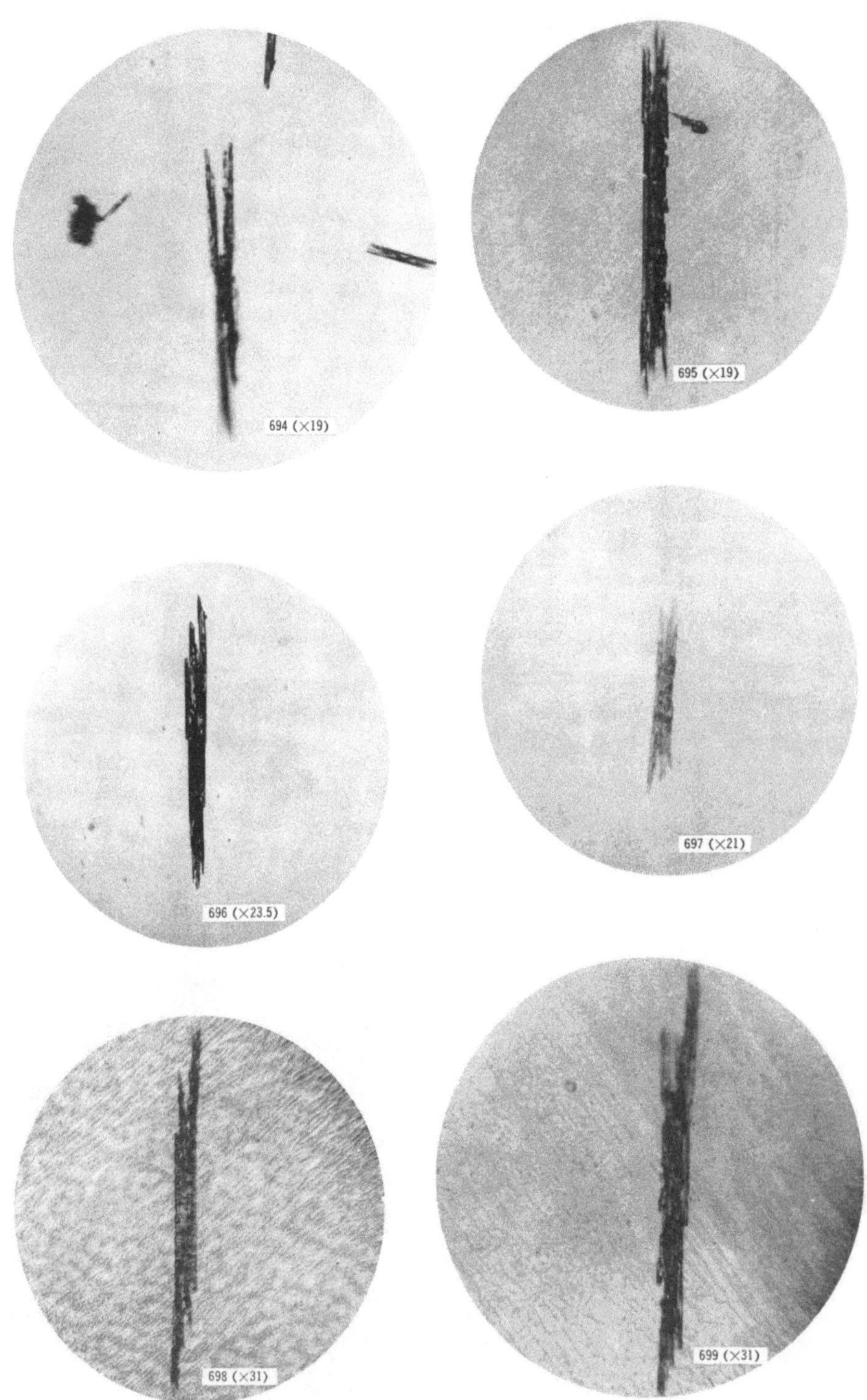

PLATE 115

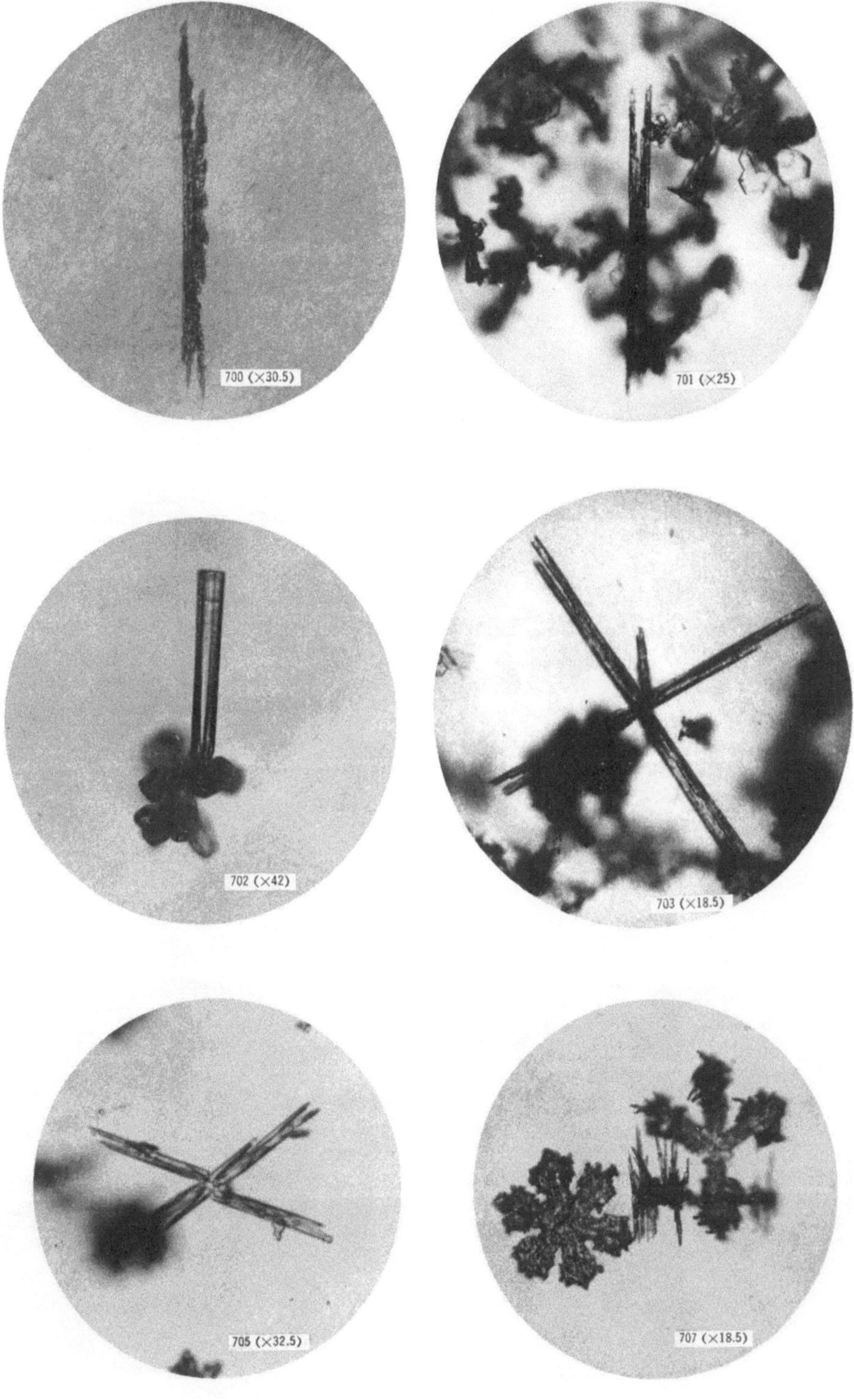

700 (×30.5)

701 (×25)

702 (×42)

703 (×18.5)

705 (×32.5)

707 (×18.5)

PLATE 116

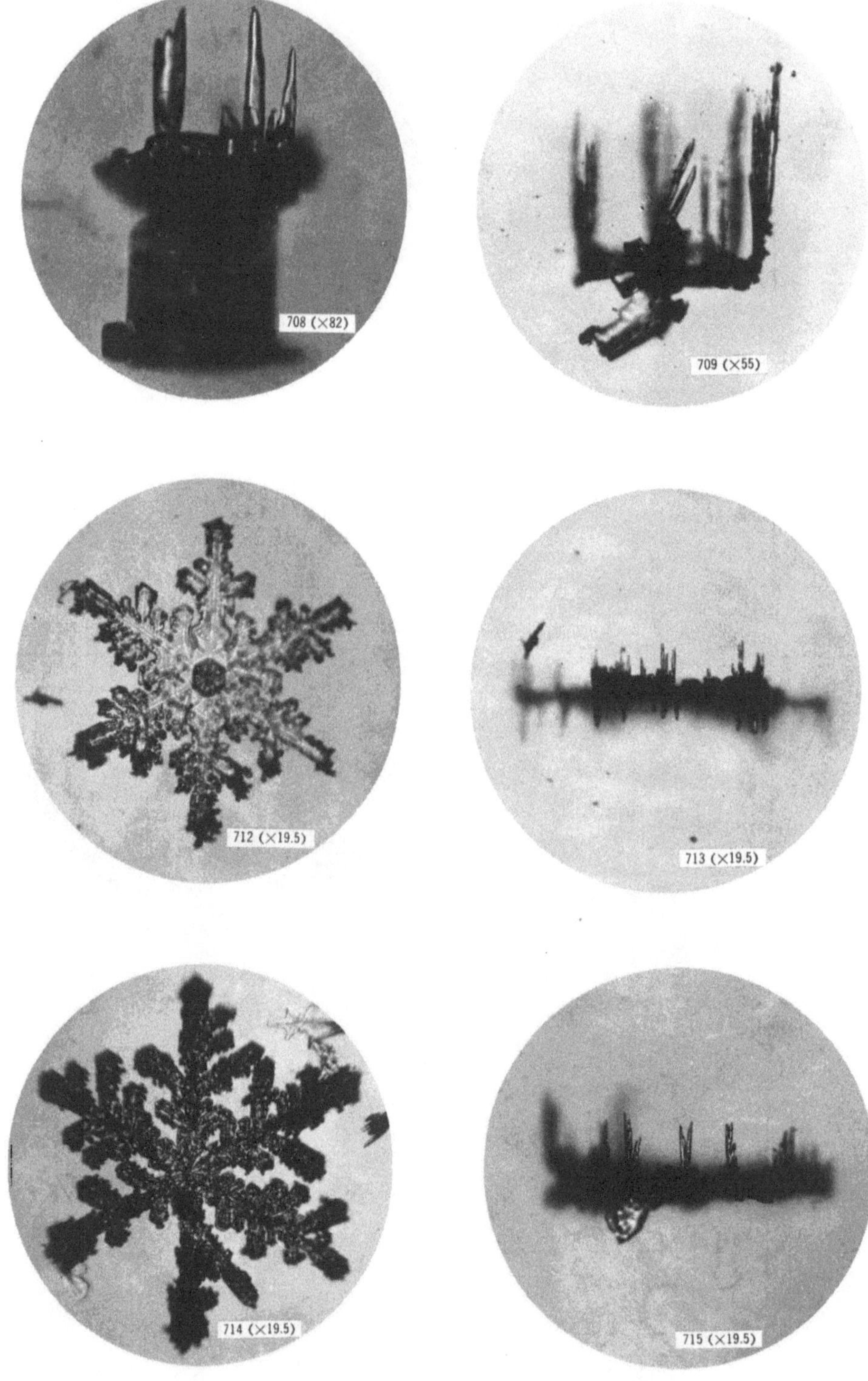

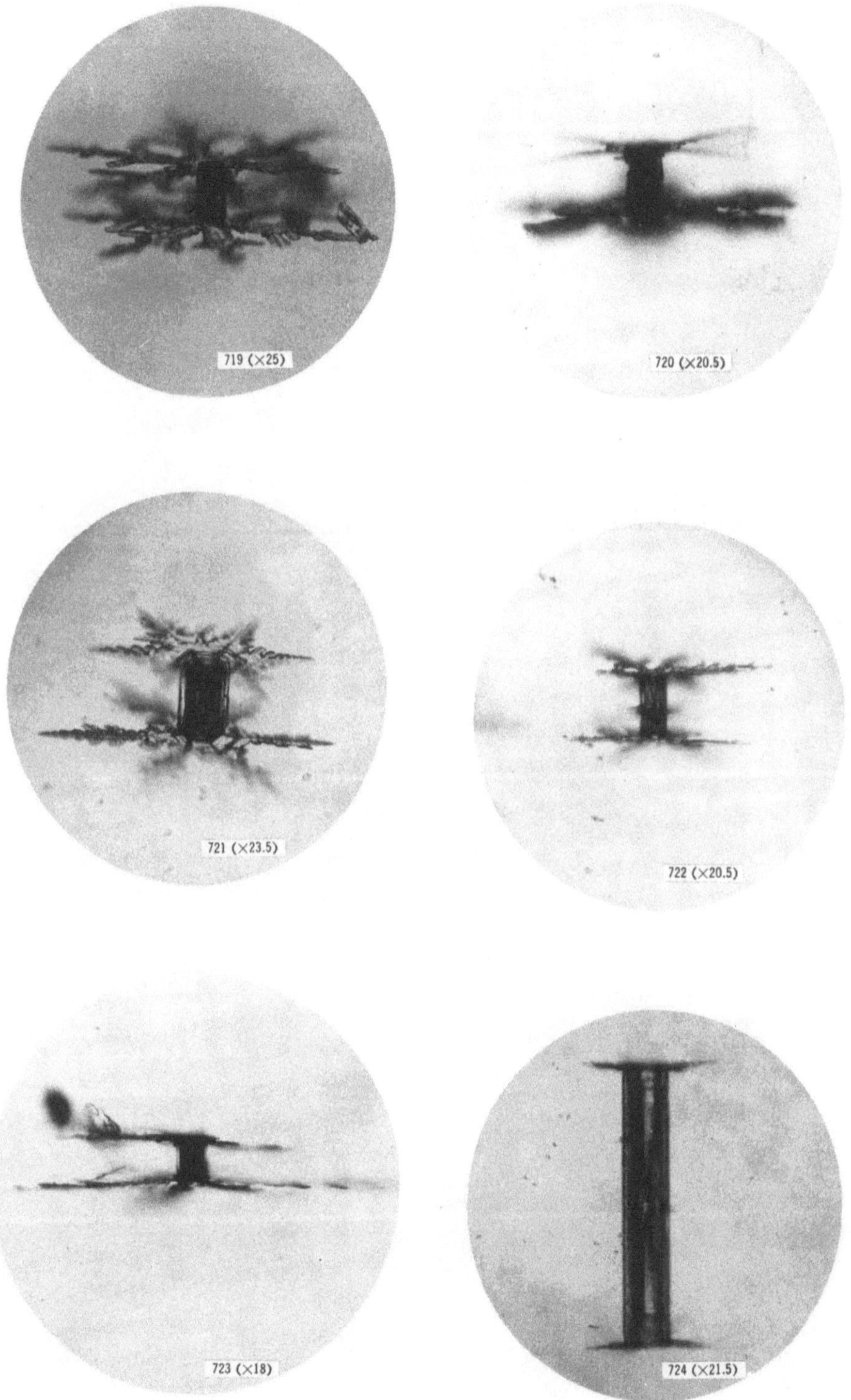
719 (×25)
720 (×20.5)
721 (×23.5)
722 (×20.5)
723 (×18)
724 (×21.5)

PLATE 118

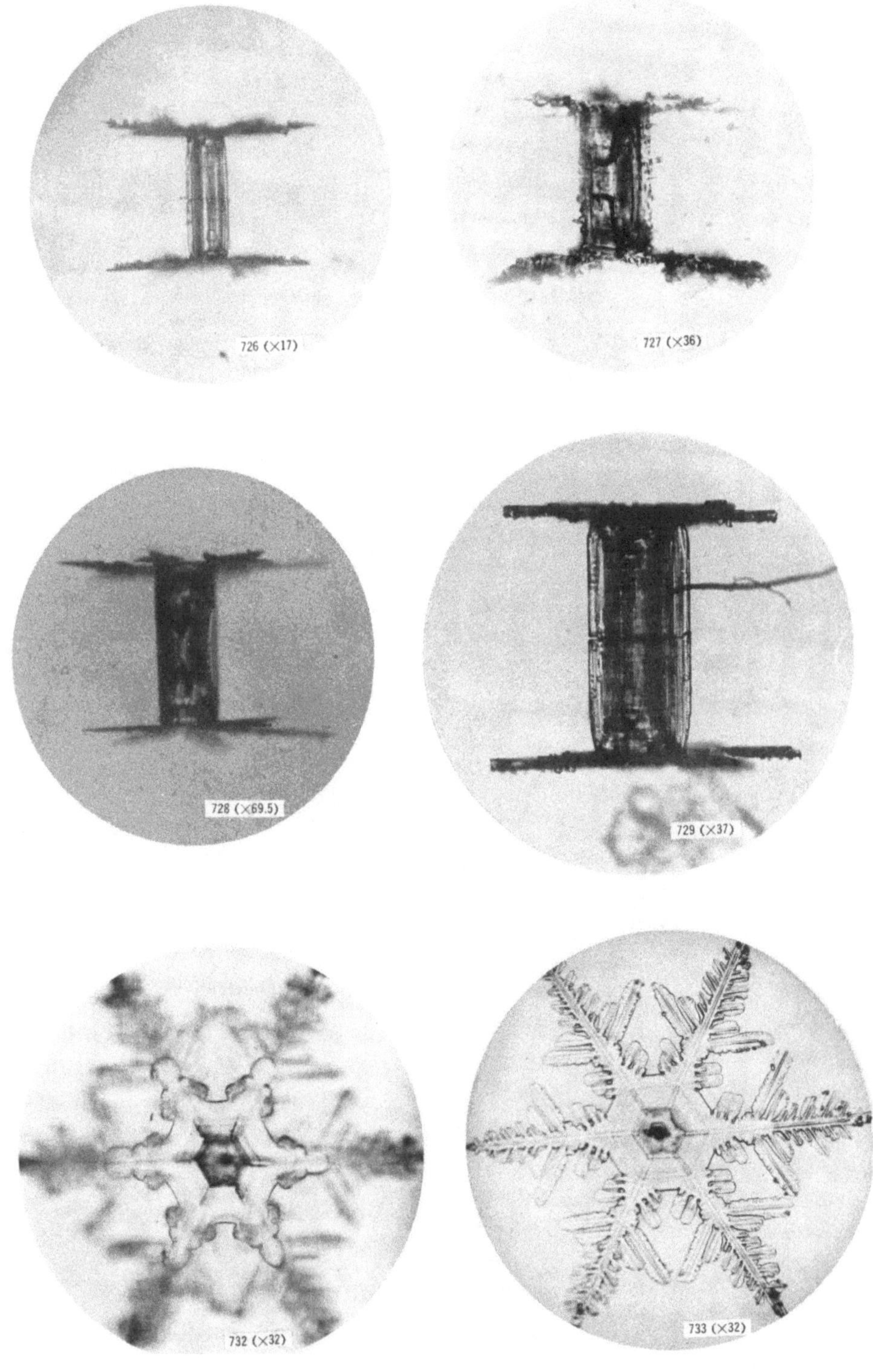

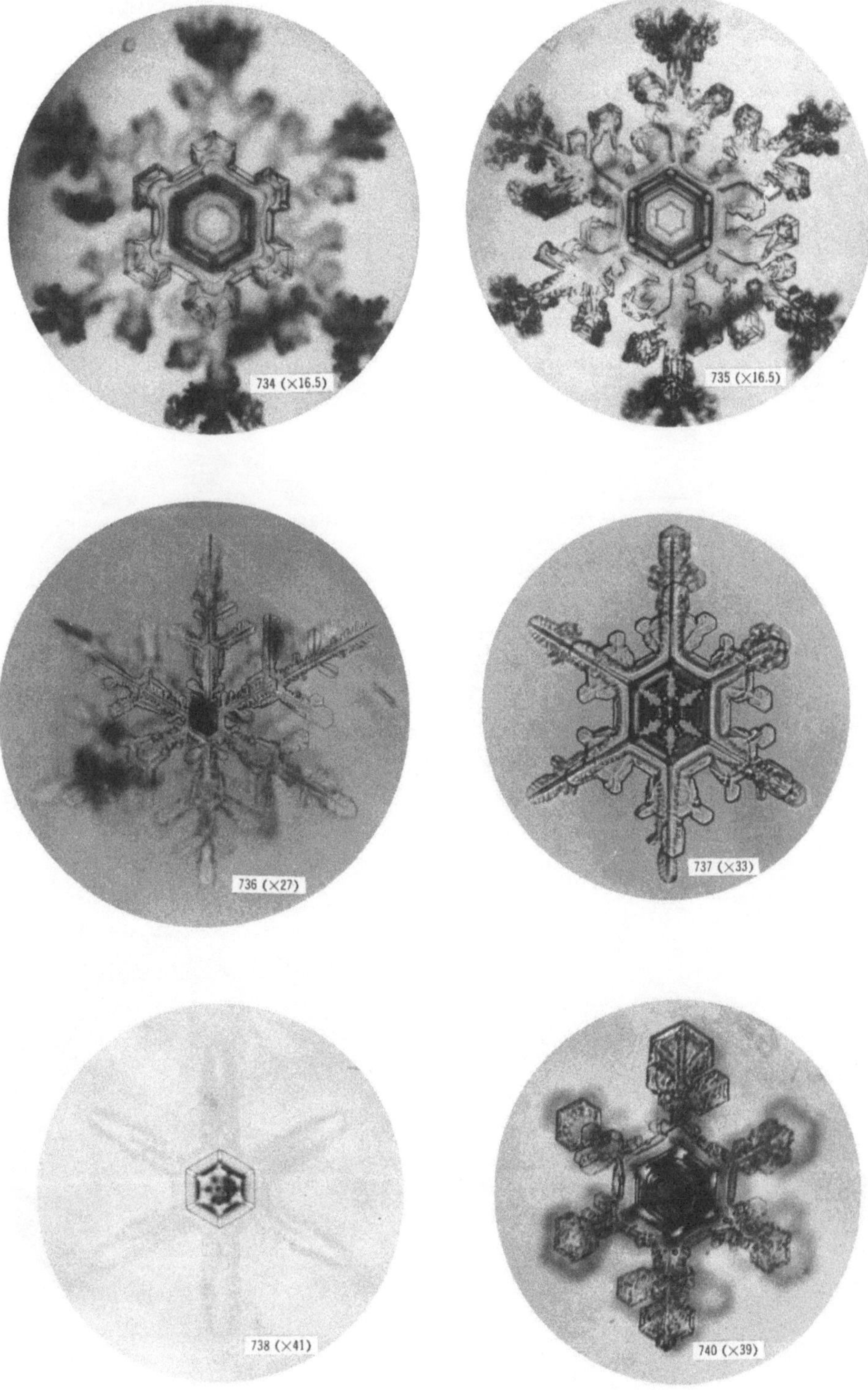
734 (×16.5)
735 (×16.5)
736 (×27)
737 (×33)
738 (×41)
740 (×39)

PLATE 120

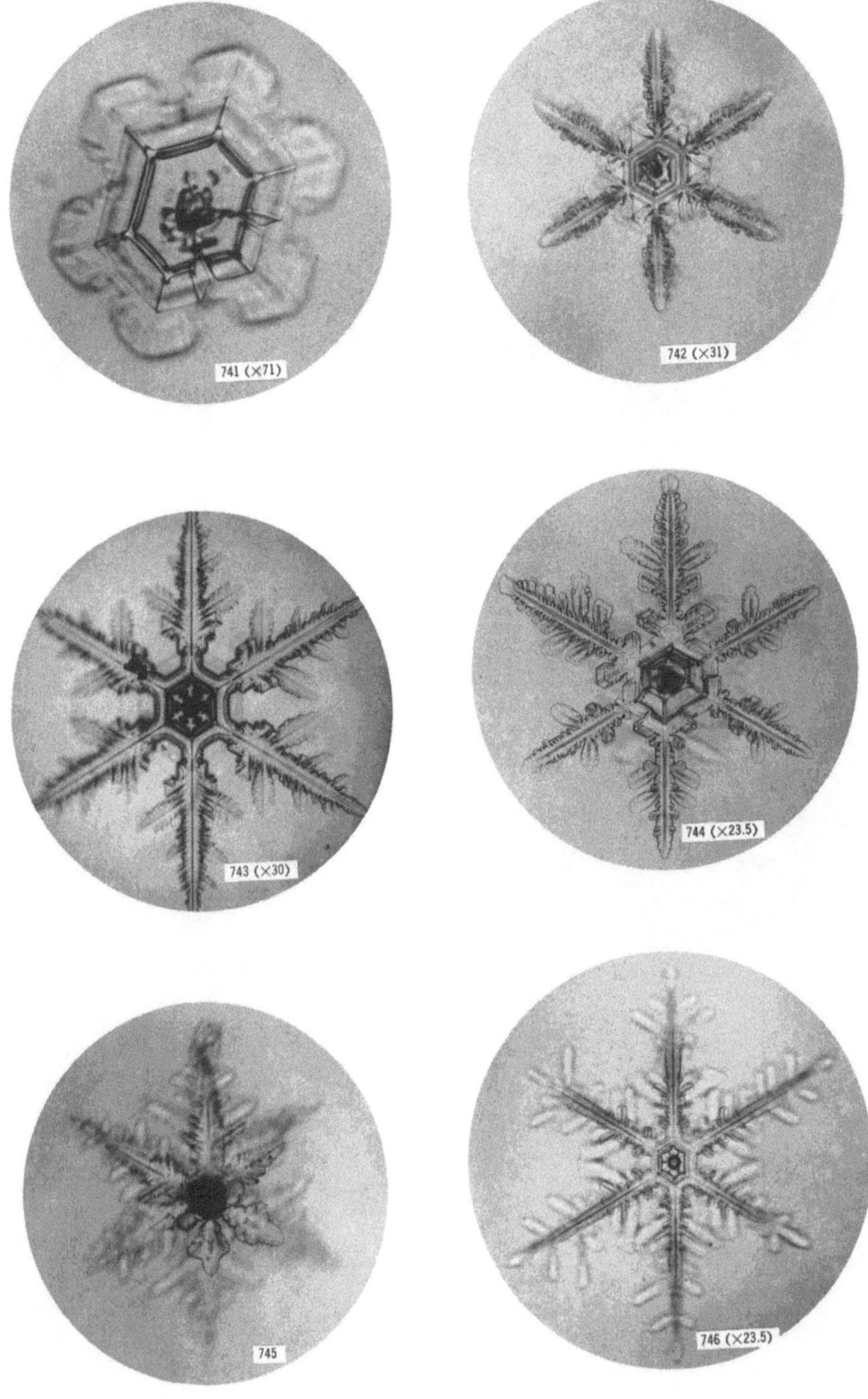

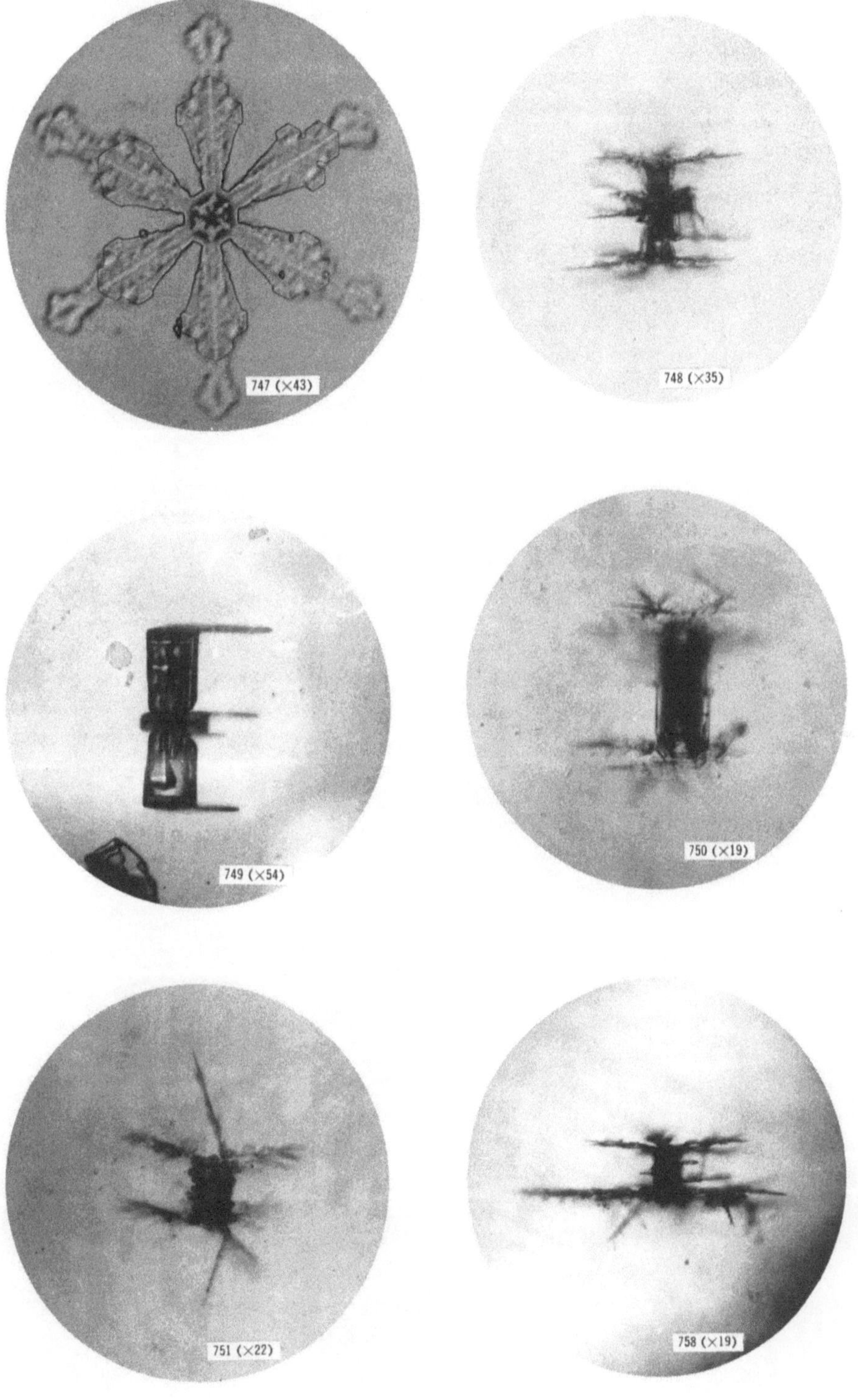
747 (×43)
748 (×35)
749 (×54)
750 (×19)
751 (×22)
758 (×19)

PLATE 122

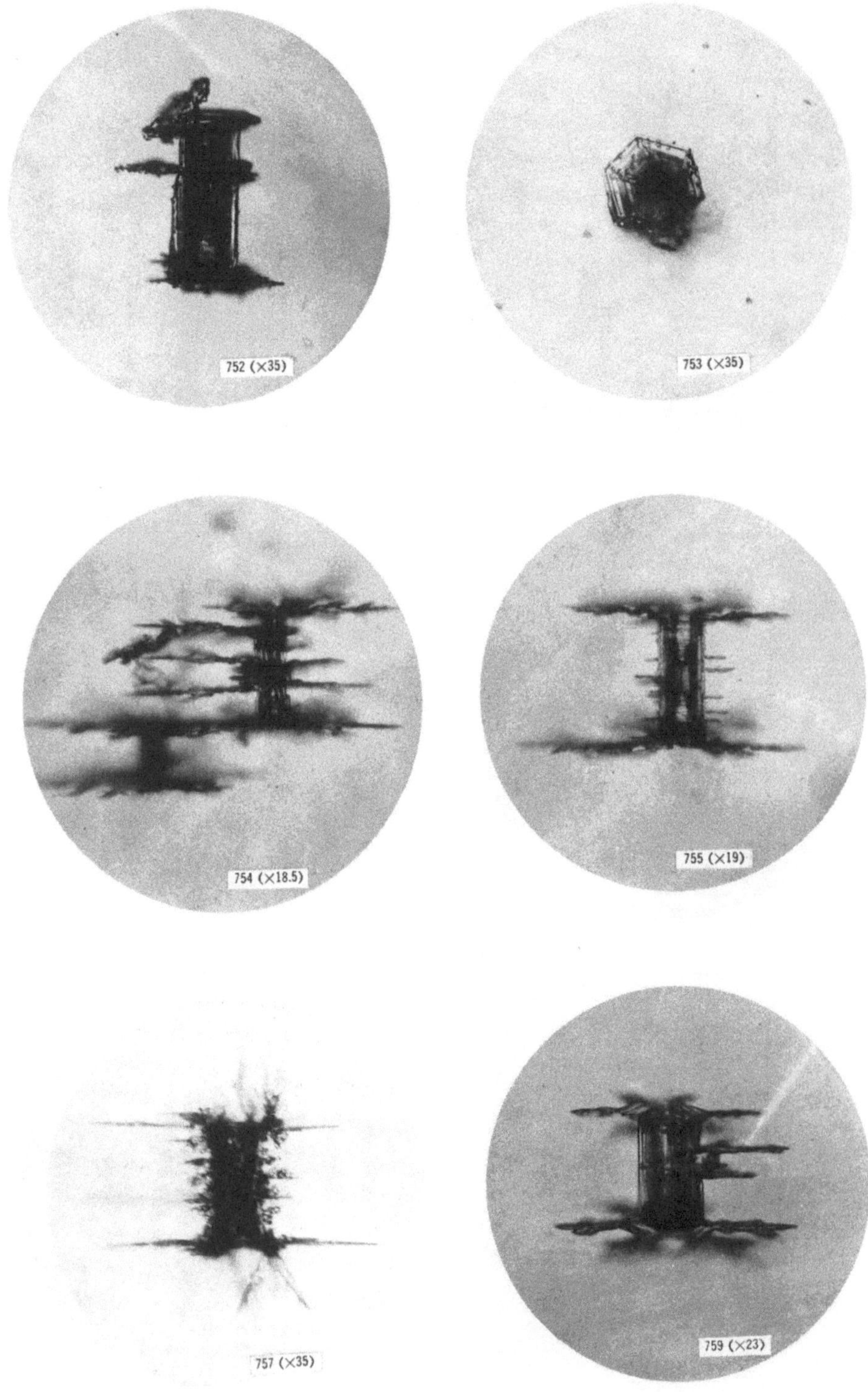

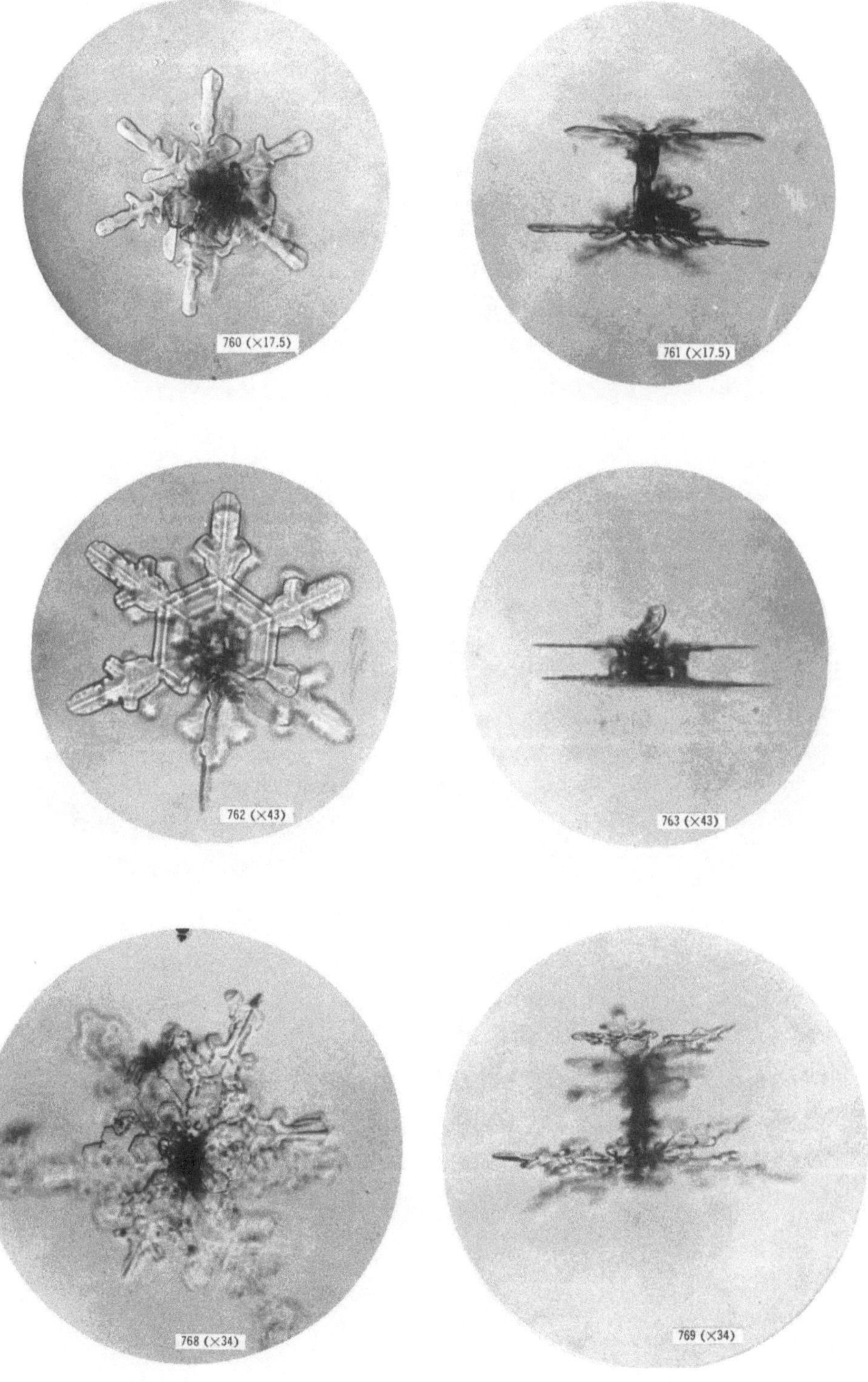
760 (×17.5)
761 (×17.5)
762 (×43)
763 (×43)
768 (×34)
769 (×34)

PLATE 124

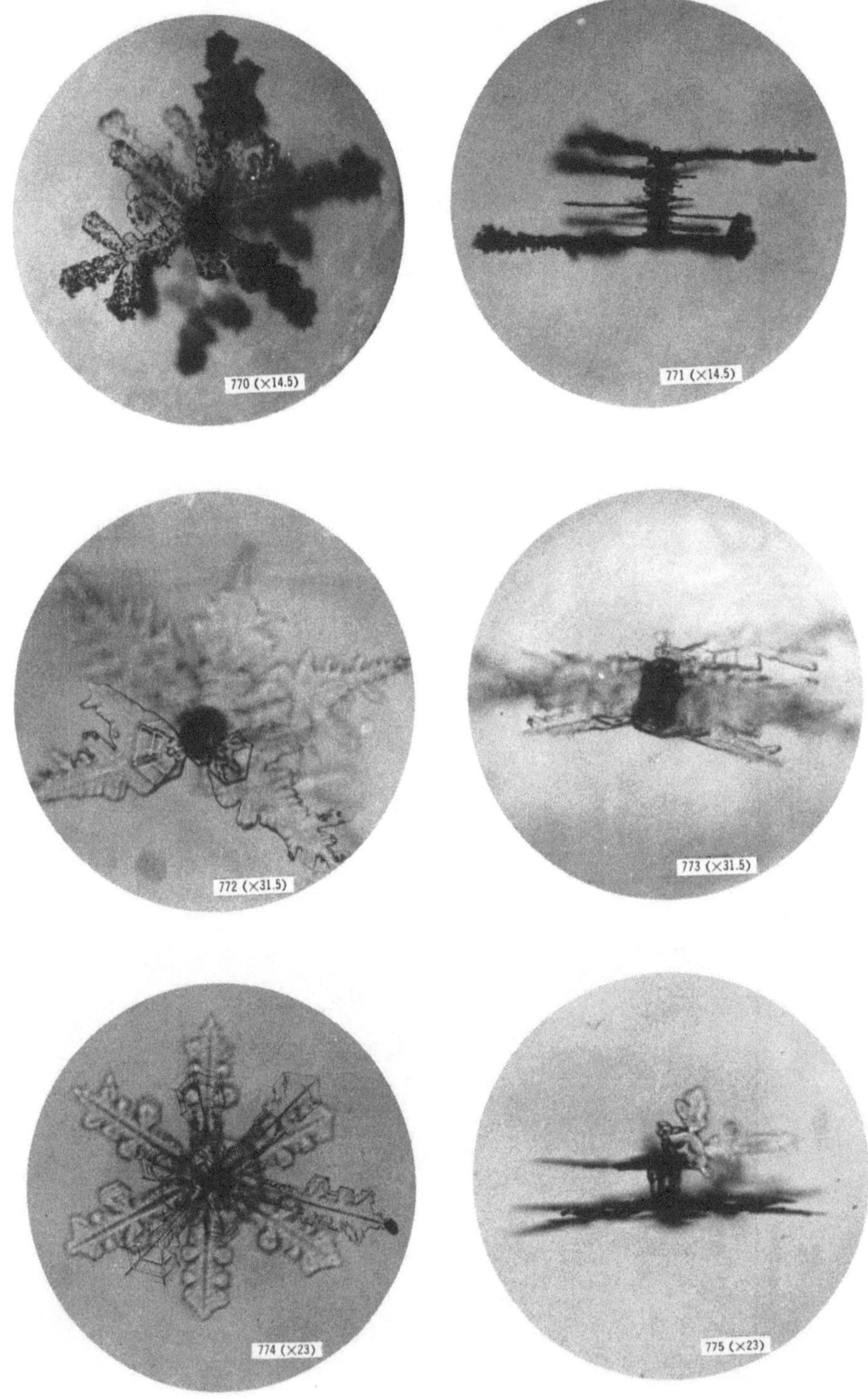

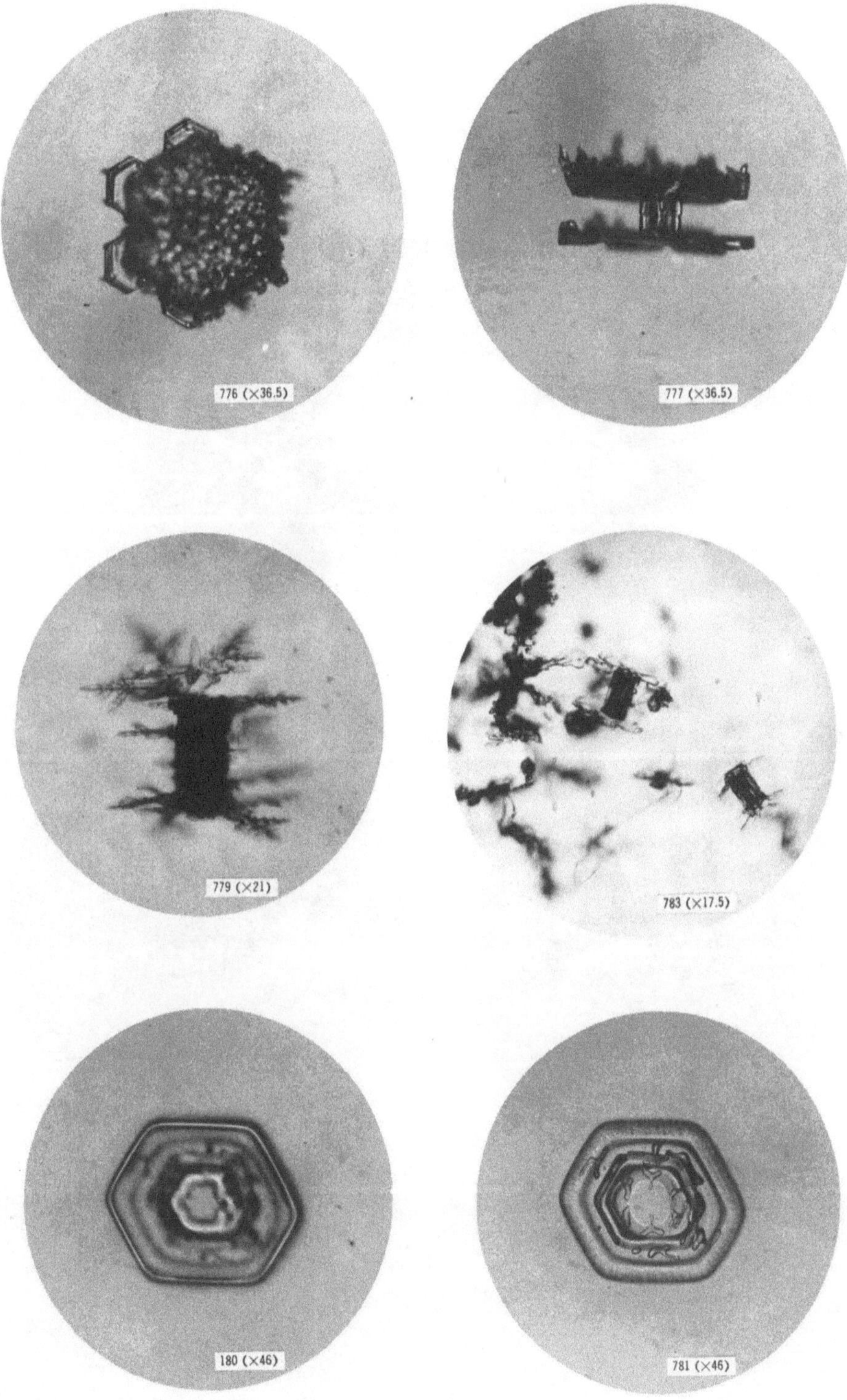
776 (×36.5)
777 (×36.5)
779 (×21)
783 (×17.5)
180 (×46)
781 (×46)

PLATE 126

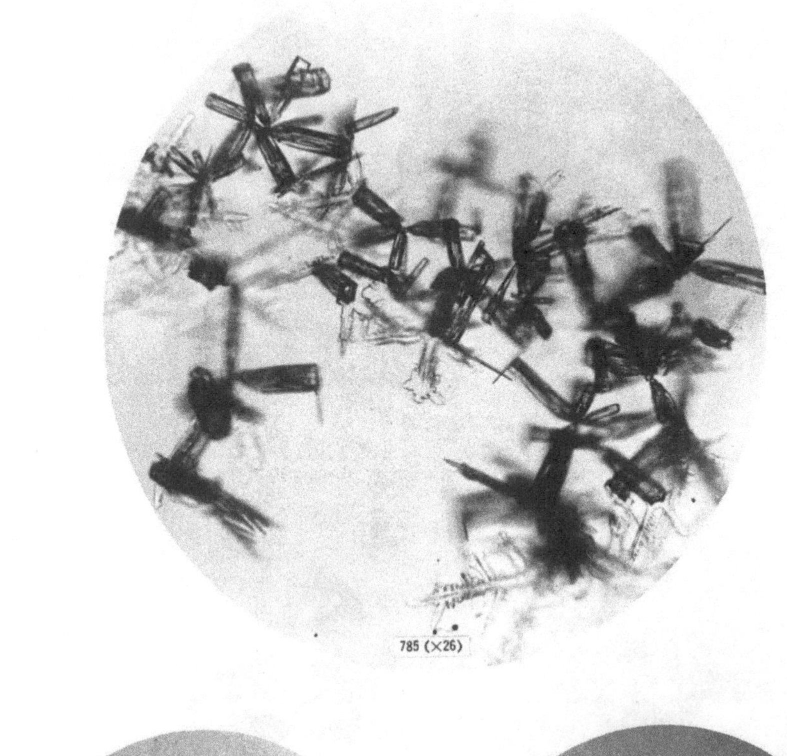

785 (×26)

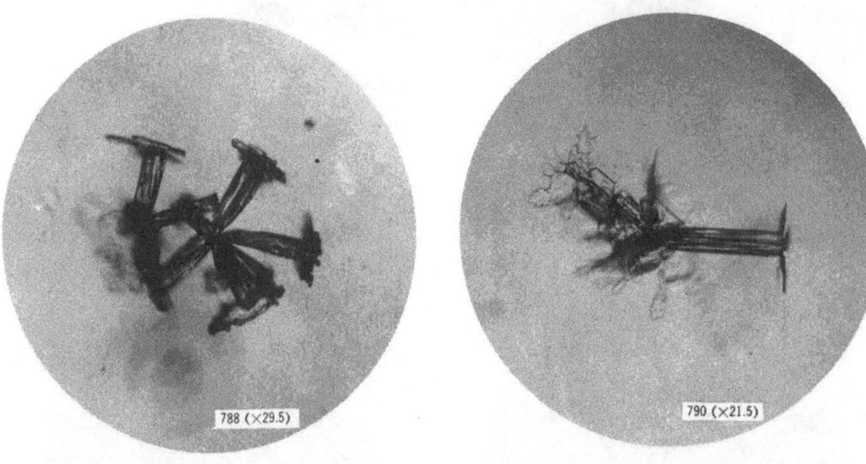

788 (×29.5)

790 (×21.5)

786 (×18.5)
789 (×25)
791
792 (×25)
794 (×25)
795 (×25)

PLATE 128

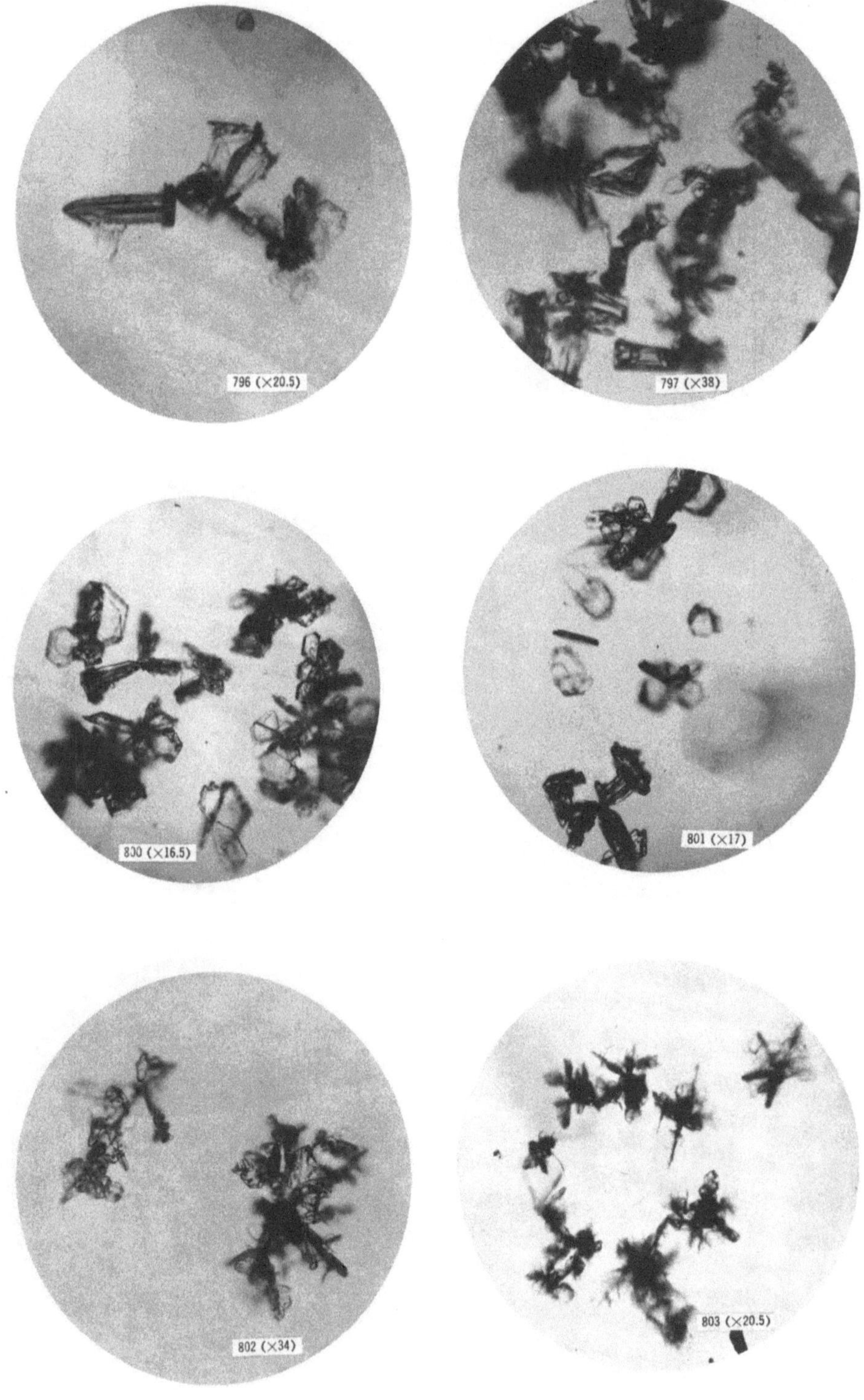

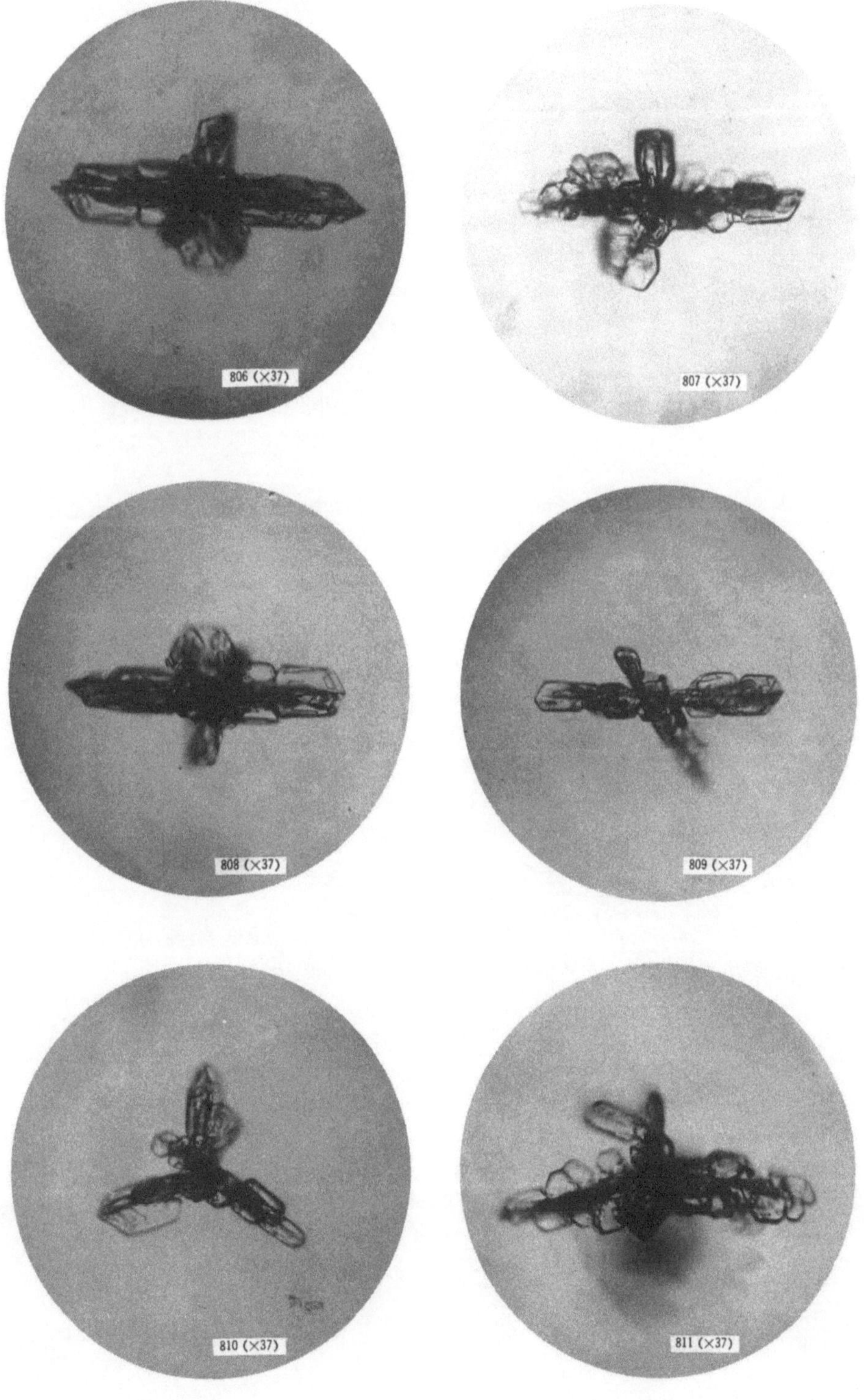
806 (×37)
807 (×37)
808 (×37)
809 (×37)
810 (×37)
811 (×37)

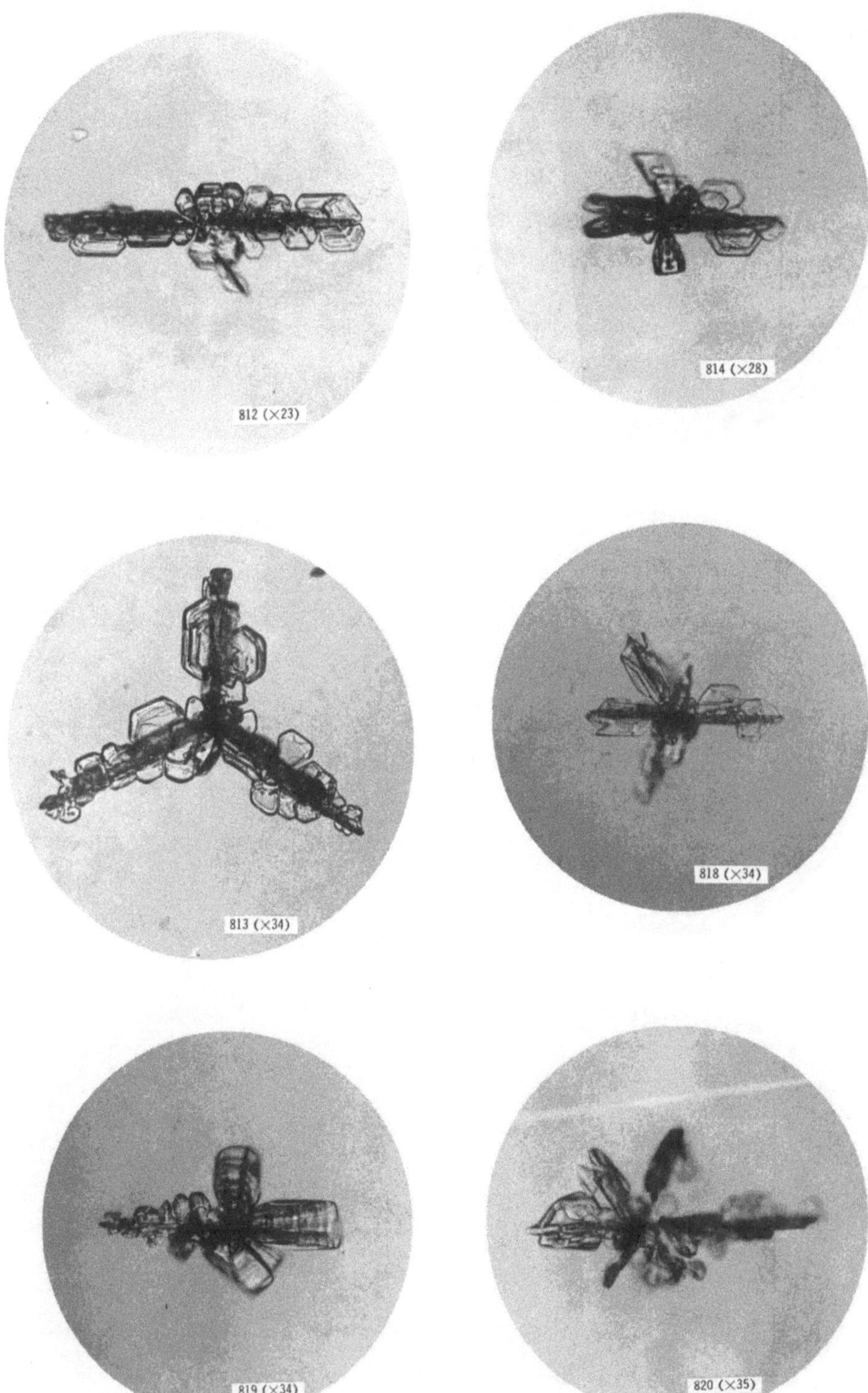
812 (×23)
814 (×28)
813 (×34)
818 (×34)
819 (×34)
820 (×35)

821 (×27.5)
822 (×40.5)
823 (×27.5)
827 (×40)
828 (×43)
829 (×33.5)

PLATE 132

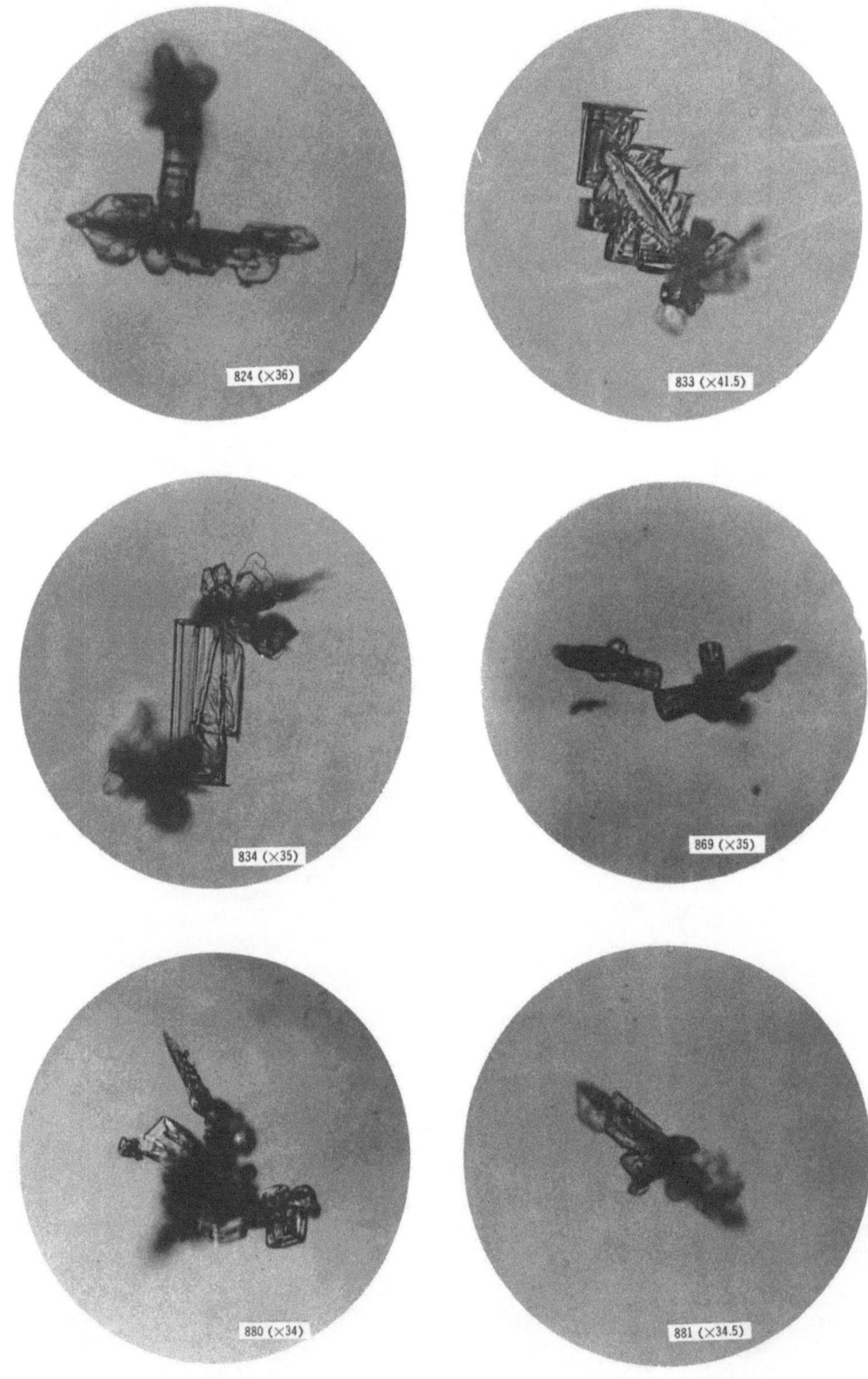

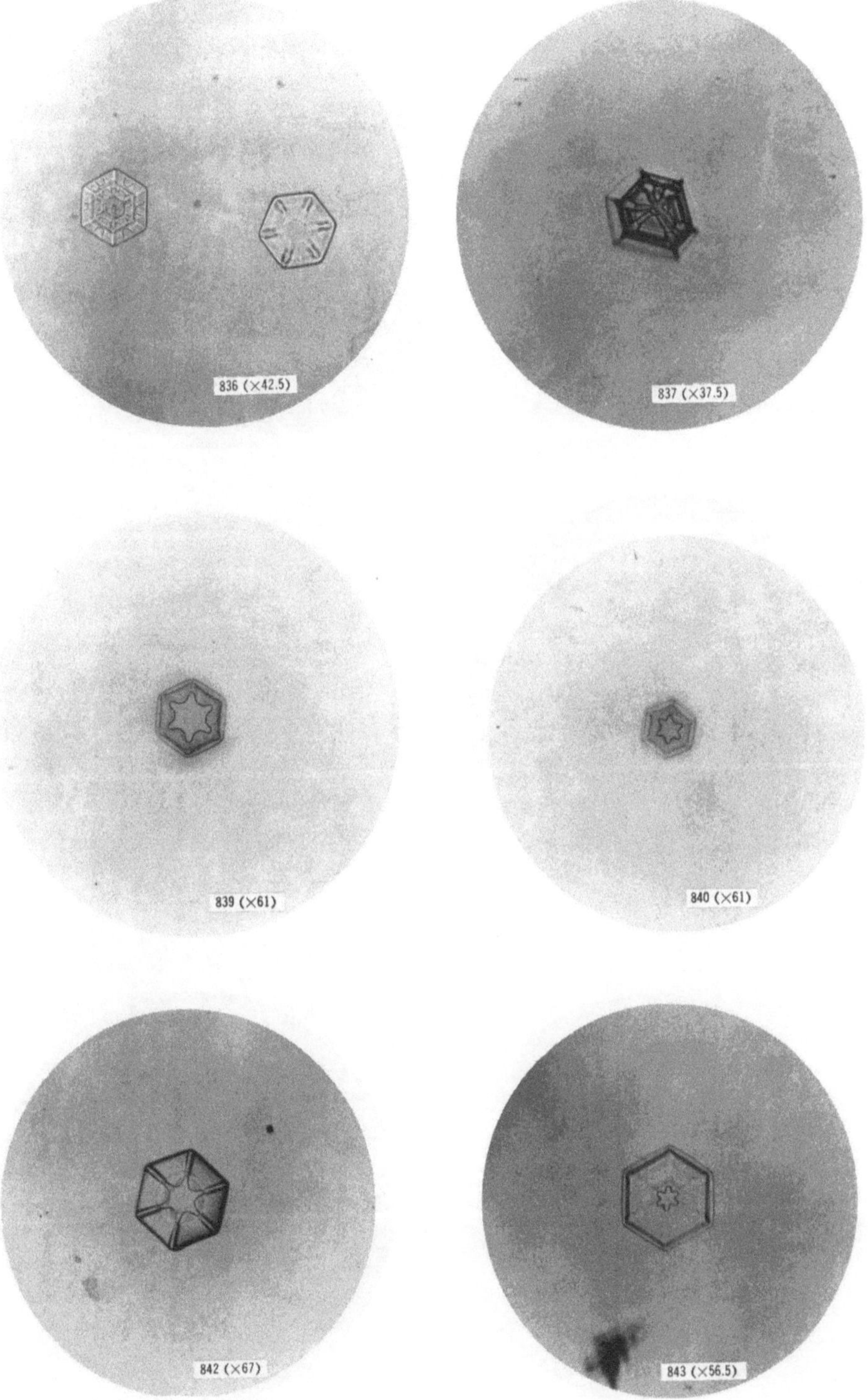
836 (×42.5)
837 (×37.5)
839 (×61)
840 (×61)
842 (×67)
843 (×56.5)

PLATE 134

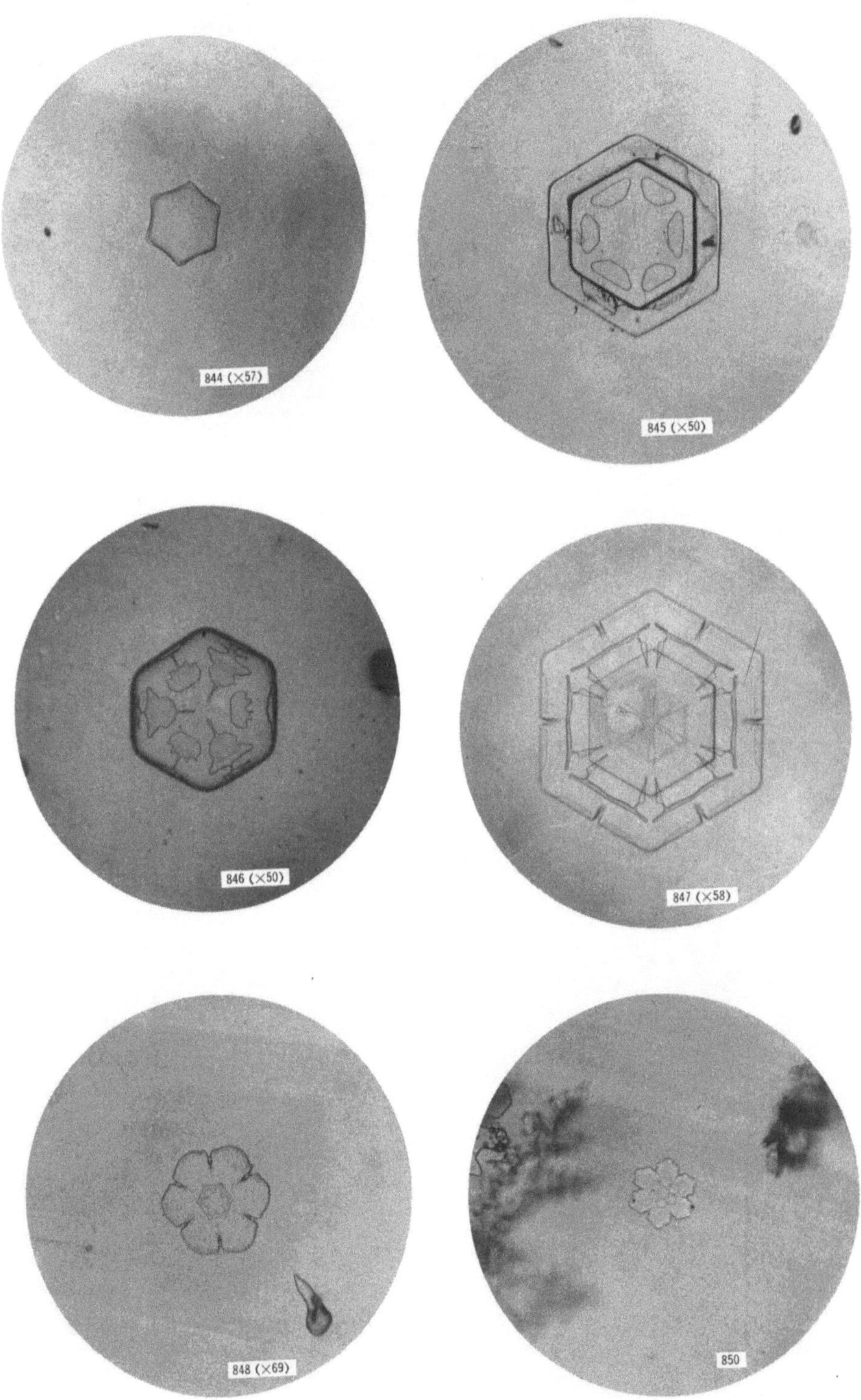

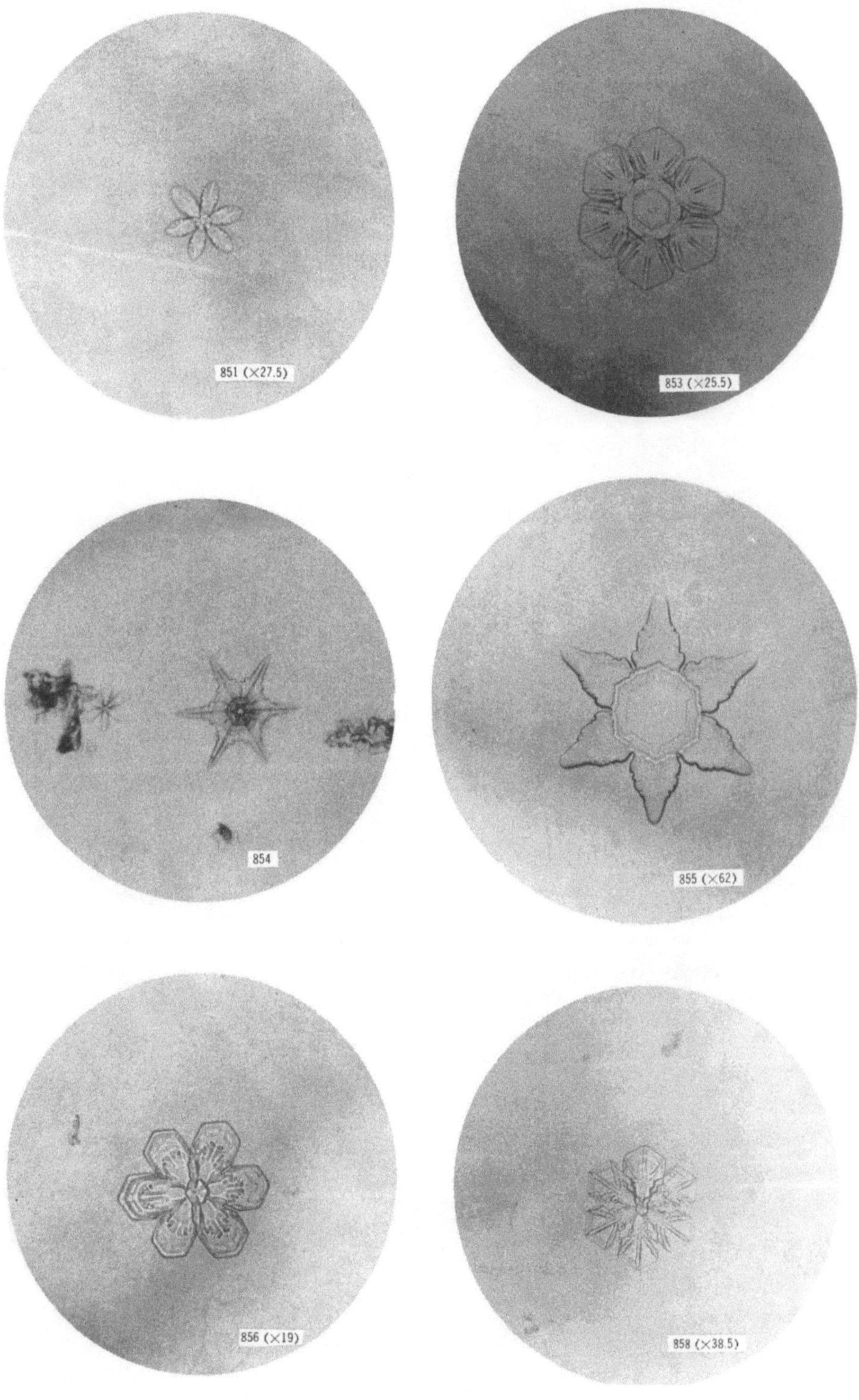
851 (×27.5)
853 (×25.5)
854
855 (×62)
856 (×19)
858 (×38.5)

PLATE 136

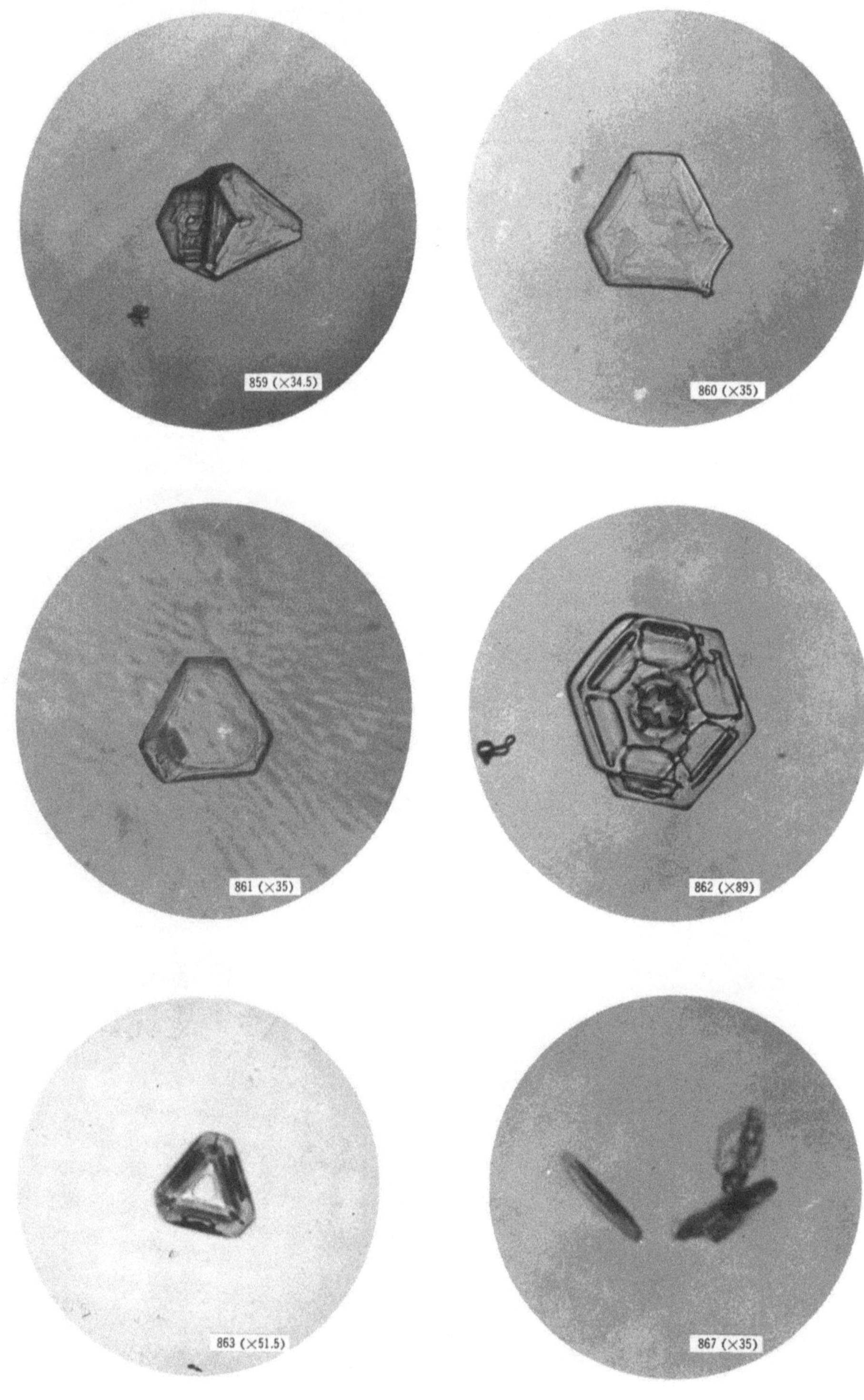

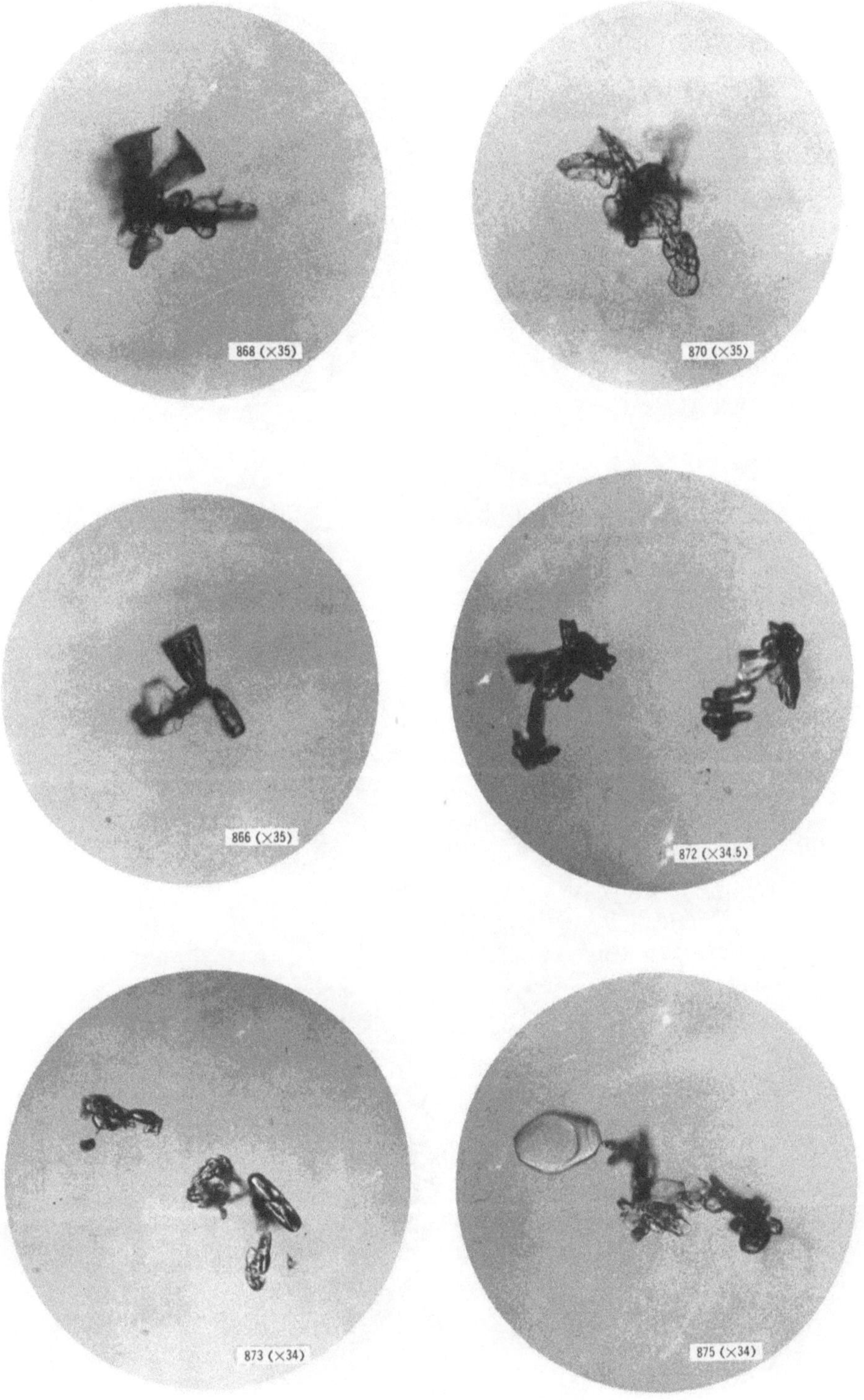
868 (×35)
870 (×35)
866 (×35)
872 (×34.5)
873 (×34)
875 (×34)

PLATE 138

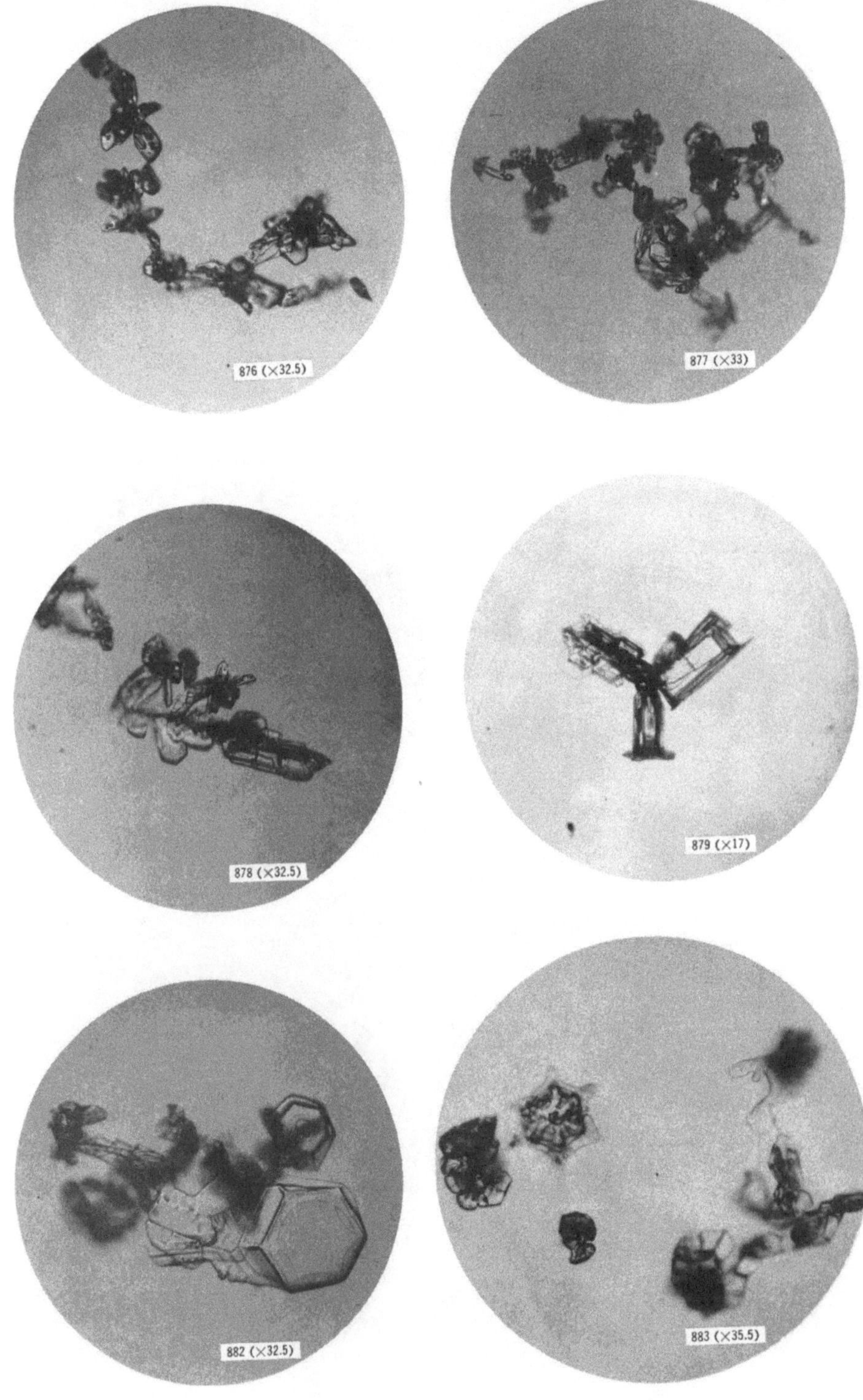

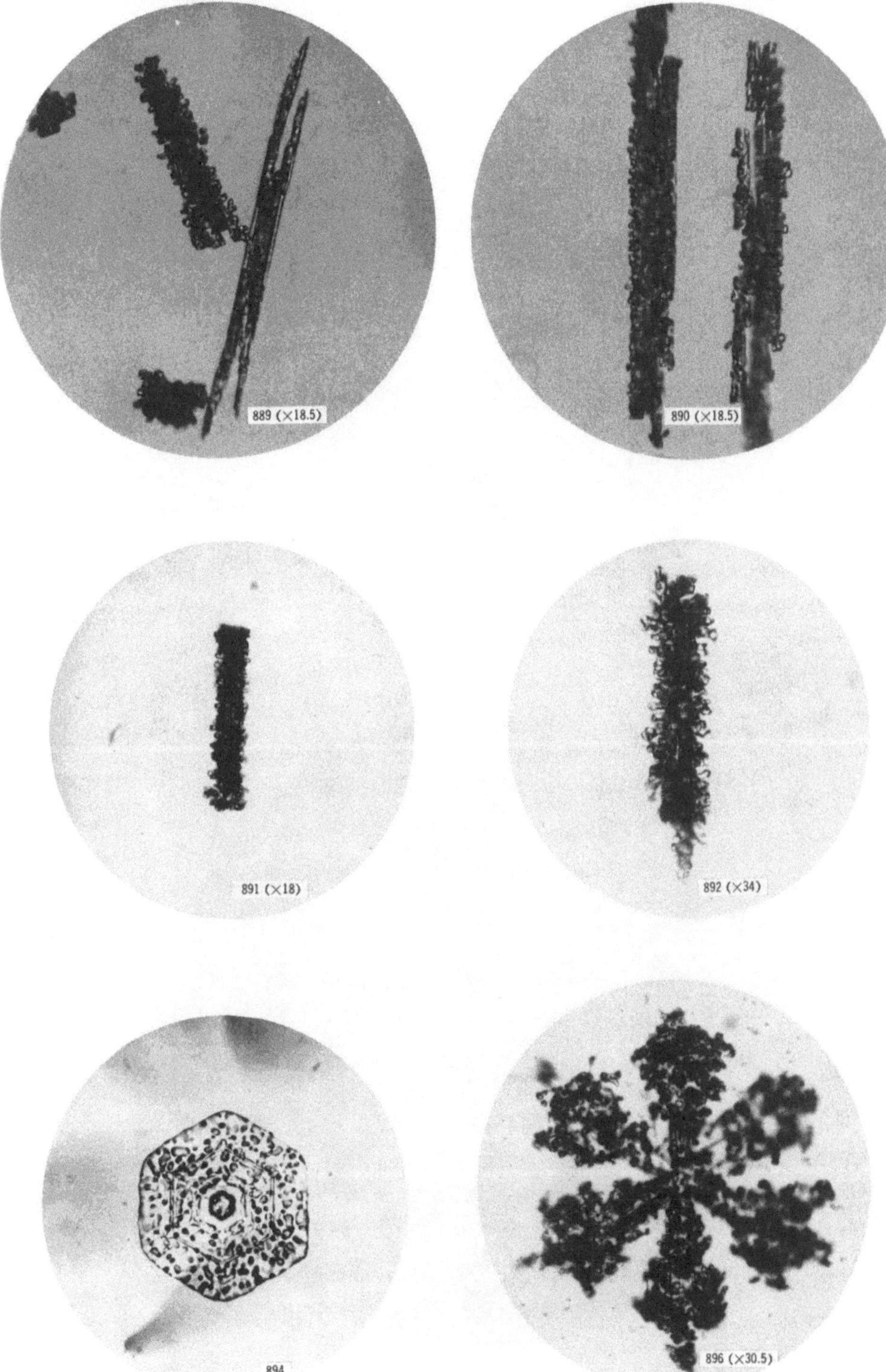
889 (×18.5)
890 (×18.5)
891 (×18)
892 (×34)
894
896 (×30.5)

PLATE 140

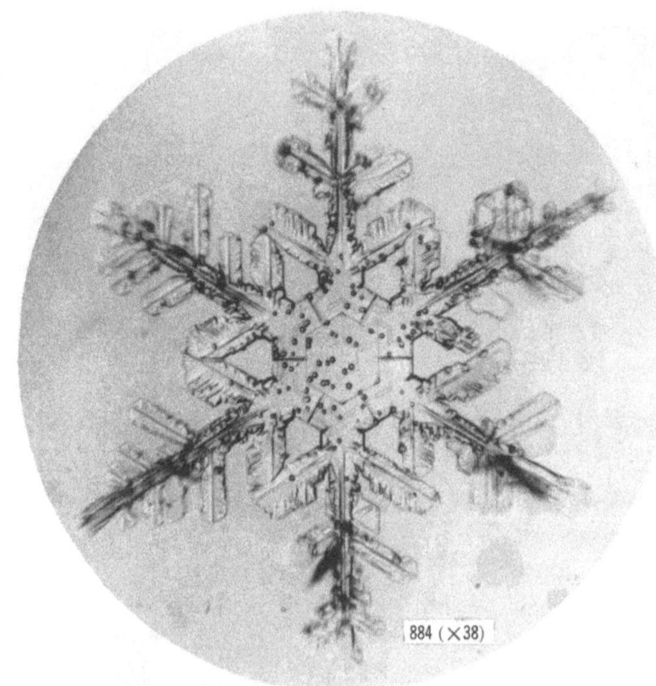

886 (×37.5)
887 (×19.5)

PLATE 142

905 (×19)

906 (×19)

907 (×36)

908 (×22)

909 (×11)

910 (×31)

PLATE 144

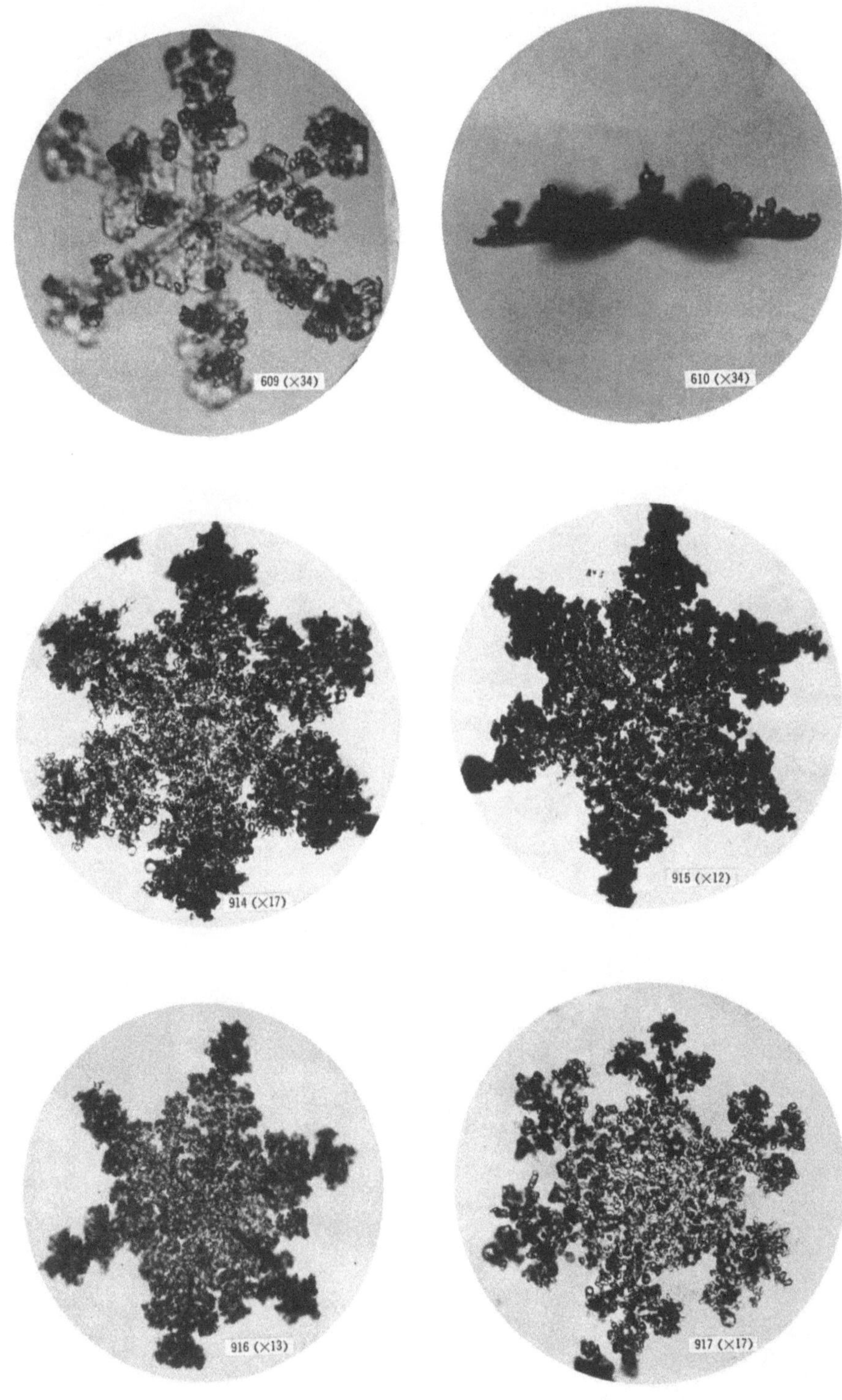

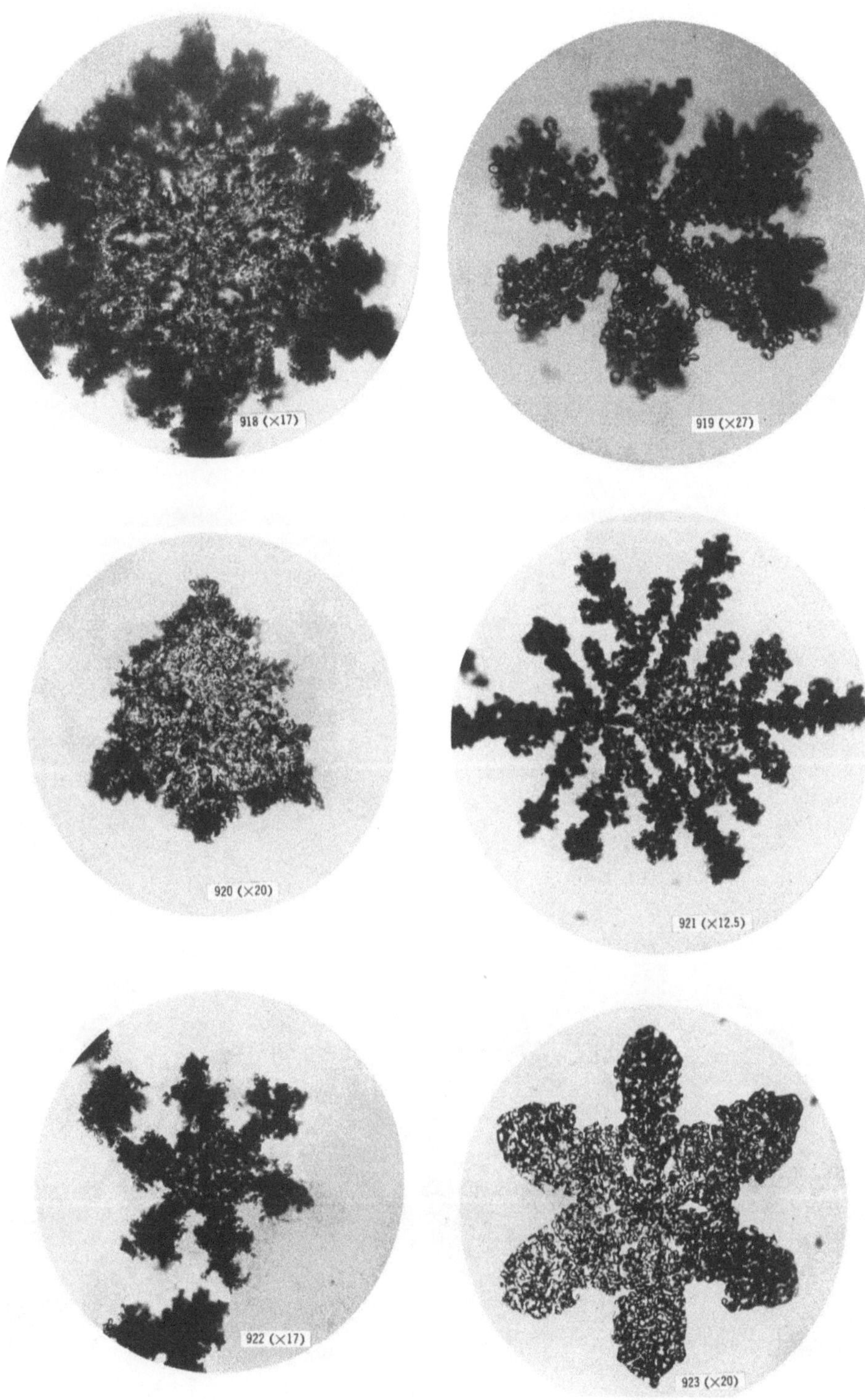

918 (×17)

919 (×27)

920 (×20)

921 (×12.5)

922 (×17)

923 (×20)

PLATE 146

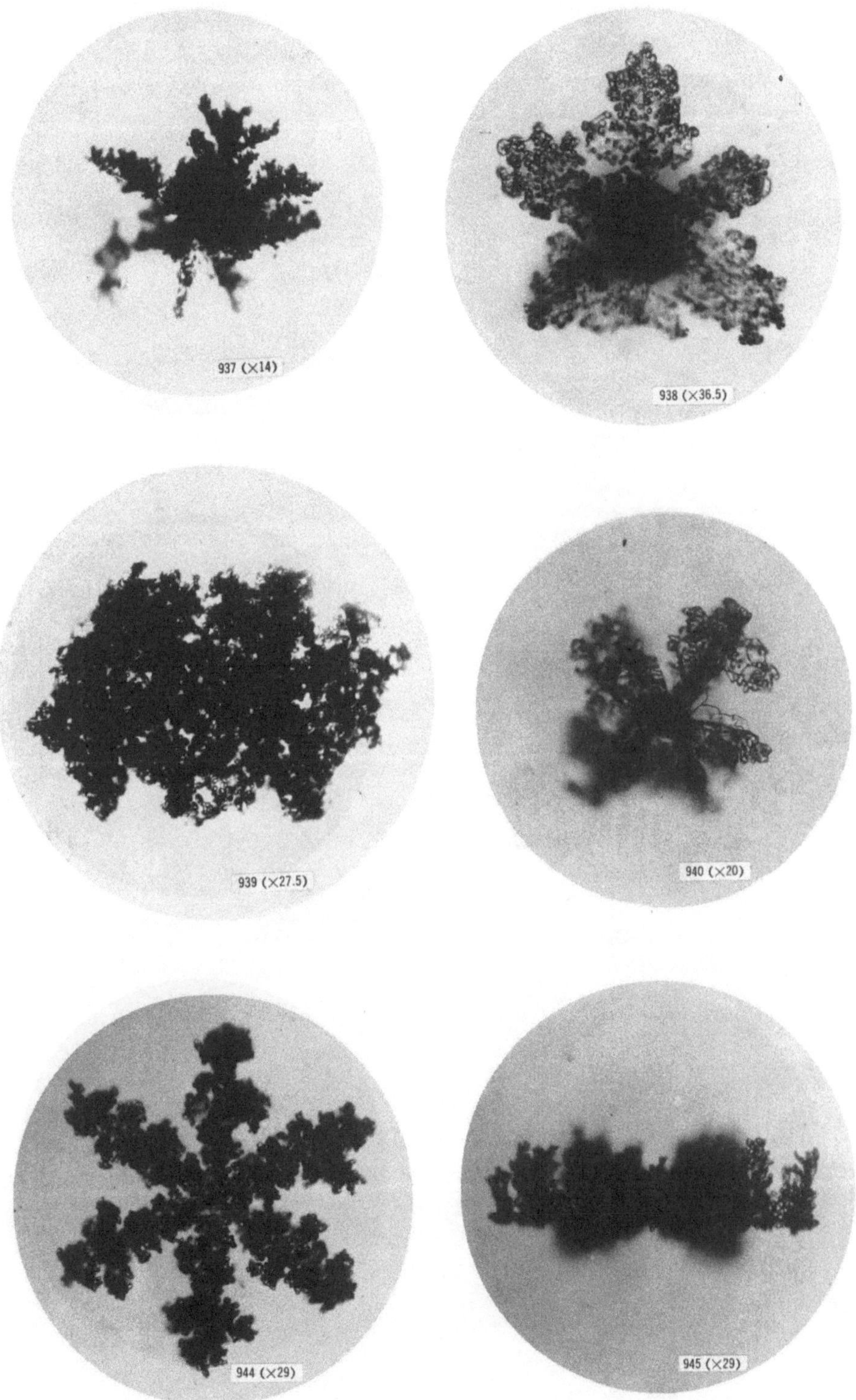

937 (×14)

938 (×36.5)

939 (×27.5)

940 (×20)

944 (×29)

945 (×29)

PLATE 148

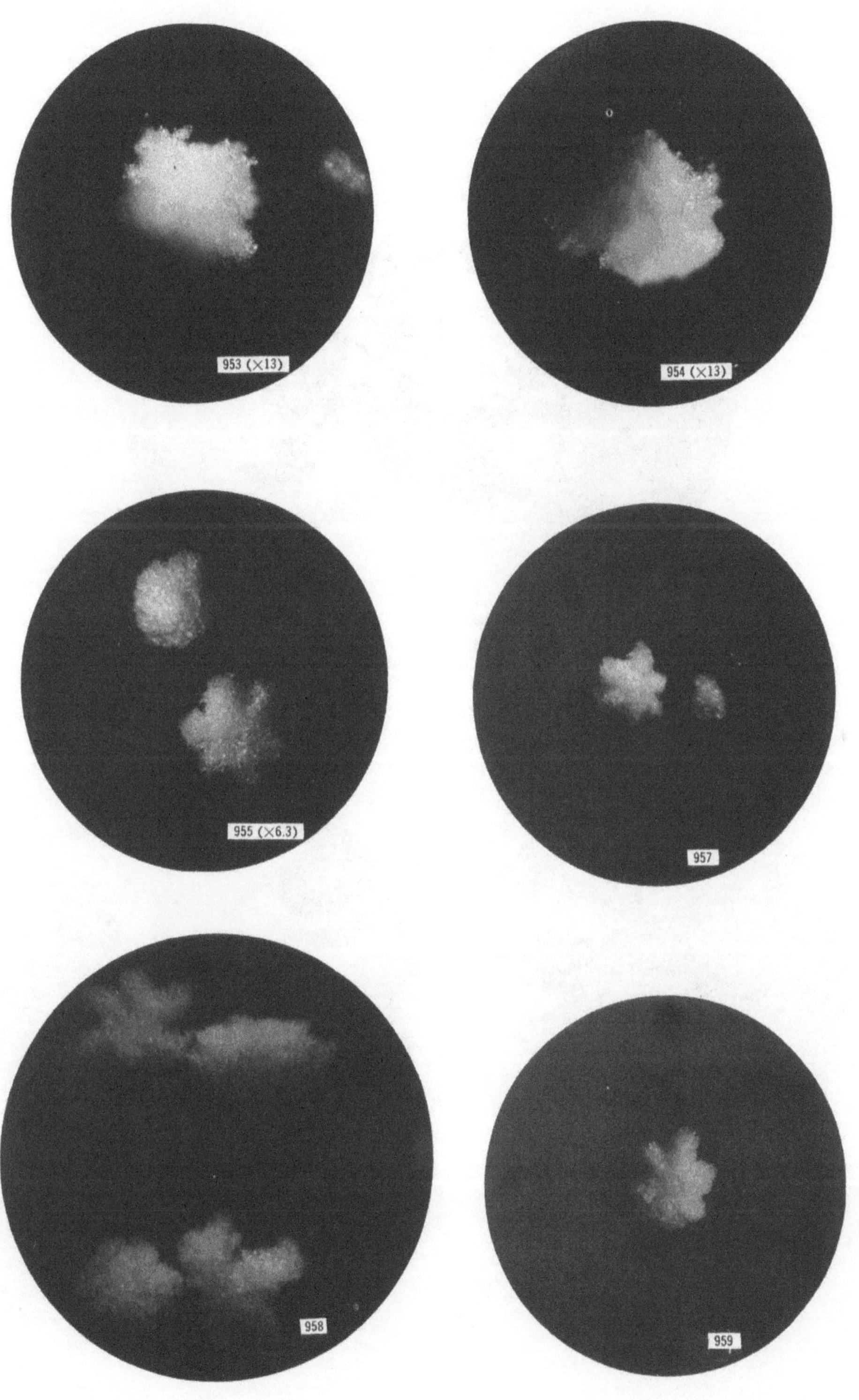
953 (×13)
954 (×13)
955 (×6.3)
957
958
959

PLATE 150

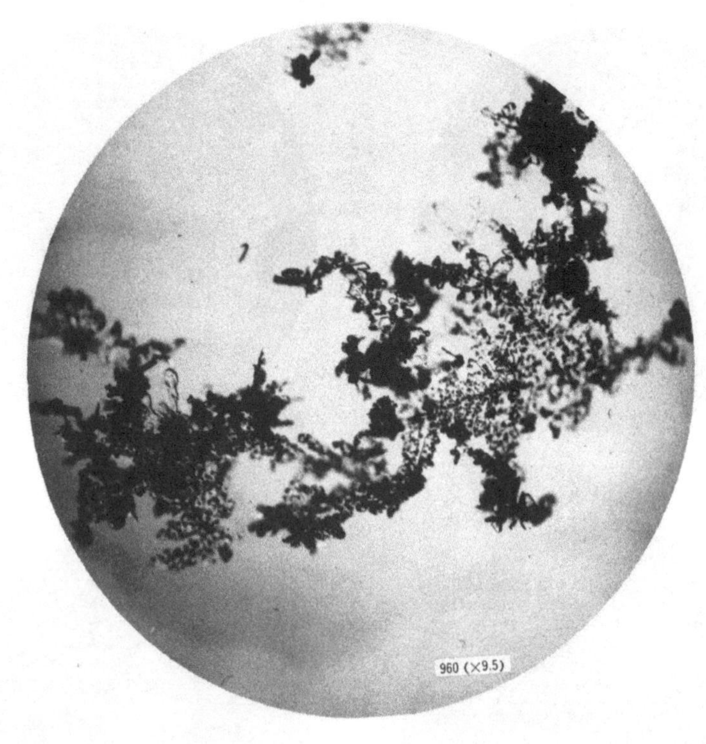

960 (×9.5)

962 (×10)

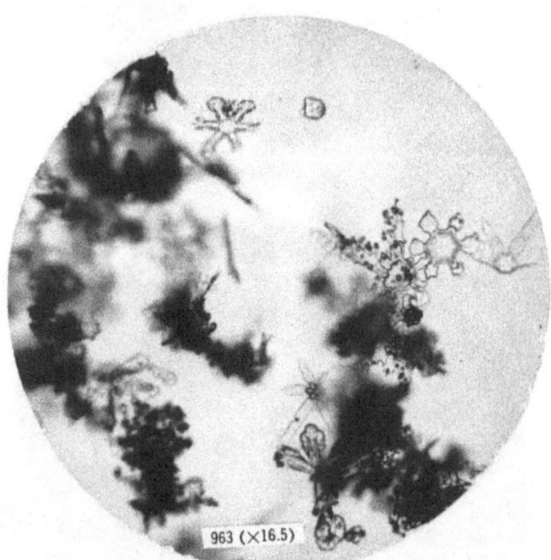

963 (×16.5)

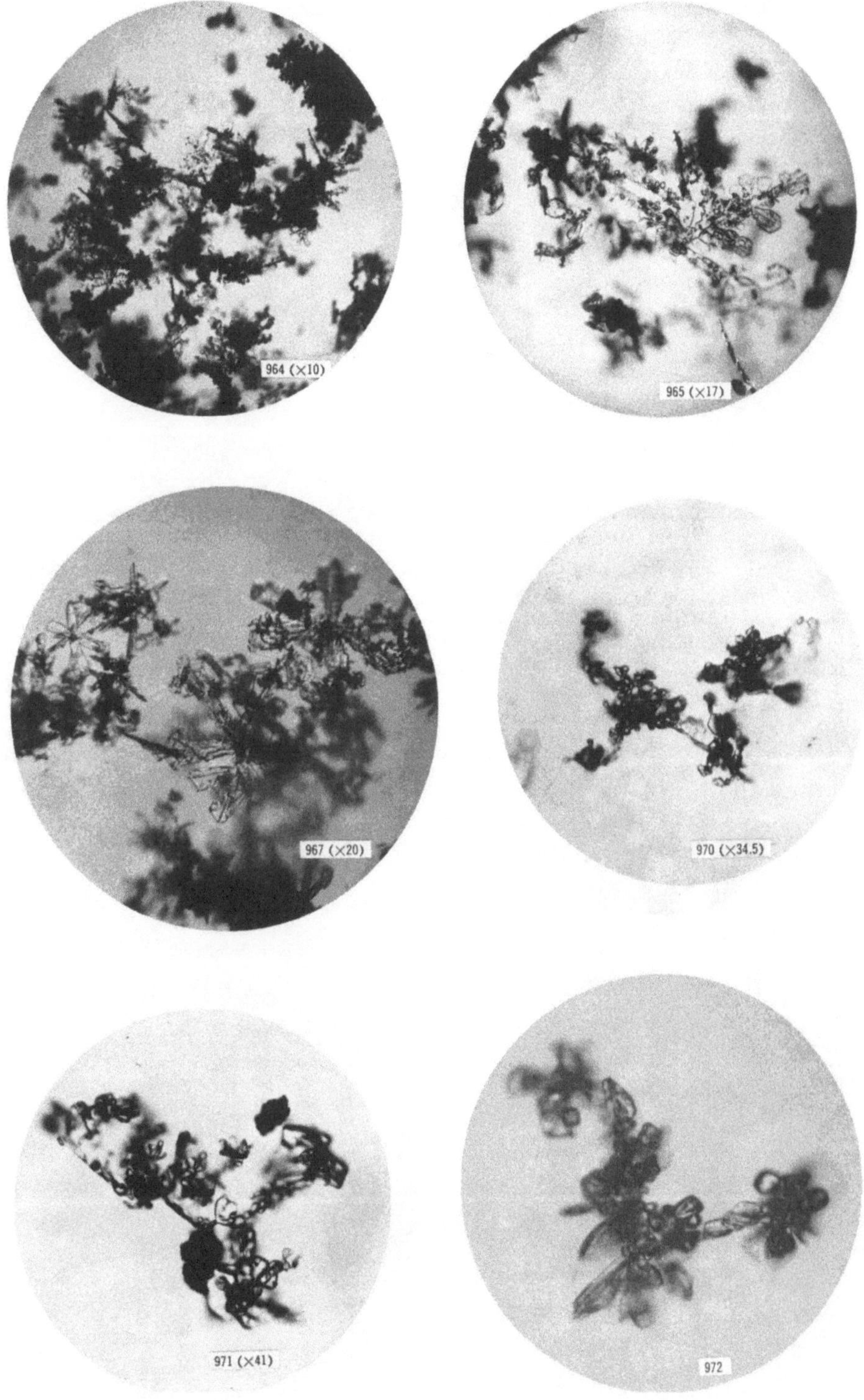
964 (×10)
965 (×17)
967 (×20)
970 (×34.5)
971 (×41)
972

PLATE 152

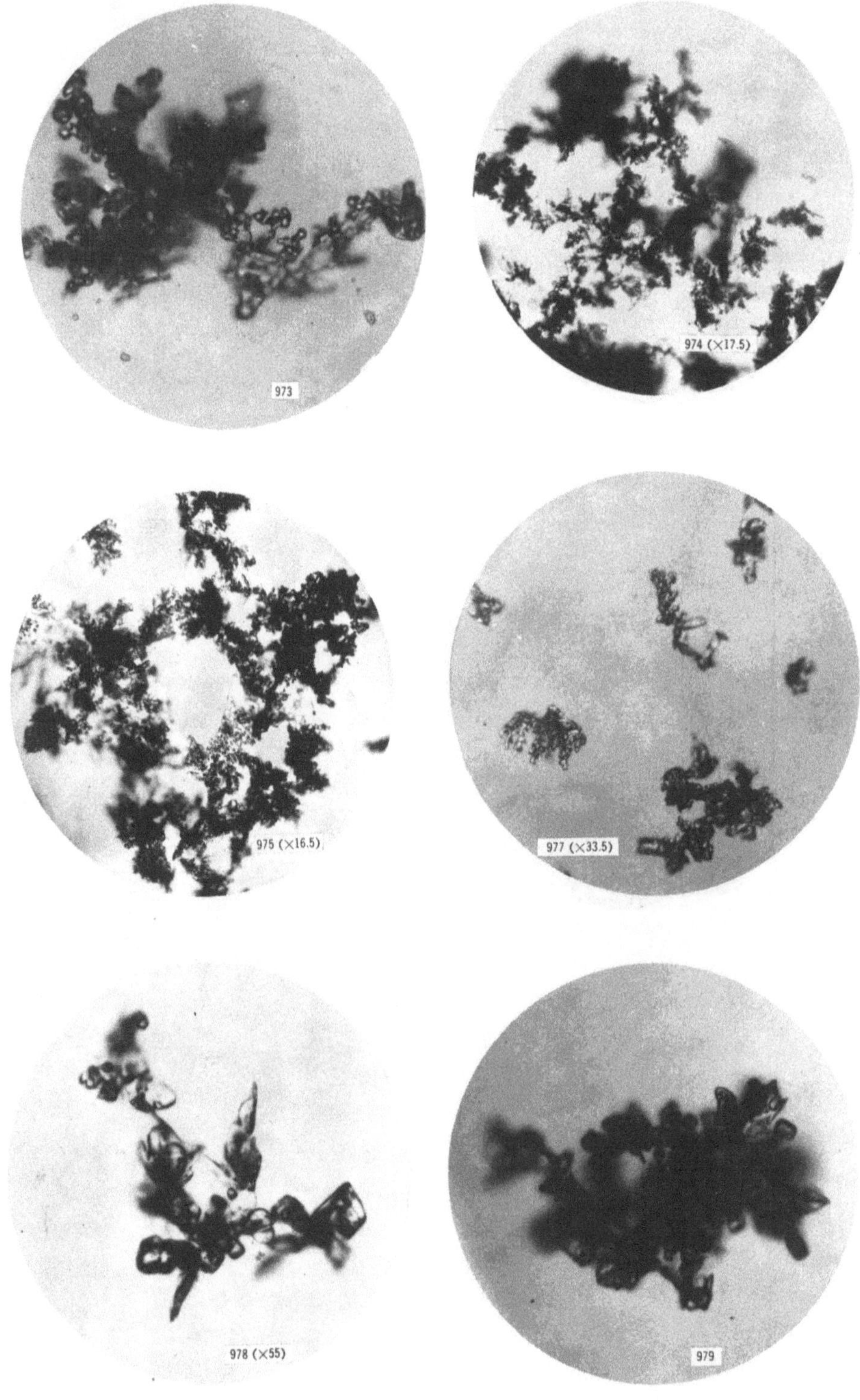

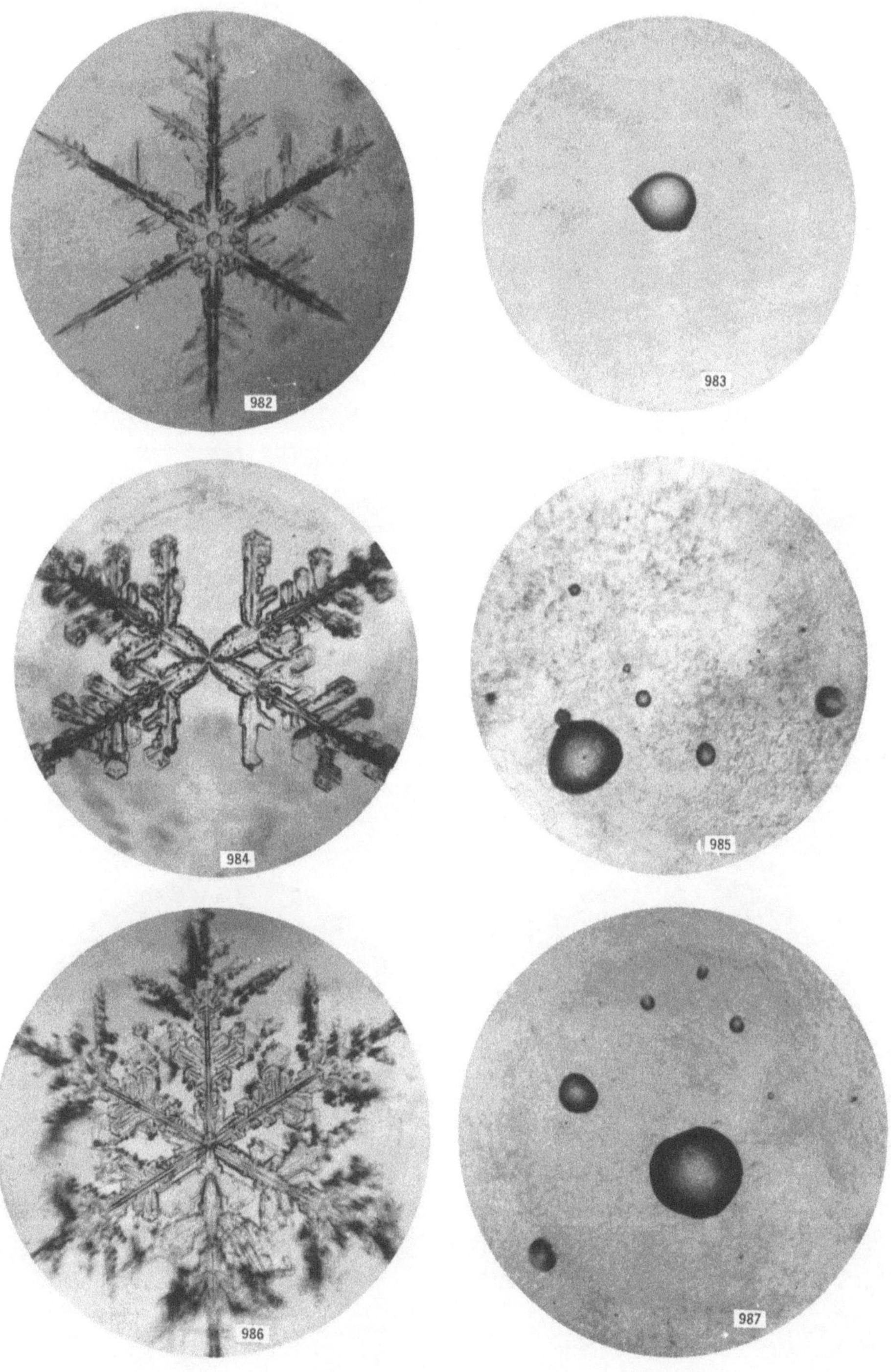

982: d = 2.9 mm; 983: m = 0.024 mg, v = 26 cm/sec; 984: d = 3.6 mm; 985: m = 0.063 mg, v = 44 cm/sec; 986: d = 3.8 mm; 987: m = 0.12 mg, v = 41 cm/sec

PLATE 154

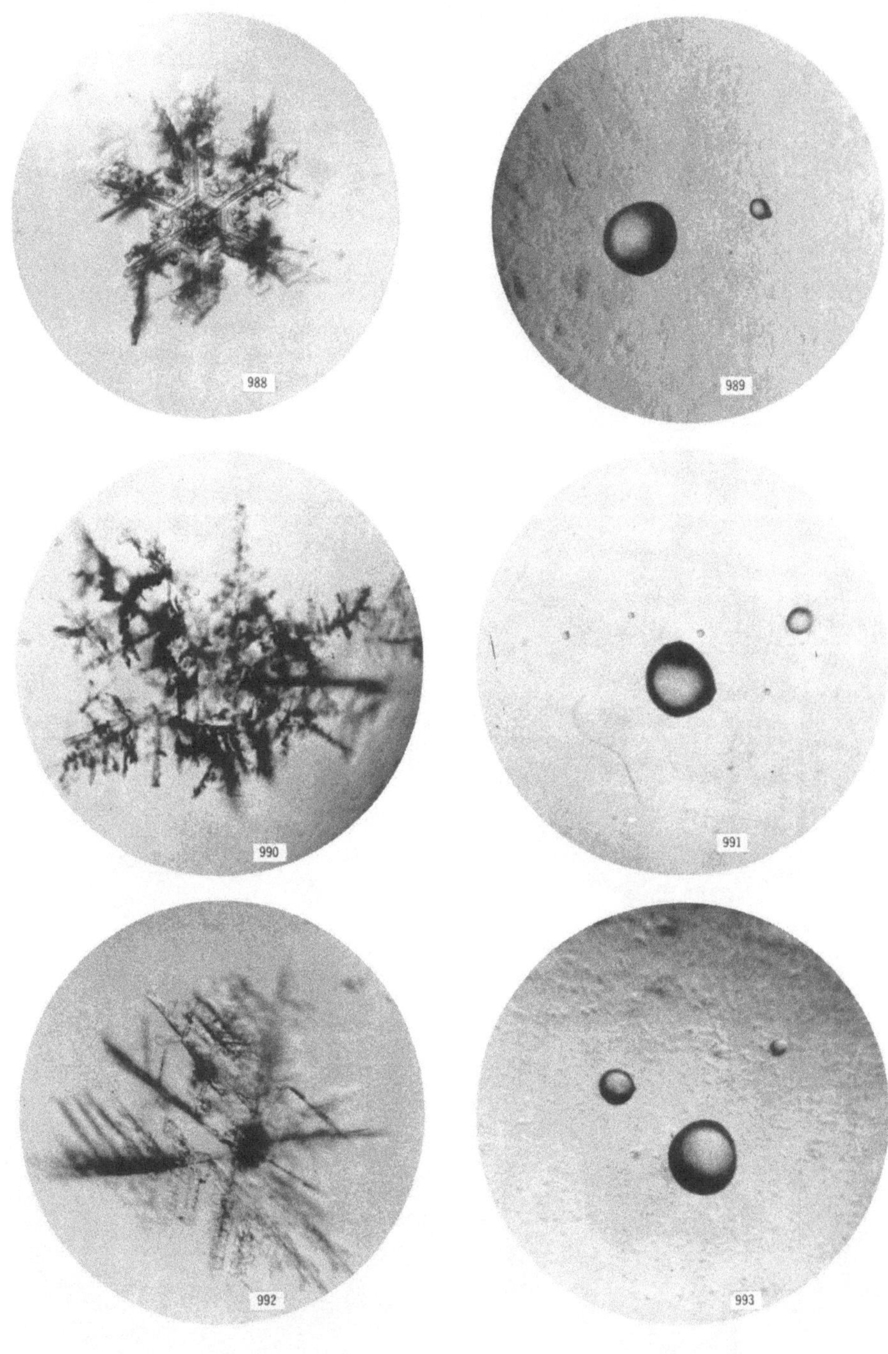

988: $d = 2.3$ mm; 989: $m = 0.058$ mg, $v = 57$ cm/sec; 990: $d = 5.3$ mm; 991: $m = 0.32$ mg, $v = 55$ cm/sec; 992: $d = 2.9$ mm; 993: $m = 0.059$ mg, $v = 49$ cm/sec

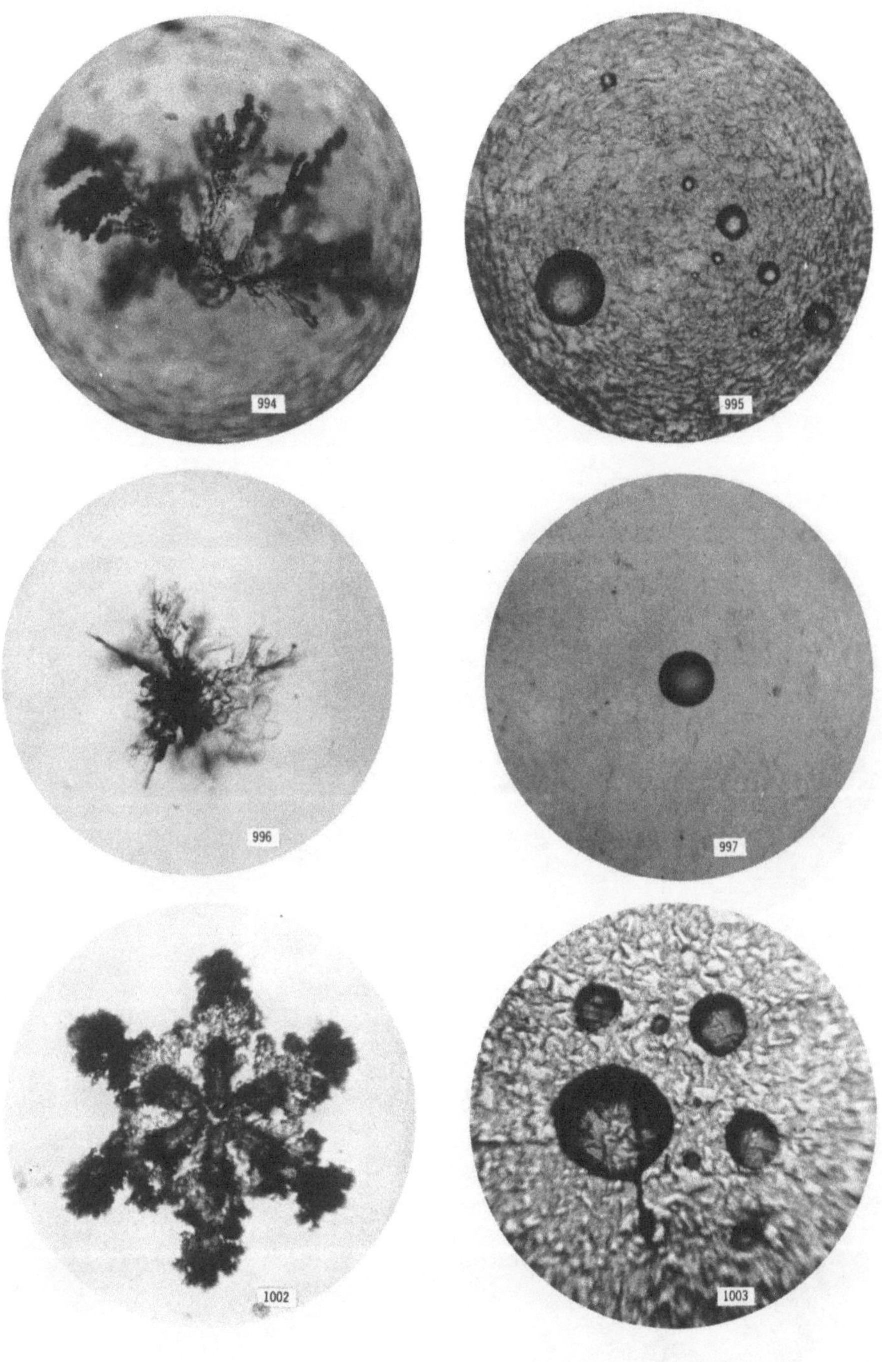

994: d = 3.4 mm; 995: m = 0.092 mg; 996: d = 1.8 mm; 997: m = 0.018 mg, v = 54 cm/sec; 1002: d = 2.5 mm; 1003: m = 0.22 mg, v = 97 cm/sec

PLATE 156

1004: d = 3.0 mm, m = 0.027 mg, v = 29 cm/sec; 1006: d = 4.6 mm, m = 0.086 mg, v = 32 cm/sec; 1007: d = 4.9 mm, v = 30 cm/sec; 1008: d = 2.9 mm, v = 33 cm/sec; 1009: d = 5.6 mm, v = 49 cm/sec; 1010: d = 2.8 mm, m = 0.027 mg, v = 29 cm/sec

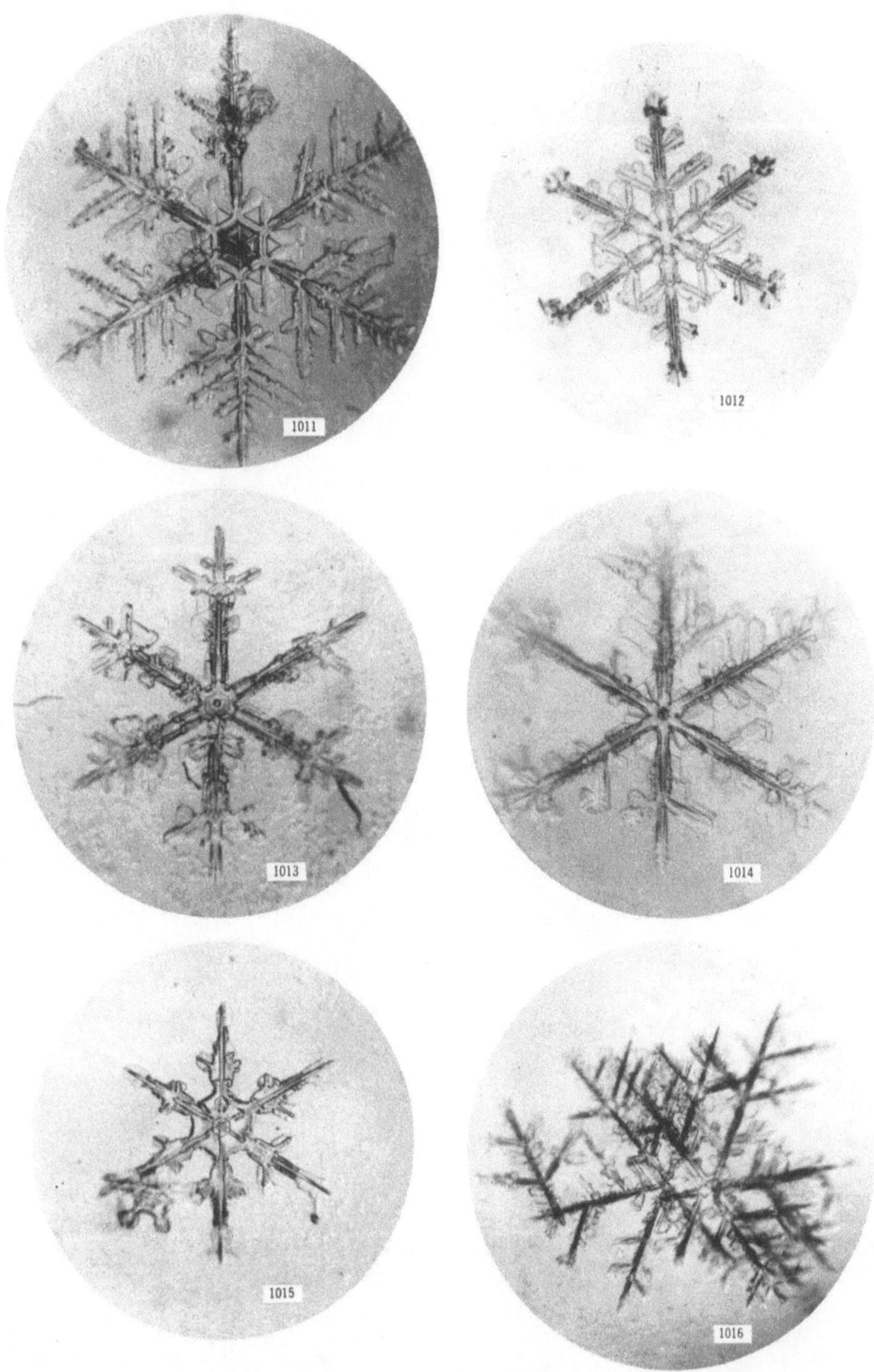

1011: $d = 3.3$ mm, $v = 29$ cm/sec; 1012: $d = 2.3$ mm, $m = 0.013$ mg, $v = 31$ cm/sec; 1013: $d = 2.5$ mm, $m = 0.017$ mg, $v = 32$ cm/sec; 1014: $d = 3.5$ mm, $m = 0.033$ mg, $v = 32$ cm/sec; 1015: $d = 1.7$ mm, $m = 0.014$ mg, $v = 28$ cm/sec; 1016: $d = 5.9$ mm, $m = 0.15$ mg, $v = 38$ cm/sec

PLATE 158

1017: $d = 4.8$ mm, $v = 52$ cm/sec; 1018: $d = 7.7$ mm, $v = 60$ cm/sec; 1019: $d = 1.8$ mm, $v = 61$ cm/sec; 1021: $d = 3.4$ mm, $m = 0.075$ mg, $v = 43$ cm/sec; 1022: $d = 2.3$ mm, $m = 0.037$ mg, $v = 49$ cm/sec; 1023: $d = 1.9$ mm, $m = 0.072$ mg, $v = 49$ cm/sec

1024: $d = 4.4$ **mm,** $v = 51$ **cm/sec; 1025:** $d = 4.8$ **mm,** $m = 0.19$ **mg; 1026:** $d = 2.9$ **mm,** $m = 0.051$ **mg,** $v = 44$ **cm/sec; 1027:** $d = 5.8$ **mm,** $m = 1.35$ **mg,** $v = 88$ **cm/sec; 1028:** $d = 1.5$ **mm,** $m = 0.041$ **mg; 1029:** $d = 2.6$ **mm,** $m = 0.045$ **mg**

PLATE 160

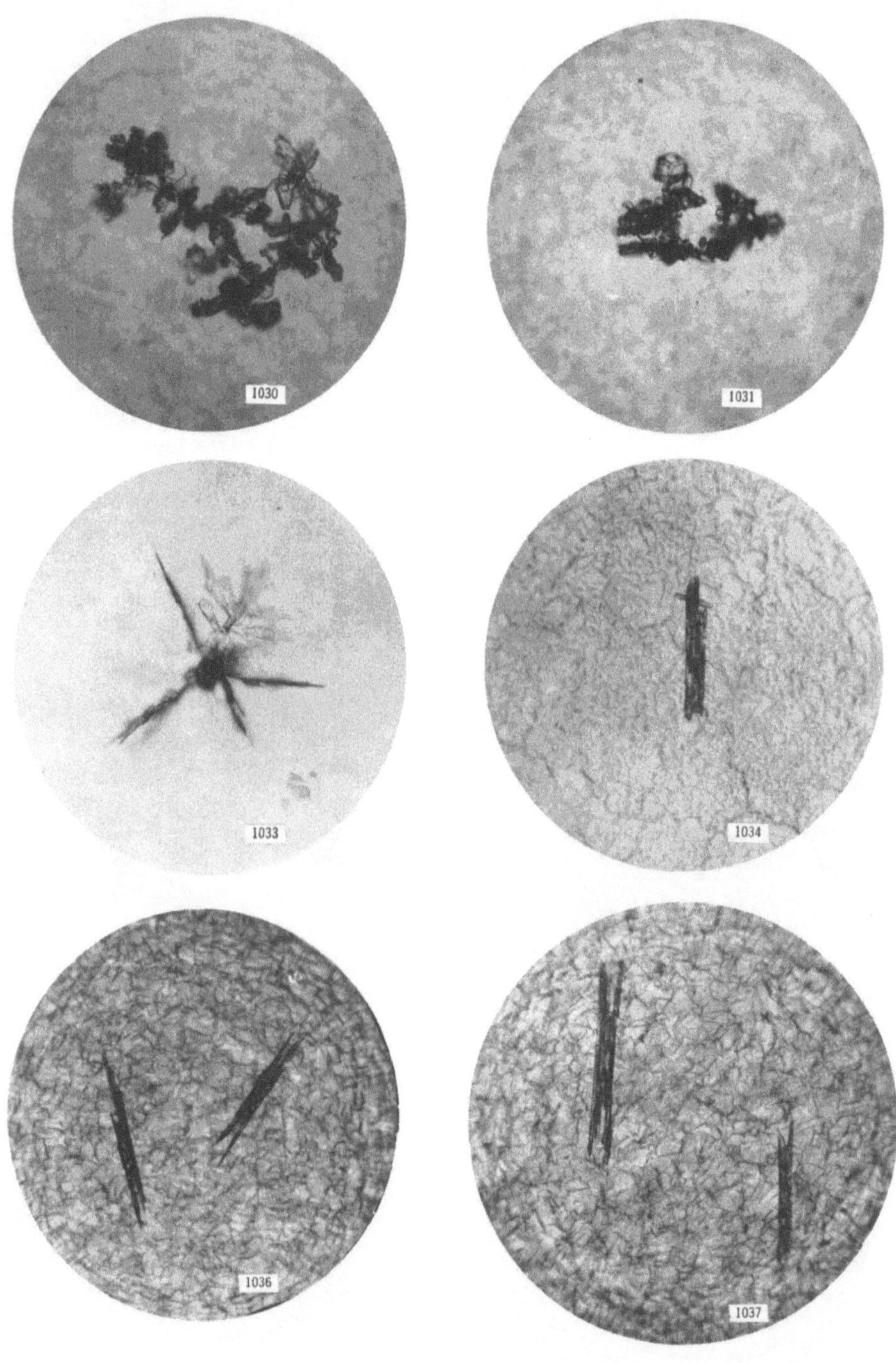

1030: $d = 2.2$ mm, $m = 0.050$ mg; **1031:** $d = 1.3$ mm, $m = 0.022$ mg; **1033:** $d = 1.6$ mm, $m = 0.014$ mg; **1034:** $l = 1.0$ mm, $v = 46$ cm/sec; **1036:** $l = 1.6$, 1.4 mm, $m = 0.0047$, 0.0037 mg; **1037:** $l = 1.9$, 1.3 mm, $m = 0.0063$, 0.0024 mg

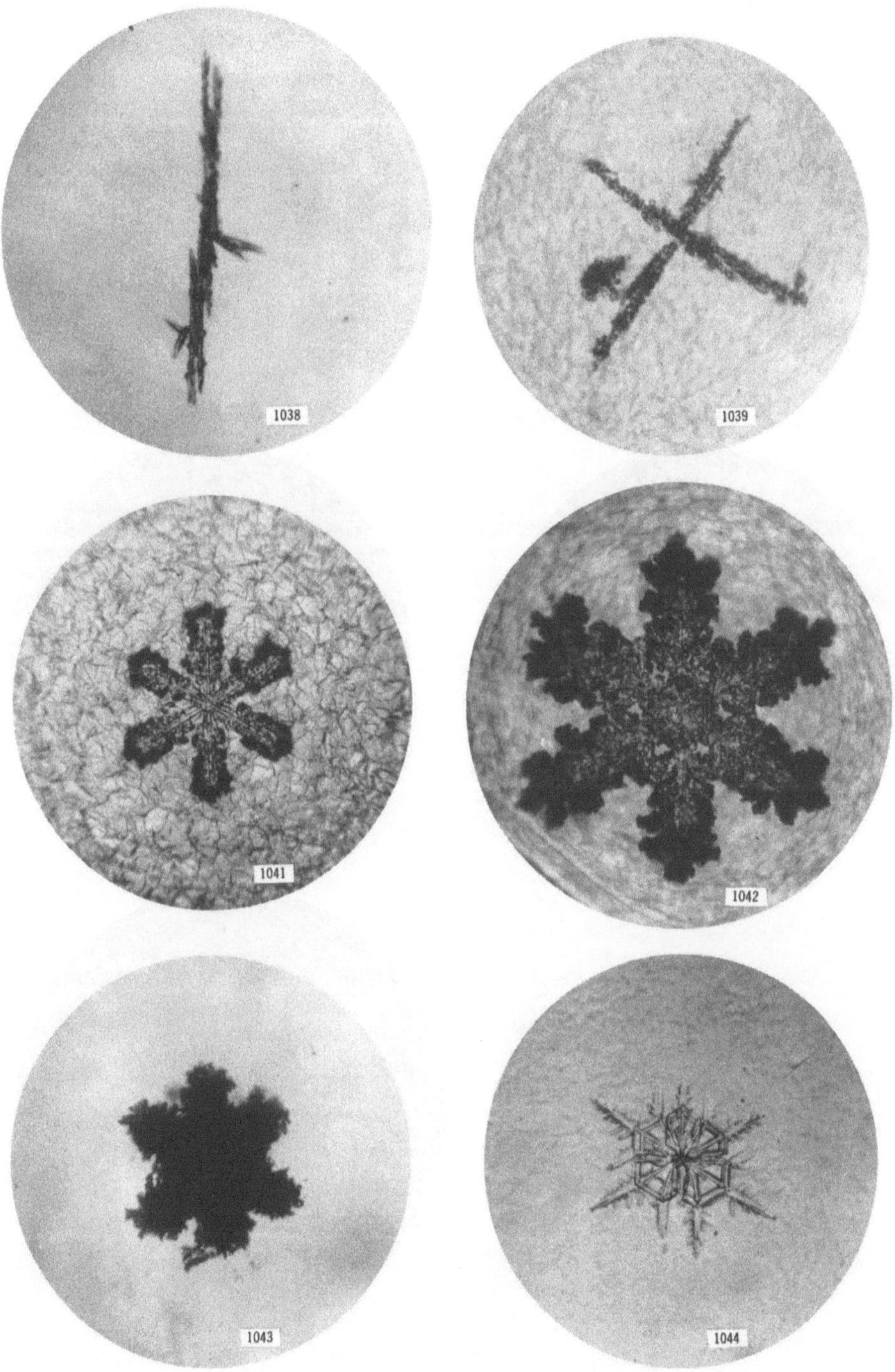

1038: $l = 2.4$ mm, $m = 0.004$ mg; 1039: $l = 2.1$ mm, $m = 0.019$ mg, 1041: $d = 1.7$ mm, $m = 0.062$ mg; 1042: $d = 3.0$ mm, $m = 0.33$ mg; 1043: $d = 1.4$ mm, $m = 0.039$ mg, $v = 97$ cm/sec; 1044: $d = 1.5$ mm, $m = 0.014$ mg, $v = 30$ cm/sec

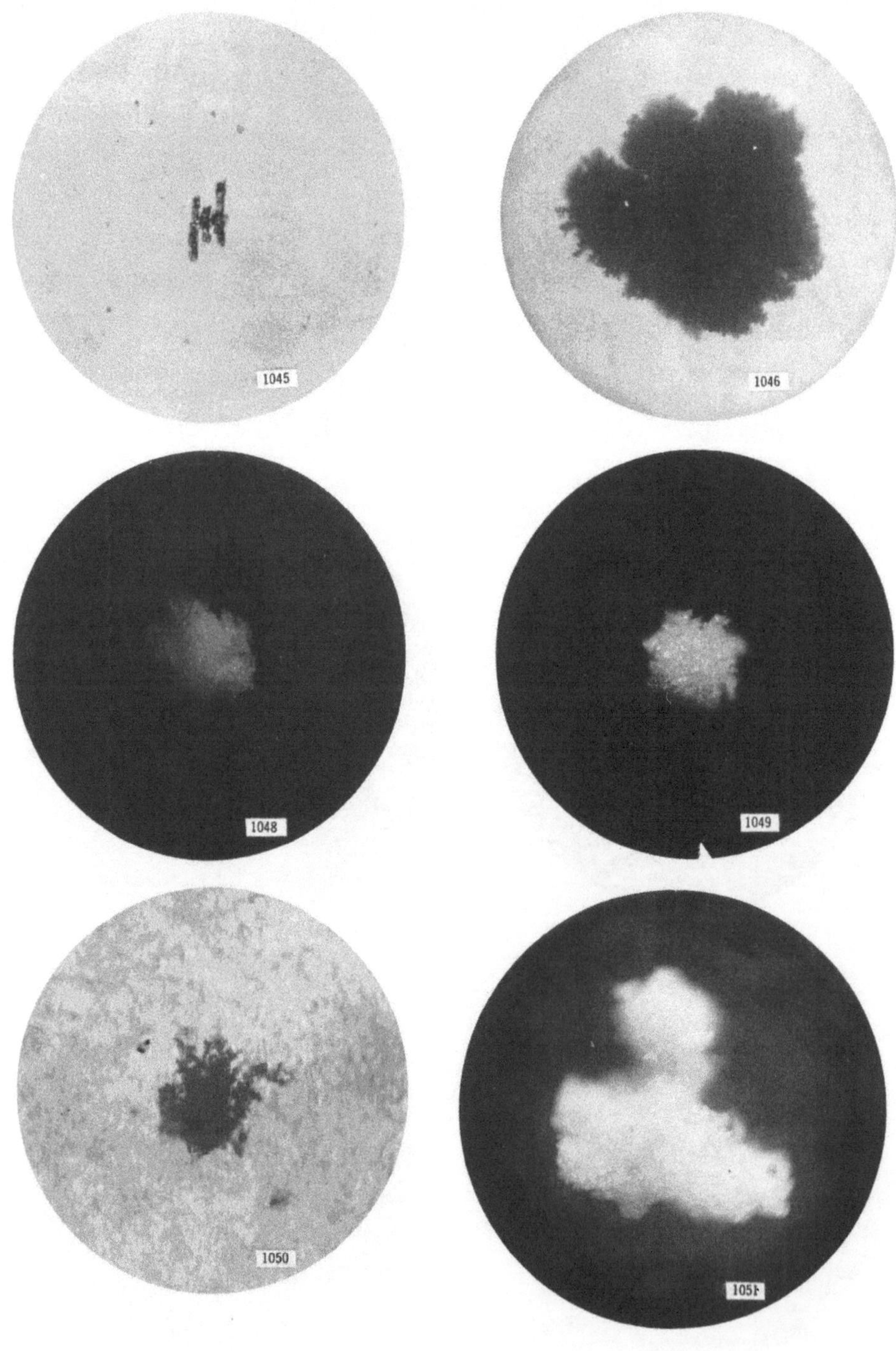

1045: l = 0.6 mm, m = 0.002 mg, v = 26 cm/sec; 1046: d = 2.8 mm, m = 1.57 mg; 1048: d = 3.0 mm, m = 1.56 mg; 1049: d = 1.5 mm, m = 0.08 mg; 1050: d = 1.1 mm, m = 0.032 mg; 1051: d = 2.7 mm, m = 0.78 mg

1064 (×32)
1066 (×32)
1067 (×32)
1068 (×12)
1070 (×55.5)
1071 (×58.5)

1073 (×55)

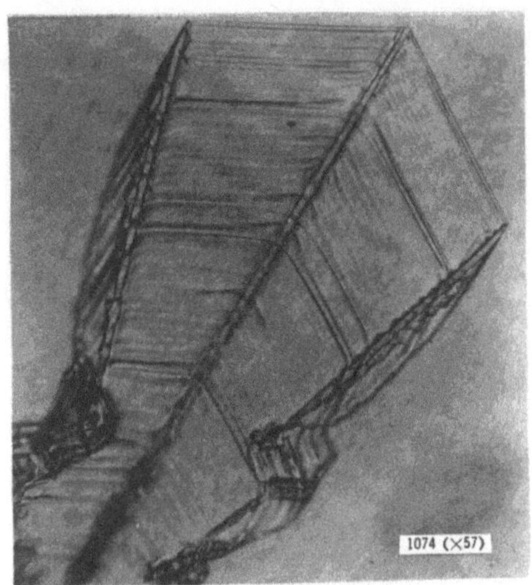

1074 (×57)

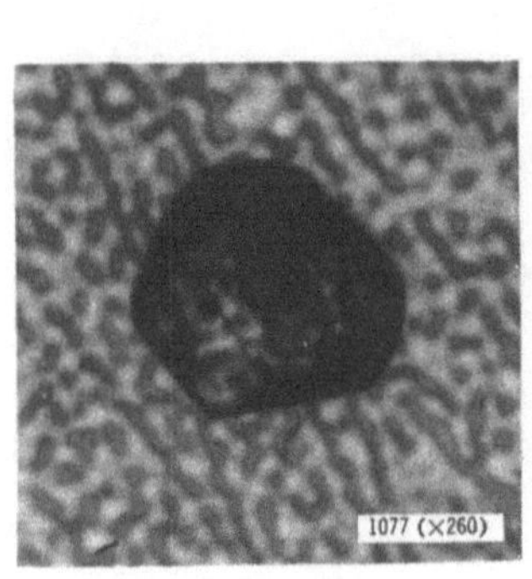

1077 (×260)

1080 (275)

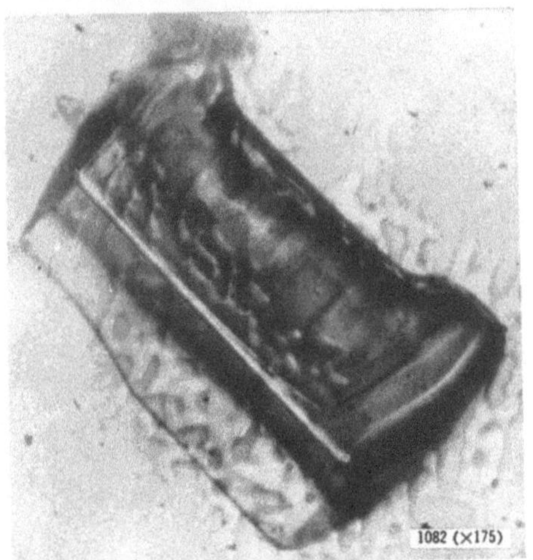

1082 (×175)

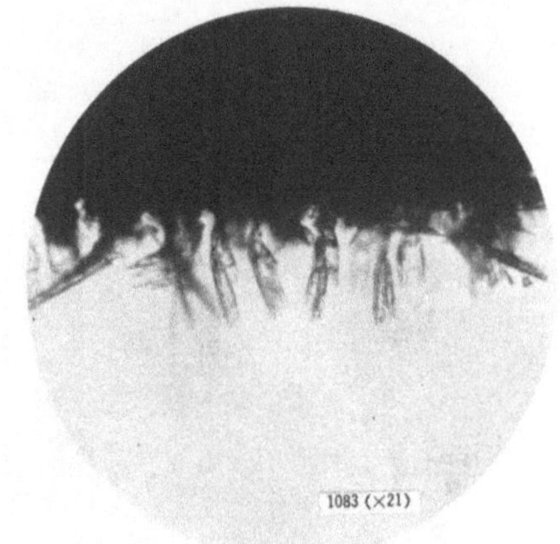

1083 (×21)

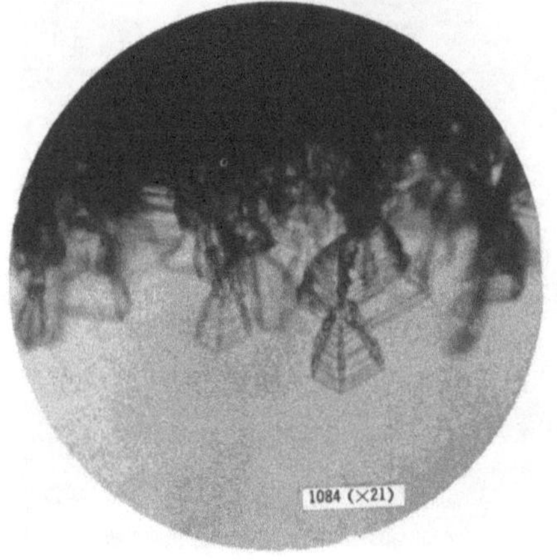

1084 (×21)

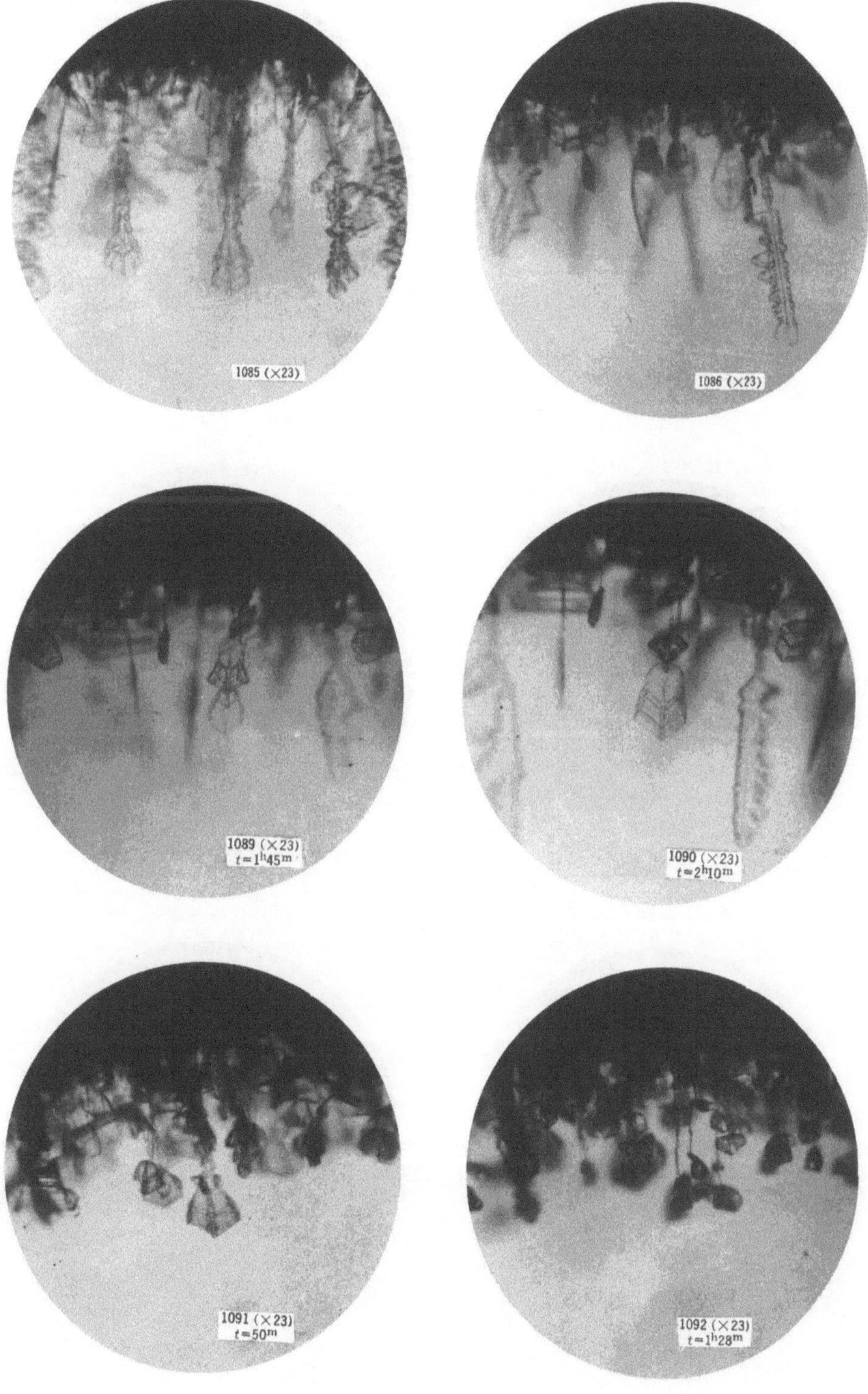
1085 (×23)
1086 (×23)
1089 (×23) $t=1^h45^m$
1090 (×23) $t=2^h10^m$
1091 (×23) $t=50^m$
1092 (×23) $t=1^h28^m$

PLATE 166

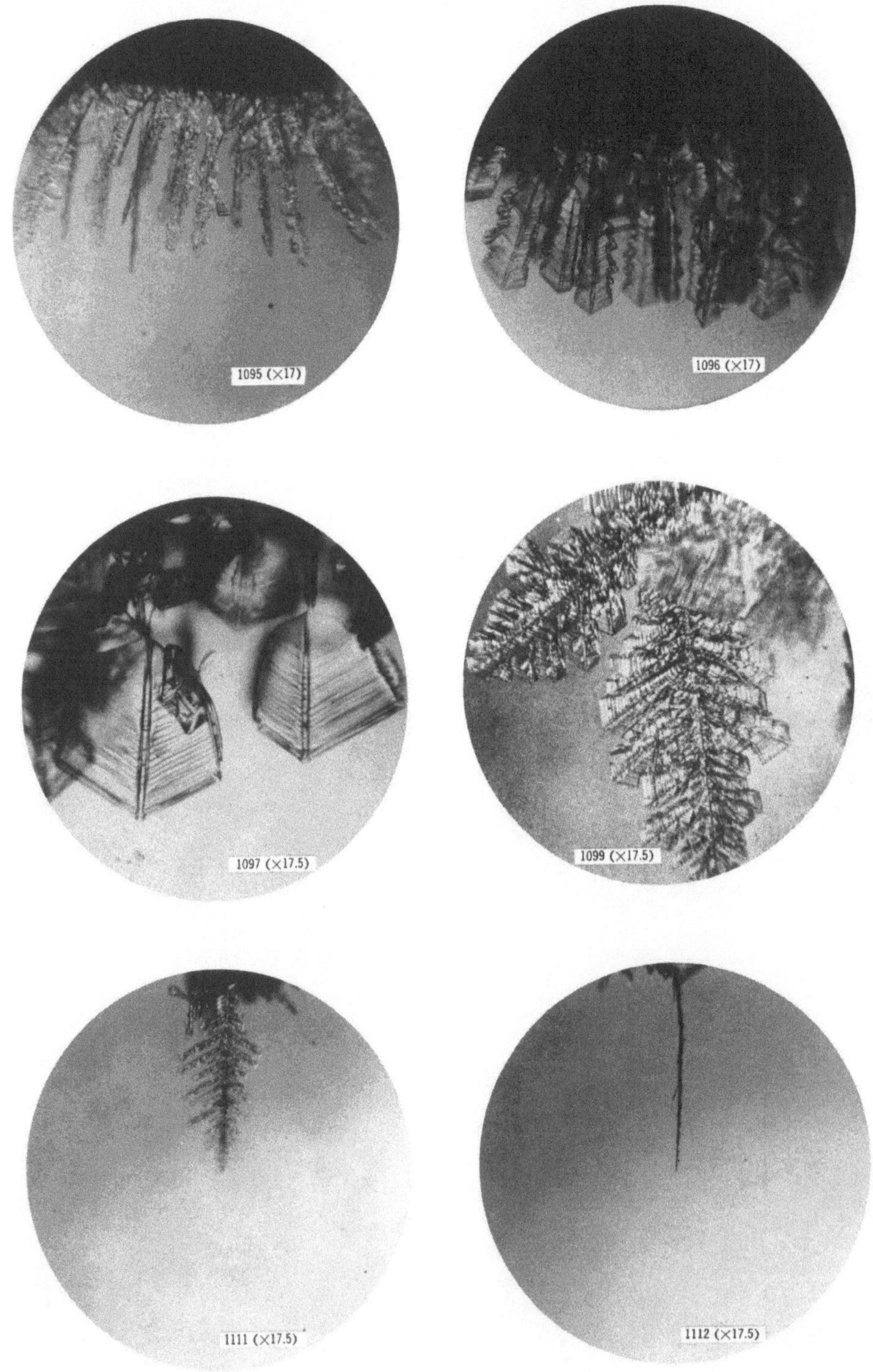

1118 (×17.5)
1119 (×17.5)
1120 (×17.5)
1123 (×15.5)
1121 (×17.5)
1122 (×17.5)

PLATE 168

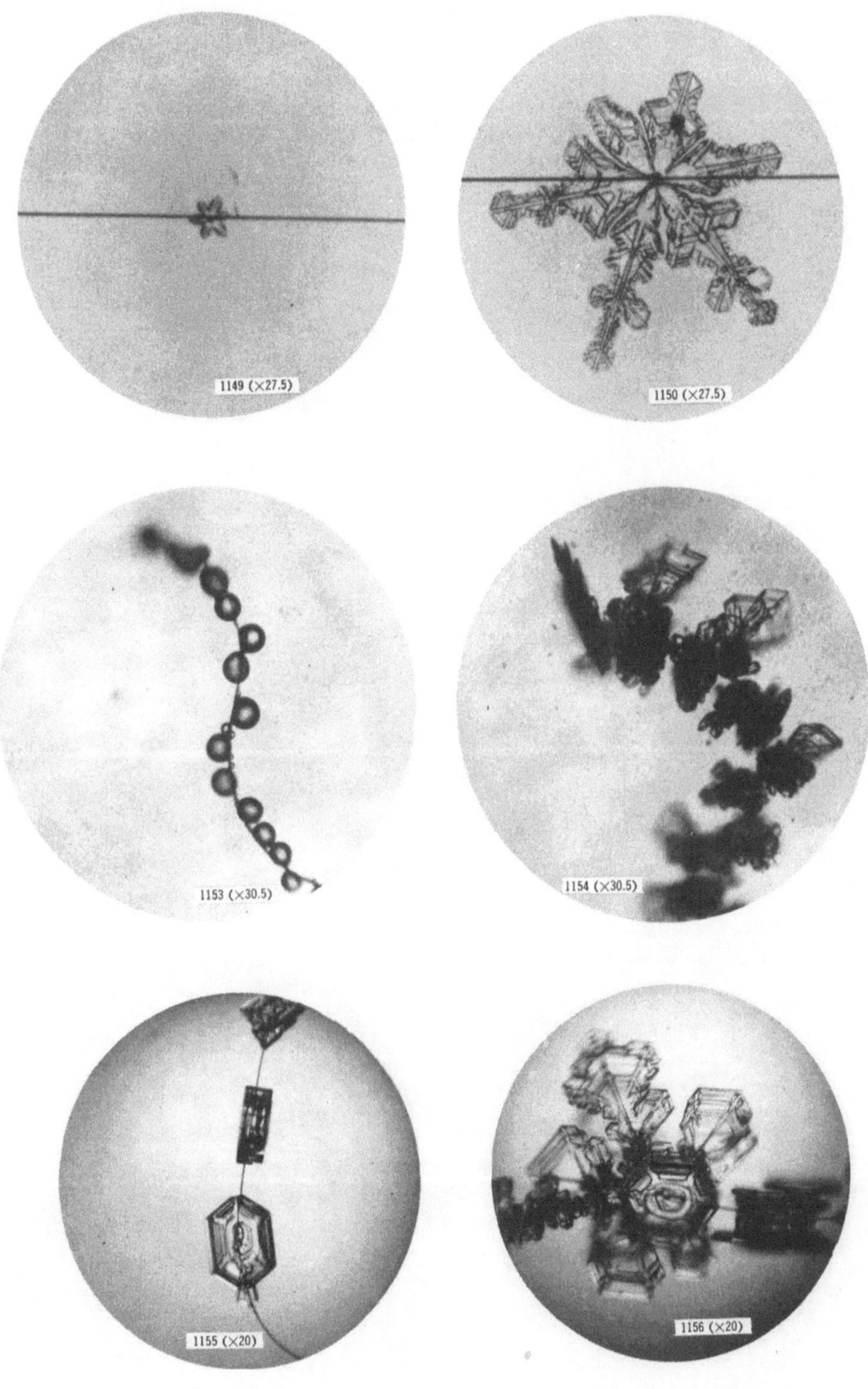
1149 (×27.5)
1150 (×27.5)
1153 (×30.5)
1154 (×30.5)
1155 (×20)
1156 (×20)

PLATE 170

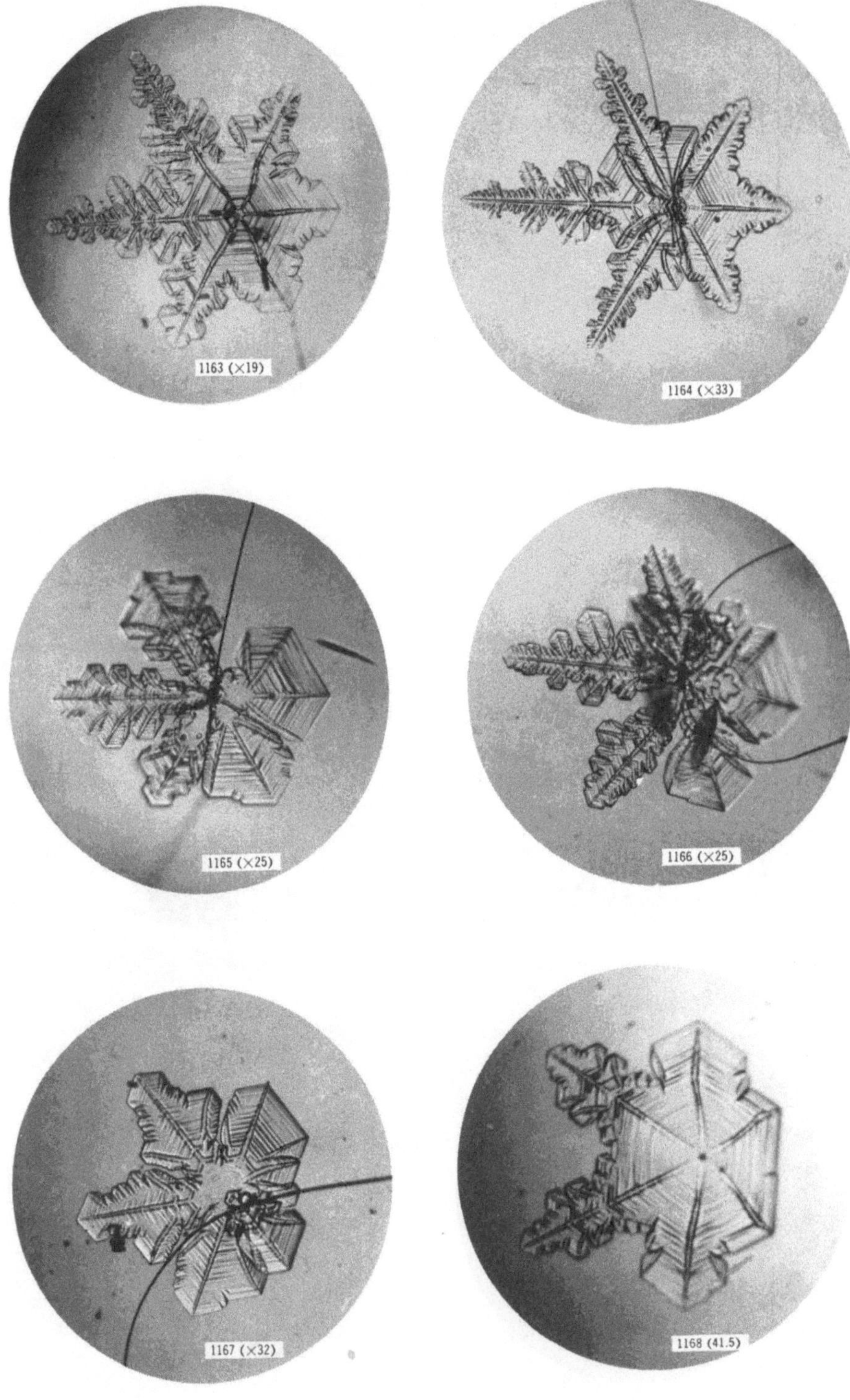

1174 (×18)

1178 (×9)

1179 (×16)

1180 (×15.5)

1181 (×18)

1182 (×18)

PLATE 172

1185 (×18.5)

1186 (×18.5)

1187 (×13)

1188 (×18.5)

1191 (×32)

1192 (×28)

1196 (×18)
1197 (×19)
1199
1200
1204 (×19)
1206 (×21)

PLATE 174

1219 (×17.5)

1222 (×23)

1223 (×26.5)

1224 (×33)

1225 (×18)

1226 (×39.5)

PLATE 176

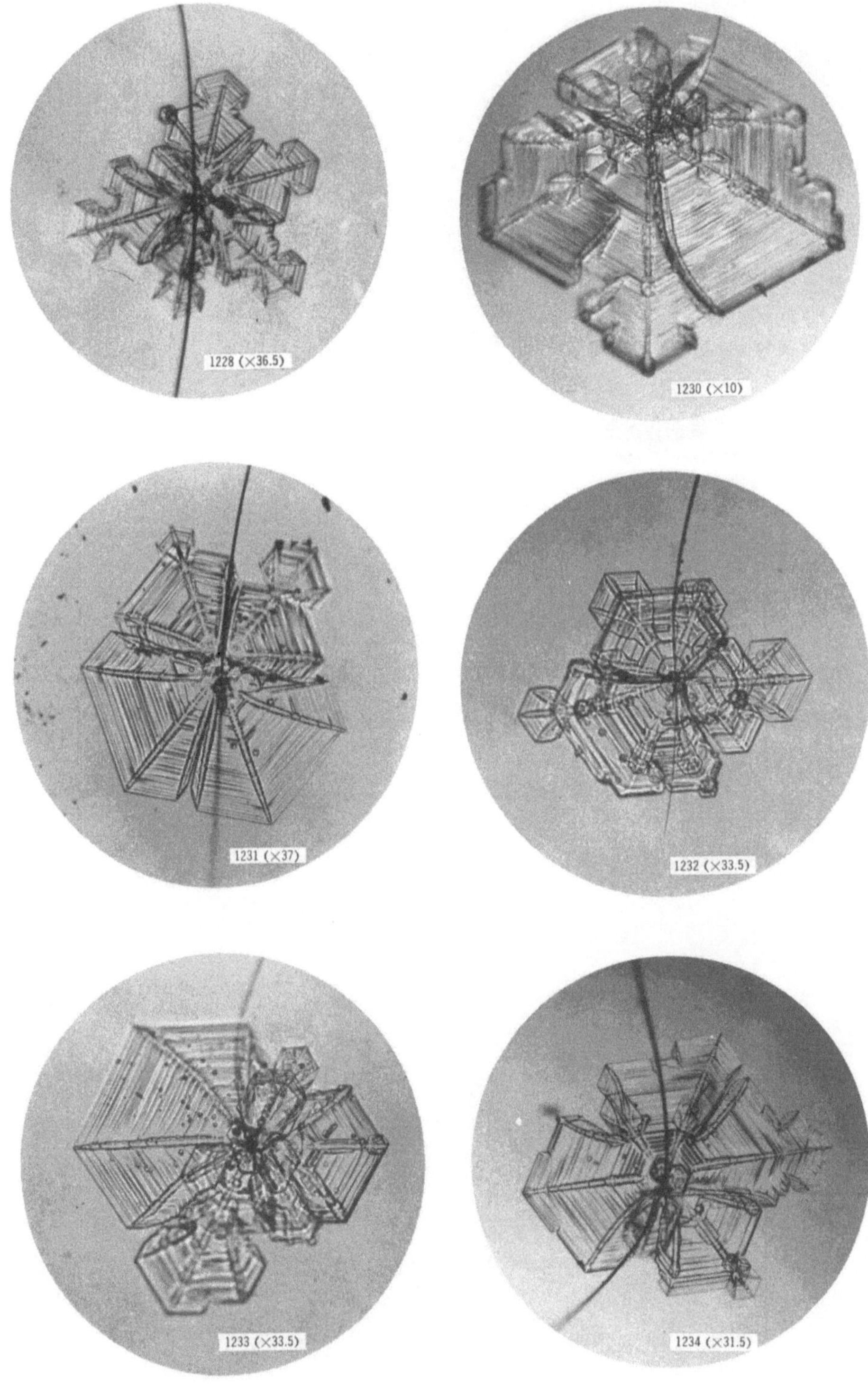

1228 (×36.5)

1230 (×10)

1231 (×37)

1232 (×33.5)

1233 (×33.5)

1234 (×31.5)

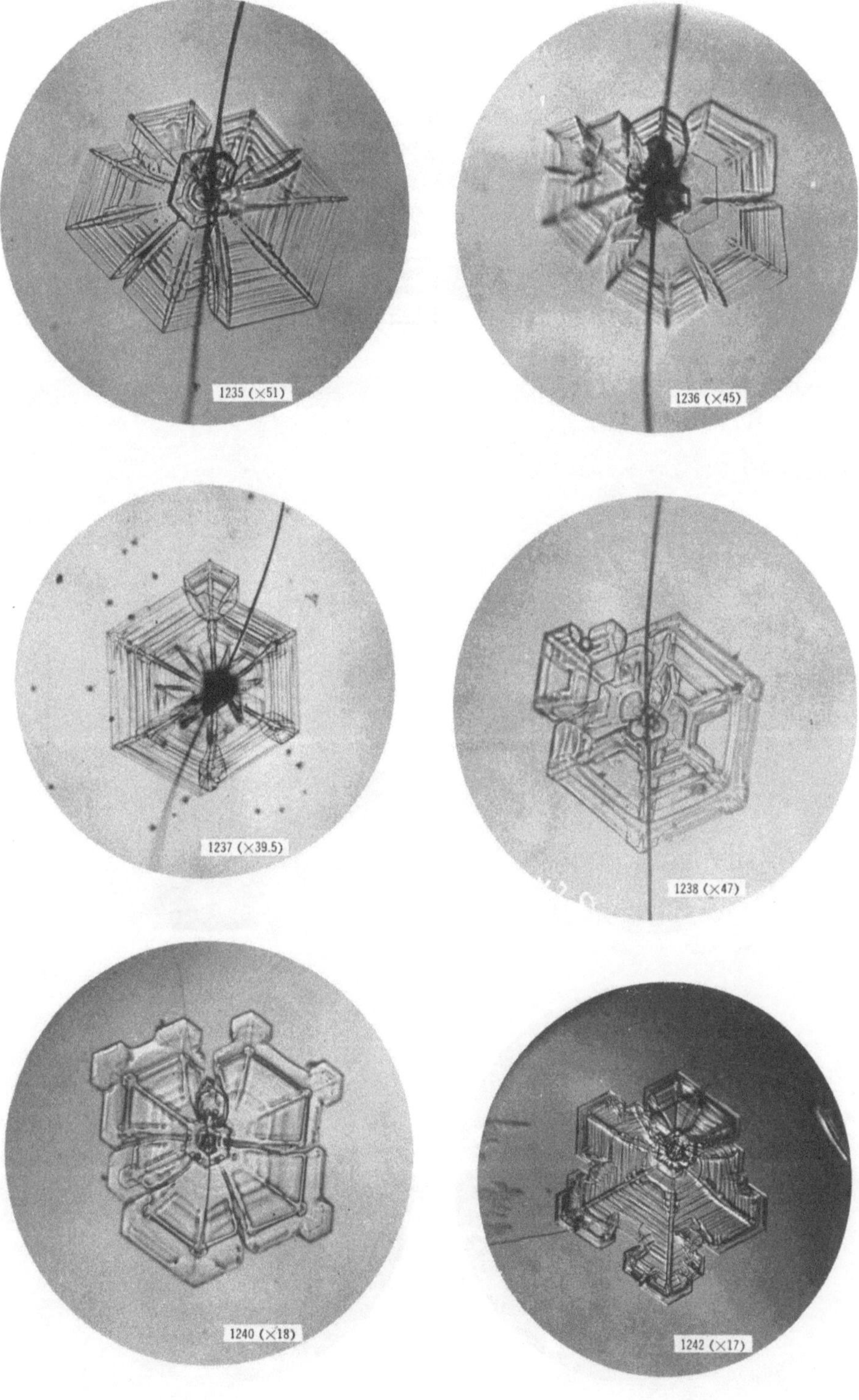
1235 (×51)
1236 (×45)
1237 (×39.5)
1238 (×47)
1240 (×18)
1242 (×17)

PLATE 178

1245 (×7)

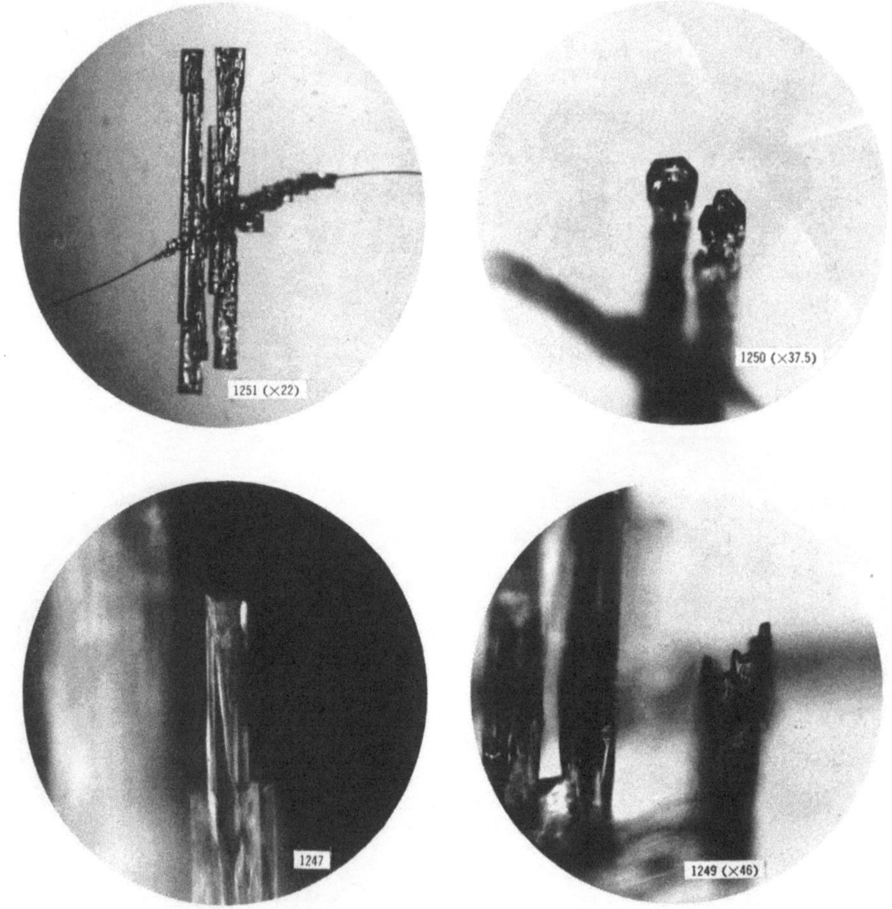

1251 (×22)

1250 (×37.5)

1247

1249 (×46)

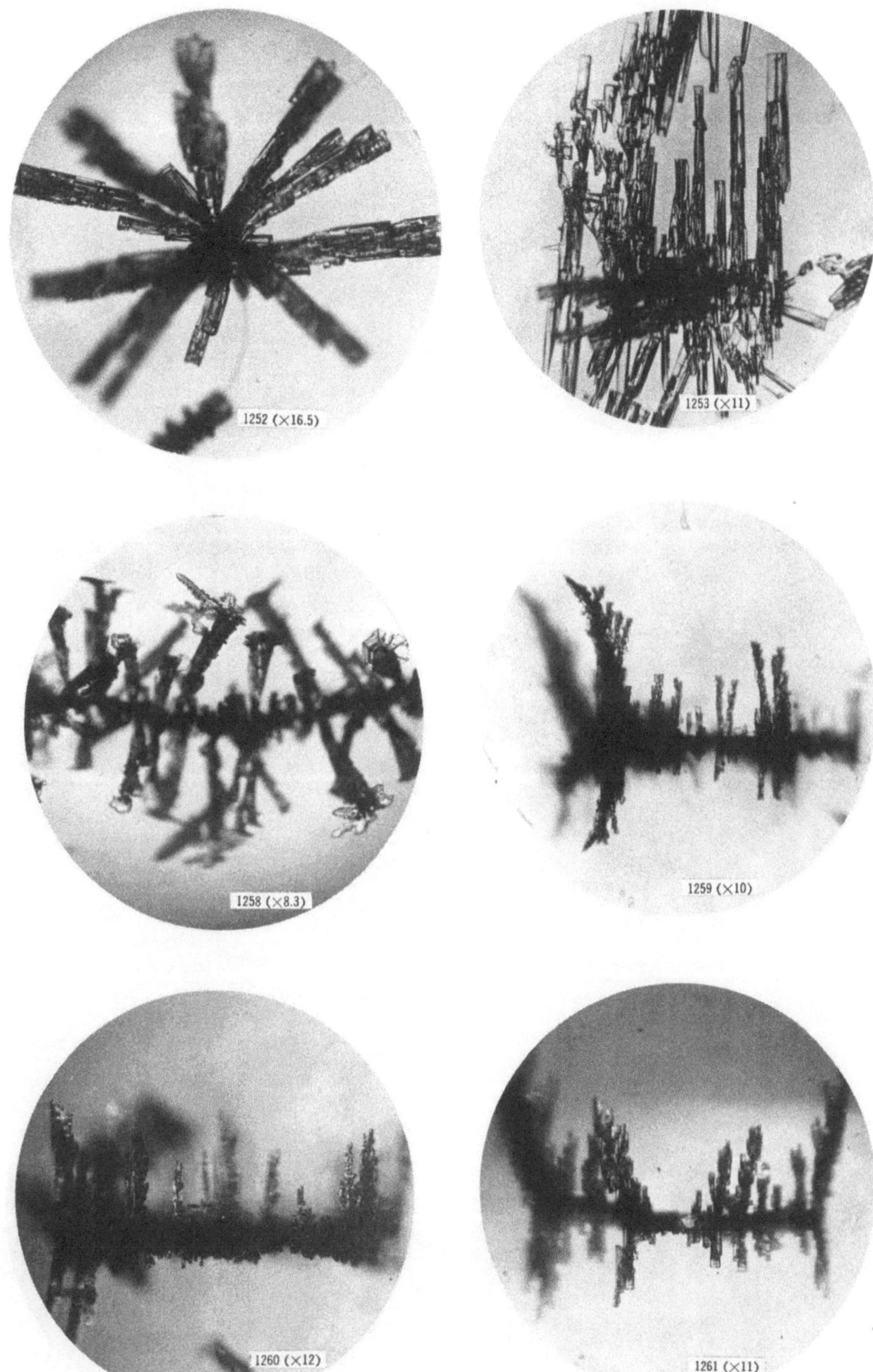
1252 (×16.5)
1253 (×11)
1258 (×8.3)
1259 (×10)
1260 (×12)
1261 (×11)

PLATE 180

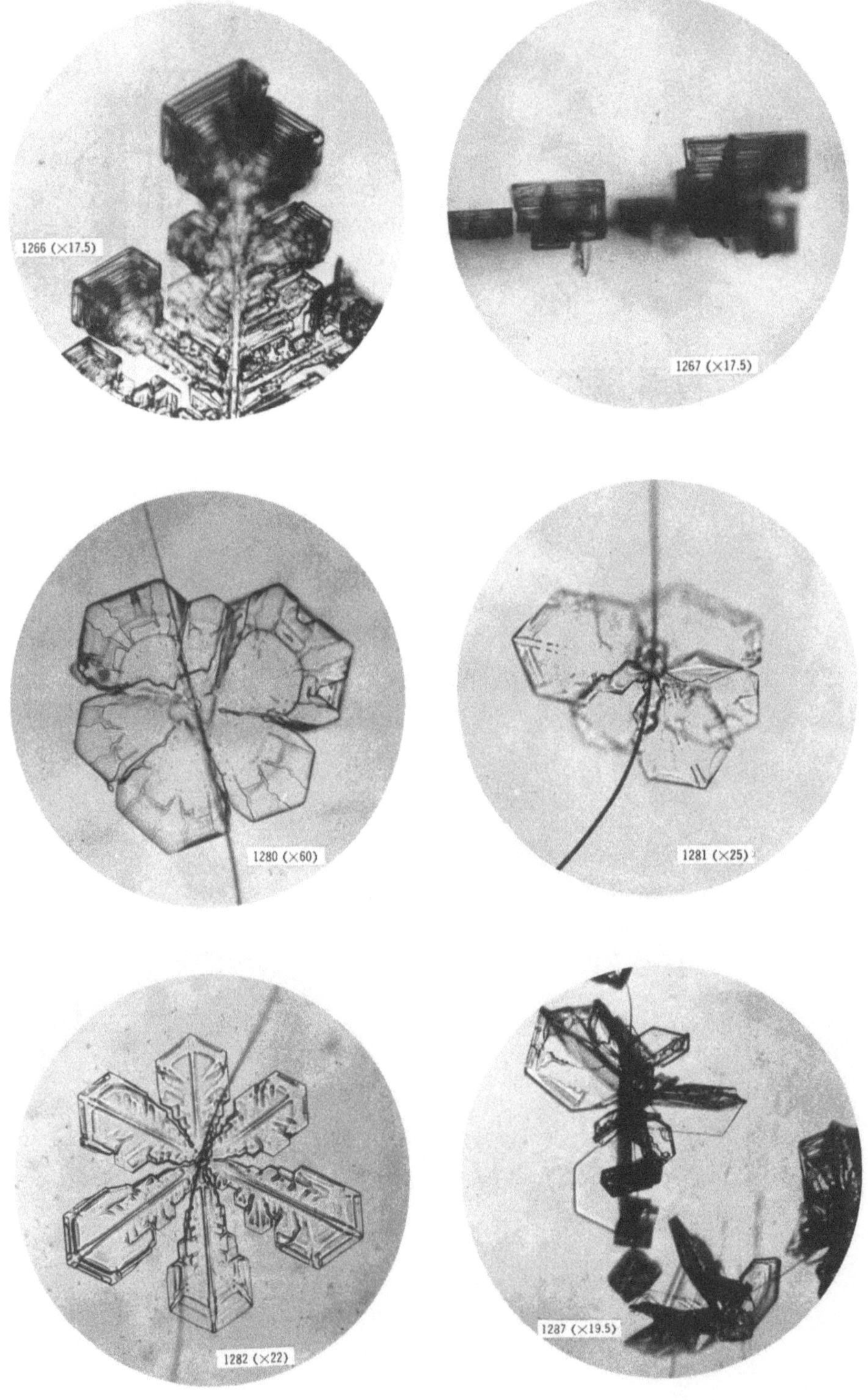

1266 (×17.5)
1267 (×17.5)
1280 (×60)
1281 (×25)
1282 (×22)
1287 (×19.5)

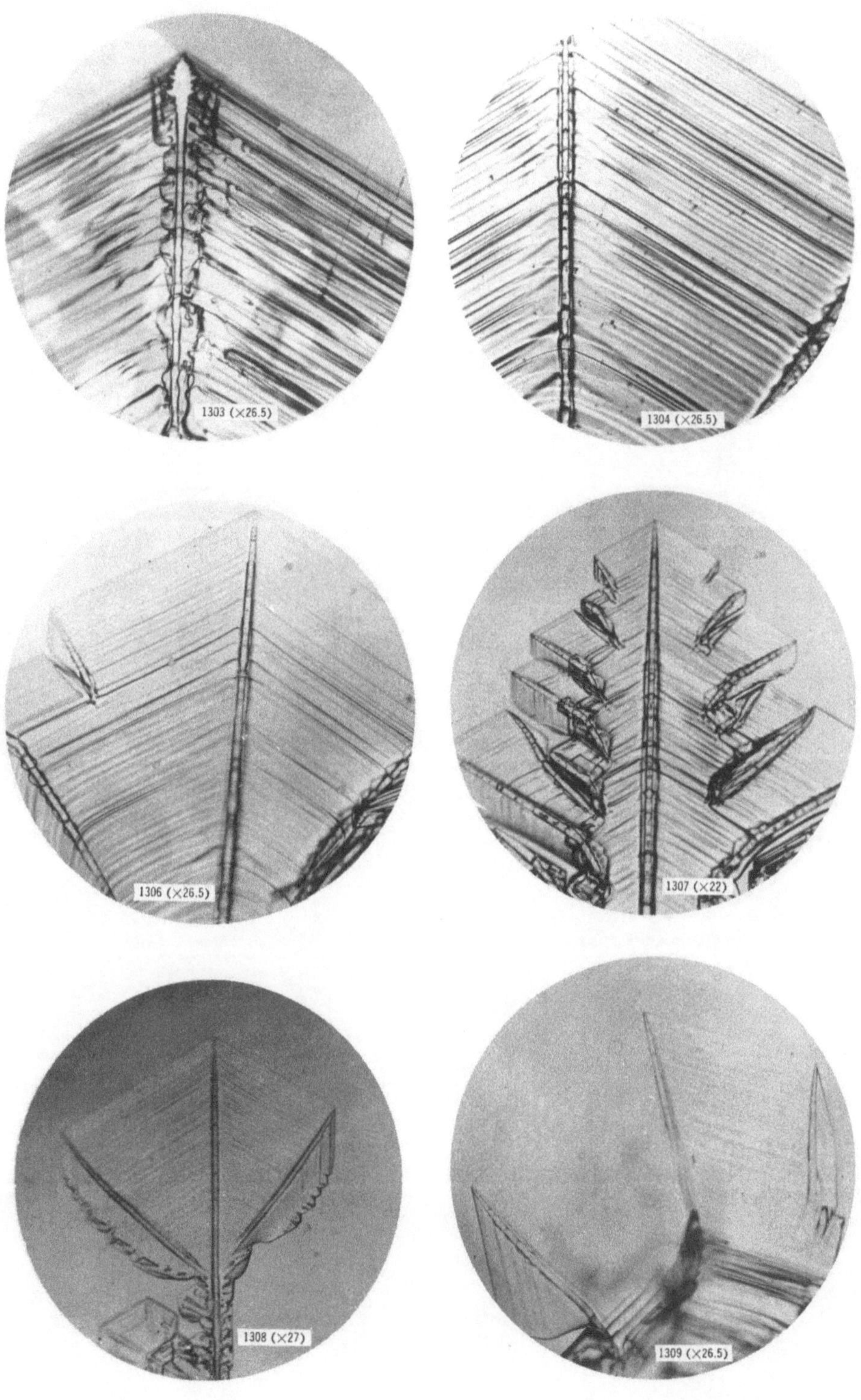
1303 (×26.5)
1304 (×26.5)
1306 (×26.5)
1307 (×22)
1308 (×27)
1309 (×26.5)

PLATE 182

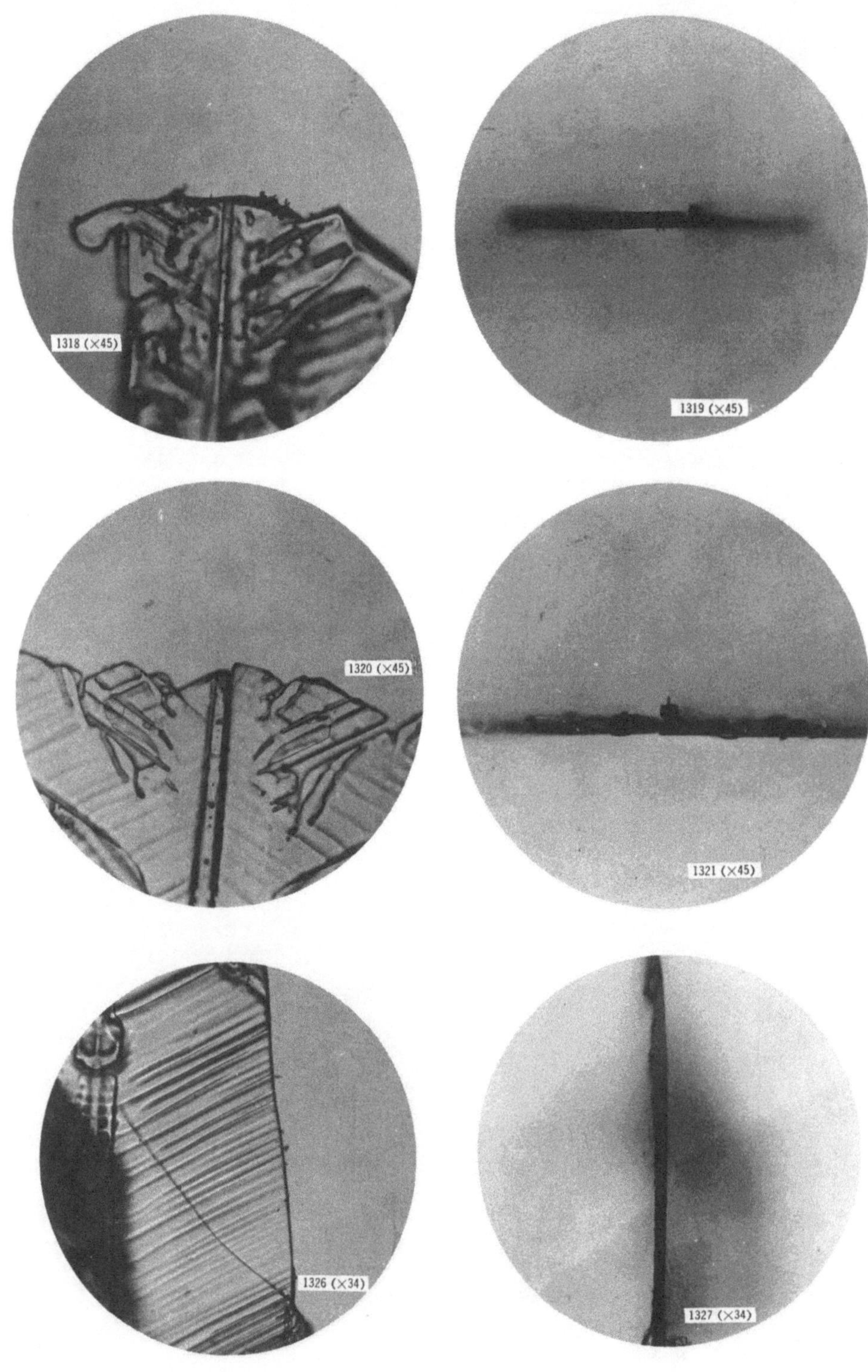

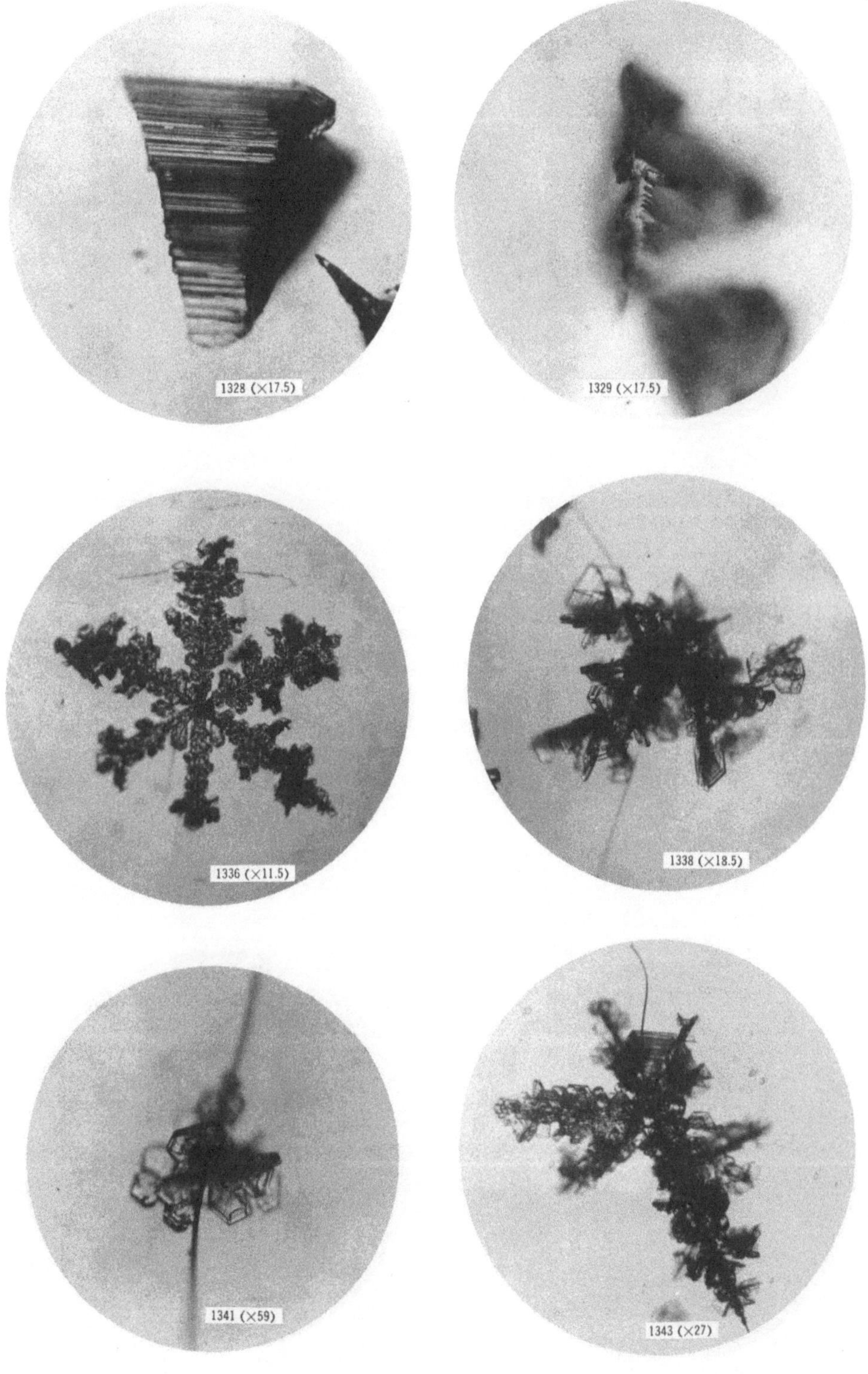

1328 (×17.5)

1329 (×17.5)

1336 (×11.5)

1338 (×18.5)

1341 (×59)

1343 (×27)

PLATE 184

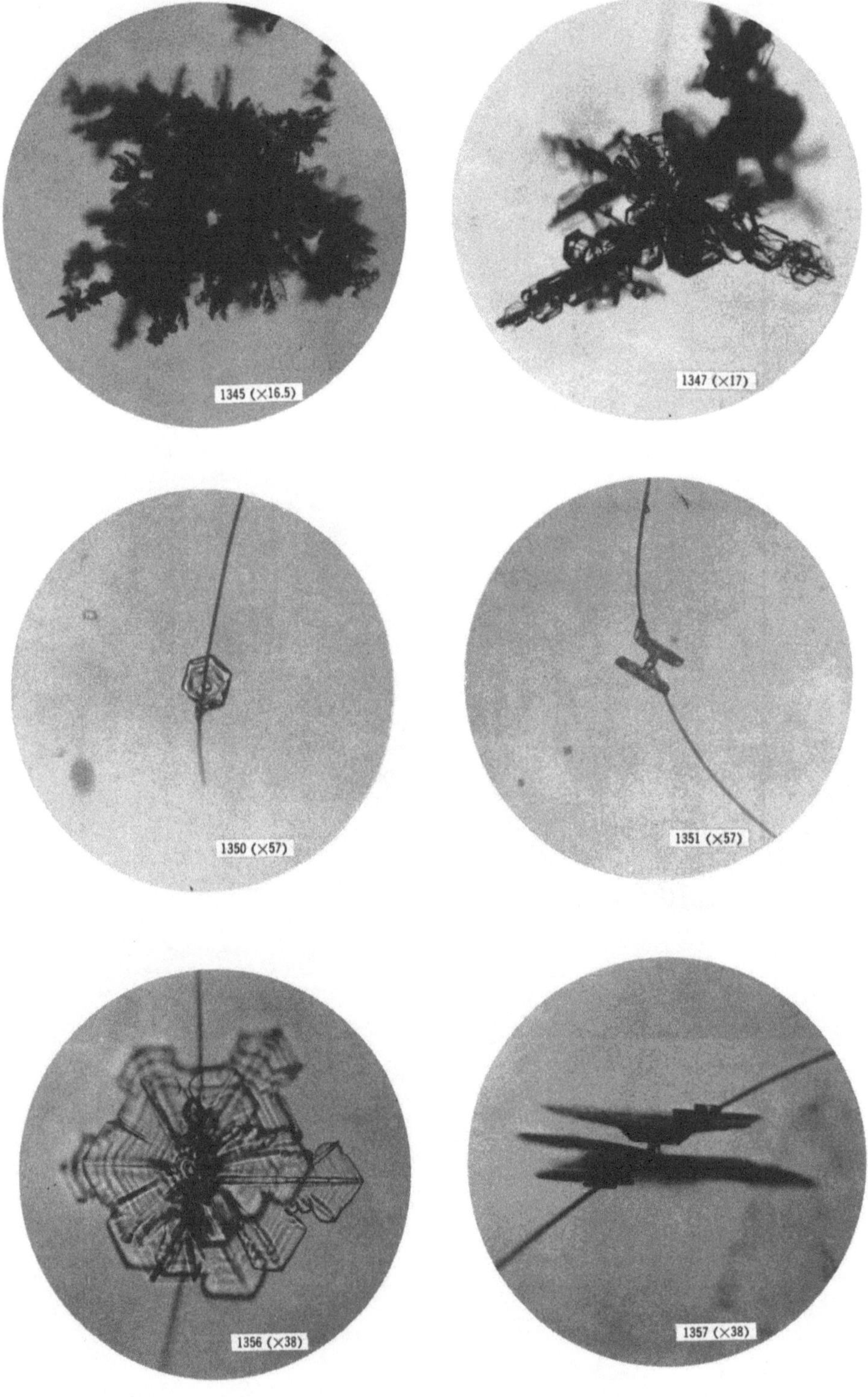

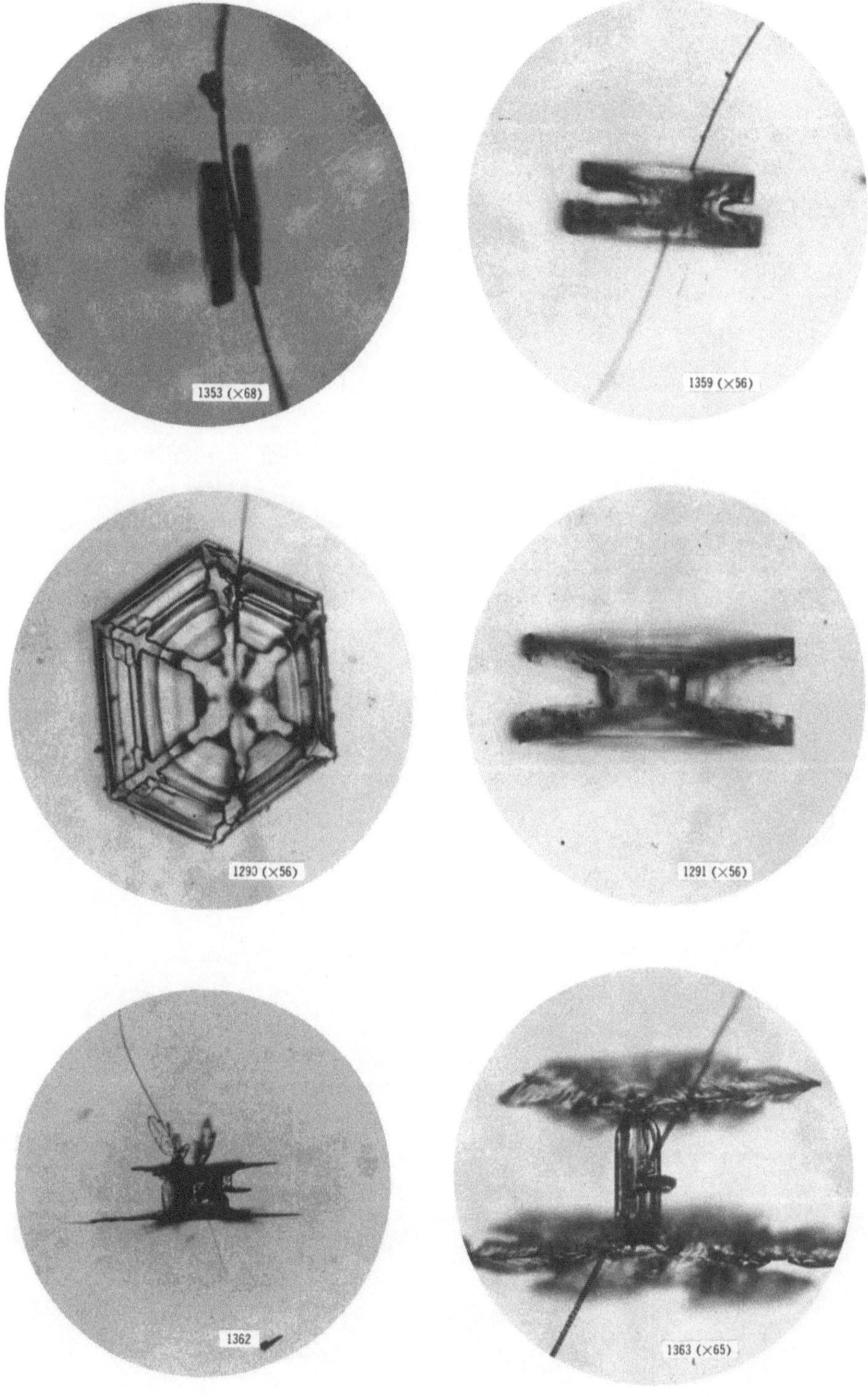

1353 (×68) 1359 (×56)

1290 (×56) 1291 (×56)

1362 1363 (×65)

PLATE 186

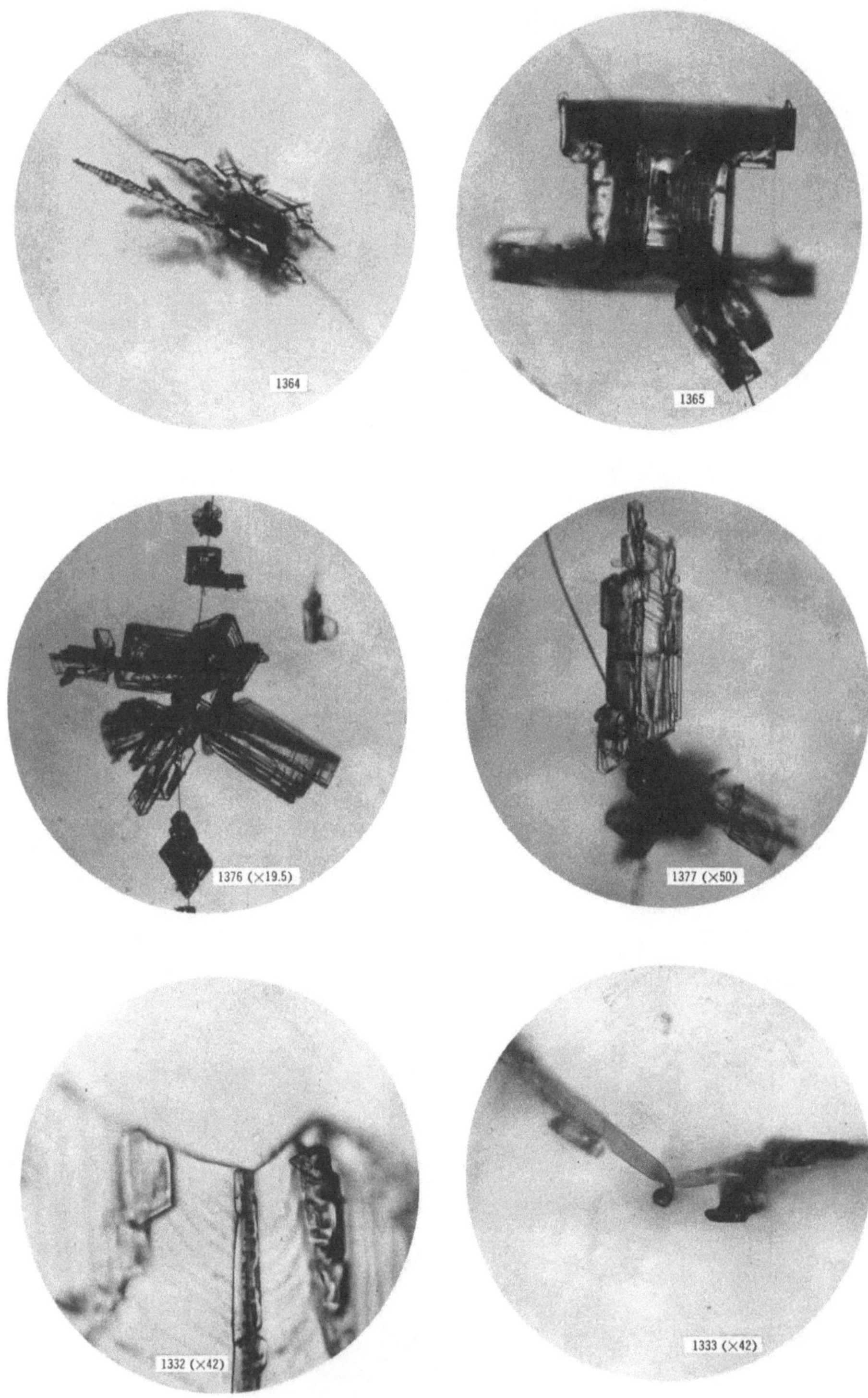

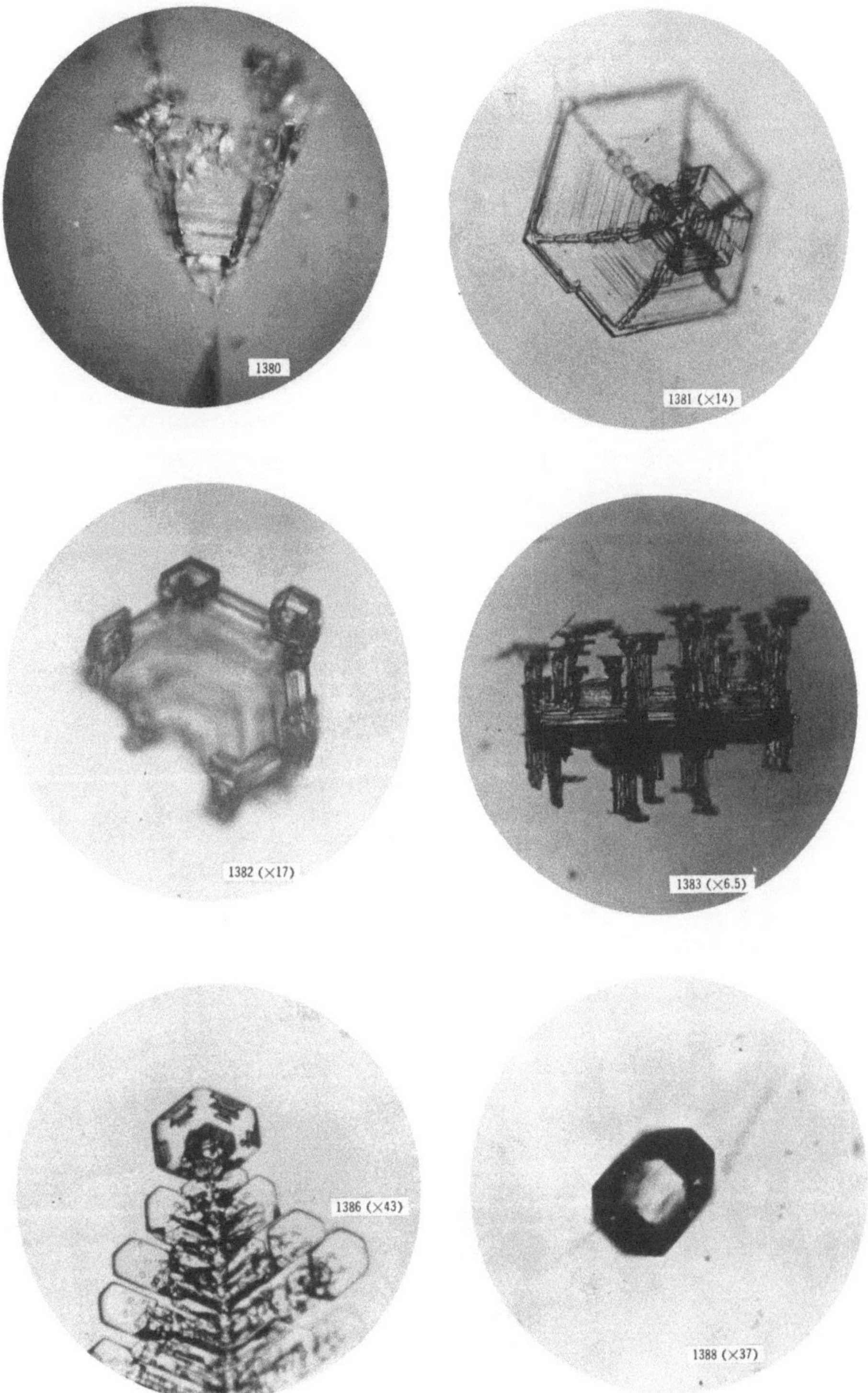
1380
1381 (×14)
1382 (×17)
1383 (×6.5)
1386 (×43)
1388 (×37)

PLATE 188

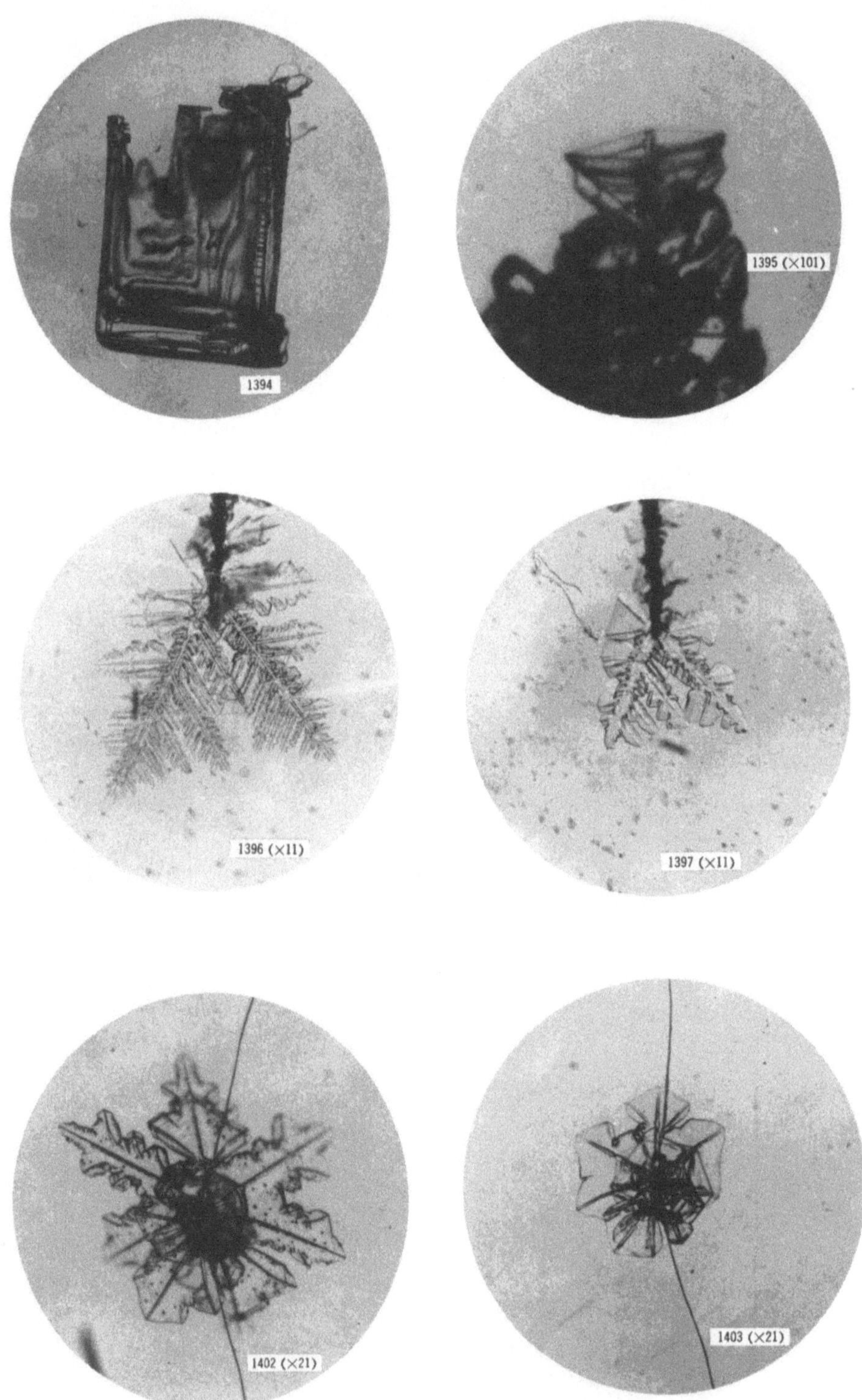

Indexes

INDEX OF NAMES

INDEX OF SUBJECTS

Bei Fragen zur Produktsicherheit wenden Sie sich bitte an:
If you have any questions regarding product safety,
please contact:

Walter de Gruyter GmbH
Genthiner Straße 13
10785 Berlin
productsafety@degruyterbrill.com